Advances in Computer Vision and Pattern Recognition

More information about this series at http://www.springer.com/series/4205

Sébastien Marcel · Mark S. Nixon
Julian Fierrez · Nicholas Evans
Editors

Handbook of Biometric Anti-Spoofing

Presentation Attack Detection

Second Edition

 Springer

Editors
Sébastien Marcel
Idiap Research Institute
Martigny, Switzerland

Julian Fierrez
Universidad Autonoma de Madrid
Madrid, Spain

Mark S. Nixon
University of Southampton
Southampton, UK

Nicholas Evans
EURECOM
Biot Sophia Antipolis, France

ISSN 2191-6586   ISSN 2191-6594 (electronic)
Advances in Computer Vision and Pattern Recognition
ISBN 978-3-319-92626-1   ISBN 978-3-319-92627-8 (eBook)
https://doi.org/10.1007/978-3-319-92627-8

Library of Congress Control Number: 2018957636

This Springer imprint is published by the registered company Springer Nature Switzerland AG
The registered company address is: Gewerbestrasse 11, 6330 Cham, Switzerland

Foreword

About 5 years ago, I had the privilege to write the Foreword for the first edition of the *Handbook of Biometric Anti-Spoofing*, edited by my good colleagues Sébastien Marcel, Mark S. Nixon, and Stan Z. Li. I was impressed with their work, and wrote that Foreword that there were four reasons that made it easy to envy what they accomplished with their Handbook. I will revisit those reasons below. I now have the privilege to write the Foreword to the second edition of the *Handbook of Biometric Anti-Spoofing*, which I enjoy even more than the first edition. The second edition is edited by good colleagues Sébastien Marcel, Mark S. Nixon, Julian Fierrez, and Nicholas Evans. The editorial team has expanded as the scope and ambition of the Handbook has expanded, and in my assessment, the editors have achieved an impressive final product.

In the Foreword to the first edition of the *Handbook of Biometric Anti-Spoofing*, I wrote that one reason to envy what the editors had accomplished is that they managed to envision a truly novel (at the time) theme for their Handbook. Theirs was the first Handbook that I am aware of to be dedicated to biometric anti-spoofing. As the advertising copy says "the first definitive study of biometric anti-spoofing". This distinction does not go away, but anti-spoofing—or "presentation attack detection" in the current lingo—is a fast-moving area of research and any work in this area can go out-of-date quickly. With the second edition, the coverage of the field has been brought up to date and also expanded to more comprehensive coverage of the field. As the scope and ambition of the field as a whole has grown, so has the scope and ambition of the Handbook.

In the Foreword to the first edition, I wrote that a second reason to envy the editors' accomplishment was that they anticipated an important emerging need. If this was not clear to the entire field 5 years ago, it certainly should be clear now. Biometric technologies continue to become more widely deployed, in consumer products such as the 3D face recognition in Apple's iPhone X, in business processes such as Yombu's fingerprint payment system, and in government applications such as Somaliland's use of iris recognition to create their national voter registration list. With bigger, broader and higher value applications, presentation attacks of more creative varieties are certain to be attempted. The need for an authoritative, broad coverage volume detailing the current state of the art in biometric anti-spoofing has only increased since the first edition, and the second edition fulfills this need.

The third reason that I outlined in the previous Foreword was that the editors had "timed the wave" well; they were on the early leading edge of the wave of popularity of research in anti-spoofing. With the second edition, I believe that they are again on the leading edge of a wave that is still to crest. I can imagine that the CTO of every business integrating biometric identity verification into one of their processes will want to study this Handbook carefully. As well, researchers wanting to begin activity in this area will find this Handbook a great place to start.

The fourth reason that I outlined previously was that the editors' efforts had resulted in a quality product. Now, to these four reasons enumerated in the Foreword to the first edition, I must add a fifth reason specific to the second edition—the editors have evolved and updated the material in a big way, and the result is that they have produced an even better, more comprehensive and more useful second edition of the *Handbook of Biometric Anti-Spoofing*.

Whereas the first edition comprised 13 chapters, the second edition has grown to 22 chapters! And the editors have been bold, and not taken the path of least resistance. They did not automatically keep a chapter corresponding to each chapter in the first edition, but instead both dropped some topics and added new topics. Where the first edition had two chapters dealing with fingerprint, two with face, and one each on iris, gait, speaker, and multimodal biometrics, the second edition has six (!) chapters dealing with face, five with fingerprint, three with iris, three with voice, and one each dealing with vein and signature. There is also coverage of the major presentation attack competitions, and of the major databases available for research. The second edition being very much up to date and globally aware, there is even a discussion of presentation attack detection and how it may be handled under the EU's new General Data Protection Regulation (GDPR). And with every chapter, the contributors are authorities on the topic, having recently published specific state-of-the-art research of their own on the topic. The editorial team has made quite significant and impressive efforts at recruiting contributors to accomplish the expansion and updating of material for the second edition.

The second edition of the *Handbook of Biometric Anti-Spoofing* by Sébastien Marcel, Mark S. Nixon, Julian Fierrez, and Nicholas Evans is the new standard in authoritative and comprehensive coverage of the current state of the art in biometric presentation attack detection. As biometric technology continues to be adapted in new large-scale applications, the wave of research attention to presentation attack detection will continue to grow. We can only hope that the editors will return in a few years with a third edition that continues in the tradition that they have set with the first two.

Notre Dame, IN, USA   Prof. Kevin W. Bowyer
July 2018   Editor-In-Chief
                                                             IEEE Transactions on Biometrics,
                                                              Behavior and Identity Science
                                                        Schubmehl-Prein Family Professor of
                                                          Computer Science and Engineering
                                                                University of Notre Dame

Preface

In the 4 years since 2014 when the TABULA RASA[1] project ended,[2] and the first edition of this Handbook was published,[3] the field of biometric anti-spoofing (term now standardized as biometric Presentation Attack Detection—PAD) has advanced significantly with large-scale industrial application. As these applications continue to grow in scale, the number of research challenges and technology requirements are also increasing significantly. The importance of the topic and the related research needs are confirmed by new highly funded research programs like the IARPA ODIN program initiated in 2016 and ongoing, aimed at advancing PAD technologies to identify known and unknown biometric presentation attacks.

The field of biometric PAD has matured significantly since the first edition, with a growing number of research groups working in the topic, various benchmarks and tools now commonly used and shared among researchers, technology competitions, and standardization activities. With the aim of updating our first edition published in 2014, heavily focused then on the research within the TABULA RASA project, we initiated this second edition in 2017 in an Open Call aiming to represent a more up-to-date and comprehensive picture of the current state of the art. We received 25 Expressions of Interest for contributions to the book, which after review resulted in a final set of 22 chapters. We are very grateful both to the authors and to the reviewers, who are listed separately.

We also thank the support provided by Springer, with special thanks to Simon Rees, who similar to the first edition has helped significantly towards this second edition.

As the body of knowledge in biometric PAD is growing in the recent years, the volume and contents in this second edition have increased significantly with respect to the first edition. Additionally, this field is attracting the interest of a growing

---

[1] Trusted Biometrics under Spoofing Attacks—http://www.tabularasa-euproject.org.
[2] A. Hadid, N. Evans, S. Marcel and J. Fierrez, "Biometrics systems under spoofing attack: an evaluation methodology and lessons learned", IEEE Signal Processing Magazine, September 2015.
[3] S. Marcel, M. S. Nixon and S. Z. Li (Eds.), Handbook of Biometric Anti-Spoofing, Springer, 2014.

number of people: from researchers to practitioners, from students to advanced researchers, and from engineers to technology consultants and marketers. In order to be useful to a wider spectrum of readers, in this second edition, we have included a number of introductory chapters for the most important biometrics. Those introductory chapters can be skipped by readers knowledgeable in the basics of biometric PAD.

With the mindset of helping researchers and practitioners, and speeding up the progress in this field, we asked authors of experimental chapters to comply with two requirements related to Reproducible Research:

- experiments are conducted on publicly available datasets;
- system scores generated with proposed PAD methods are openly available.

Additionally, some chapters and more particularly chapters 2, 4, 7, 11, 12, 13, 16, 17, 18, 19 and 20, also include code for generating performance plots and figures, open source codes for the presented methods, and detailed instructions on how to reproduce the reported results. All this Reproducible Research material is available here: https://gitlab.idiap.ch/biometric-resources.

As researchers in the field for many years, we trust you find this text of use as guidance and as reference in a topic that will continue to inspire and challenge many researchers.

Martigny, Switzerland   Sébastien Marcel
Southampton, England   Mark S. Nixon
Madrid, Spain   Julian Fierrez
Biot Sophia Antipolis, France   Nicholas Evans
July 2018

List of Reviewers

Zahid Akhtar, INRS-EMT, University of Quebec, Canada
Jos Luis Alba Castro, Universidad de Vigo, Spain
André Anjos, Idiap Research Institute, Switzerland
Sushil Bhattacharjee, Idiap Research Institute, Switzerland
Christophe Champod, University of Lausanne, Switzerland
Adam Czajka, University of Notre Dame, USA
Héctor Delgado, EURECOM, France
Nesli Erdogmus, Izmir Institute of Technology, Turkey
Nicholas Evans, EURECOM, France
Jiangjiang Feng, Tsinghua University, China
Julian Fierrez, Universidad Autonoma de Madrid, Spain
Javier Galbally, European Commission, Joint Research Centre, Italy
Anjith George, Idiap Research Institute, Switzerland
Luca Ghiani, University of Cagliari, Italy
Marta Gomez-Barrero, Hochschule Darmstadt, Germany
Abdenour Hadid, University of Oulu, Finland
Guillaume Heusch, Idiap Research Institute, Switzerland
Ivan Himawan, Queensland University of Technology, Brisbane, Australia
Els J. Kindt, KU Leuven, Belgium
Tomi Kinnunen, University of Eastern Finland, Finland
Jukka Komulainen, University of Oulu, Finland
Stan Z. Li, Chinese Academy of Sciences, China
Sébastien Marcel, Idiap Research Institute, Switzerland
Gian Luca Marcialis, University of Cagliari, Italy
Amir Mohammadi, Idiap Research Institute, Switzerland
Mark S. Nixon, University of Southampton, UK
Jonathan Phillips, NIST, USA
Hugo Proenca, University of Beira Interior, Portugal
Kiran B. Raja, Norwegian University of Science and Technology, Norway
Raghavendra Ramachandra, Norwegian University of Science and Technology, Norway

Arun Ross, Michigan State University, USA
Md Sahidullah, Inria, France
Richa Singh, IIIT-Delhi, India
Massimiliano Todisco, EURECOM, France

Contents

Contributors

Zahid Akhtar INRS-EMT, University of Quebec, Quebec City, Canada

José Luis Alba-Castro Universidade de Vigo, Vigo, Spain

André Anjos Biometrics Security and Privacy Group, Idiap Research Institute, Martigny, Switzerland

Benedict Becker University of Notre Dame, Notre Dame, IN, USA

Sushil Bhattacharjee Biometrics Security and Privacy Group, Idiap Research Institute, Martigny, Switzerland

Zinelabidine Boulkenafet Center for Machine Vision and Signal Analysis, University of Oulu, Oulu, Finland

Kevin Bowyer Notre Dame University, Notre Dame, IN, France

Christoph Busch Hochschule Darmstadt and CRISP (Center for Research in Security and Privacy), Darmstadt, Germany; Norwegian Biometrics Laboratory, Norwegian University of Science and Technology (NTNU), Trondheim, Norway

Raffaele Cappelli Università di Bologna, Cesena, Italy

Milos Cernak Logitech, Lausanne, Switzerland

Ivana Chingovska Idiap Research Institute, Martigny, Switzerland

Artur Costa-Pazo GRADIANT, CITEXVI, Vigo, Spain

Adam Czajka Research and Academic Computer Network (NASK), Warsaw, Poland; University of Notre Dame, Notre Dame, IN, USA; Warsaw University of Technology, Warsaw, Poland

Luke Darlow Council for Scientific and Industrial Research, Pretoria, South Africa

Héctor Delgado Department of Digital Security, EURECOM, Biot Sophia Antipolis, France

Nicholas Evans Department of Digital Security, EURECOM, Biot Sophia Antipolis, France

Julian Fierrez Universidad Autonoma de Madrid, Madrid, Spain

Clinton Fookes Queensland University of Technology, Brisbane, Australia

Javier Galbally European Commission - DG Joint Research Centre, Ispra, Italy

Luca Ghiani Department of Electrical and Electronic Engineering, University of Cagliari, Cagliari, Italy

Marta Gomez-Barrero da/sec Biometrics and Internet Security Research Group, Hochschule Darmstadt, Darmstadt, Germany

Daniel González-Jiménez GRADIANT, CITEXVI, Vigo, Spain

Javier Hernandez-Ortega Biometrics and Data Pattern Analytics - BiDA Lab, Universidad Autonoma de Madrid, Madrid, Spain

Guillaume Heusch Idiap Research Institute, Martigny, Switzerland

Ivan Himawan Queensland University of Technology, Brisbane, Australia

Els J. Kindt KU Leuven – Law Faculty – Citip – iMec, Leuven, Belgium; Universiteit Leiden - Law Faculty - eLaw, Leiden, The Netherlands

Tomi Kinnunen School of Computing, University of Eastern Finland, Kuopio, Finland

Naman Kohli West Virginia University, Morgantown, WV, USA

Jukka Komulainen Center for Machine Vision and Signal Analysis, University of Oulu, Oulu, Finland

Pavel Korshunov Idiap Research Institute, Martigny, Switzerland

Kong-Aik Lee Data Science Research Laboratories, NEC Corporation (Japan), Tokyo, Japan

Xiaobai Li Center for Machine Vision and Signal Analysis, University of Oulu, Oulu, Finland

Si-Qi Liu Department of Computer Science, Hong Kong Baptist University, Kowloon, Hong Kong

Srikanth Madikeri Idiap Research Institute, Martigny, Switzerland

Sébastien Marcel Biometrics Security and Privacy Group, Idiap Research Institute, Martigny, Switzerland

Gian Luca Marcialis Department of Electrical and Electronic Engineering, University of Cagliari, Cagliari CA, Italy

Amir Mohammadi Biometrics Security and Privacy Group, Idiap Research Institute, Martigny, Switzerland

Yaseen Moolla Council for Scientific and Industrial Research, Pretoria, South Africa

Aythami Morales School of Engineering, Universidad Autonoma de Madrid, Madrid, Spain

Petr Motlicek Idiap Research Institute, Martigny, Switzerland

V. Mura Department of Electrical and Electronic Engineering, University of Cagliari, Cagliari, Italy

Afzel Noore West Virginia University, Morgantown, WV, USA

Javier Ortega-Garcia Biometrics and Data Pattern Analytics - BiDA Lab, Escuela Politecnica Superior, Universidad Autonoma de Madrid, Madrid, Spain

R. Raghavendra Norwegian Biometrics Laboratory, Norwegian University of Science and Technology (NTNU), Trondheim, Norway

Kiran B. Raja Norwegian Biometrics Laboratory, Norwegian University of Science and Technology (NTNU), Trondheim, Norway

Christian Rathgeb da/sec Biometrics and Internet Security Research Group, Hochschule Darmstadt, Darmstadt, Germany

Fabio Roli Department of Electrical and Electronic Engineering, University of Cagliari, Cagliari CA, Italy

Md Sahidullah School of Computing, University of Eastern Finland, Kuopio, Finland

Stephanie Schuckers Clarkson University, Potsdam, NY, USA

Ameeth Sharma Council for Scientific and Industrial Research, Pretoria, South Africa

Ann Singh Council for Scientific and Industrial Research, Pretoria, South Africa

Richa Singh IIIT-Delhi Okhla Industrial Estate, New Delhi, India

Sridha Sridharan Queensland University of Technology, Brisbane, Australia

Massimiliano Todisco Department of Digital Security, EURECOM, Biot Sophia Antipolis, France

Ruben Tolosana Biometrics and Data Pattern Analytics - BiDA Lab, Escuela Politecnica Superior, Universidad Autonoma de Madrid, Madrid, Spain

Pedro Tome Universidad Autonoma de Madrid, Madrid, Spain

Pierliugi Tuveri Department of Electrical and Electronic Engineering, University of Cagliari, Cagliari, Italy

Johan van der Merwe Council for Scientific and Industrial Research, Pretoria, South Africa

Mayank Vatsa IIIT-Delhi Okhla Industrial Estate, New Delhi, India

Esteban Vazquez-Fernandez GRADIANT, CITEXVI, Vigo, Spain

Sushma Venkatesh Norwegian Biometrics Laboratory, Norwegian University of Science and Technology (NTNU), Trondheim, Norway

Ruben Vera-Rodriguez Biometrics and Data Pattern Analytics - BiDA Lab, Escuela Politecnica Superior, Universidad Autonoma de Madrid, Madrid, Spain

Daksha Yadav West Virginia University, Morgantown, WV, USA

Junichi Yamagishi National Institute of Informatics, Tokyo, Japan; University of Edinburgh, Edinburgh, Scotland

David Yambay Clarkson University, Potsdam, NY, USA

Pong C. Yuen Department of Computer Science, Hong Kong Baptist University, Kowloon, Hong Kong

Guoying Zhao Center for Machine Vision and Signal Analysis, University of Oulu, Oulu, Finland

Mikel Zurutuza Department of Electrical and Electronic Engineering, University of Cagliari, Cagliari, Italy

Part I
Fingerprint Biometrics

Chapter 1
An Introduction to Fingerprint Presentation Attack Detection

Javier Galbally, Julian Fierrez and Raffaele Cappelli

Abstract This chapter provides an introduction to Presentation Attack Detection (PAD), also coined anti-spoofing, in fingerprint biometrics, and summarizes key developments for that purpose in the last two decades. After a review of selected literature in the field, we also revisit the potential of quality assessment for presentation attack detection. We believe that, beyond the interest that the described techniques may intrinsically have by themselves, the case study presented may serve as an example of how to develop and validate fingerprint PAD techniques based on common and publicly available benchmarks and following a systematic and replicable protocol.

1.1 Introduction

"Fingerprints cannot lie, but liars can make fingerprints". Unfortunately, this paraphrase of an old quote attributed to Mark Twain[1] has been proven right on many occasions now.

J. Galbally
European Commission, Joint Research Centre, Ispra, Italy
e-mail: javier.galbally@ec.europa.eu

J. Fierrez (✉)
Universidad Autonoma de Madrid, Madrid, Spain
e-mail: julian.fierrez@uam.es

R. Cappelli
Università di Bologna, Cesena, Italy
e-mail: raffaele.cappelli@unibo.it

[1]Figures do not lie, but liars do figure.

© Springer Nature Switzerland AG 2019
S. Marcel et al. (eds.), *Handbook of Biometric Anti-Spoofing*,
Advances in Computer Vision and Pattern Recognition,
https://doi.org/10.1007/978-3-319-92627-8_1

As the deployment of fingerprint systems keeps growing year after year in such different environments as airports, laptops, or mobile phones, people are also becoming more familiar to their use in everyday life and, as a result, the security weaknesses of fingerprint sensors are becoming better known to the general public. Nowadays, it is not difficult to find websites or even tutorial videos, which give detail guidance on how to create fake fingerprints which may be used for spoofing biometric systems.

As a consequence, the fingerprint stands out as one of the biometric traits which has arisen the most attention not only from researchers and vendors, but also from the media and users, regarding its vulnerabilities to Presentation Attacks (PAs) (aka spoofing). This increasing interest of the biometric community in the security evaluation of fingerprint recognition systems against presentation attacks has led to the creation of numerous and very diverse initiatives in this field: the publication of many research works disclosing and evaluating different fingerprint presentation attack approaches [1–4]; the proposal of new Presentation Attack Detection (PAD) (aka anti-spoofing) methods [5–7]; related book chapters [8, 9]; PhD and MSc Theses which propose and analyze different fingerprint PA and PAD techniques [10–13]; several patented fingerprint PAD mechanisms both for touch-based and contactless systems [14–18]; the publication of Supporting Documents and Protection Profiles in the framework of the security evaluation standard Common Criteria for the objective assessment of fingerprint-based commercial systems [19, 20]; the organization of competitions focused on vulnerability assessment to fingerprint presentation attacks [21, 22]; the acquisition of specific datasets for the evaluation of fingerprint protection methods against direct attacks [23, 24], the creation of groups and laboratories which have the evaluation of fingerprint security as one of their major tasks [25–27]; or of several European Projects on fingerprint PAD as one of their main research interests [28, 29].

The aforementioned initiatives and other analogue studies have shown the importance given by all parties involved in the development of fingerprint-based biometrics to the improvement of the systems security and the necessity to propose and develop specific protection methods against PAs in order to bring this rapidly emerging technology into practical use. This way, researchers have focused on the design of specific countermeasures that enable fingerprint recognition systems to detect fake samples and reject them, improving this way the robustness of the applications.

In the fingerprint field, besides other PAD approaches such as the use of multi-biometrics or challenge–response methods, special attention has been paid by researchers and industry to the so-called *liveness detection* techniques. These algorithms use different physiological properties to distinguish between real and fake traits. Liveness assessment methods represent a challenging engineering problem as they have to satisfy certain demanding requirements [30]: (i) noninvasive, the technique should in no case be harmful for the individual or require an excessive contact with the user; (ii) user-friendly, people should not be reluctant to use it; (iii) fast, results have to be produced in a very reduced interval as the user cannot be asked to interact with the sensor for a long period of time; (iv) low cost, a wide use cannot be expected if the cost is excessively high; (v) performance, in addition to having a good fake detection rate, the protection scheme should not degrade the recognition performance (i.e., false rejection) of the biometric system.

Liveness detection methods are usually classified into one of two groups: (i) *Hardware-based* techniques, which add some specific device to the sensor in order to detect particular properties of a living trait (e.g., fingerprint sweat, blood pressure, or odor); (ii) *Software-based* techniques, in this case, the fake trait is detected once the sample has been acquired with a standard sensor (i.e., features used to distinguish between real and fake traits are extracted from the biometric sample, and not from the trait itself).

The two types of methods present certain advantages and drawbacks over the other and, in general, a combination of both would be the most desirable protection approach to increase the security of biometric systems. As a coarse comparison, hardware-based schemes usually present a higher fake detection rate, while software-based techniques are in general less expensive (as no extra device is needed), and less intrusive since their implementation is transparent to the user. Furthermore, as they operate directly on the acquired sample (and not on the biometric trait itself), software-based techniques may be embedded in the feature extractor module which makes them potentially capable of detecting other types of illegal break-in attempts not necessarily classified as presentation attacks. For instance, software-based methods can protect the system against the injection of reconstructed or synthetic samples into the communication channel between the sensor and the feature extractor [31, 32].

Although, as shown above, a great amount of work has been done in the field of fingerprint PAD and big advances have been reached over the last decade, the attacking methodologies have also evolved and become more and more sophisticated. This way, while many commercial fingerprint readers claim to have some degree of PAD embedded, many of them are still vulnerable to presentation attack attempts using different artificial fingerprint samples. Therefore, there are still big challenges to be faced in the detection of fingerprint direct attacks.[2]

This chapter represents an introduction to the problem of fingerprint PAD, including an example of experimental methodology [33], and example results extracted from [34]. More comprehensive and up to date surveys of recent advances can be found elsewhere [35–37]. After a review of early works in fingerprint PAD, we analyze and evaluate the potential of quality assessment for liveness detection purposes. In particular, we consider two different sets of features: (i) one based on fingerprint-specific quality measures (i.e., quality measures which may only be extracted from a fingerprint image); (ii) a second set based on general image quality measures (i.e., quality measures which may be extracted from any image). Both techniques are tested on publicly available fingerprint spoofing databases where they have reached results fully comparable to those obtained on the same datasets and following the same experimental protocols by top-ranked approaches from the state of the art.

In addition to their very competitive performance, as they are software-based, both methods present the usual advantages of this type of approaches: fast, as they only

---

[2]https://www.iarpa.gov/index.php/research-programs/odin/

need one image (i.e., the same sample acquired for verification) to detect whether it is real or fake; nonintrusive; user-friendly (transparent to the user); cheap and easy to embed in already functional systems (as no new piece of hardware is required).

The rest of the chapter is structured as follows. A review of relevant early works in the field of fingerprint PAD is given is Sect. 1.2. A brief description of large and publicly available fingerprint spoofing databases is presented in Sect. 1.3. A case study based on the use of quality assessment as PAD tool is introduced in Sect. 1.4 where we give some key concepts about image quality assessment and the rationale behind its use for biometric protection. The two fingerprint PAD approaches studied in the chapter based on fingerprint-specific and general quality features are described respectively in Sects. 1.5 and 1.6. The evaluation of the methods and experimental results are given in Sect. 1.7. Conclusions are finally drawn in Sect. 1.8.

1.2   Early Works in Fingerprint Presentation Attack Detection

The history of fingerprint forgery in the forensic field is probably almost as old as that of fingerprint development and classification itself. In fact, the question of whether or not fingerprints could be forged was positively answered [38] several years before it was officially posed in a research publication [39].

Regarding modern automatic fingerprint recognition systems, although other types of attacks with dead [40] or altered [41] fingers have been reported, almost all the available vulnerability studies regarding presentations attacks are carried out either by taking advantage of the residual fingerprint left behind on the sensor surface, or by using some type of gummy fingertip (or even complete prosthetic fingers) manufactured with different materials (e.g., silicone, gelatin, plastic, clay, dental molding material, or glycerin). In general, these fake fingerprints may be generated with the cooperation of the user, from a latent fingerprint or even from a fingerprint image reconstructed from the original minutiae template [1–3, 23, 42–46].

These very valuable works and other analogue studies have highlighted the necessity to develop efficient protection methods against presentation attacks. One of the first efforts in fingerprint PAD initiated a research line based on the analysis of the skin perspiration pattern which is very difficult to be faked in an artificial finger [5, 47]. These pioneer studies, which considered the periodicity of sweat and the sweat diffusion pattern, were later extended and improved in two successive works applying a wavelet-based algorithm and adding intensity-based perspiration features [48, 49]. These techniques were finally consolidated and strictly validated on a large database of real, fake, and dead fingerprints acquired under different conditions in [24]. More recently, a novel region-based liveness detection approach also based on perspiration parameters and another technique analyzing the valley noise have been proposed by the same group [50, 51]. Part of these approaches has been implemented in commercial products [52], and has also been combined with other morphological features [53, 54] in order to improve the presentation attack detection rates [55].

A second group of fingerprint liveness detection techniques has appeared as an application of the different fingerprint distortion models described in the literature [56–58]. These models have led to the development of a number of liveness detection techniques based on the flexibility properties of the skin [6, 59–61]. In most of these works the user is required to move his finger while pressing it against the scanner surface, thus deliberately exaggerating the skin distortion. When a real finger moves on a scanner surface, it produces a significant amount of distortion, which can be observed to be quite different from that produced by fake fingers which are usually more rigid than skin. Even if highly elastic materials are used, it seems very difficult to precisely emulate the specific way a real finger is distorted, because the behavior is related to the way the external skin is anchored to the underlying derma and influenced by the position and shape of the finger bone.

Other liveness detection approaches for fake fingerprint detection include: the combination of both perspiration and elasticity-related features in fingerprint image sequences [62]; fingerprint-specific quality-related features [7, 34]; the combination of the local ridge frequency with other multiresolution texture parameters [53]; techniques which, following the perspiration-related trend, analyze the skin sweat pores visible in high definition images [63, 64]; the use of electric properties of the skin [65]; using several image processing tools for the analysis of the finger tip surface texture such as wavelets [66], or three very related works using Gabor filters [67], ridgelets [68] and curvelets [69]; analyzing different characteristics of the Fourier spectrum of real and fake fingerprint images [70–74].

A critical review of some of these solutions for fingerprint liveness detection was presented in [75]. In a subsequent work [76], the same authors gave a comparative analysis of the PAD methods efficiency. In this last work, we can find an estimation of some of the best performing static (i.e., measured on one image) and dynamic (i.e., measured on a sequence of images) features for liveness detection, that were later used together with some fake-finger specific features in [77] with very good results. Different static features are also combined in [78], significantly improving the results of the individual parameters. Other comparative results of different fingerprint PAD techniques are available in the results of the Fingerprint Liveness Detection Competitions (LivDet series) [21, 22].

In addition, some very interesting hardware-based solutions have been proposed in the literature applying: multispectral imaging [79, 80], an electrotactile sensor [81], pulse oximetry [82], detection of the blood flow [14], odor detection using a chemical sensor [83], or another trend based on Near Infrared (NIR) illumination and Optical Coherence Tomography (OCT) [84–89].

More recently, a third type of protection methods which fall out of the traditional two-type classification software- and hardware-based approaches has been started to be analyzed in the field of fingerprint PAD. These protection techniques focus on the study of biometric systems under direct attacks at the *score level*, in order to propose and build more robust matchers and fusion strategies that increase the resistance of the systems against presentation attack attempts [90–94].

Outside the research community, some companies have also proposed different methods for fingerprint liveness detection such as the ones based on ultrasounds

[95, 96], light measurements [97], or a patented combination of different unimodal experts [98]. A comparative study of the PAD capabilities of different commercial fingerprint sensors appears in [99].

Although the vast majority of the efforts dedicated by the biometric community in the field of fingerprint presentation attacks and PAD are focused on touch-based systems, some preliminary works have also been conducted to study the vulnerabilities of contactless fingerprint systems against direct attacks and some protection methods to enhance their security level have been proposed [17, 47, 100].

The approaches mentioned above represent the main historical developments in fingerprint PAD until ca. 2012–2013. For a survey of more recent and advanced methods in the last 5 years we refer the reader to [36, 37], and the ODIN program.[3]

1.3 Fingerprint Spoofing Databases

The availability of public datasets comprising real and fake fingerprint samples and of associated common evaluation protocols is basic for the development and improvement of fingerprint PAD methods.

However, in spite of the large amount of works addressing the challenging problem of fingerprint protection against direct attacks (as shown in Sect. 1.2), in the great majority of them, experiments are carried out on proprietary databases which are not distributed to the research community.

Currently, the two largest fingerprint spoofing databases publicly available for researchers to test their PAD algorithms are:

- LivDet DBs [21, 22]: These datasets were generated for the different campaigns of the Fingerprint Liveness Detection Competition series (in 2009, 2011, 2013, 2015, and 2017). Most of the data can be found in the LivDet series website.[4] Each dataset is complemented with specific training and testing protocols and most campaigns contain over 10,000 samples from over 100 fingers generated with materials such as: silicone, gelatine, latex, wood glue, ecoflex, and playdoh.
- ATVS-Fake Fingerprint DB (ATVS-FFp DB) [34]: This database is available from the website.[5] It contains over 3,000 real and fake fingerprint samples coming from 68 different fingers acquired using a flat optical sensor, a flat capacitive sensor, and a thermal sweeping sensor. The gummy fingers were generated with and without the cooperation of the user (i.e., recovered from a latent fingerprint) using modeling silicone.

---

[3]https://www.iarpa.gov/index.php/research-programs/odin/

[4]http://livdet.org/

[5]http://atvs.ii.uam.es/index.jsp

1.4   A Case Study: Quality Assessment Versus Fingerprint Spoofing

The problem of presentation attack detection can be seen as a two-class classification problem where an input biometric sample has to be assigned to one of two classes: real or fake (Fig. 1.1).

Simple visual inspection of an image of a real fingerprint and a fake sample of the same trait shows that the two images can be very similar and even the human eye may find it difficult to make a distinction between them after a short inspection. Yet, some differences between the real and fake fingerprints may become evident once the images are translated into a proper feature space.

Therefore, the key point of the process is to find a set of discriminant features which permits to build an appropriate classifier which gives the probability of the image "liveness" given the extracted set of features.

In the present chapter, we explore and evaluate the potential of quality assessment for fingerprint liveness detection. In particular, we consider two different sets of features: (i) one based on fingerprint-specific quality measures (i.e., quality measures which may only be extracted from a fingerprint image); (ii) a second set based on general image quality measures (i.e., quality measures which may be extracted from any image).

The use of quality assessment for PAD purposes is promoted by the assumption that: "*It is expected that a fake image captured in an attack attempt will have a different quality than a real sample acquired in the normal operation scenario for which the sensor was designed.*"

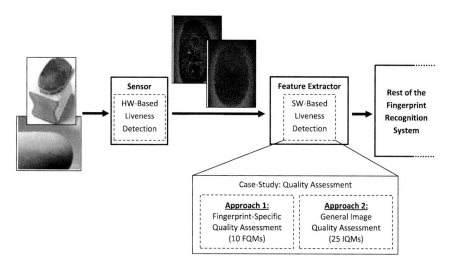

Fig. 1.1 General diagram of the fingerprint PAD case study considered in Sect. 1.4. Approach 1 and Approach 2 are described in Sects. 1.5 and 1.6, respectively. FQMs stands for Fingerprint Quality Measures, while IQMs stands for Image Quality Measures

Expected quality differences between real and fake samples may include: degree of sharpness, color and luminance levels, local artifacts, amount of information found in both types of images (entropy), structural distortions, or natural appearance. For example, it is not rare that fingerprint images captured from a gummy finger present local acquisition artifacts such as spots and patches, or that they have a lower definition of ridges and valleys due to the lack of moisture.

In the current state of the art, the rationale behind the use of quality assessment features for liveness detection is supported by three factors:

- Image quality has been successfully used in previous works for image manipulation detection [101, 102] and steganalysis [103–105] in the forensic field. To a certain extent, many fingerprint presentation attacks may be regarded as a type of image manipulation which can be effectively detected, as shown in the present research work, by the use of different quality features.
- Human observers very often refer to the "different appearance" of real and fake samples to distinguish between them. The different metrics and methods implemented here for quality assessment intend to estimate in an objective and reliable way the perceived appearance of fingerprint images.
- Moreover, different quality measures present different sensitivity to image artifacts and distortions. For instance, measures like the mean squared error respond more to additive noise, whereas others such as difference measured in the spectral domain are more sensitive to blur; while gradient-related features react to distortions concentrated around edges and textures. Therefore, using a wide range of quality measures exploiting complimentary image quality properties should permit to detect the aforementioned quality differences between real and fake samples expected to be found in many attack attempts.

All these observations lead us to believe that there is sound proof for the "quality difference" hypothesis and that quality measures have the potential to achieve success in biometric protection tasks.

In the next sections, we describe two particular software-based implementations for fingerprint PAD. Both methods use only one input image (i.e., the same sample acquired for authentication purposes) to distinguish between real and fake fingerprints. The difference between the two techniques relies on the sets of quality-based features used to solve the classification problem: (i) the first PAD method uses a set of 10 fingerprint-specific quality measures (see Sect. 1.5); (ii) the second uses a set of 25 general image quality measures (see Sect. 1.6). Later, both techniques are evaluated on two publicly available databases and their results are compared to other well-known techniques from the state of the art (see Sect. 1.7).

1.5 Approach 1: Fingerprint-Specific Quality Assessment (FQA)

The parameterization proposed in this section comprises ten Fingerprint-specific Quality Measures (FQMs). A number of approaches for fingerprint image quality computation have been described in the literature [110]. Fingerprint image quality can be assessed by measuring one of the following properties: ridge strength or directionality, ridge continuity, ridge clarity, integrity of the ridge–valley structure, or estimated verification performance when using the image at hand. A number of information sources are used to measure these properties: (i) angle information provided by the direction field, (ii) Gabor filters, which represent another implementation of the direction angle [111], (iii) pixel intensity of the gray-scale image, (iv) power spectrum, and (v) neural networks. Fingerprint quality can be assessed either analyzing the image in a holistic manner, or combining the quality from local non-overlapped blocks of the image.

In the following, we give some details about the ten fingerprint-specific quality measures used in this PAD method. The features implemented have been selected in order to cover the different fingerprint quality assessment approaches mentioned above so that the maximum degree of complementarity among them may be achieved. This way, the protection method presents a high generality and may be successfully

Table 1.1 Summary of the 10 Fingerprint-specific Quality Measures (FQMs) implemented in Sect. 1.5 for fingerprint PAD. All features were either directly taken or adapted from the references given. For each feature, the fingerprint property measured and the information source used for its estimation is given. For a more detailed description of each feature, we refer the reader to Sect. 1.5

List of 10 FQMs implemented					
#	Acronym	Name	Ref.	Property measured	Source
1	OCL	Orientation Certainty Level	[106]	Ridge strength	Local angle
2	PSE	Power Spectrum Energy	[107]	Ridge strength	Power spectrum
3	LOQ	Local Orientation Quality	[108]	Ridge continuity	Local angle
4	COF	Continuity of the Orientation Field	[106]	Ridge continuity	Local angle
5	MGL	Mean Gray Level	[76]	Ridge clarity	Pixel intensity
6	SGL	Standard Deviation Gray Level	[76]	Ridge clarity	Pixel intensity
7	LCS1	Local Clarity Score 1	[108]	Ridge clarity	Pixel intensity
8	LCS2	Local Clarity Score 2	[108]	Ridge clarity	Pixel intensity
9	SAMP	Sinusoid Amplitude	[109]	Ridge clarity	Pixel intensity
10	SVAR	Sinusoid Variance	[109]	Ridge clarity	Pixel intensity

used to detect a wide range of presentation attacks. A classification of the ten features and of the information source exploited by each of them is given in Table 1.1.

As the features used in this approach evaluate fingerprint-specific properties, prior to the feature extraction process, it is necessary to segment the actual fingerprint from the background. For this preprocessing step, the same method proposed in [112] is used.

1.5.1 Ridge Strength Measures

- **Orientation Certainty Level (OCL)** [106] measures the energy concentration along the dominant direction of ridges using the intensity gradient. It is computed as the ratio between the two eigenvalues of the covariance matrix of the gradient vector. A relative weight is given to each region of the image based on its distance from the centroid, since regions near the centroid are supposed to provide more reliable information [107]. An example of Orientation Certainty Level computation for a real and fake fingerprints is shown in Fig. 1.2.
- **Power Spectrum Energy (PSE)** [107] is computed using ring-shaped bands. For this purpose, a set of bandpass filters is employed to extract the energy in each frequency band. High quality images will have the energy concentrated in few bands while poor ones will have a more diffused distribution. The energy concentration is measured using the entropy. An example of quality estimation using the global quality index PSE is shown in Fig. 1.3 for fake and real fingerprints.

(a)   **(b)**

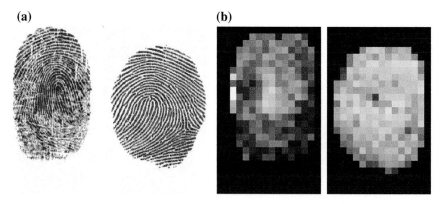

Fig. 1.2 Computation of the Orientation Certainty Level (OCL) for fake and real fingerprints. Panel **a** are the input fingerprint (left is fake, right is real). Panel **b** are the block-wise values of the OCL; blocks with brighter color indicate higher quality in the region

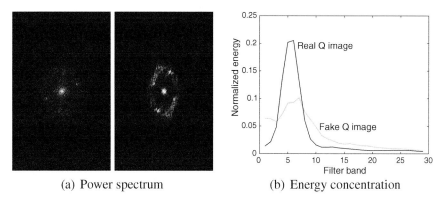

(a) Power spectrum   (b) Energy concentration

Fig. 1.3 Computation of the energy concentration in the power spectrum for fake and real finger-prints. Panel **a** are the power spectra of the images shown in Fig. 1.2. Panel **b** shows the energy distributions in the region of interest

1.5.2 Ridge Continuity Measures

- **Local Orientation Quality (LOQ)** [108] is computed as the average absolute difference of direction angle with the surrounding image blocks, providing information about how smoothly direction angle changes from block to block. Quality of the whole image is finally computed by averaging all the Local Orientation Quality scores of the image. In high quality images, it is expected that ridge direction changes smoothly across the whole image. An example of Local Orientation Quality computation is shown in Fig. 1.4 for fake and real fingerprints.
- **Continuity of the Orientation Field (COF)** [106]. This method relies on the fact that, in good quality images, ridges and valleys must flow sharply and smoothly in a locally constant direction. The direction change along rows and columns of the image is examined. Abrupt direction changes between consecutive blocks are then accumulated and mapped into a quality score. As we can observe in Fig. 1.4, ridge direction changes smoothly across the whole image in case of high quality.

1.5.3 Ridge Clarity Measures

- **Mean Gray Level (MGL)** and **Standard Deviation Gray Level (SGL)**, computed from the segmented foreground only. These two features had already been considered for liveness detection in [76].
- **Local Clarity Score (LCS1 and LCS2)** [108]. The sinusoidal-shaped wave that models ridges and valleys [109] is used to segment ridge and valley regions (see Fig. 1.5). The clarity is then defined as the overlapping area of the gray level distributions of segmented ridges and valleys. For ridges/valleys with high clarity,

(a) **(b)**

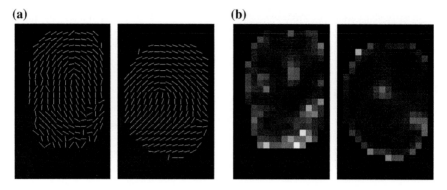

Fig. 1.4 Computation of the Local Orientation Quality (LOQ) for fake and real fingerprints. Panel **a** are the direction fields of the images shown in Fig. 1.2a. Panel **b** are the block-wise values of the average absolute difference of local orientation with the surrounding blocks; blocks with brighter color indicate higher difference value and thus, lower quality

Fig. 1.5 Modeling of ridges
and valleys as a sinusoid

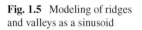

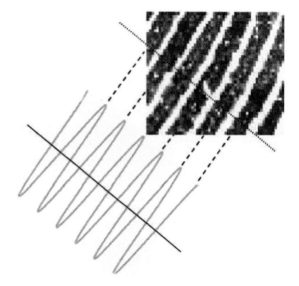

both distributions should have a very small overlapping area. An example of quality estimation using the Local Clarity Score is shown in Fig. 1.6 for two fingerprint blocks coming from fake and real fingerprints. It should be noted that sometimes the sinusoidal-shaped wave cannot be extracted reliably, specially in bad quality regions of the image. The quality measure LCS1 discards these regions, therefore being an optimistic measure of quality. This is compensated with LCS2, which does not discard these regions, but they are assigned the lowest quality level.

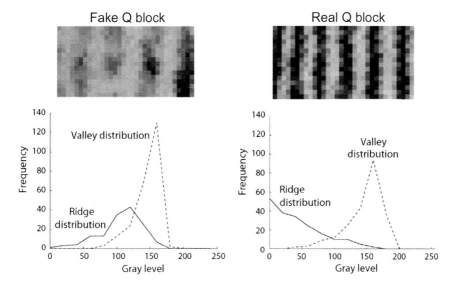

Fig. 1.6 Computation of the Local Clarity Score for two blocks coming from real and fake finger-prints. The fingerprint blocks appear on top, while below we show the gray level distributions of the segmented ridges and valleys. The degree of overlapping for the real and fake blocks is 0.22 and 0.10, respectively

- **Amplitude and Variance of the Sinusoid that models Ridges and Valleys (SAMP and SVAR)** [109]. Based on these parameters, blocks are classified as *good* and *bad*. The quality of the fingerprint is then computed as the percentage of foreground blocks marked as *good*.

1.6   Approach 2: General Image Quality Assessment (IQA)

The goal of an objective Image Quality Measure (IQM) is to provide a quantitative score that describes the degree of fidelity or, conversely, the level of distortion of a given image. Many different approaches for objective Image Quality Assessment (IQA) have been described in the literature [113]. From a general perspective, IQ metrics can be classified according to the availability of an original (distortion-free) image, with which the distorted image is to be compared. Thus, objective IQA methods can fall in one of two categories: (i) *full reference* techniques, which include the majority of traditional automatic image estimation approaches, and where a complete reference image is assumed to be known (e.g., with a large use in the field of image compression algorithms) [114]; (ii) *no-reference* techniques (also referred as *blind*), which assess the quality of the test image without any reference to the original sample, generally using some pretrained statistical model [115].

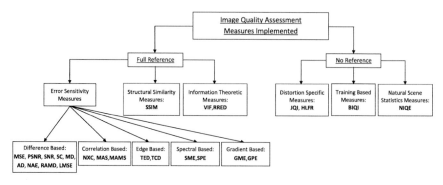

Fig. 1.7 Classification of the 25 image quality measures implemented in Sect. 1.6. Acronyms (in bold) of the different measures are explained in Table 1.2

The parameterization proposed in this section and applied to fingerprint liveness detection comprises 25 image quality measures (IQMs) both full reference and blind. In order to generate a system as general as possible in terms of number of attacks detected, we have given priority to IQMs which evaluate complementary properties of the image (e.g., sharpness, entropy or structure). In addition, to assure a user-friendly nonintrusive system, big importance has been given to the complexity and the feature extraction time of each IQM, so that the overall speed of the final fake detection algorithm allows it to operate in real-time environments.

Furthermore, as the method operates on the whole image without searching for any fingerprint-specific properties, it does not require any preprocessing steps (e.g., fingerprint segmentation) prior to the computation of the IQ features. This characteristic minimizes its computational load.

The final 25 selected image quality measures are summarized in Table 1.2. Details about each of these 25 IQMs are given in Sects. 1.6.1 and 1.6.2. For clarity, in Fig. 1.7, we show a diagram with the general IQM classification followed in these sections. Acronyms of the different features are highlighted in bold in the text and in Fig. 1.7.

1.6.1 Full Reference IQ Measures

As described previously, Full Reference (FR) IQA methods rely on the availability of a clean undistorted reference image to estimate the quality of the test sample. In the problem of fake detection addressed in this work such a reference image is unknown, as the detection system only has access to the input sample. In order to circumvent this limitation, the same strategy already successfully used for image manipulation detection in [101] and for steganalysis in [103] is implemented here.

The input gray-scale image $\mathbf{I}$ (of size $N \times M$) is filtered with a low-pass Gaussian kernel ($\sigma = 0.5$ and size 3×3) in order to generate a distorted version $\hat{\mathbf{I}}$. Then, the quality between both images ($\mathbf{I}$ and $\hat{\mathbf{I}}$) is computed according to the corresponding

Table 1.2 List of the 25 Image Quality Measures (IQMs) implemented in Sect. 1.6 for fingerprint PAD. All the features were either directly taken or adapted from the references given. In the table: $\mathbf{I}$ denotes the reference clean image (of size $N \times M$) and $\hat{\mathbf{I}}$ the distorted version of the reference image. For other notation specifications and undefined variables or functions, we refer the reader to the description of each particular feature in Sect. 1.6. Also, for those features with no mathematical definition, the exact details about their computation may be found in the given references

List of the 25 IQMs implemented

#	Type	Acronym	Name	Ref.	Description
1	FR	MSE	Mean Squared Error	[118]	$\text{MSE}(\mathbf{I}, \hat{\mathbf{I}}) = \frac{1}{NM} \sum_{i=1}^{N} \sum_{j=1}^{M} (\mathbf{I}_{i,j} - \hat{\mathbf{I}}_{i,j})^2$
2	FR	PSNR	Peak Signal-to-Noise Ratio	[119]	$\text{PSNR}(\mathbf{I}, \hat{\mathbf{I}}) = 10 \log \left(\frac{\max(\mathbf{I}^2)}{MSE(\mathbf{I},\hat{\mathbf{I}})} \right)$
3	FR	SNR	Signal-to-Noise Ratio	[120]	$\text{SNR}(\mathbf{I}, \hat{\mathbf{I}}) = 10 \log \left(\frac{\sum_{i=1}^{N} \sum_{j=1}^{M} (\mathbf{I}_{i,j})^2}{N \cdot M \cdot MSE(\mathbf{I},\hat{\mathbf{I}})} \right)$
4	FR	SC	Structural Content	[121]	$\text{SC}(\mathbf{I}, \hat{\mathbf{I}}) = \frac{\sum_{i=1}^{N} \sum_{j=1}^{M} (\mathbf{I}_{i,j})^2}{\sum_{i=1}^{N} \sum_{j=1}^{M} (\hat{\mathbf{I}}_{i,j})^2}$
5	FR	MD	Maximum Difference	[121]	$\text{MD}(\mathbf{I}, \hat{\mathbf{I}}) = \max |\mathbf{I}_{i,j} - \hat{\mathbf{I}}_{i,j}|$
6	FR	AD	Average Difference	[121]	$\text{AD}(\mathbf{I}, \hat{\mathbf{I}}) = \frac{1}{NM} \sum_{i=1}^{N} \sum_{j=1}^{M} (\mathbf{I}_{i,j} - \hat{\mathbf{I}}_{i,j})$
7	FR	NAE	Normalized Absolute Error	[121]	$\text{NAE}(\mathbf{I}, \hat{\mathbf{I}}) = \frac{\sum_{i=1}^{N} \sum_{j=1}^{M} |\mathbf{I}_{i,j} - \hat{\mathbf{I}}_{i,j}|}{\sum_{i=1}^{N} \sum_{j=1}^{M} |\mathbf{I}_{i,j}|}$
8	FR	RAMD	R-Averaged MD	[118]	$\text{RAMD}(\mathbf{I}, \hat{\mathbf{I}}, R) = \frac{1}{R} \sum_{r=1}^{R} \max_r |\mathbf{I}_{i,j} - \hat{\mathbf{I}}_{i,j}|$
9	FR	LMSE	Laplacian MSE	[121]	$\text{LMSE}(\mathbf{I}, \hat{\mathbf{I}}) = \frac{\sum_{i=1}^{N-1} \sum_{j=2}^{M-1} (h(\mathbf{I}_{i,j}) - h(\hat{\mathbf{I}}_{i,j}))^2}{\sum_{i=1}^{N-1} \sum_{j=2}^{M-1} h(\mathbf{I}_{i,j})^2}$
10	FR	NXC	Normalized Cross-Correlation	[121]	$\text{NXC}(\mathbf{I}, \hat{\mathbf{I}}) = \frac{\sum_{i=1}^{N} \sum_{j=1}^{M} (\mathbf{I}_{i,j} \cdot \hat{\mathbf{I}}_{i,j})}{\sum_{i=1}^{N} \sum_{j=1}^{M} (\mathbf{I}_{i,j})^2}$
11	FR	MAS	Mean Angle Similarity	[118]	$\text{MAS}(\mathbf{I}, \hat{\mathbf{I}}) = 1 - \frac{1}{NM} \sum_{i=1}^{N} \sum_{j=1}^{M} \left(\frac{2}{\pi} \cos^{-1} \frac{\langle \mathbf{I}_{i,j}, \hat{\mathbf{I}}_{i,j} \rangle}{\|\mathbf{I}_{i,j}\| \|\hat{\mathbf{I}}_{i,j}\|} \right)$

(continued)

Table 1.2 (continued)

#	Type	Acronym	Name	Ref.	Description
			List of the 25 IQMs implemented		
12	FR	MAMS	Mean Angle Magnitude Similarity	[118]	$MAMS(\mathbf{I}, \hat{\mathbf{I}}) = \frac{1}{NM}\sum_{i=1}^{N}\sum_{j=1}^{M}(1-[1-\alpha_{i,j}][1-\frac{\|\mathbf{I}_{i,j}-\hat{\mathbf{I}}_{i,j}\|}{255}])$
13	FR	TED	Total Edge Difference	[122]	$TED(\mathbf{I_E}, \hat{\mathbf{I}}_E) = \frac{1}{NM}\sum_{i=1}^{N}\sum_{j=1}^{M}|\mathbf{I}_{E_{i,j}} - \hat{\mathbf{I}}_{E_{i,j}}|$
14	FR	TCD	Total Corner Difference	[122]	$TCD(N_{cr}, \hat{N}_{cr}) = \frac{|N_{cr}-\hat{N}_{cr}|}{\max(N_{cr},\hat{N}_{cr})}$
15	FR	SME	Spectral Magnitude Error	[123]	$SME(\mathbf{F}, \hat{\mathbf{F}}) = \frac{1}{NM}\sum_{i=1}^{N}\sum_{j=1}^{M}(|\mathbf{F}_{i,j}| - |\hat{\mathbf{F}}_{i,j}|)^2$
16	FR	SPE	Spectral Phase Error	[123]	$SPE(\mathbf{F}, \hat{\mathbf{F}}) = \frac{1}{NM}\sum_{i=1}^{N}\sum_{j=1}^{M}|(\mathbf{F}_{i,j}) - (\hat{\mathbf{F}}_{i,j})|^2$
17	FR	GME	Gradient Magnitude Error	[124]	$SME(\mathbf{G}, \hat{\mathbf{G}}) = \frac{1}{NM}\sum_{i=1}^{N}\sum_{j=1}^{M}(|\mathbf{G}_{i,j}| - |\hat{\mathbf{G}}_{i,j}|)^2$
18	FR	GPE	Gradient Phase Error	[124]	$SPE(\mathbf{G}, \hat{\mathbf{G}}) = \frac{1}{NM}\sum_{i=1}^{N}\sum_{j=1}^{M}|(\mathbf{G}_{i,j}) - (\hat{\mathbf{G}}_{i,j})|^2$
19	FR	SSIM	Structural Similarity Index	[125]	See [125] and practical implementation available in [126]
20	FR	VIF	Visual Information Fidelity	[127]	See [127] and practical implementation available in [126]
21	FR	RRED	Reduced Ref. Entropic Difference	[128]	See [128] and practical implementation available in [126]
22	NR	JQI	JPEG Quality Index	[129]	See [129] and practical implementation available in [126]
23	NR	HLFI	High-Low Frequency Index	[130]	$SME(\mathbf{I}) = \frac{\sum_{i=1}^{il}\sum_{j=1}^{jl}|\mathbf{I}_{i,j}| - \sum_{i=ih}^{N}\sum_{j=jh+1}^{M}|\mathbf{I}_{i,j}|}{\sum_{i=1}^{N}\sum_{j=1}^{M}|\mathbf{I}_{i,j}|}$
24	NR	BIQI	Blind Image Quality Index	[131]	See [131] and practical implementation available in [126]
25	NR	NIQE	Naturalness Image Quality Estimator	[132]	See [132] and practical implementation available in [126]

full reference IQA metric. This approach assumes that the loss of quality produced by Gaussian filtering differs between real and fake biometric samples. Assumption which is confirmed by the experimental results given in Sect. 1.7.

1.6.1.1   FR-IQMs: Error Sensitivity Measures

Traditional perceptual image quality assessment approaches are based on measuring the errors (i.e., signal differences) between the distorted and the reference images, and attempt to quantify these errors in a way that simulates human visual error sensitivity features.

Although their efficiency as signal fidelity measures is somewhat controversial [116], up to date, these are probably the most widely used methods for IQA as they conveniently make use of many known psychophysical features of the human visual system [117], they are easy to calculate and usually have very low computational complexity.

Several of these metrics have been included in the 25-feature parameterization applied in the present work. For clarity, these features have been classified here into five different categories (see Fig. 1.7) according to the image property measured [118]:

- **Pixel Difference Measures** [118, 121]. These features compute the distortion between two images on the basis of their pixelwise differences. Here we include: Mean Squared Error (**MSE**), Peak Signal-to-Noise Ratio (**PSNR**), Signal-to-Noise Ratio (**SNR**), Structural Content (**SC**), Maximum Difference (**MD**), Average Difference (**AD**), Normalized Absolute Error (**NAE**), R-Averaged Maximum Difference (**RAMD**) and Laplacian Mean Squared Error (**LMSE**). The formal definitions for each of these features are given in Table 1.2.
  In the RAMD entry in Table 1.2, $\max_r$ is defined as the r-highest pixel difference between two images. For the present implementation, $R = 10$.
  In the LMSE entry in Table 1.2, $h(\mathbf{I}_{i,j}) = \mathbf{I}_{i+1,j} + \mathbf{I}_{i-1,j} + \mathbf{I}_{i,j+1} + \mathbf{I}_{i,j-1} - 4\mathbf{I}_{i,j}$.
- **Correlation-Based Measures** [118, 121]. The similarity between two digital images can also be quantified in terms of the correlation function. A variant of correlation-based measures can be obtained by considering the statistics of the angles between the pixel vectors of the original and distorted images. These features include (also defined in Table 1.2): Normalized Cross-Correlation (**NXC**), Mean Angle Similarity (**MAS**), and Mean Angle Magnitude Similarity (**MAMS**).
  In the MAMS entry in Table 1.2, $\alpha_{i,j} = \frac{2}{\pi} \cos^{-1} \frac{\langle \mathbf{I}_{i,j}, \hat{\mathbf{I}}_{i,j} \rangle}{||\mathbf{I}_{i,j}|| ||\hat{\mathbf{I}}_{i,j}||}$
- **Edge-Based Measures**. Edges and other two-dimensional features such as corners are some of the most informative parts of an image, which play a key role in the human visual system and in many computer vision algorithms including quality assessment applications [122].
  Since the structural distortion of an image is tightly linked with its edge degradation, here we have considered two edge-related quality measures: Total Edge Difference (**TED**) and Total Corner Difference (**TCD**).

In order to implement both features, which are computed according to the corresponding expressions given in Table 1.2, we use: (i) the Sobel operator to build the binary edge maps $\mathbf{I_E}$ and $\mathbf{\hat{I}_E}$; (ii) the Harris corner detector [133] to compute the number of corners N_{cr} and $\hat{N}_{cr}$ found in $\mathbf{I}$ and $\mathbf{\hat{I}}$.

- **Spectral Distance Measures**. The Fourier transform is another traditional image processing tool which has been applied to the field of image quality assessment [118, 123]. In this work, we will consider as IQ spectral-related features: the Spectral Magnitude Error (**SME**) and the Spectral Phase Error (**SPE**), defined in Table 1.2 (where $\mathbf{F}$ and $\mathbf{\hat{F}}$ are the respective Fourier transforms of $\mathbf{I}$ and $\mathbf{\hat{I}}$).

- **Gradient-Based Measures**. Gradients convey important visual information which can be of great use for quality assessment. Many of the distortions that can affect an image are reflected by a change in its gradient. Therefore, using such information, structural and contrast changes can be effectively captured [124].

  Two simple gradient-based features are included in the biometric protection system studied here: Gradient Magnitude Error (**GME**) and Gradient Phase Error (**GPE**), defined in Table 1.2 (where $\mathbf{G}$ and $\mathbf{\hat{G}}$ are the gradient maps of $\mathbf{I}$ and $\mathbf{\hat{I}}$ defined as $\mathbf{G} = \mathbf{G}_x + i\mathbf{G}_y$, where $\mathbf{G}_x$ and $\mathbf{G}_y$ are the gradients in the x and y directions).

1.6.1.2 FR-IQMs: Structural Similarity Measures

Although being very convenient and widely used, the aforementioned image quality metrics based on error sensitivity present several problems which are evidenced by their mismatch (in many cases) with subjective human-based quality scoring systems [116]. In this scenario, a recent new paradigm for image quality assessment based on structural similarity was proposed following the hypothesis that the human visual system is highly adapted for extracting structural information from the viewing field [125]. Therefore, distortions in an image that come from variations in lighting, such as contrast or brightness changes (nonstructural distortions), should be treated differently from structural ones.

Among these recent objective perceptual measures, the Structural Similarity Index Measure (**SSIM**) has the simplest formulation and has gained widespread popularity in a broad range of practical applications [125, 134]. In view of its very attractive properties, the SSIM has been included in the 25-feature parameterization.

1.6.1.3 FR-IQMs: Information Theoretic Measures

The quality assessment problem may also be understood, from an information theory perspective, as an information fidelity problem (rather than a signal fidelity problem). The core idea behind these approaches is that an image source communicates to a receiver through a channel that limits the amount of information that could flow through it, thereby introducing distortions. The goal is to relate the visual quality of the test image to the amount of information shared between the test and the reference signals, or more precisely, the mutual information between them. Under this general framework, image quality measures based on information fidelity exploit the (in some cases unprecise) relationship between statistical image information and visual quality [127, 128].

In the present work, we consider two of these information theoretic features: the Visual Information Fidelity (**VIF**) which measures quality fidelity as the ratio between the total information ideally extracted by the brain from the distorted image and that from the reference sample [127]; and the Reduced Reference Entropic Difference index (**RRED**), which approaches the problem of QA from the perspective of measuring distances between the reference image and the projection of the distorted image onto the space of natural images [128].

1.6.2   No-Reference IQ Measures

Unlike the objective reference IQA methods, in general, the human visual system does not require of a reference sample to determine the quality level of an image. Following this same principle, automatic no-reference image quality assessment (NR-IQA) algorithms try to handle the very complex and challenging problem of assessing the visual quality of images in the absence of a reference. Presently, NR-IQA methods generally estimate the quality of the test image according to some pretrained statistical model. Depending on the images used to train this model and on the a priori knowledge required, the methods are coarsely divided into one of three trends [115]:

- **Distortion-Specific Approaches**. These techniques rely on previously acquired knowledge about the type of visual quality loss caused by a specific distortion. The final quality measure is computed according to a model trained on clean images and on images affected by this particular distortion. Two of these measures have been included in the biometric protection method studied in the present work.
  The JPEG Quality Index (**JQI**) evaluates the quality in images affected by the usual block artifacts found in many compression algorithms running at low bit rates such as the JPEG [129].
  The High-Low Frequency Index (**HLFI**) is formally defined in Table 1.2. It was inspired by previous work which considered local gradients as a blind metric to detect blur and noise [130]. Similarly, the HLFI feature is sensitive to the sharpness of the image by computing the difference between the power in the lower and upper frequencies of the Fourier Spectrum. In the HLFI entry in Table 1.2, i_l, i_h, j_l, j_h are respectively the indices corresponding to the lower and upper frequency thresholds considered by the method. In the current implementation, $i_l = i_h = 0.15N$ and $j_l = j_h = 0.15M$.
- **Training-Based Approaches**. Similarly to the previous class of NR-IQA methods, in this type of techniques a model is trained using clean and distorted images. Then, the quality score is computed based on a number of features extracted from the test image and related to the general model [131]. However, unlike the former

approaches, these metrics intend to provide a general quality score not related to a specific distortion. To this end, the statistical model is trained with images affected by different types of distortions.

This is the case of the Blind Image Quality Index (**BIQI**) described in [131], which is part of the 25 feature set used in the present work. The BIQI follows a two-stage framework in which the individual measures of different distortion-specific experts are combined to generate one global quality score.

- **Natural Scene Statistic Approaches**. These blind IQA techniques use a priori knowledge taken from natural scene distortion-free images to train the initial model (i.e., no distorted images are used). The rationale behind this trend relies on the hypothesis that undistorted images of the natural world present certain *regular* properties which fall within a certain subspace of all possible images. If quantified appropriately, deviations from the regularity of natural statistics can help to evaluate the perceptual quality of an image [132].

This approach is followed by the Natural Image Quality Evaluator (**NIQE**) used in the present work [132]. The NIQE is a completely blind image quality analyzer based on the construction of a quality aware collection of statistical features (derived from a corpus of natural undistorted images) related to a multi-variate Gaussian natural scene statistical model.

1.7 Results

In order to achieve reproducible results, we have used in the experimental validation two of the largest publicly available databases for fingerprint spoofing (introduced in Sect. 1.3): (i) the LivDet 2009 DB [21] and (ii) the ATVS-FFp DB [34]. This has allowed us to compare, in an objective and fair way, the performance of the proposed system with other existing state-of-the-art liveness detection solutions.

According to their associated protocols, the databases are divided into a: train set, used to train the quadratic classifier (i.e., based on Quadratic Discriminant Analysis, QDA); and test set, used to evaluate the performance of the protection method. In order to generate unbiased results, there is no overlap between both sets (i.e., samples corresponding to each user are just included in the train or the test set).

The task in *all* the scenarios and experiments described in the next sections is to automatically distinguish between real and fake fingerprints. Therefore, in all cases, results are reported in terms of: the False Genuine Rate (FGR), which accounts for the number of false samples that were classified as real; and the False Fake Rate (FFR), which gives the probability of an image coming from a genuine sample being considered as fake. The Half Total Error Rate (HTER) is computed as HTER $= (FGR + FFR)/2$.

Table 1.3 Results obtained in the ATVS-FFp DB by the two biometric protection methods described in Sects. 1.5 and 1.6

| | Results: ATVS-FFp DB | | | | | | | | |
| | Biometrika | | | Precise | | | Yubee | | |
	FFR	FGR	HTER	FFR	FGR	HTER	FFR	FGR	HTER
IQF-based	4.9	7.6	5.8	1.8	7.0	4.4	2.2	9.7	5.9
IQA-based	9.2	4.0	6.6	6.8	1.5	4.2	7.9	1.9	4.9

1.7.1 Results: ATVS-FFp DB

Both the development and the test set of the ATVS-FFp DB contain half of the fingerprint images acquired with and without the cooperation of the user, following a twofold cross validation protocol. In Table 1.3, we show the detection results of the two systems described in Sects. 1.5 (top row) and 1.6 (bottom row).

The performance of both algorithms is similar, although in the overall, the method based on general image quality assessment is slightly better in two of the three datasets (Precise and Yubee). In addition, thanks to its simplicity and lack of image preprocessing steps, the IQA-based method is around 30 times faster than the one using fingerprint-specific quality features (tested on the same Windows-based platform). This gives the IQA-based scheme the advantage of being usable in practical real-time applications, without losing any accuracy.

1.7.2 Results: LivDet 2009 DB

The train and test sets selected for the evaluation experiments on this database are the same as the ones used in the LivDet 2009 competition, so that the results obtained by the two described methods based on quality assessment may be directly compared to the participants of the contest. Results are shown in the first two rows of Table 1.4. For comparison, the best results achieved in LivDet 2009 for each of the individual datasets are given in the third row.

Rows four to seven show post-competition results over the same dataset and protocol. In [55], a novel fingerprint liveness detection method combining perspiration and morphological features was presented and evaluated on the LivDet 2009 DB following the same protocol (training and test sets) used in the competition. In that work, comparative results were reported with particular implementations of the techniques proposed in: [66], based on wavelet analysis; [69], based on curvelet analysis; and [53], based on the combination of local ridge frequencies and multiresolution texture analysis. In the last four rows of Table 1.4, we also present those results so that they can be compared with the two quality-based methods described in Sects. 1.5 (first row) and 1.6 (second row).

Table 1.4 Results obtained in the LivDet 2009 DB by: the two biometric protection methods described in Sects. 1.5 and 1.6 (IQF-based and IQA-based, top two rows); each of the best approaches participating in LivDet 2009 [21] (third row); the method proposed in [55] which combines perspiration and morphological features (fourth row); the method proposed in [66] based on wavelet analysis, according to an implementation from [55] (fifth row); the method proposed in [69] based on curvelet analysis, according to an implementation from [55] (sixth row); and the method proposed in [53] based on the combination of local ridge frequencies and multiresolution texture analysis, according to an implementation from [55] (bottom row)

| | Results: LivDet 2009 DB | | | | | | | | |
| | Biometrika | | | CrossMatch | | | Identix | | |
	FFR	FGR	HTER	FFR	FGR	HTER	FFR	FGR	HTER
IQF-based	3.1	71.8	37.4	8.8	20.8	13.2	4.8	5.0	6.7
IQA-based	14.0	11.6	12.8	8.6	12.8	10.7	1.1	1.4	1.2
LivDet 2009	15.6	20.7	18.2	7.4	11.4	9.4	2.7	2.8	2.8
Marasco et al.	12.2	13.0	12.6	17.4	12.9	15.2	8.3	11.0	9.7
Moon et al.	20.8	25.0	23.0	27.4	19.6	23.5	74.7	1.6	38.2
Nikam et al.	14.3	42.3	28.3	19.0	18.4	18.7	23.7	37.0	30.3
Abhyankar et al.	24.2	39.2	31.7	39.7	23.3	31.5	48.4	46.0	47.2

The results given in Table 1.4 show that the method based on general image quality assessment outperforms all the contestants in LivDet 2009 in two of the datasets (Biometrika and Identix), while its classification error is just slightly worse than the best of the participants for the Crossmatch data. Although the results are not as good for the case of the IQF-based method, its performance is still competitive compared to that of the best LivDet 2009 participants.

The classification rates of the two quality-based approaches are also clearly lower than those reported in [55] for the different liveness detection solutions tested.

1.8   Conclusions

The study of the vulnerabilities of biometric systems against presentation attacks has been a very active field of research in recent years [36, 37, 135]. This interest has led to big advances in the field of security-enhancing technologies for fingerprint-based applications. However, in spite of this noticeable improvement, the development of efficient protection methods against known threats (usually based on some type of self-manufactured gummy finger) has proven to be a challenging task.

Simple visual inspection of an image of a real fingerprint and its corresponding fake sample shows that the two images can be very similar and even the human eye may find it difficult to make a distinction between them after a short inspection. Yet, some disparities between the real and fake images may become evident once the images are translated into a proper feature space. These differences come from the fact that fingerprints, as 3-D objects, have their own optical qualities (absorption, reflection, scattering, refraction), which other materials (silicone, gelatin, glycerin) or synthetically produced samples do not possess. Furthermore, fingerprint acquisition devices are designed to provide good quality samples when they interact, in a normal operation environment, with a real 3-D trait. If this scenario is changed, or if the trait presented to the scanner is an unexpected fake artifact, the characteristics of the captured image may significantly vary.

In this context, it is reasonable to assume that the image quality properties of real accesses and fraudulent attacks will be different. Following this *"quality difference"* hypothesis, in this chapter, after an overview of early works and main research lines in fingerprint PAD methods, we have explored the potential of quality assessment as a protection tool against fingerprint direct attacks.

For this purpose, we have considered two different feature sets which we have combined with simple classifiers to detect real and fake access attempts: (i) a set of 10 fingerprint-specific quality measures which requires of some preprocessing steps (e.g., fingerprint segmentation); (ii) a set of 25 complementary general image quality measures which may be computed without any image preprocessing.

The two PAD methods have been evaluated on two large publicly available databases following their associated protocols. This way, the results are reproducible and may be fairly compared with other past or future fingerprint PAD solutions.

Several conclusions can be extracted from the evaluation results presented in the experimental sections of the chapter: (i) The proposed methods, especially the one based on general image quality assessment, are able to generalize well performing consistently well for different databases, acquisition conditions, and spoofing scenarios. (ii) The error rates achieved by the described protection schemes are in many cases lower than those reported by other related fingerprint PAD systems which have been tested in the framework of different independent competitions. (iii) In addition to its very competitive performance, the IQA-based approach presents some other very attractive features such as: its simple, fast, nonintrusive, user-friendly and cheap, all of them very desirable properties in a practical protection system.

All the previous results validate the "different-quality" hypothesis formulated in Sect. 1.4, and show the great potential of quality assessment as a PAD tool to secure fingerprint recognition systems.

Overall, the chapter has tried to give an introduction to fingerprint PAD, including an overview of early works, main research lines, and selected results. For more recent and advanced developments occurred in the last 5 years we refer the reader to [36, 37]. In addition, the experimental evaluation carried out in the chapter has been performed following a clear and standard methodology [33] based on common protocols, metrics, and benchmarks, which may serve as a good baseline starting point for the validation of future fingerprint PAD methods.

Acknowledgements This work was done in the context of the TABULA RASA and BEAT projects funded under the 7th Framework Programme of EU. This work was supported by project Cogni-Metrics from MINECO/FEDER under Grant TEC2015-70627-R, and the COST Action CA16101 (Multi-Foresee).

References

1. van der Putte T, Keuning J (2000) Biometrical fingerprint recognition: don't get your fingers burned. In: Proceedings of the IFIP conference on smart card research and advanced applications, pp 289–303
2. Matsumoto T, Matsumoto H, Yamada K, Hoshino S (2002) Impact of artificial gummy fingers on fingerprint systems. In: Proceedings of the SPIE optical security and counterfeit deterrence techniques IV, vol 4677, pp 275–289
3. Thalheim L, Krissler J (2002) Body check: biometric access protection devices and their programs put to the test. ct magazine, pp 114–121
4. Sousedik C, Busch C (2014) Presentation attack detection methods for fingerprint recognition systems: a survey. IET Biom 3(14):219–233. http://digital-library.theiet.org/content/journals/10.1049/iet-bmt.2013.0020
5. Derakhshani R, Schuckers S, Hornak L, O'Gorman L (2003) Determination of vitality from non-invasive biomedical measurement for use in fingerprint scanners. Pattern Recognit 36:383–396
6. Antonelli A, Capelli R, Maio D, Maltoni D (2006) Fake finger detection by skin distortion analysis. IEEE Trans Inf Forensics Secur 1:360–373
7. Galbally J, Alonso-Fernandez F, Fierrez J, Ortega-Garcia J (2012) A high performance fingerprint liveness detection method based on quality related features. Future Gener Comput Syst 28:311–321
8. Franco A, Maltoni D (2008) Advances in biometrics: sensors, algorithms and systems, chap. fingerprint synthesis and spoof detection. Springer, Berlin, pp 385–406
9. Li SZ (ed) (2009) Encyclopedia of biometrics. Springer, Berlin
10. Coli P (2008) Vitality detection in personal authentication systems using fingerprints. PhD thesis, Universita di Cagliari
11. Sandstrom M (2004) Liveness detection in fingerprint recognition systems. Master's thesis, Linkoping University
12. Lane M, Lordan L (2005) Practical techniques for defeating biometric devices. Master's thesis, Dublin City University
13. Blomme J (2003) Evaluation of biometric security systems against artificial fingers. Master's thesis, Linkoping University
14. Lapsley P, Less J, Pare D, Hoffman N (1998) Anti-fraud biometric sensor that accurately detects blood flow
15. Setlak DR (1999) Fingerprint sensor having spoof reduction features and related methods
16. Kallo I, Kiss A, Podmaniczky JT (2001) Detector for recognizing the living character of a finger in a fingerprint recognizing apparatus
17. Diaz-Santana E, Parziale G (2008) Liveness detection method
18. Kim J, Choi H, Lee W (2011) Spoof detection method for touchless fingerprint acquisition apparatus
19. Centro Criptologico Nacional (CCN) (2011) Characterizing attacks to fingerprint verification mechanisms CAFVM v3.0. Common Criteria Portal
20. Bundesamt fur Sicherheit in der Informationstechnik (BSI) (2008) Fingerprint spoof detection protection profile FSDPP v1.8. Common Criteria Portal
21. Marcialis GL, Lewicke A, Tan B, Coli P, Grimberg D, Congiu A, Tidu A, Roli F, Schuckers S (2009) First international fingerprint liveness detection competition – livdet 2009. In: Proceedings of the IAPR international conference on image analysis and processing (ICIAP). LNCS, vol 5716, pp 12–23
22. Ghiani L, Yambay DA, Mura V, Marcialis GL, Roli F, Schuckers SA (2017) Review of the fingerprint Liveness Detection (LivDet) competition series: 2009 to 2015. Image Vis Comput 58:110–128
23. Galbally J, Fierrez J, Alonso-Fernandez F, Martinez-Diaz M (2011) Evaluation of direct attacks to fingerprint verification systems. J Telecommun Syst Special Issue Biom Syst Appl 47:243–254

24. Abhyankar A, Schuckers S (2009) Integrating a wavelet based perspiration liveness check with fingerprint recognition. Pattern Recognit 42:452–464
25. Biometrics Institute: Biometric Vulnerability Assessment Expert Group (2011). http://www.biometricsinstitute.org/pages/biometric-vulnerability-assessment-expert-group-bvaeg.html
26. NPL: National Physical Laboratory: Biometrics (2010). http://www.npl.co.uk/biometrics
27. CESG: Communications-Electronics Security Group - Biometric Working Group (BWG) (2001). https://www.cesg.gov.uk/policyguidance/biometrics/Pages/index.aspx
28. BEAT: Biometrics Evaluation and Testing (2016). http://www.beat-eu.org/
29. TABULA RASA: Trusted biometrics under spoofing attacks (2014). http://www.tabularasa-euproject.org/
30. Maltoni D, Maio D, Jain A, Prabhakar S (2009) Handbook of fingerprint recognition. Springer, Berlin
31. Cappelli R, Maio D, Lumini A, Maltoni D (2007) Fingerprint image reconstruction from standard templates. IEEE Trans Pattern Anal Mach Intell 29:1489–1503
32. Cappelli R (2009) Handbook of fingerprint recognition, chapter, synthetic fingerprint generation. Springer, Berlin, pp 270–302
33. Hadid A, Evans N, Marcel S, Fierrez J (2015) Biometrics systems under spoofing attack: an evaluation methodology and lessons learned. IEEE Signal Process Mag 32(5):20–30
34. Galbally J, Marcel S, Fierrez J (2014) Image quality assessment for fake biometric detection: application to iris, fingerprint and face recognition. IEEE Trans on Image Process 23(2):710–724
35. Sousedik C, Busch C (2014) Presentation attack detection methods for fingerprint recognition systems: a survey. IET Biometrics 3(4):219–233
36. Marasco E, Ross A (2015) A survey on anti-spoofing schemes for fingerprint recognition systems. ACM Comput Surv 47(2):1–36
37. Pinto A, Pedrini H, Krumdick M, Becker B, Czajka A, Bowyer KW, Rocha A (2018) Counteracting presentation attacks in face, fingerprint, and iris recognition. In: Vatsa M, Singh R, Majumdar A (eds) Deep learning in biometrics. CRC Press
38. Wehde A, Beffel JN (1924) Fingerprints can be forged. Tremonia Publish Co, Chicago
39. de Water MV (1936) Can fingerprints be forged? Sci News-Lett 29:90–92
40. Sengottuvelan P, Wahi A (2007) Analysis of living and dead finger impressions identification for biometric applications. In: Proceedings of the international conference on computational intelligence and multimedia applications
41. Yoon S, Feng J, Jain AK (2012) Altered fingerprints: analysis and detection. IEEE Trans Pattern Anal Mach Intell 34:451–464
42. Willis D, Lee M (1998) Biometrics under our thumb. Netw Comput http://www.networkcomputing.com/
43. Sten A, Kaseva A, Virtanen T (2003) Fooling fingerprint scanners - biometric vulnerabilities of the precise biometrics 100 SC scanner. In: Proceedings of the australian information warfare and IT security conference
44. Wiehe A, Sondrol T, Olsen K, Skarderud F (2004) Attacking fingerprint sensors. Technical report NISlab, Gjovik University College
45. Galbally J, Cappelli R, Lumini A, de Rivera GG, Maltoni D, Fierrez J, Ortega-Garcia J, Maio D (2010) An evaluation of direct and indirect attacks using fake fingers generated from ISO templates. Pattern Recognit Lett 31:725–732
46. Barral C, Tria A (2009) Fake fingers in fingerprint recognition: glycerin supersedes gelatin. In: Formal to Practical Security. LNCS, vol 5458, pp 57–69
47. Parthasaradhi S, Derakhshani R, Hornak L, Schuckers S (2005) Time-series detection of perspiration as a liveness test in fingerprint devices. IEEE Trans Syst Man Cybern - Part C: Appl Rev 35:335–343
48. Schuckers S, Abhyankar A (2004) A wavelet based approach to detecting liveness in fingerprint scanners. In: Proceeding of the biometric authentication workshop (BioAW). LNCS, vol 5404. Springer, Berlin, pp 278–386

49. Tan B, Schuckers S (2006) Comparison of ridge- and intensity-based perspiration liveness detection methods in fingerprint scanners. In: Proceeding of the SPIE biometric technology for human identification III (BTHI III), vol 6202, p 62020A
50. Tan B, Schuckers S (2009) A new approach for liveness detection in fingerprint scanners based on valley noise analysis. J Electron Imaging 17:011,009
51. DeCann B, Tan B, Schuckers S (2009) A novel region based liveness detection approach for fingerprint scanners. In: Proceeding of the IAPR/IEEE international conference on biometrics. LNCS, vol 5558. Springer, Berlin, pp 627–636
52. NexIDBiometrics: (2012). http://nexidbiometrics.com/
53. Abhyankar A, Schuckers S (2006) Fingerprint liveness detection using local ridge frequencies and multiresolution texture analysis techniques. In: Proceedings of the IEEE international conference on image processing (ICIP)
54. Marasco E, Sansone C (2010) An anti-spoofing technique using multiple textural features in fingerprint scanners. In: Proceeding of the IEEE workshop on biometric measurements and systems for security and medical applications (BIOMS), pp 8–14
55. Marasco E, Sansone C (2012) Combining perspiration- and morphology-based static features for fingerprint liveness detection. Pattern Recognit Lett 33:1148-1156
56. Cappelli R, Maio D, Maltoni D (2001) Modelling plastic distortion in fingerprint images. In: Proceedings of the international conference on advances in pattern recognition (ICAPR). LNCS, vol 2013. Springer, Berlin, pp 369–376
57. Bazen AM, Gerez SH (2003) Fingerprint matching by thin-plate spline modelling of elastic deformations. Pattern Recognit 36:1859–1867
58. Chen Y, Dass S, Ross A, Jain AK (2005) Fingerprint deformation models using minutiae locations and orientations. In: Proceeding of the IEEE workshop on applications of computer vision (WACV), pp 150–156
59. Chen Y, Jain AK (2005) Fingerprint deformation for spoof detection. In: Proceeding of the IEEE biometric symposium (BSym), pp 19–21
60. Zhang Y, Tian J, Chen X, Yang X, Shi P (2007) Fake finger detection based on thin-plate spline distortion model. In: Proceeding of the IAPR international conference on biometrics. LNCS, vol 4642. Springer, Berlin, pp 742–749
61. Yau WY, Tran HT, Teoh EK, Wang JG (2007) Fake finger detection by finger color change analysis. In: Proceedings of the international conference on biometrics (ICB). LNCS, vol 4642. Springer, Berlin, pp 888–896
62. Jia J, Cai L (2007) Fake finger detection based on time-series fingerprint image analysis. In: Proceedings of the IEEE international conference on intelligent computing (ICIC). LNCS, vol 4681. Springer, Berlin, pp 1140–1150
63. Marcialis GL, Roli F, Tidu A (2010) Analysis of fingerprint pores for vitality detection. In: Proceedings of the IEEE international conference on pattern recognition (ICPR), pp 1289–1292
64. Memon S, Manivannan N, Balachandran W (2011) Active pore detection for liveness in fingerprint identification system. In: Proceeding of the IEEE Telecommuncations Forum (TelFor), pp 619–622
65. Martinsen OG, Clausen S, Nysather JB, Grimmes S (2007) Utilizing characteristic electrical properties of the epidermal skin layers to detect fake fingers in biometric fingerprint systems-a pilot study. IEEE Trans Biomed Eng 54:891–894
66. Moon YS, Chen JS, Chan KC, So K, Woo KC (2005) Wavelet based fingerprint liveness detection. Electron Lett 41
67. Nikam SB, Agarwal S (2009) Feature fusion using gabor filters and cooccrrence probabilities for fingerprint antispoofing. Int J Intell Syst Technol Appl 7:296–315
68. Nikam SB, Argawal S (2009) Ridgelet-based fake fingerprint detection. Neurocomputing 72:2491–2506
69. Nikam S, Argawal S (2010) Curvelet-based fingerprint anti-spoofing. Signal Image Video Process 4:75–87

70. Coli P, Marcialis GL, Roli F (2007) Power spectrum-based fingerprint vitality detection. In: Proceedings of the IEEE workshop on automatic identification advanced technologies (AutoID), pp 169–173

71. Jin C, Kim, H, Elliott S (2007) Liveness detection of fingerprint based on band-selective Fourier spectrum. In: Proceedings of the international conference on information security and cryptology (ICISC). LNCS, vol 4817. Springer, Berlin, pp 168–179

72. Jin S, Bae Y, Maeng H, Lee H (2010) Fake fingerprint detection based on image analysis. In: Proceedings of the SPIE, Sensors, cameras, and systems for industrial/scientific applications XI, vol 7536, p 75360C

73. Lee H, Maeng H, Bae Y (2009) Fake finger detection using the fractional Fourier transform. In: Proceedings of the biometric ID management and multimodal communication (BioID). LNCS, vol 5707. Springer, Berlin, pp 318–324

74. Marcialis GL, Coli P, Roli F (2012) Fingerprint liveness detection based on fake finger characteristics. Int J Digit Crime Forensics 4

75. Coli P, Marcialis GL, Roli F (2007) Vitality detection from fingerprint images: a critical survey. In: Proceedings of the international conference on biometrics (ICB). LNCS, vol 4642. Springer, Berlin, pp 722–731

76. Coli P, Marcialis GL, Roli F (2008) Fingerprint silicon replicas: static and dynamic features for vitality detection using an optical capture device. Int J Image Graph, pp 495–512

77. Marcialis GL, Coli P, Roli F (2012) Fingerprint liveness detection based on fake finger characteristics. Int J Digit Crime Forensics 4:1–19

78. Choi H, Kang R, Choi K, Jin ATB, Kim J (2009) Fake-fingerprint detection using multiple static features. Optic Eng 48:047, 202

79. Nixon KA, Rowe RK (2005) Multispectral fingerprint imaging for spoof detection. In: Proceedings of the SPIE, biometric technology for human identification II (BTHI), vol 5779, pp 214–225

80. Rowe RK, Nixon KA, Butler PW (2008) Advances in biometrics: Sensors, algorithms and systems, Chapter, multispectral fingerprint image acquisition. Springer, Berlin, pp 3–23

81. Yau WY, Tran HL, Teoh EK (2008) Fake finger detection using an electrotactile display system. In: Proceedings of the international conference on control, automation, robotics and vision (ICARCV), pp 17–20

82. Reddy PV, Kumar A, Rahman SM, Mundra TS (2008) A new antispoofing approach for biometric devices. IEEE Trans Biomed Circuits Syst 2:328–337

83. Baldisserra D, Franco A, Maio D, Maltoni D (2006) Fake fingerprint detection by odor analysis. In: Proceedings of the IAPR international conference on biometrics (ICB). LNCS, vol 3832. Springer, Berlin, pp 265–272

84. Cheng Y, Larin KV (2006) Artificial fingerprint recognition using optical coherence tomography with autocorrelation analysis. Appl Opt 45:9238–9245

85. Manapuram RK, Ghosn M, Larin KV (2006) Identification of artificial fingerprints using optical coherence tomography technique. Asian J Phys 15:15–27

86. Cheng Y, Larin KV (2007) In vivo two- and three-dimensional imaging of artificial and real fingerprints with optical coherence tomography. IEEE Photonics Technol Lett 19:1634–1636

87. Larin KV, Cheng Y (2008) Three-dimensional imaging of artificial fingerprint by optical coherence tomography. In: Proceedings of the SPIE biometric technology for human identification (BTHI), vol 6944, p 69440M

88. Chang S, Larin KV, Mao Y, Almuhtadi W, Flueraru C (2011) State of the art in biometrics, chap. fingerprint spoof detection using near infrared optical analysis, Intechopen, pp 57–84

89. Nasiri-Avanaki MR, Meadway A, Bradu A, Khoshki RM, Hojjatoleslami A, Podoleanu AG (2011) Anti-spoof reliable biometry of fingerprints using en-face optical coherence tomography. Opt Photonics J 1:91–96

90. Rattani A, Poh N, Ross A (2012) Analysis of user-specific score characteristics for spoof biometric attacks. In: Proceedings of the IEEE computer society workshop on biometrics at the international conference on computer vision and pattern recognition (CVPR), pp 124–129

91. Marasco E, Ding Y, Ross A (2012) Combining match scores with liveness values in a fingerprint verification system. In: Proceedings of the IEEE international conference on biometrics: theory, applications and systems (BTAS), pp 418–425
92. Hariri M, Shokouhi SB (2011) Possibility of spoof attack against robustness of multibiometric authentication systems. SPIE J Opt Eng 50:079, 001
93. Akhtar Z, Fumera G, Marcialis GL, Roli F (2011) Robustness analysis of likelihood ratio score fusion rule for multi-modal biometric systems under spoof attacks. In: Proceedings of the IEEE international carnahan conference on security technology (ICSST), pp 237–244
94. Akhtar Z, Fumera G, Marcialis GL, Roli F (2012) Evaluation of serial and parallel multi-biometric systems under spoofing attacks. In: Proceedings of the international conference on biometrics: theory, applications and systems (BTAS)
95. Ultra-Scan: (2012). http://www.ultra-scan.com/
96. Optel: (2012). http://www.optel.pl/
97. PosID: (2012). http://www.posid.co.uk/
98. VirdiTech: (2012). http://www.virditech.com/
99. Kang H, Lee B, Kim H, Shin D, Kim J (2003) A study on performance evaluation of the liveness detection for various fingerprint sensor modules. In: Proceedings of the international conference on knowledge-based intelligent information and engineering systems (KES). LNAI, vol 2774. Springer, Berlin, pp 1245–1253
100. Wang L, El-Maksoud RA, Sasian JM, William Kuhn P, Gee K, 2009, V.S.V (2009) A novel contactless aliveness-testing fingerprint sensor. In: Proceedings of the SPIE novel optical systems design and optimization XII, vol 7429, pp 742–915
101. Bayram S, Avcibas I, Sankur B, Memon N (2006) Image manipulation detection. J Electron Imaging 15:041,102
102. Stamm MC, Liu KJR (2010) Forensic detection of image manipulation using statistical intrinsic fingerprints. IEEE Trans Inf Forensics Secur 5:492–496
103. Avcibas I, Memon N, Sankur B (2003) Steganalysis using image quality metrics. IEEE Trans Image Process 12:221–229
104. Avcibas I, Kharrazi M, Memon N, Sankur B (2005) Image steganalysis with binary similarity measures. EURASIP J Appl Signal Process 1:2749–2757
105. Lyu S, Farid H (2006) Steganalysis using higher-order image statistics. IEEE Trans Inf Forensics Secur 1:111–119
106. Lim E, Jiang X, Yau W (2002) Fingerprint quality and validity analysis. In: Proceeding of the IEEE international conference on image processing (ICIP), vol 1, pp 469–472
107. Chen Y, Dass S, Jain A (2005) Fingerprint quality indices for predicting authentication performance. In: Proceedings of the IAPR audio- and video-based biometric person authentication (AVBPA). LNCS, vol 3546. Springer, Berlin, pp 160–170
108. Chen T, Jiang X, Yau W (2004) Fingerprint image quality analysis. In: Proceeding of the IEEE international conference on image processing (ICIP), vol 2, pp 1253–1256
109. Hong L, Wan Y, Jain AK (1998) Fingerprint image enhancement: algorithm and performance evaluation. IEEE Trans Pattern Anal Mach Intell 20(8):777–789
110. Alonso-Fernandez F, Fierrez J, Ortega-Garcia J, Gonzalez-Rodriguez J, Fronthaler H, Kollreider K, Bigun, J (2008) A comparative study of fingerprint image quality estimation methods. IEEE Trans Inf Forensics Secur 2(4):734–743
111. Bigun J (2006) Vision with direction. Springer, Berlin
112. Shen L, Kot A, Koo W (2001) Quality measures of fingerprint images. In: Proceedings of the IAPR audio- and video-based biometric person authentication (AVBPA). LNCS, vol 2091. Springer, Berlin, pp 266–271
113. Wong PW, Pappas TN, Safranek RJ, Chen J, Wang Z, Bovik AC, Simoncelli EP, Sheikh HR (2005) Handbook of image and video processing, Chapter, Sect. VIII: image and video rendering and assessment. Academic Press, New York, pp 925–989
114. Sheikh HRS, Sabir MF, Bovik AC (2006) A statistical evaluation of recent full reference image quality assessment algorithms. IEEE Trans Image Process 15:3440–3451

115. Saad MA, Bovik AC, Charrier C (2012) Blind image quality assessment: a natural scene statatistics approach in the DCT domain. IEEE Trans Image Process 21:3339–3352
116. Wang Z, Bovik AC (2009) Mean squared error: love it or leave it? IEEE Signal Process Mag 26:98–117
117. Teo PC, Heeger DJ (1994) Perceptual image distortion. In: Proceedings of the international conference on image processing, pp 982–986
118. Avcibas I, Sankur B, Sayood K (2002) Statistical evaluation of image quality measures. J Electron Imaging 11:206–223
119. Huynh-Thu Q, Ghanbari M (2008) Scope of validity of PSNR in image/video quality assessment. Electron Lett 44:800–801
120. Yao S, Lin W, Ong E, Lu Z (2005) Contrast signal-to-noise ratio for image quality assessment. In: Proceedings of the international conference on image processing (ICIP), pp 397–400
121. Eskicioglu AM, Fisher PS (1995) Image quality measures and their performance. IEEE Trans Commun 43:2959–2965
122. Martini MG, Hewage CT, Villarini B (2012) Image quality assessment based on edge preservation. Signal Process Image Commun 27:875–882
123. Nill NB, Bouzas B (1992) Objective image quality measure derived from digital image power spectra. Opt Eng 31:813–825
124. Liu A, Lin W, Narwaria M (2012) Image quality assessment based on gradient similarity. IEEE Trans Image Process 21:1500–1511
125. Wang Z, Bovik AC, Sheikh HR, Simoncelli EP (2004) Image quality assessment: from error visibility to structural similarity. IEEE Trans Image Process 13:600–612
126. LIVE: (2012). http://live.ece.utexas.edu/research/Quality/index.htm
127. Sheikh HR, Bovik AC (2006) Image information and visual quality. IEEE Trans Image Process 15:430–444
128. Soundararajan R, Bovik AC (2012) RRED indices: reduced reference entropic differencing for image quality assessment. IEEE Trans Image Process 21:517–526
129. Wang Z, Sheikh HR, Bovik AC (2002) No-reference perceptual quality assessment of JPEG compressed images. In: Proceedings of the IEEE international conference on image processing (ICIP), pp 477–480
130. Zhu X, Milanfar P (2009) A no-reference sharpness metric sensitive to blur and noise. In: Proceedings of the international workshop on quality of multimedia experience (QoMEx), pp 64–69
131. Moorthy AK, Bovik AC (2010) A two-step framework for constructing blind image quality indices. IEEE Signal Process Lett 17:513–516
132. Mittal A, Soundararajan R, Bovik AC (2012) Making a completely blind image quality analyzer. IEEE Signal Process Lett. https://doi.org/10.1109/LSP.2012.2227726
133. Harris C, Stephens M (1988) A combined corner and edge detector. In: Proceeding of the alvey vision conference (AVC), pp 147–151
134. Brunet D, Vrscay ER, Wang Z (2012) On the mathematical properties of the structural similarity index. IEEE Trans Image Process 21:1488–1499
135. Nixon KA, Aimale V, Rowe RK (2008) Handbook of biometrics, Chapter, Spoof detection schemes. Springer, Berlin, pp 403–423

Chapter 2
A Study of Hand-Crafted and Naturally Learned Features for Fingerprint Presentation Attack Detection

Kiran B. Raja, R. Raghavendra, Sushma Venkatesh, Marta Gomez-Barrero, Christian Rathgeb and Christoph Busch

Abstract Fingerprint-based biometric systems have shown reliability in terms of accuracy in both biometric and forensic scenarios. Although fingerprint systems are easy to use, they are susceptible to presentation attacks that can be carried out by employing lifted or latent fingerprints. This work presents a systematic study of the fingerprint presentation attack detection (PAD aka., spoofing detection) using textural features. To this end, this chapter reports an evaluation of both hand-crafted features and naturally learned features via deep learning techniques for fingerprint presentation attack detection. The evaluation is presented on publicly available fake fingerprint database that consists of both bona fide (i.e., real) and presentation attack fingerprint samples captured by capacitive, optical and thermal sensors. The results indicate the need for further approaches that can detect attacks across data from different sensors.

K. B. Raja and R. Raghavendra—Equal contribution of authors.

K. B. Raja (✉) · R. Raghavendra · S. Venkatesh · C. Busch
Norwegian Biometrics Laboratory, Norwegian University of Science and Technology (NTNU),
Trondheim, Norway
e-mail: kiran.raja@ntnu.no

R. Raghavendra
e-mail: raghavendra.ramachandra@ntnu.no

C. Busch
e-mail: christoph.busch@ntnu.no

M. Gomez-Barrero · C. Rathgeb
da/sec Biometrics and Internet Security Research Group, Hochschule Darmstadt, Darmstadt,
Germany
e-mail: marta.gomez-barrero@h-da.de

C. Rathgeb
e-mail: christian.rathgeb@h-da.de

© Springer Nature Switzerland AG 2019
S. Marcel et al. (eds.), *Handbook of Biometric Anti-Spoofing*,
Advances in Computer Vision and Pattern Recognition,
https://doi.org/10.1007/978-3-319-92627-8_2

2.1 Introduction

Fingerprint-based identification or verification of an individual has been in use in ubiquitous scenarios of authentication such as civilian border crossing, forensic analysis, smartphone authentication among a long list of other applications. The preference for such fingerprint-based systems can mainly be attributed to the long proven reliability in terms of accuracy and ability to reinforce the authentication decision by employing multiple fingerprints. While the accuracy and verification performance speak for themselves, there is a stronger concern for the reliability. The key factors for such a concern as outlined in the earlier chapters of this book are the ability to generate artefacts in a simple and cost-effective manner by exploiting the latent fingerprints or lifted fingerprints. It has to be noted that these problems are in addition to the existing challenges emerging due to traditional problem of cuts, abrasions and burns on the fingerprint, which can lead to a degradation of the pattern. These in turn result in low-quality fingerprint captured/acquired with on the optical sensors. Recent research has resulted in a number of newer techniques to tackle presentation attacks through the use of advanced sensing techniques (for instance, optical coherent tomography, full-field optical coherence tomography and Multispectral Imaging (MSI)), Short Wave Infra-Red (SWIR), Laser Contrast Speckle Imaging (LSCI) [1–6] to detect artefacts.

Although newer sensing techniques provide reliable artefact detection, it has to be noted that the cost of sensor production is very high and the technology by itself is at infancy. However, the challenge is to make the traditional and existing fingerprint systems reliable by making them attack resistant using their primary output to detect the attacks (e.g. with software-based methods or by challenge–response approaches). The prominence of the problem is exemplified by the number of works reported on Fake Fingerprint Database (ATVS-FFp database) [7] and a series of ongoing fingerprint Presentation Attack Detection (PAD) competitions (such as LiveDet 2009, LiveDet 2011, LiveDet 2013, LiveDet 2015, LiveDet 2017) [8].

In the process of addressing this challenge, a number of works have been proposed, which rely on minutiae, texture, or quality-based approaches, including the latest techniques involving deep learning [8–15]. This chapter presents an exhaustive summary of techniques dedicated for presentation attack detection on fingerprint recognition systems leveraging the texture-based approaches. Further, we present a comparison using deep learning-based technique against the list of comprehensive techniques. To benchmark the techniques against a common baseline, we present our evaluation results on a publicly available Fake Fingerprint database from ATVS (ATVS-FFp DB) [7] whose details are presented in the later sections of this chapter. The ATVS-FFp database is a large-scale fingerprint presentation attack database which is publicly available at the ATVS-Biometric Recognition Group website.[1] The choice of the database was based on three major components: (1) to report the results on a publicly available database; (2) to study both cooperative and non-cooperative

---

[1] https://atvs.ii.uam.es/atvs/databases.jsp.

presentation attacks; (3) to evaluate different kinds of sensors that include capacitive, optical and thermal sensors for both real and artefact presentation.

2.1.1 Related Works

A number of recent approaches that have been proposed for presentation attack detection in fingerprint recognition systems are presented in recent surveys [8, 16]. General categorization of the presentation attack detection can be provided into hardware or software-based approaches. While the hardware-based methods are discussed in other chapters of this book, we restrict the scope of this chapter to software-based techniques/algorithms. These approaches can broadly be classified as approaches based on dynamic feature and static features [17]. Dynamic features are typically derived by processing multiple frames of the same fingerprint to analyse the dynamic characteristics such as perspiration over time or ridge distortion [17]. Static features are commonly extracted from a single the fingerprint or single image/impression of fingerprint. Taxonomic classes of the fingerprint features have been reported under:

- *Level 1 (Global features)*—which consists of dense singular points, the main orientation of ridges that include arch, tented arch, left loop, right loop and whorl;
- *Level 2 (Local features)*—which consists of dense minutiae details that include ridge ending, ridge bifurcation, lake, independent ridges, island ridge, spur and crossover of two ridges;
- *Level 3 (Fine details)*—which constitutes of concrete details of ridges such as width, shapes, contours and strength of sweat pores.

The attack detection techniques based on the fingerprint descriptors described above were further extended by the use of complementary physical properties such as elasticity of skin from fingerprint, perspiration-based features or a combination of these [17]. A key factor for the fingerprint systems is the quality of the sample captured and it was noted that the quality of artefact significantly differed from quality of bona-fide sample (aka real presentation) [18, 19]. Thus, one of the early works proposed measuring the quality of fingerprint features including strength of ridge, directionality, ridge continuity, clarity of ridges and ridge integrity, structure to differentiate the presentation attack instruments (PAIs) from bona fide samples [10, 18]. Another work identified the use of perspiration pores (sweat pores) to detect the bona fide (i.e., live) fingerprint as compared to the artefacts which have almost no sweat pores or sparsely seen pore-like structures [15].

In many practical operational scenarios, fingerprint acquisition results in insufficient number of Level 1 and Level 2 features. The challenge is addressed in number of works by employing image-based features, such as analysing orientation field and minutiae distribution [9, 10]. Subsequent works have explored other image features such as various texture descriptors [9, 11–15].

Texture is typically characterized by a set of patterns or local variations in image intensity due to structural changes within the image. The classical texture extraction builds upon mathematical calculations on the pixel intensities of the images in a specific manner. The texture features can be further classified as *local texture features* and *global texture features* which are both expected to be invariant to monotonic transformations of grey levels, robust to rotation and translation, illumination invariant. The texture of the image can be expressed in terms of the description of gradients, orientation, local and global statistical features based on mathematical formulations in either local or global neighbourhood. The texture features can therefore be extracted in hand-crafted manner or through the use of filter banks inspired in a natural manner (e.g. Gabor filter banks, Directional Filter banks). Recent works have also proposed texture extraction using bio-inspired approaches of Deep Convolutional Neural Networks (D-CNN). These properties of texture descriptors have resulted in a number of works that have employed various texture features for both fingerprint recognition [9, 11, 20] and attack detection til date [10, 15, 21, 22]. Although the texture descriptors can be used for fingerprint recognition and PAD, in this chapter, we restrict the scope of texture feature descriptors for PAD within fingerprint recognition.

Of the number of earlier works, histograms and wavelets features were used to determine the liveness of the presented finger in [12–15]. Local Phase Quantization was explored by representing all spectrum characteristics in a compact histogram feature by [23]. Weber Local Image Descriptor (WLD) was employed to detect attacks in [24] which was demonstrated to be well suited to high-contrast patterns such as the ridges and valleys of fingerprints images. Multi-scale Block Local Ternary Patterns (MBLTP) (in the family of Local Binary Patterns (LBP)) was explored for fingerprint PAD [25]. The recent works have further explored Binarized Statistical Image Features (BSIF) [26, 27] which is based on extracting texture features using a set of naturally learned features. Recent works have further employed features from Deep Convolutional Neural Networks for fingerprint PAD [28–30]. Motivated by these earlier works, in this chapter, we present an evaluation of texture-based approaches that include both hand-crafted and deeply learned features for fingerprint PAD. Specifically, we evaluate three popular texture- based approaches—Local Binary Pattern (LBP) [31], Local Phase Quantization (LPQ) [32] and Binarized Statistical Image Features (BSIF) [27]. Along with the set of aforementioned texture features, we evaluate three different deep learning (D-CNN) based features such as VGG16, VGG19 and AlexNet [33, 34]. Further, to give the reader a brief overview of working principles of textural features, we provide a brief summary of all the above mentioned techniques in the next section.

2.2 Hand-Crafted Texture Descriptors

This section of the chapter presents three hand-crafted texture descriptors employed for evaluation and a brief description of the corresponding working principles.

2.2.1 Local Binary Pattern

The local binary pattern (LBP) [31] thresholds the intensity values of a pixel around a specified neighbourhood in an image. The threshold is computed based on the intensity of central pixel intensity in a chosen window around the selected pixel. The new binary value of the neighbourhood is computed in a circular symmetric manner by interpolating the locations and checking against the value of the central pixel. If a particular value in the neighbourhood is greater than the chosen central value, 1 is assigned and 0 otherwise. The set of values in a particular chosen block encoded to form the compact pixel value f in the range of 0–255 by using simple binary to decimal conversion strategy as given by Eq. 2.1.

$$f = \sum_{j=1}^{8} Q(i - c) * (2^{(j)})$$
(2.1)

where Q represents the quantized values (such as central pixel value $Q(c)$ and considered pixel $Q(i)$ in a neighbourhood).

2.2.2 Local Phase Qunatization

Local Phase Qunatization (LPQ) [32] is obtained by employing Short-Term Fourier Transform (STFT). In particular, the local time–frequency responses are computed using the discrete STFT in a local window ω in the neighbourhood of n given by

$$F(u, x) = I(x, y)\omega_R(y - x) \exp\{-j2\pi U^T y\}$$
(2.2)

where I is the fingerprint image. The local Fourier coefficients are computed for the frequency points u_1, u_2, u_3 and u_4, which correspond to four points $[a, 0]^T$, $[0, a]^T$, $[a, a]^T$, $[a, -a]^T$ such that the spectral response $H(u_i) > 0$ [32]. The spectral information present in the form of Fourier coefficients is further separated into real and imaginary parts of each component in the Fourier response $[Re\{F\}, Im\{F\}]$ to form a final vector $R = [Re\{F\}, Im\{F\}]$. The Fourier response in the final vector is binarized (Q_i for ith) bit and assigned a value of 1 for all components with response greater than 1 and 0 otherwise as given in Eq. 2.3.

$$Q_i = \begin{cases} 1, & \text{if } R_i > 0 \\ 0, & \text{otherwise} \end{cases}$$
(2.3)

Finally, these are encoded to form the compact pixel value f in the range of 0–255 by using simple binary to decimal conversion strategy as given by Eq. 2.4.

$$f = \sum_{j=1}^{8} Q_j \times 2^{(j-1)} \qquad (2.4)$$

2.2.3 Binarized Statistical Image Features

Binarized Statistical Image Features (BSIF) is motivated by the texture extraction methods such as LBP and LPQ [27]. Instead of the hand-crafted filter configuration, BSIF automatically learns a fixed set of filters from a set of natural images. The BSIF-based technique consists of applying learned textural filters to obtain statistically meaningful representation of the fingerprint sample, which in turn enables an efficient information encoding using binary quantization. A set of filters of patch size $l \times l$ are learned using natural images and Independent Component Analysis (ICA) [27] where the patch size l is defined as:

$$l = (2 * n + 1)$$

such that n ranges from $\{1, 2, \ldots, 8\}$. The set of pre-learned filters from natural images are used to extract the texture features from fingerprint images. If a fingerprint image is represented using $I(x, y)$ and the filter is represented by $H_i(x, y)$ where i represents the basis of the filter, the linear response of the filter s_i can be given as [27]:

$$s_i = \sum_{x,y} I(x, y) H_i(x, y) \qquad (2.5)$$

where x, y represent the dimensions of image and filter, and i the basis of the filter. The response is further binarized based on the obtained response value. If the linear filter response is greater than the threshold, a binarized value of 1 is assigned as given by [27]:

$$b_i = \begin{cases} 1, & \text{if } s_i > 0 \\ 0, & \text{otherwise} \end{cases} \qquad (2.6)$$

The obtained responses b are encoded to form the compact pixel value f in the range of 0–255 by using binary to decimal conversion as provided by Eq. 2.7.

$$f = \sum_{j=1}^{8} b_j \times 2^{(j-1)} \qquad (2.7)$$

2.3 Naturally Learned Features Using Transfer Learning Approaches

Along with the set of hand-crafted features, we evaluate the applicability of naturally learned features using the transfer learning approaches on pre-trained Deep Convolutional Neural Networks (D-CNN). Specifically, we have employed three popular pre-trained D-CNNs such as D-CNNs, namely: VGG16, VGG19, and AlexNet, which are pre-trained on the large-scale ImageNet database [33, 34]. The key motivation to choose the pre-trained networks is due to the ability to handle a small volume of data unlike CNN who rely on large-scale data. Another factor resulting in our choice of transfer learning is their proven good performance for various biometric applications including PAD [28–30, 35]. Given the image I, we crop the Region of Interest (ROI) I_c, corresponding to the fingerprint image I, which is further normalized to have 227×227 pixels for AlexNet and 224×224 pixels for VGG16 and VGG19, in order to comply with the input layer requirements of each network. Further, to perform the fine-tuning of each network, we carried out data augmentation using random cropping that can preserve the semantics of the fingerprint images. As the fine-tuning process is intended to control the learning rate, we have boosted the learning rate of the last layers in all three networks such that they change faster than the rest of the network and thereby, we do not modify the previous layers in AlexNet or VGG net while quickly learning the weights of the newer layer. Thus, for all three networks, we have used the *weight learning rate factor* equalling 10 and *bias learning rate factor* equalling 20.

Given the training set, we tune the pre-trained AlexNet/VGG16 or VGG19 and extract the features f from the last fully connected layer, which is of dimension 1×4096 to train a SoftMax classifier. Considering the non-engineered efforts, these set of features obtained from the set of three different networks are learned purely on the basis of natural features within each class of images (bona fide and artefact).

2.4 Experiments and Results

This section presents the details of the database employed for the evaluation and the corresponding experiments and results.

2.4.1 Database

The ATVS-FFp DB [7] is publicly available at the ATVS-Biometric Recognition Group website.[2] The database comprises bona-fide and artefact fingerprint images

---

[2]https://atvs.ii.uam.es/atvs/databases.jsp.

coming from the index and middle fingers of both hands of 17 subjects ($17 \times 4 = 68$ different fingers). A sample of the images in the database is presented in Fig. 2.1. The database contains over 3,000 bona- fide and artefact fingerprint samples captured from 68 different fingers. Corresponding to each bona-fide finger presentation, two gummy finger artefacts are created by modelling silicone following a cooperative and non-cooperative process. Under the cooperative artefact generation scenario, the subject is asked to place his finger on a mouldable and stable material in order to obtain the negative of the fingerprint. The actual fingerprint is further obtained by recovering the negative mould. Whereas, in non-cooperative artefact generation, a latent fingerprint is used to recover the fingerprint unnoticeably by using a specialized fingerprint development toolkit and then digitized with a scanner. The scanned image is then enhanced through image processing and finally printed on a PCB from which the gummy finger is generated.

Four samples of each fingerprint (bona fide and artefact) were captured in one acquisition session with three different kinds of sensors which include:

- Flat optical sensor Biometrika (Model FX2000, Resolution—569 dpi, Image size—312×372).
- Flat capacitive sensor by Precise Biometrics (Model Precise100SC, 500 dpi, Image size—300×300).
- Sweeping thermal sensor by Yubee with Atmels Fingerchip (Resolution—500 dpi, Image size—232×412).

2.4.2 Performance Evaluation Protocol

The database presented in the previous section is augmented in this section for the experimental protocols. As provided in Table 2.1, the database is divided into training and testing subsets for each of the different types of sensor data. In the case of data captured with cooperation, data from each sensor comprises training samples corresponding to 128 fingers and testing samples corresponding to 144 bona-fide fingers. A similar distribution of dataset is followed for artefact subset for all three sensors: capacitive, optical and thermal sensors. Along the lines of division of dataset for cooperative artefacts generation, the dataset is subdivided for training and testing set for all three sensors. The training and testing set consists of samples from 128 fingers in both bona-fide and artefact samples.

2.4.3 Results on Cooperative Data

This section presents the results on the presentation attack detection for the bona-fide samples and artefacts with cooperation. The results are presented in Table 2.2

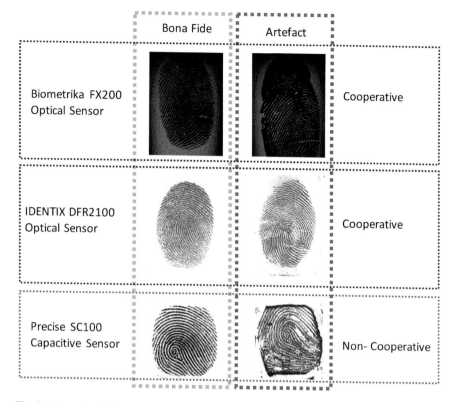

Fig. 2.1 Samples of fingerprint from ATVS database

Table 2.1 Statistics of the training and testing samples

Data collection	Sensor type	Data partition			
		Bona-fide samples		Artefact samples	
		Training samples	Testing samples	Training samples	Testing samples
With cooperation	Capacitive	128	144	128	144
	Optical	128	144	128	144
	Thermal	128	144	128	144
Without cooperation	Capacitive	128	128	128	128
	Optical	128	128	128	128
	Thermal	128	128	128	128

Table 2.2 Quantitative performance of the algorithms on cooperative data

Sensor	Algorithms		D-EER (%)
Capacitive	Hand-crafted features	LBP-SVM	2.77
		LPQ-SVM	6.25
		BSIF-SVM	0.69
	Naturally learned features	AlexNet	0
		VGG16	0
		VGG19	0
Optical	Hand-crafted features	LBP-SVM	29.77
		LPQ-SVM	22.04
		BSIF-SVM	24.26
	Naturally learned features	AlexNet	0.69
		VGG16	0
		VGG19	0.69
Thermal	Hand-crafted features	LBP-SVM	6.98
		LPQ-SVM	12.50
		BSIF-SVM	24.26
	Naturally learned features	AlexNet	0
		VGG16	2.77
		VGG19	2.08

for each different set of attack detection schemes. The performance for both hand-crafted and naturally learned filters are listed together. The Detection Error Trade-off (DET) curves are presented in Figs. 2.2, 2.3 and 2.4 for data captured using capacitive sensor, optical sensor and thermal sensors. The DET illustrates the trade-off between the Attack Presentation Classification Error Rate (APCER) and the Bona-Fide Classification Error Rate (BPCER). A set of the observations can be deduced from the study on applicability of both features for detecting artefacts as listed below:

- In the case of fingerprints captured from capacitive sensor, the best performance is obtained with an Detection-Equal Error Rate (D-EER) of 0.69% using the features. On the other hand, in the case of naturally learned filter, it is reached using deep learning techniques.
- All three different deep learning-based approaches involving AlexNet, VGG16 and VGG19 have achieved an D-EER of 0% in detecting the PAIs, thereby indicating high efficiency/performance with naturally learned features.
- Unlike the data captured from a capacitive sensor, the data captured using optical sensor has been observed as challenging and thereby results in higher error rate in detecting artefacts using hand-crafted features. The highest accuracy of D-EER=22.04% is obtained for LPQ-SVM technique.

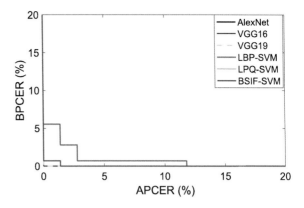

Fig. 2.2 Capacitive sensor—with cooperation

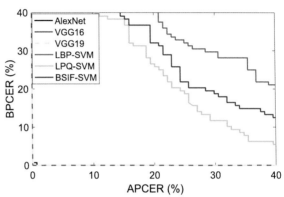

Fig. 2.3 Optical sensor—with cooperation

- Naturally learned features using deep learning techniques such as AlexNet, VGG16 and VGG19 have achieved an D-EER of 0.69%, 0% and 0.69%, respectively. The naturally learned features have drastically reduced the error rates close to 0%.
- For the data captured from the thermal sensor, LBP-SVM outperforms both BSIF- and LPQ-based attack detection schemes. While the lowest error rate obtained is close to 7%, naturally learned features from AlexNet results in D-EER=0% and similar low error rates (<3%) can be observed for VGG net-based features.
- A key observation from the set of results is that the naturally learned features have resulted in very low error rates as compared to hand-crafted features in detecting the attacks stemming from cooperative scenarios.

2.4.4 Results on Non-cooperative Data

This section presents the results obtained on the artefact detection for the data collected without cooperation. This set of experiments correspond to realistic scenarios

Fig. 2.4 Thermal
sensor—with cooperation

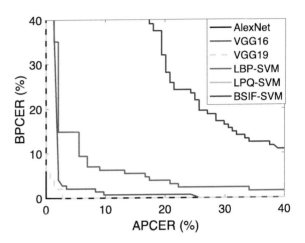

of attacks where lifted fingerprints can be used to perform the attacks. The results
pertaining to this set of experiments are listed in Table 2.3, and the corresponding
DET curves are presented in Figs. 2.5, 2.6 and 2.7 for data captured using capacitive
sensor, optical sensor and thermal sensor, respectively. An analysis of the obtained
results is presented in the following section:

Table 2.3 Quantitative performance of the algorithms on non-cooperative data

Sensor	Algorithms		EER (%)
Capacitive	Hand-crafted features	LBP-SVM	0
		LPQ-SVM	28.90
		BSIF-SVM	0
	Naturally learned features	AlexNet	0
		VGG16	0
		VGG19	0
Optical	Hand-crafted features	LBP-SVM	32.81
		LPQ-SVM	33.20
		BSIF-SVM	47.65
	Naturally learned features	AlexNet	3.12
		VGG16	0
		VGG19	1.56
Thermal	Hand-crafted features	LBP-SVM	48.63
		LPQ-SVM	39.84
		BSIF-SVM	0
	Naturally learned features	AlexNet	33.39
		VGG16	33.39
		VGG19	33.39

Fig. 2.5 Capacitive
sensor—without cooperation

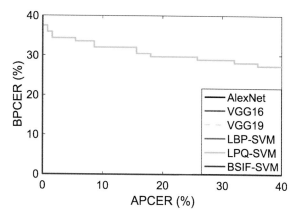

Fig. 2.6 Optical
sensor—without cooperation

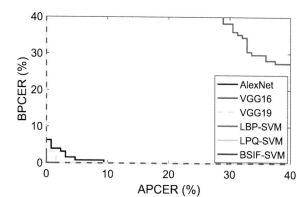

Fig. 2.7 Thermal
sensor—without cooperation

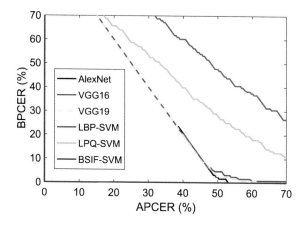

- The data captured from capacitive sensor can be effectively detected using both hand-crafted features like BSIF-SVM and naturally learned features using all three different deep learning networks resulting in D-EER=0%.
- It can be noted that the data captured from optical sensor is not optimal to detect the presentation attacks using the hand-crafted features which results in lowest error rates of D-EER=32.81%. The high error rates suggest the non-applicability of the hand-crafted features to detect attacks, while the naturally learned features using deep learning models result in low error rates. It can be further noted that the VGG16 obtains an D-EER=0% implying the robustness in detecting the attacks.
- Further, the highest rates of error can be seen in the data captured using thermal sensor. While the lowest error rate can be observed as D-EER=0% using BSIF-SVM under hand-crafted features, the lowest error from deep learning models is D-EER=33.39%. An important observation can be deduced is that the artefact data from thermal sensor cannot be easily classified using deep learning models unlike for rest of the artefact data.

2.5   Conclusions

Long-standing use of fingerprint is mainly attributed to ease of capture using the sensors and the robustness in authentication performance. However, the ease of acquiring the fingerprints by lifting the latent prints presents the threat of presentation attacks. In this chapter, we have analysed the techniques based on naturally learned features and hand-crafted features to gauge their ability to detect the presentation attacks. We have systematically demonstrated the applicability of hand-crafted features to detect the artefacts captured under the scenarios of cooperative and non-cooperative scenarios. It was noted that the hand-crafted features work reasonably well for capacitive sensor while not optimal for optical and thermal sensors. Despite the challenge in detecting the PAIs from thermal sensor under non-cooperative attacks, naturally learned features have proven very efficient in detecting the artefacts and have resulted in an D-EER=0%. The set of results obtained on different subsets of artefact data suggests the suitability of naturally learned features. A detailed analysis of naturally learned features using deep learning-based approaches and large training sets need to be carried out in future to devise strategies for reliable presentation attack detection.

Acknowledgements This work was carried out under the funding for SWAN project from the Research Council of Norway under Grant No. IKTPLUSS-248030/O70. This work was partially supported by the German Federal Ministry of Education and Research (BMBF) as well as by the Hessen State Ministry for Higher Education, Research and the Arts (HMWK) within the Center for Research in Security and Privacy (CRISP, www.crisp-da.de).

References

1. Auksorius E, Boccara AC (2015) Fingerprint imaging from the inside of a finger with full-field optical coherence tomography. Biomed Opt Express 6(11)
2. Bicz A, Bicz W (2016) Development of ultrasonic finger reader based on ultrasonic holography having sensor area with 80 mm diameter. In: 2016 international conference of the biometrics special interest group (BIOSIG). IEEE, pp 1–6
3. Yu X, Xiong Q, Luo Y, Wang N, Wang L, Tey HL, Liu L (2016) Contrast enhanced subsurface fingerprint detection using high-speed optical coherence tomography. IEEE Photonics Technol Lett 29(1):70–73
4. Harms F, Dalimier E, Boccara AC (2014) En-face full-field optical coherence tomography for fast and efficient fingerprints acquisition. In: SPIE Defense+ Security, pp 90,750E–90,750E (International society for optics and photonics)
5. Raja KB, Auksorius E, Raghavendra R, Boccara AC, Busch C (2017) Robust verification with subsurface fingerprint recognition using full field optical coherence tomography. In: Proceedings of the IEEE conference on computer vision and pattern recognition workshops, pp 144–152
6. Sousedik C, Breithaupt R, Busch C (2013) Volumetric fingerprint data analysis using optical coherence tomography. In: 2013 international conference of the biometrics special interest group (BIOSIG). IEEE, pp 1–6
7. Galbally J (2015) Anti-spoofing, fingerprint databases. Encyclopedia of biometrics, pp 79–86
8. Ghiani L, Yambay DA, Mura V, Marcialis GL, Roli F, Schuckers SA (2017) Review of the fingerprint liveness detection (LivDet) competition series: 2009 to 2015. Image Vis Comput 58:110–128
9. Jain A, Ross A, Prabhakar S (2001) Fingerprint matching using minutiae and texture features. In: 2001 international conference on image processing, 2001 proceedings, vol 3. IEEE, pp 282–285
10. Yoon S, Feng J, Jain AK (2012) Altered fingerprints: analysis and detection. IEEE Trans Pattern Anal Mach Intell 34(3):451–464
11. Jain AK, Prabhakar S, Hong L, Pankanti S (2000) Filterbank-based fingerprint matching. IEEE Trans Image Process 9(5):846–859
12. Gottschlich C, Marasco E, Yang AY, Cukic B (2014) Fingerprint liveness detection based on histograms of invariant gradients. In: 2014 IEEE international joint conference on biometrics (IJCB). IEEE, pp 1–7
13. Gragnaniello D, Poggi G, Sansone C, Verdoliva L (2014) Wavelet-Markov local descriptor for detecting fake fingerprints. Electron Lett 50(6):439–441
14. Gragnaniello D, Poggi G, Sansone C, Verdoliva L (2015) Local contrast phase descriptor for fingerprint liveness detection. Pattern Recognit 48(4):1050–1058
15. Marasco E, Sansone C (2012) Combining perspiration-and morphology-based static features for fingerprint liveness detection. Pattern Recognit Lett 33(9):1148–1156
16. Sousedik C, Busch C (2014) Presentation attack detection methods for fingerprint recognition systems: a survey. IET Biom 3(4):219–233
17. Marasco E, Ross A (2015) A survey on antispoofing schemes for fingerprint recognition systems. ACM Comput Surv (CSUR) 47(2):28
18. Galbally J, Alonso-Fernandez F, Fierrez J, Ortega-Garcia J (2009) Fingerprint liveness detection based on quality measures. In: 2009 international conference on biometrics, identity and security (BIdS). IEEE, pp 1–8
19. Galbally J, Marcel S, Fierrez J (2014) Image quality assessment for fake biometric detection: application to iris, fingerprint, and face recognition. IEEE Trans Image Process 23(2):710–724
20. Maltoni D, Maio D, Jain AK, Prabhakar S (2009) Handbook of fingerprint recognition. Springer Science & Business Media, New York
21. Menotti D, Chiachia G, Pinto A, Schwartz WR, Pedrini H, Falcão AX, Rocha A (2015) Deep representations for iris, face, and fingerprint spoofing detection. IEEE Trans Inf Forensics Secur 10(4):864–879

22. Nogueira RF, de Alencar Lotufo R, Machado RC (2016) Fingerprint liveness detection using convolutional neural networks. IEEE Trans Inf Forensics Secur 11(6):1206–1213

23. Ghiani L, Marcialis GL, Roli F (2012) Fingerprint liveness detection by local phase quantization. In: 2012 21st international conference on pattern recognition (ICPR). IEEE, pp 537–540

24. Gragnaniello D, Poggi G, Sansone C, Verdoliva L (2013) Fingerprint liveness detection based on weber local image descriptor. In: 2013 IEEE workshop on biometric measurements and systems for security and medical applications (BIOMS). IEEE, pp 46–50

25. Jia X, Yang X, Zang Y, Zhang N, Dai R, Tian J, Zhao J (2013) Multi-scale block local ternary patterns for fingerprints vitality detection. In: 2013 international conference on biometrics (ICB). IEEE, pp 1–6

26. Ghiani L, Hadid A, Marcialis GL, Roli F (2013) Fingerprint liveness detection using binarized statistical image features. In: 2013 IEEE sixth international conference on biometrics: theory, applications and systems (BTAS). IEEE, pp 1–6

27. Kannala J, Rahtu E (2012) BSIF: binarized statistical image features. In: 21st international conference on pattern recognition (ICPR) 2012. IEEE, pp 1363–1366

28. Gottschlich C (2016) Convolution comparison pattern: an efficient local image descriptor for fingerprint liveness detection. PloS One 11(2), e0148,552

29. Kim S, Park B, Song BS, Yang S (2016) Deep belief network based statistical feature learning for fingerprint liveness detection. Pattern Recognit Lett 77, 58–65. https://doi.org/10.1016/j.patrec.2016.03.015. http://www.sciencedirect.com/science/article/pii/S0167865516300198

30. Nogueira RF, de Alencar Lotufo R, Machado RC (2014) Evaluating software-based fingerprint liveness detection using convolutional networks and local binary patterns. In: 2014 IEEE workshop on biometric measurements and systems for security and medical applications (BIOMS) proceedings. IEEE, pp 22–29

31. Ojala T, Pietikainen M, Maenpaa T (2002) Multiresolution gray-scale and rotation invariant texture classification with local binary patterns. IEEE Trans Pattern Anal Mach Intell 24(7):971–987

32. Ojansivu V, Heikkilä J (2008) Blur insensitive texture classification using local phase quantization. In: Elmoataz A, Lezoray O, Nouboud F, Mammass D (eds) Image and signal processing, vol 5099. Springer, Berlin, pp 236–243

33. Krizhevsky A, Sutskever I, Hinton GE (2012) Imagenet classification with deep convolutional neural networks. In: Pereira F, Burges C, Bottou L, Weinberger K (eds) Advances in neural information processing systems, vol 25. Curran Associates, Inc, pp 1097–1105

34. Simonyan K, Zisserman A (2014) Very deep convolutional networks for large-scale image recognition. CoRR. arXiv:1409.1556

35. Raghavendra R, Raja KB, Venkatesh S, Busch C (2017) Transferable deep-CNN features for detecting digital and print-scanned morphed face images. In: 2017 IEEE conference on computer vision and pattern recognition workshops (CVPRW). IEEE, pp 1822–1830

36. Tolosana R, Gomez-Barrero M, Kolberg J, Morales A, Busch C, Ortega-Garcia J (2018) Towards fingerprint presentation attack detection based on convolutional neural networks and short wave infrared imaging. In: proceedings of the IEEE 17th international conference of the biometrics special interest group (BIOSIG), Darmstadt, Germany, September 2018

37. Keilbach P, Kolberg J, Gomez-Barrero M, Busch C, Langweg H (2018) Fingerprint presentation attack detection using laser speckle contrast imaging. In: proceedings of the IEEE 17th international conference of the biometrics special interest group (BIOSIG), Darmstadt, Germany, September 2018

38. Gomez-Barrero M, Kolberg J, Busch C (2018) Towards fingerprint presentation attack detection based on short wave infrared imaging and spectral signatures. In: proceedings Norwegian Information Security Conference (NISK), Svalbard, Norway, September 2018

Chapter 3
Optical Coherence Tomography for Fingerprint Presentation Attack Detection

Yaseen Moolla, Luke Darlow, Ameeth Sharma, Ann Singh and Johan van der Merwe

Abstract New research in fingerprint biometrics uses optical coherence tomography (OCT) technology to acquire fingerprints from where they originate below the surface of the skin. The penetrative nature of this technology means that rich information is available regarding the structure of the skin. This access, in turn, enables new techniques in detecting spoofing attacks, and therefore also introduces mitigation steps against current presentation attack methods. These techniques include the ability to detect fake fingers; fake layers applied above the skin; differentiate between fakes and surface skin conditions; and liveness detection based on, among others, the analysis of eccrine glands and capillary blood flow from below the surface of the skin. Through advances in the OCT hardware and processing techniques, one has increased capabilities to capture large fingerprint volumes at a reasonable speed at the relevant necessary resolution to detect current known attempts at spoofing. The nature of OCT and the data it produces means that a truly high-security fingerprint acquisition system may exist in the future. This work serves to detail current research in this domain.

Y. Moolla (✉) · L. Darlow · A. Sharma · A. Singh · J. van der Merwe
Council for Scientific and Industrial Research, Pretoria, South Africa
e-mail: ymoolla@csir.co.za

L. Darlow
e-mail: s1739461@sms.ed.ac.uk

A. Sharma
e-mail: asharma@csir.co.za

A. Singh
e-mail: asingh1@csir.co.za

J. van der Merwe
e-mail: jvdmerwe3@csir.co.za

© Springer Nature Switzerland AG 2019
S. Marcel et al. (eds.), *Handbook of Biometric Anti-Spoofing*,
Advances in Computer Vision and Pattern Recognition,
https://doi.org/10.1007/978-3-319-92627-8_3

3.1 Introduction

Fingerprints are one of the oldest, most reliable and most widely applied biometrics used in secure identity authentication systems. The fingerprint of any given finger is statistically unique enough to be used to recognise an individual. Since the end of the nineteenth century, law enforcement agencies and forensic departments have used fingerprints for the identification of criminals. This provided a means by which criminals using aliases to avoid the law could still be identified. Over time, with the birth of the digital era, the acquisition and recording of fingerprints has moved from a paper-based system to electronic systems. Comparisons of fingerprints have shifted from a manual to an automated process, giving rise to automated fingerprint identification systems [1]. These systems have since found use in a wider range of applications, such as national identification databases [2], social security agencies [2, 3], border control [4], healthcare management [5], home security systems [6], mobile devices and the banking sector [7].

However, as the technology to identify individuals has advanced, so too have the efforts of criminals advanced in attempts to outwit these systems. Such deception by impostors may grant unauthorised access to the property of another individual, leading to the theft of information, money or even an entire identity. There are many points at which a criminal can attack an identity authentication system. These include

- circumventing the security measures of the servers on which the information is stored;
- intersecting the communication between a fingerprint acquisition device and the server to insert false information, i.e. a man-in-the-middle attack; or
- presenting the stolen identity of another individual at the point of acquisition, i.e. presentation attack [8].

The moment when a user presents their finger to a system is thus a crucial point at which the security of a fingerprint system must be ensured. By early detection of an attempt at unauthorised access, the wealth, possessions and personal information of a system's users can be protected. Optical coherence tomography (OCT) technology provides new means by which a presented fingerprint may be assessed for authenticity and liveness. This chapter will briefly describe the forms of presentation attacks, followed by a discussion in the advances of OCT as a means of detecting these presentation attacks.

3.2 Background

This section will serve to provide background regarding the history and properties of OCT, skin physiology and presentation attack detection. These three ideas contextualise this chapter in that the skin physiology has well-defined physical and optical properties on the surface and below the surface of the skin, OCT has the capability of accurately measuring this physiology, and current fingerprint presentation attack techniques do not perfectly reproduce this physiology. It is this decoupling between skin physiology and presentation attacks that makes OCT well-suited as a fingerprint acquisition device that is capable of consistent and reliable presentation attack detection. The following section details the history and properties of OCT.

3.2.1 History and Properties of OCT

OCT is a non-invasive, non-contact, optical imaging technique that is able to yield volumetric subsurface morphology, both 2D and 3D, of scattering samples in situ and in real time. OCT is often described as the optical analogue to ultrasound. However, the back scattered light cannot be measured electronically due to the high speed of light. Therefore, OCT uses the technique of low coherence interferometry and was first demonstrated by Huang in 1991 [9] and later for a different configuration by Fercher [10]. Since then it has been applied extensively in biomedical applications especially ophthalmology [11], oncology [12], dermatology [13] and cardiology [14] as well as applications in material structure analysis [15], artwork [16] and biometrics [17].

The principle of OCT is shown in Fig. 3.1. The light is split between a sample and a reference mirror (reference path). When the difference between the distance travelled by the light for the sample and the reference paths is within the coherence length of the light source, then interference will occur at the detector. OCT measures the echo time delay and intensity of backscattered light.

Regarding the use of OCT in presentation attack detection, various approaches have been taken to acquire the relevant data. However, the representation of the data afforded by OCT follows a standard structure.

An OCT scan is constructed by measuring the 1D internal structure of a material at some point, or spot. This produces a depth profile, or intensity graph, of this point. This is referred to as an A-scan. Multiple consecutive lateral points are obtained to produce a B-scan, which can be represented as a single image or 'slice' of a volume. Multiple B-scans can then be combined to create a complete volumetric

Fig. 3.1 Diagram depicting the principle of OCT

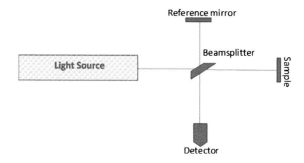

representation of a material. Figure 3.2 shows the progression from an A-scan to a B-scan, to a 3D volume. Figure 3.3 shows the general axes labelling standard for volumetric OCT data. This figure is also an example of a 3D scan of a fingerprint.

The axial width resolution of the light source is referred to as the spot size. The distance between each lateral spot is referred to as the step size. This may be measured in μm/pixel or dots per inch (dpi). Dpi is relevant since it is the standard metric used for measuring fingerprint resolution.

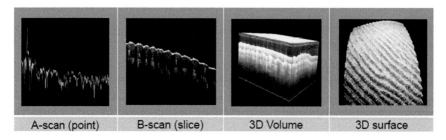

| A-scan (point) | B-scan (slice) | 3D Volume | 3D surface |

Fig. 3.2 The progression from an A-scan, to the concatenation to form a B-scan, to the construction of a 3D volume, to the extraction of a fingerprint [18]

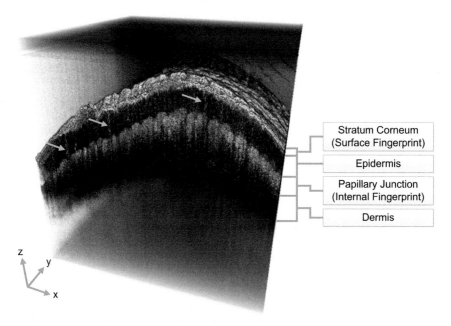

Fig. 3.3 Example OCT volume. Each cross-section slice in the X–Z plane is known as a B-scan. Each 1D signal in the Z direction is read by the OCT scanner. The various layers of the skin are labeled. The yellow arrows indicate the location of some eccrine sweat glands, which are also visible in an OCT scan. ©2016 IEEE. Reprinted with permission from [19]

The axial depth resolution is dependent on the spectral bandwidth of the light source, i.e. the broader the bandwidth, the lower the coherence length and the higher the depth resolution [8, 9], as shown in Eq. 3.1.

$$\Delta_z = \frac{2ln2}{\pi} \frac{\lambda^2}{\Delta\lambda} = 0.44 \frac{\lambda^2}{\Delta\lambda} \tag{3.1}$$

For a refractive index of n, the axial resolution becomes $\Delta\frac{z}{n}$ where Δz is the full-width-at-half-maximum of the autocorrelation function, $\Delta\lambda$ is the spectral bandwidth of the power spectrum and λ is the centre wavelength of the light source. Transverse resolution is determined by

$$\Delta_x = \frac{4\lambda f}{\pi d} \tag{3.2}$$

where d is the spot size of the beam on the objective lens and f is the focal length of the lens. Increasing the transverse resolution decreases the penetration depth. For OCT systems, an objective with low numerical aperture is used. The penetration depth is given by

$$z_{max} = \frac{\lambda_0^2}{4n\delta\lambda} \tag{3.3}$$

OCT systems operate primarily either in the near infrared (NIR), i.e. 1250–1350 nm or 800–900 nm depending on the application. The NIR wavelengths are preferable when imaging non-transparent tissue, or similar samples, due to the better penetration depth. Tissue has low absorption of the beam by proteins, haemoglobin, water and lipids in this region and hence better penetration. Ophthalmology is performed at 800–850 nm due to the transparent nature of the eye. However, this wavelength limits the penetration depth for other tissues, such as skin.

The most common type of OCT systems are Time Domain (TD) and Fourier Domain (FD) OCT. In TD OCT, A-scans are acquired by scanning the reference mirror back and forth to match different depths in the sample to within the coherence length of the light source. The moving mirror limits the acquisition speed and also makes the data susceptible to motion artefacts.

FD OCT, first demonstrated in 1995 [10], obtains A-scans with a fixed reference mirror and measures the spectral response of the interferogram. The interferogram is encoded in an optical frequency space and undergoes a Fourier transform to yield the reflectivity profile of the sample. This configuration is divided into Spectral Domain (SD) OCT, which requires a broadband light source for illumination and separating the spectral components with a spectrometer; and Swept Source (SS) OCT, which uses a light source which probes the sample with different optical frequencies sequentially. The power is then measured with a single photon detector.

The main push for OCT systems for fingerprint comparisons would be resolution, acquisition speed and compactness and cost. At present, the price of OCT systems remain relatively high due to the specialised optical components and this is a challenge that will change as productions costs decrease. Acquisition speed is another challenging performance parameter and, at present, post-processing of OCT data is computationally intensive, requiring high bandwidth electronics. In addition, the scan rate in swept source OCT systems poses another limitation. However, most manufacturers are slowly increasing their offered scan rates. Alternatively, in FD systems, the frame rate, i.e. line speed, of line CCD and CMOS sensors has increased and such systems will soon become competitive in terms of both cost and performance. Researchers at MIT have attempted to push forward the performance of OCT systems using sub-sampling and Fabry–Perót swept sources, which have received interest in recent years [20].

3.2.2 Skin Physiology

The human skin consists of several functional and distinct layers. Most notably, the three layers of the skin are the epidermis, dermis and the hypodermis. The uppermost layer is the epidermis, and the uppermost and external-facing sub-layer of the epidermis is the stratum corneum. The dermis is the layer below the epidermis and the hypodermis is below that. The upper sub-layer of the dermis is the papillary dermis. Thus, the junction between the epidermis and the dermis is known as the papillary junction. Figure 3.3 shows the fingerprint skin layer structure as scanned by an OCT system.

On areas such as fingertips, palms and the soles of the feet, the papillary junction forms friction ridge patterns which result in fingerprints, palmprints and footprints, respectively, on the surfaces of the skin [21]. The stratum corneum and the papillary junction sub-layers are of particular interest in this domain because they present high-contrast regions when scanned using OCT. The particular undulations of these skin layers are the ridges and valleys of the surface and subsurface fingerprints. The surface fingerprint, which is acquired by conventional fingerprint scanners, is a copy of the subsurface fingerprint that exists at the papillary junction. The skin cells grow outwards from the papillary junction and, owing to this relationship, the subsurface fingerprint is the 'master' copy of the surface fingerprint. These two layers can be seen clearly in Fig. 3.3.

Since OCT is able to measure the internal reflectivity, and thus structure of the skin, it grants access to this master fingerprint and other subsurface structures. It is this property that makes OCT a tool of high potential in presentation attack detection. Section 3.2.3 gives a brief outline of presentation attack detection and Sect. 3.3 details presentation attack detection capabilities using OCT technology.

3.2.3 Presentation Attack Detection

A presentation attack is an attempt by an impostor to assume the identity of another individual to obtain unauthorised access to a system. In the context of fingerprint presentation attacks, this usually takes one of the following forms:

1. **Thin layered fakes**: Usually constructed from a master impression and can be made using a variety of materials, such as silicon, these are placed on an attackers finger as an additional layer to be presented to an acquisition device. The master impression could be obtained from a mold of an individual's fingerprint, a latent fingerprint left on a surface, or even from a photograph [22].
2. **Full finger fakes**: Constructed from the same materials as thin layered fakes, full finger fakes are similar to prosthetics and are presented as such.
3. **Severed fingers**: Dismembered fingers and fingers of deceased individuals do not lose their fingerprint pattern immediately and can be used to deceive acquisition devices.
4. **Masking**: A substance without a fingerprint pattern is used to obscure the fingerprint pattern, rather than imposing a false identity.
5. **False claims**: Apart from actual presentation attacks, fraudulent behaviour may include instances where an authorised individual accesses a system and then denies doing so. They may claim to be the victim of a presentation attack. Such denials may require evidence for nonrepudiation.

Fingerprint presentation attack detection is thus the act of detecting such attacks before they succeed. There are many manners in which this may be achieved. From hardware solutions such as pulse, moisture, heat or conductivity detection to software solutions that use trained classifiers to detect the differences in images obtained from fingerprint scanners [23]. These all achieve varying success. OCT technology has been recently explored for presentation attack detection. Owing to the inherent capabilities therein, it has the potential to provide a uniquely well-suited solution in that it can look beyond the presented media and into the internal structure. The following section details the research undertaken thus far.

3.3 Existing and Ongoing Research

Numerous research groups have approached the used of OCT in presentation attack detection. This section will discuss in detail the research carried out by each of these groups in a chronological order.

3.3.1 University of Houston

From 2006 to 2008, the researchers from the Biomedical Optics Laboratory at the University of Houston introduced the use of OCT for fingerprint presentation attack detection. Cheng and Larin [24] suggested enhancing existing fingerprint scanning systems using OCT. They averaged B-scan slices to reduce speckle noise and arrived at a single 1D curve that represented the distribution of light into the skin. Figure 3.4 shows the information they used.

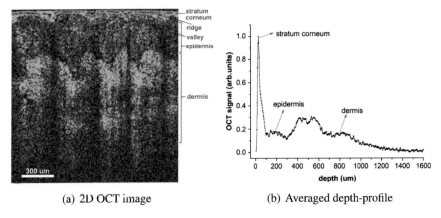

(a) 2D OCT image   (b) Averaged depth-profile

Fig. 3.4 a B-scans were averaged into **b** depth profiles, and autocorrelation analysis was then applied on this. Image reprinted with permission from [24], OSA Publishing

Autocorrelation analysis was then applied to these 1D signals. This technique is used in signal analysis to detect repeating structures. Thus, highly homogeneous signals yield high absolute autocorrelation coefficients, while inhomogeneous signals yield autocorrelation coefficients close to zero. Their assumption was that real human skin exhibits inhomogeneity while fake fingerprints do not. Several materials were tested, namely, gelatin, silicon, wax and agar. Additionally, layer fakes were created from eight fingers, and 10–20 impressions of each fake on a different individuals hand were taken.

The authors used time-domain OCT with a wavelength of 1300 ± 15 nm and a power output of $375\,\mu$W. Single B-scans were performed at a depth of 2.2 mm in air and lateral scan length of 2.4 mm. The B-scans obtained were 450×450 pixels. A single B-scan took 3 s to acquire.

The authors assume that the 1D depth signal of a real finger exhibits changes that are greater than those changes exhibited in materials used to construct fingerprint fakes. This assumption is both a strength and weakness. A strength of this system is that it provides a simple yet effective means of analysing presentation attacks based on the natural physiological layered structure of human skin. Conversely, a weakness is that human skin may not exhibit this behaviour under all circumstances. For instance, in the case of skin damage to the fingerprint [17] the epidermis may be eroded to an extent that autocorrelation analysis begins to fail.

This autocorrelation technique introduced OCT to the domain of presentation attack detection and showed how a simple signal processing technique could use the valuable depth information granted by OCT technology to detect fake fingerprints. In a later work by the same research team [25], the group extended their previous work from a single lateral B-scan of a finger towards a collection of lateral scans to create a volumetric representation of the finger. This development allowed visual analysis of topography of the fake layer and underlying real fingerprint.

This paper also pointed to the possibility of recording and analysing the real fingerprint pattern that is behind the dummy fingerprint. This information could be used to determine the true identity of a perpetrator in a presentation attack. However, the overall area of the finger that was scanned was very small (2.4 mm × 10 mm) with anisotropic resolution of 4762 dpi in one direction and 254 dpi in the other direction. Thus, while this technology was still immature for fingerprint comparisons, the results showed promise for another method of visual analysis to detect presentation attacks.

In a further extension of this work [26], the researchers collaborated with the Optics Group at the Institute for Microstructural Science at the National Research Council of Canada. Instead of a time domain OCT device, a full field OCT device was used. This parallelizes the capture of information to increase the speed of the system and collect data over a wider area and in a shorter time. This allowed for capturing of volumetric information of the fingerprint without movement artefacts, thus providing clearer resolution of 3D fingerprints.

Through the use of a full field OCT device, the investigators were able to render a graphical representation of the fingerprint pattern of a fake fingerprint. This system provided: information of the upper surface of the layered fake, with a fingerprint pattern; the internal structure of the layered fake, which differs from the structure of human tissue; and the bottom surface of the layered fake, which does not hold a fingerprint pattern, is smooth, and does not exist in a real finger.

3.3.2 Bern University of Applied Sciences

In 2010, Bossen et al. [27] from the Bern University of Applied Sciences used a frequency domain OCT system to obtain volumetric information from live fingers. The system was also able to detect differences between fake layers placed on a finger and the real finger below. This was done by visually analysing the internal fingerprint from a fixed depth. When an additional layer is present, the internal fingerprint is obscured. They also discussed the detection of eccrine glands through visual inspection. Further, the paper reported the first biometric comparison study of automatically extracted subsurface fingerprint patterns using the index fingers of 51 individuals. These comparisons showed promising reliability in using the subsurface for fingerprint verification.

The system required ±20 s to capture an area of 14 mm × 14 mm and a depth of 3 mm in air, producing a volume of 512 × 512 × 512 volumetric pixels. Although the performance results were promising, the speed of acquisition of this system was too slow for commercial application. Although OCT systems are capable of contactless fingerprint acquisition, the above- mentioned study required participants to press their fingers against a glass slide when scanning.

3.3.3   University of Delaware

In 2010, Liu and Buma [28] from the Optics and Ultrasonics Research Laboratory at the University of Delaware presented a technique for mapping eccrine sweat glands on the fingertip, using a spectral domain OCT system. The eccrine glands present a 3D helical structure under the surface skin and are the conduits for perspiration that extrude sweat out through the sweat pores on the surface. Thus, there is a direct correlation between eccrine glands and sweat pores which have been shown as effective third-level minutiae detail for biometric comparison, in addition to the conventional comparison of ridge ending and bifurcation second-level minutiae details [29]. Figure 3.5 shows the eccrine glands, the subsurface fingerprint from the papillary junction layer, the mapping between these two, and a fake finger that has no eccrine glands. The use of third-level minutiae features have the potential to improve the confidence of fingerprint comparisons.

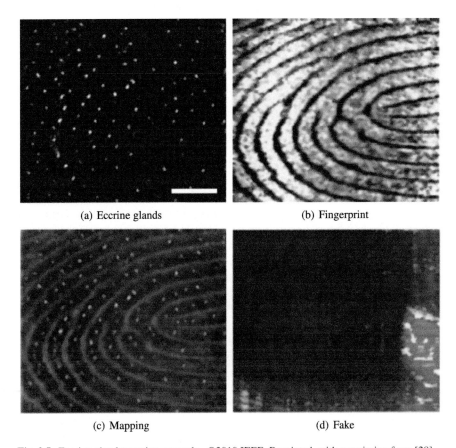

(a) Eccrine glands   (b) Fingerprint

(c) Mapping   (d) Fake

Fig. 3.5 Eccrine gland mapping example. ©2010 IEEE. Reprinted, with permission from [28]

The internal fingerprint from the papillary junction was extracted with the eccrine glands. A number of unique details can be afforded in the following manner:

1. the **internal fingerprint pattern** in the papillary junction is as unique as a surface fingerprint, enabling biometric comparisons;
2. the **global pattern of the eccrine glands** are as unique as the sweat pore pattern of the surface, enabling further reliability to biometric comparisons;
3. the existence of the **helical structures of the eccrine glands** in the epidermis and their optical properties are very nuanced, making it very difficult to reproduce in a fake finger and
4. the distinguishing optical properties of the **papillary junction** between the dermis and epidermis, which is due to the layered structure of skin growth and is also difficult to reproduce in a fake finger.

The data extracted from real fingerprints was compared to full finger fakes that were created to model the optical scattering properties of a real finger, using a mixture of polydimethylsiloxane and titanium dioxide. However, since the full finger fakes do no possess the same underlying subsurface structure as a real finger, the fake and real fingers were easily differentiated. The downsides of the presented system were a slow scanning time and a small captured area. However, the authors proposed means of improving on these limitations.

3.3.4 University of Kent

In 2011, Nasiri-Avanaki et al. [30] of the Applied Optics Group of the University of Kent showed how *en-face* OCT can be used to detect additional layers placed on top of the skin. They used a combination of a dynamic focus *en-face* OCT system and a time-domain OCT system, and showed how an additional masking layer made of sellotape was detected while the true fingerprint below it was extracted. They achieved a high resolution using this type of OCT setup and the sweat pores were well-defined. Figure 3.6 shows the difference between fingers with and without a sellotape mask. Once more, this work evidenced the strength of OCT to image multilayer objects for presentation attack detection.

They also discussed another avenue regarding how OCT can be configured to measure blood flow for liveness detection by measuring differences caused by the Doppler effect. Analysing the movement of blood below the surface of the skin would allow the system to detect if a severed finger or the hand of a deceased individual is presented. Furthermore, they postulate that a single B-scan OCT image may be sufficient for liveness detection. The idea of blood flow analysis was actualized by another group [31] which is discussed in Sect. 3.3.5. They also mentioned that owing to the clear visibility of the eccrine glands, an OCT system could be configured to give a measure of stress by analysing the rate of sweat production.

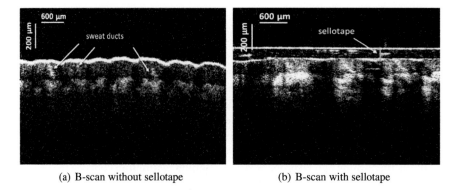

(a) B-scan without sellotape   (b) B-scan with sellotape

Fig. 3.6 Two B-scans showing the **a** absence and **b** presence of sellotape. The sweat ducts are also visible as helical structures in **a**. Images taken from [30]

3.3.5 University of California

In 2013, Liu and Chen [31] of the University of California showed the detection of subsurface fingerprint patterns, eccrine gland patterns, and micro-circulation blood flow patterns using a swept source OCT system. While this work did not explicitly use fake fingerprints to test presentation attack detection, it showed repeatability in the capability of OCT technology to extract reliable subsurface information from a real finger, which are not present in fakes.

Intensity amplitude autocorrelation analysis between adjacent B-scans was used to determine micro-variances that were indicative of blood flow. This technique is known as inter-frame intensity-based Doppler variance. As discussed in the previous section, this provides a means of liveness detection in that blood does not flow in a severed or dead finger. Figure 3.7 shows the micro-circulation patterns for a fingerprint volume. These will not be present in a dead or fake finger.

3.3.6 National University of Ireland

In 2013, through a collaboration between the National University of Ireland and the University of Houston, Dsouza et al. [32] imaged and mapped the micro-circulation of the subsurface fingerprint using a technique called correlation mapping OCT (cmOCT) [33]. This technique applied a correlation mapping algorithm to swept source OCT scans. The 3D scans were processed in sub-stacks of eight *en-face*/X-Z slices: each sub-stack correlation map was generated by passing a 7 × 7 window across the eight slices and measuring the average correlation. The maximum intensity projection map was then calculated. The result thereof was vascular patterns that represented the micro-circulation of the scanned fingerprint. Although their study was small and lacks suitable automation, they demonstrated the utility of OCT regarding

(a) 3D OCT volume   (b) Subsurface fingerprint

(c) Micro-circulation pattern of dermal papilla   (d) Micro-circulation pattern of the entire depth
     region

Fig. 3.7 Vascular patterns obtained using Doppler OCT. Image reprinted with permission from
[31], OSA Publishing

liveness detection through the measurement of blood flow. To simulate an attack that
would use a severed or dead finger, they inhibited the blood flow. This resulted in a
micro-circulation pattern sufficiently different from a normal live finger Fig. 3.8.

McNamara et al. [34] went on to develop a first-generation cost-effective multiple
reference OCT system. The ongoing endeavour by this team is to reduce the cost of
OCT through the use of off-the-shelf components and by the envisioned leveraging
of economies of scale. The reduced cost could pave the way to wider usage of OCT
in commercial applications.

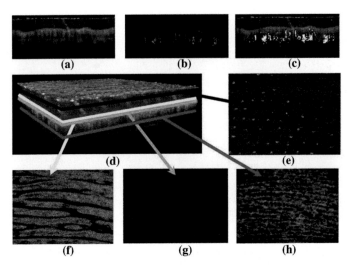

Fig. 3.8 A demonstration of cmOCT. **a** is a B-scan; **b** is after applying the cmOCT algorithm; **c** is an overlay of **a** and **b**; **d** is the OCT volume used; **e** shows the eccrine glands; **f** is the subsurface fingerprint; **g** shows the rising capillary loops at a shallower depth than **h**, which shows the micro-circulation pattern deeper into the skin [32]

3.3.7 OCT Ingress Project

The Ingress project [35] is a collaboration between 10 European partners towards developing technology for fingerprint security including subsurface fingerprint acquisition. This sections will detail some of the research from this project with is relevant to fingerprint presentation attack detection using OCT.

Meissner et al. [36] demonstrated how OCT is a useful tool for high-security biometric control systems regarding presentation attacks using layered fakes. Three-dimensional volume stacks were obtained using a standard swept source OCT system. The dimensions of the scans were 4.5 mm × 4 mm × 2 mm. They manually classified these into real and fake fingerprints by visual assessment of the OCT volumes. Eccrine gland detection was performed and used to classify these scans. Figure 3.9 demonstrates the scans used in this study.

The study was larger than those before it: 7458 scans of live persons, 330 scans of bodies, and 2970 scans of fakes. They achieved a success rate of almost 100% when performing manual detection. The eccrine glands were detected in all live scans but the number differed between persons. The high success rate was attributed to the ease of observing the abnormal layer arrangement that is caused by fake additional layers. Although the technique was not expounded upon, they reported that an 'automatic analysis' of the scans resulted in a misinterpretation of 7% for real fingers and a failure rate of 26% for additional layer fakes.

Sousedik et al. [37] analysed OCT fingerprint volumes towards creating a strong presentation attack detection system. Although research continues in this field, they

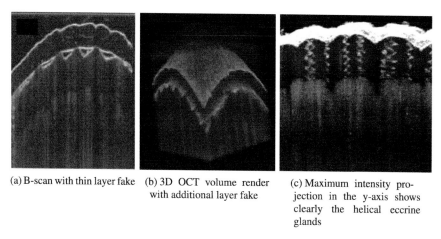

(a) B-scan with thin layer fake

(b) 3D OCT volume render with additional layer fake

(c) Maximum intensity projection in the y-axis shows clearly the helical eccrine glands

Fig. 3.9 **a** B-scan, **b** OCT volume, and **c** maximum intensity projection for presentation attack detection. Images taken from [36]

stated that it remains relatively simple to fool most state-of-the-art fingerprint sensors using low-cost and widely available materials. Even with pattern recognition to distinguish the subtle differences in elasticity or perspiration between real and fake fingerprints, for example, 2D scans acquired by conventional fingerprint sensors are limited and do not provide a level of attack detection to satisfy a high-security application space. Moreover, any single hardware solution, such as moisture detection can usually be fooled by using new materials or production techniques to produce the fingerprint fakes.

Once more, the work in Sousedik et al. [37] stated that because of the penetrative nature of OCT, access is granted to information that can greatly improve presentation attack detection. From eccrine glands to the 'master' subsurface fingerprint, these structures are exceedingly difficult to reproduce in a fake. These do, however, create challenges to overcome when producing an automated presentation attack detection solution using OCT. These challenges include processing large quantities of data (OCT scans can easily exceed 512 MB in size), speckle noise that differs from scanner to scanner, and high intra-class variability in real fingerprints. An automated system must be able to account for the above-mentioned in a reasonable time (a few seconds) and still accurately and reliably detect presentation attacks.

The data used in Meissner et al. [36] and Sousedik et al. [37] remains the most diverse and comprehensive regarding real fingerprints and fake fingerprint classes. With 7458 scans of real fingerprints, 2970 fake fingerprint scans (representing a total of 30 classes considering different combinations of mold material and artefact material compositions), and 330 cadaver fingerprints, this study is noteworthy.

Further work by Sousedik and Busch [38] involved detailing various presentation attack tools and methods, and also prevention techniques. They discussed the potential of OCT for presentation attack detection because it can image the sweat glands under the skin and also grant access to the internal layers of skin. Additionally, they discussed that their work is under the supervision of a member of the German Federal Office for Information Security. Recent work by Breithaupt et al. [39] detailed the full fingerprint OCT-based scanner developed by this team, and Sousedik and Busch [40] detailed some of the challenges related to non-compliant behaviour during capture.

The system developed by Sousedik and Breithaupt in [41] is capable of delivering a full fingerprint scan of a 2 cm × 2 cm area that clearly shows subsurface details, such as sweat glands, that could be used for presentation attack detection. The techniques they used to detect the fingerprint layers, i.e. surface and subsurface fingerprints are detailed in Sousedik et al. [42]. They utilised a novel edge-detection procedure that was implemented on a GPU setup where each edge-detection 'core' was capable of processing an entire A-line scan without memory swaps. In this manner, edge-detection, which is a vital component to 3D to 2D OCT fingerprint processing, could be carried out in near real-time.

In [41], Sousedik and Breithaupt discussed three different resolution configurations for their system. First, at a resolution of 512 dpi, which is close to the standard fingerprint resolution of 500 dpi, an area of 2 cm × 2 cm was scanned in 1.63 s. This was to measure the speed of their system for fingerprint acquisition. Second, samples of a 'medium' resolution of 1408 × 1408 pixels over an area of 2 cm × 2 cm were taken to assess this system as a fingerprint acquisition device. This data was analysed in Sousedik et al. [42] by comparing the extracted OCT fingerprints against fingerprints from the same individuals which were collected using a conventional 2D scanner. Equal error rates (EER) of 0.7 and 1.0% were obtained when comparing the external OCT fingerprint and subsurface OCT fingerprint to 2D surface fingerprints, respectively.

Finally, a small area of higher resolution was intentionally rescanned immediately after the medium resolution scan for each sample. These scans were of an area of 3.58 mm × 3.58 mm and 512 × 512 pixels. These were visualised to show the presence of eccrine glands for presentation attack detection, but not automatically assessed. The work by this research team in the OCT Ingress project is towards a fingerprint acquisition tool that is impervious to presentation attacks. Figure 3.10 shows the surface and subsurface fingerprints that are extracted from the high-resolution B-scans.

3.3.8 Council for Scientific and Industrial Research

The team from the Council for Scientific and Industrial Research in South Africa have also been working on OCT for fingerprint acquisition. They developed two alternative approaches to automated presentation attack detection. The first of these analysed the depth profile of an OCT scan in order to determine if the expected real physiological layer structure is present, which would be different with a fingerprint fake [43]. This is made possible because of the clear differences OCT technology is able to measure, as shown in Fig. 3.11.

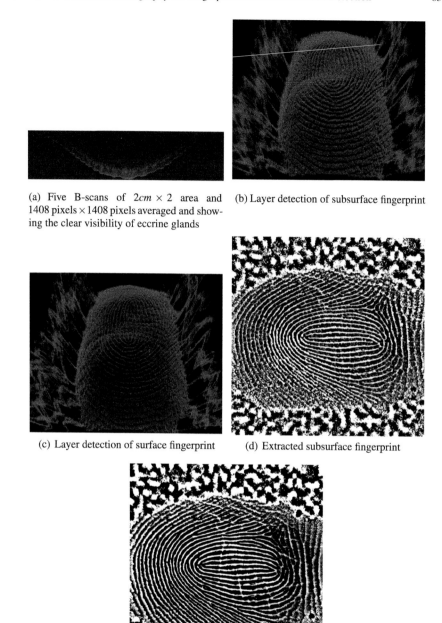

(a) Five B-scans of $2cm \times 2$ area and 1408 pixels $\times$ 1408 pixels averaged and showing the clear visibility of eccrine glands

(b) Layer detection of subsurface fingerprint

(c) Layer detection of surface fingerprint

(d) Extracted subsurface fingerprint

(e) Extracted surface fingerprint

Fig. 3.10 Subsurface and surface layers and corresponding fingerprints. Reprinted from [42] with permission from Springer

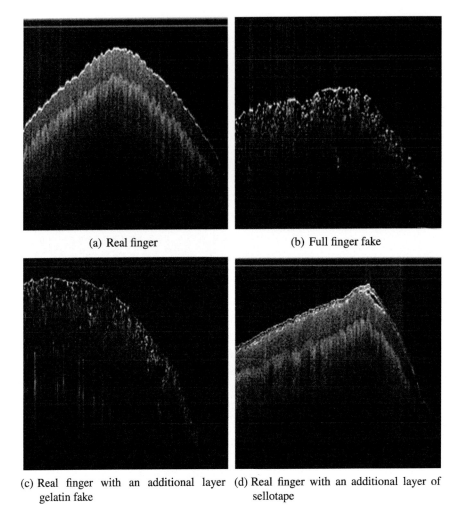

(a) Real finger   (b) Full finger fake

(c) Real finger with an additional layer   (d) Real finger with an additional layer of
    gelatin fake   sellotape

Fig. 3.11 B-scans of several different attack types, demonstrating the obvious qualitative and quantitative differences between real fingers, fake fingers, and real fingers with artefacts present. Images taken from [43]

They computed two independent features to determine (1) the presence of an additional layer fake, and (2) the presence of a thin layer such as sellotape. The first feature was computed by applying autocorrelation analysis to an averaged depth profile. The deviation of the gradient at the inflection point of the autocorrelation analysis was determined to be greater for real fingers when compared to additional layer fakes. This was because the homogeneous repetitive structure of real fingers yields high absolute autocorrelation [24].

The second feature they computed was a measure of 'double' peaks at the surface of the skin. By determining the location of the surface skin through scaling resolution

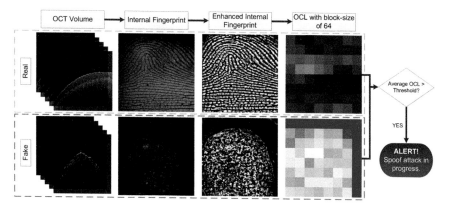

Fig. 3.12 Presentation attack detection by assessment of the quality of subsurface fingerprints. Image taken from [17]

zone detection, as per Darlow et al. [44], and then probing for additional 1D spikes in the depth profiles, they were able to detect additional thin sellotape layers.

The second approach which they developed involved analysis of the extracted subsurface fingerprint [17]. The assumption made therein was that the subsurface fingerprint will only be accessible when there is no additional layer occluding the laser from measuring the skin structure. Although it may be theoretically possible to construct a fingerprint fake that possesses similar layer characteristics, this remains to be accomplished in practice, to our knowledge, as it is a difficult undertaking. Figure 3.12 outlines the procedure for detecting fingerprint presentation attacks through quality assessment of the subsurface fingerprint. The orientation certainty level (OCL) is a simple fingerprint quality assessment technique that measures the consistency of the ridge frequency in non-overlapping blocks.

As is evident in Fig. 3.12, when an additional layer fake is presented to an OCT scanner, the optical properties of the media used to create such a fake can obscure the structure beneath. Thus, the subsurface fingerprint component extracted is meaningless as a fingerprint and results in a very poor quality when assessed. This is also true for a fake finger. This is a straightforward but effective means of detecting presentation attacks. Furthermore, even if the substances used to create the fakes allow for the internal structure to be measured, the actual fingerprint of the perpetrator will be available and the presentation attack will be rendered useless.

Both approaches by this research team yielded a 100% success rate. However, the data was limited and this needs to be confirmed in the future. This research also performed a case study on damaged fingerprints and demonstrated that the internal fingerprints still persisted even when the surface fingerprints were badly eroded. This could be used to reduce the false rejection rate in presentation attack detection systems under these circumstances. Moreover, intentionally damaged fingerprints may still be successfully scanned using OCT.

3.4   Other Advantages and Future Work

OCT has the additional functionality of being able to image a fingerprint even when the surface skin is damaged. Darlow et al. [17] evidenced how OCT could be used to acquire fingerprints even when the surface skin was damaged. In the context of presentation attacks, this damage would be intentional and the effects thereof should be tested in a future work.

Giving the assurance that somebody cannot deny something, or nonrepudiation, is another subtle yet invaluable advantage afforded through OCT technology. Consider the circumstance that an individual, having accessed their personal bank accounts using their fingerprint, claims after the fact that this was actually a fraudulent transaction. With current fingerprint acquisition technology this results in a difficult situation in that there is a low level of assurance that the presented fingerprint was, in fact, real. With OCT, however, the high level of detail and assurance afforded makes this situation less foreseeable.

The greatest existing hurdle to the commercialization of this technology is the cost of such systems. However, research into reducing the cost of such systems is also ongoing [34].

Advances in the bioprinting of human skin [45] and other organs [46] may eventually allow the creation of fake fingers which can mimic fingerprint minutiae patterns, the structure of eccrine glands and blood flow. However, such an attack would require significant advances in artificial organ growth, and has thus not yet been tested to our knowledge.

3.5   Conclusion

This chapter has served to familiarise the reader with OCT as a technology suited to fingerprint presentation attack detection. We have summarised the existing research in the overlapping domains of OCT for fingerprint acquisition and presentation attack detection, describing the research undertaken and ongoing by various groups from around the world.

With numerous approaches, assumptions, and advances demonstrated by these contributors, OCT as a modality for fingerprint acquisition and presentation attack detection may be a reality in the near future.

References

1. Maltoni D, Maio D, Jain A K, Prabhakar S (2009) Handbook of fingerprint recognition. Springer Science & Business Media, New York
2. Breckenridge K (2005) The biometric state: the promise and peril of digital government in the new South Africa. J S Afr Stud 31(2):267–282

3. Mthethwa S, Barbour G, Thinyane M (2016) An improved smartcard for the South African social security agency (SASSA): a proof of life based solution. In: 2016 international conference on information science and security (ICISS). IEEE, pp 1–4
4. Anand A, Labati RD, Genovese A, Munoz E, Piuri V, Scotti F, Sforza G (2016) Enhancing fingerprint biometrics in automated border control with adaptive cohorts. In: 2016 IEEE symposium series on computational intelligence (SSCI). IEEE, pp 1–8
5. Jain AK, Cao K, Arora SS (2014) Recognizing infants and toddlers using fingerprints: increasing the vaccination coverage. In: 2014 IEEE international joint conference on biometrics (IJCB). IEEE, pp 1–8
6. Yukawa M (2004) Home security system. US Patent App. 10/909,354
7. British Broadcasting Corporation (BBC): credit card with a fingerprint sensor revealed by Mastercard (2017). http://www.bbc.com/news/technology-39643453. Accessed 15 August 2017
8. Sussman A, Cukic B, McKeown P, Becker K, Zektser G, Bataller C (2012) IEEE certified biometrics professional (CBP) learning system. Module 3: biometric system design and evaluation. IEEE
9. Huang D, Swanson EA, Lin CP, Schuman JS, Stinson WG, Chang W, Hee MR, Flotte T, Gregory K, Puliafito CA et al (1991) Optical coherence tomography. Science (New York, NY) 254(5035):1178
10. Fercher AF, Hitzenberger CK, Kamp G, El-Zaiat SY (1995) Measurement of intraocular distances by backscattering spectral interferometry. Opt Commun 117(1–2):43–48
11. Gabriele ML, Wollstein G, Ishikawa H, Kagemann L, Xu J, Folio LS, Schuman JS (2011) Optical coherence tomography: history, current status, and laboratory work. Investig Ophthalmol Vis Sci 52(5):2425–2436
12. Verga N, Mirea D, Busca I, Poroschianu M, Zarma S, Grînişteanu L, Gheorghe C, Stan C, Verga M, Vasilache R (2014) Optical coherence tomography in oncological imaging. Romanian Rep Phys 66(1):75–86
13. Welzel J (2001) Optical coherence tomography in dermatology: a review. Skin Res Technol 7(1):1–9
14. Hamdan R, Gonzalez RG, Ghostine S, Caussin C (2012) Optical coherence tomography: from physical principles to clinical applications. Arch Cardiovasc Dis 105(10):529–534
15. Ju MJ, Lee SJ, Min EJ, Kim Y, Kim HY, Lee BH (2010) Evaluating and identifying pearls and their nuclei by using optical coherence tomography. Opt Express 18(13):13,468–13,477
16. Liang H, Cid MG, Cucu RG, Dobre G, Podoleanu AG, Pedro J, Saunders D (2005) En-face optical coherence tomography-a novel application of non-invasive imaging to art conservation. Opt Express 13(16):6133–6144
17. Darlow LN, Singh A, Moolla Y (2016) Damage invariant and high security acquisition of the internal fingerprint using optical coherence tomography. In: World congress on internet security
18. Sharma A, Singh A, Roberts T, Ramokolo R, Strauss H (2016) A high speed OCT system developed at the CSIR national laser centre. In: The 61st annual conference of the South African institute of physics. University of Cape Town
19. Darlow LN, Connan J, Singh A (2016) Performance analysis of a hybrid fingerprint extracted from optical coherence tomography fingertip scans. In: 2016 international conference on biometrics (ICB). IEEE, pp 1–8
20. Siddiqui M, Tozburun S, Vakoc BJ (2016) Simultaneous high-speed and long-range imaging with optically subsampled OCT (conference presentation). In: SPIE BiOS, pp 96,970–96,970 (International society for optics and photonics)
21. Babler WJ (1991) Embryologic development of epidermal ridges and their configurations. Birth Defects Orig Artic Ser 27(2):95–112
22. Cable News Network (CNN): Hackers recreate fingerprints using public photos (2014). http://money.cnn.com/2014/12/30/technology/security/fingerprint-hack/index.html. Accessed 7 August 2017
23. Galbally J, Fierrez J, Ortega-Garcia J, Cappelli R (2014) Fingerprint anti-spoofing in biometric systems. Handbook of biometric anti-spoofing. Springer, Berlin, pp 35–64

24. Cheng Y, Larin KV (2006) Artificial fingerprint recognition by using optical coherence tomography with autocorrelation analysis. Appl Opt 45(36):9238–9245
25. Cheng Y, Larin KV (2007) In vivo two- and three-dimensional imaging of artificial and real fingerprints with optical coherence tomography. IEEE Photonics Technol Lett 19(20):1634–1636
26. Chang S, Cheng Y, Larin KV, Mao Y, Sherif S, Flueraru C (2008) Optical coherence tomography used for security and fingerprint-sensing applications. IET Image Process 2(1):48–58
27. Bossen A, Lehmann R, Meier C (2010) Internal fingerprint identification with optical coherence tomography. IEEE Photonics Technol Lett 22(7):507–509
28. Liu M, Buma T (2010) Biometric mapping of fingertip eccrine glands with optical coherence tomography. IEEE Photonics Technol Lett 22(22):1677–1679
29. Jain AK, Chen Y, Demirkus M (2007) Pores and ridges: high-resolution fingerprint matching using level 3 features. IEEE Trans Pattern Anal Mach Intell 29(1):15–27
30. Nasiri-Avanaki MR, Meadway A, Bradu A, Khoshki RM, Hojjatoleslami A, Podoleanu AG (2011) Anti-spoof reliable biometry of fingerprints using en-face optical coherence tomography. Opt Photonics J 1(03):91
31. Liu G, Chen Z (2013) Capturing the vital vascular fingerprint with optical coherence tomography. Appl Opt 52(22):5473–5477
32. Dsouza RI, Zam A, Subhash HM, Larin KV, Leahy M (2013) In vivo microcirculation imaging of the sub surface fingertip using correlation mapping optical coherence tomography (cmOCT). In: SPIE BiOS, pp 85,800M–1–85,800M–5 (International society for optics and photonics)
33. Jonathan E, Enfield J, Leahy MJ (2011) Correlation mapping method for generating microcirculation morphology from optical coherence tomography (OCT) intensity images. J Biophotonics 4(9):583–587
34. McNamara PM, Dsouza R, ORiordan C, Collins S, OBrien P, Wilson C, Hogan J, Leahy MJ (2016) Development of a first-generation miniature multiple reference optical coherence tomography imaging device. J Biomed Opt 21(12):126,020–126,020
35. Ingress: innovative technology for fingerprint live scanners (2017). http://www.ingress-project.eu/. Accessed 27 July 2017
36. Meissner S, Breithaupt R, Koch E (2013) Defense of fake fingerprint attacks using a swept source laser optical coherence tomography setup. In: SPIE LASE, pp 86,110L–1–86,110L–4 (International society for optics and photonics)
37. Sousedik C, Breithaupt R, Busch C (2013) Volumetric fingerprint data analysis using optical coherence tomography. In: 2013 international conference of the biometrics special interest group (BIOSIG). IEEE, pp 1–6
38. Sousedik C, Busch C (2014) Presentation attack detection methods for fingerprint recognition systems: a survey. IET Biom 3(4):219–233
39. Breithaupt R, Sousedik C, Meissner S (2015) Full fingerprint scanner using optical coherence tomography. In: International workshop on biometrics and forensics. IEEE, pp 1–6
40. Sousedik C, Busch C (2014) Quality of fingerprint scans captured using optical coherence tomography. In: International joint conference on biometrics. IEEE, pp 1–8
41. Sousedik C, Breithaupt R (2017) Full-fingerprint volumetric subsurface imaging using Fourier-domain optical coherence tomography. In: 2017 5th international workshop on biometrics and forensics (IWBF). IEEE, pp 1–6
42. Sousedik C, Breithaupt R, Bours P (2017) Classification of fingerprints captured using optical coherence tomography. In: Scandinavian conference on image analysis. Springer, pp 326–337
43. Darlow LN, Webb L, Botha N (2016) Automated spoof-detection for fingerprints using optical coherence tomography. Appl Opt 55:3387–3396
44. Darlow LN, Connan J, Akhoury SS (2015) Internal fingerprint zone detection in optical coherence tomography fingertip scans. J Electron Imaging **24**, 24 – 24 – 14 (2015). https://doi.org/10.1117/1.JEI.24.2.023027
45. Cubo N, Garcia M, del Cañizo JF, Velasco D, Jorcano JL (2016) 3D bioprinting of functional human skin: production and in vivo analysis. Biofabrication **9**(1), 015,006
46. Murphy SV, Atala A (2014) 3D bioprinting of tissues and organs. Nat Biotechnol 32(8):773–785

Chapter 4
Interoperability Among Capture Devices for Fingerprint Presentation Attacks Detection

Pierliugi Tuveri, L. Ghiani, Mikel Zurutuza, V. Mura and G. L. Marcialis

Abstract A fingerprint verification system is vulnerable to attacks led through the fingertip replica of an enrolled user. The countermeasure is a software/hardware module called fingerprint presentation attacks detector (FPAD) that is able to detect images coming from a real (live) and a spoof (fake) fingertip. We focused our work on the so-called software-based solutions that use a classifier trained with a collection of live and fake fingerprint images in order to determine the liveness level of a finger, that is, the probability that the submitted fingerprint image is not a replica. The chapter goal is to give an overview of FPAD systems by focusing on the problem of the interoperability among different capture devices. In other words, the FPAD performance variation arises when the capture device is substituted by another one, for example, due to upgrading reasons. After a brief summary of the main and most effective state-of-the-art approaches to feature extraction, we introduce the interoperability FPAD problem from the image captured by the fingerprint sensor to the impact on the related feature space and classifier. In particular, we take into account the so-called textural descriptors used for FPAD. We review the state of the art in order to see if and how this problem has been already treated. Finally, a possible solution is suggested and a set of experiments is done to investigate its effectiveness.

We thank Mikel Zurutuza who contributed to this research work during his visiting period for the Global Training Program.

P. Tuveri · L. Ghiani (✉) · M. Zurutuza · V. Mura · G. L. Marcialis
Department of Electrical and Electronic Engineering, University of Cagliari, Cagliari, Italy
e-mail: luca.ghiani@diee.unica.it

P. Tuveri
e-mail: pierluigi.tuveri@diee.unica.it

M. Zurutuza
e-mail: mikelzuru@gmail.com

V. Mura
e-mail: valerio.mura@diee.unica.it

G. L. Marcialis
e-mail: marcialis@diee.unica.it

4.1 Introduction

The personal recognition based on fingerprints has been studied since 1686. Many strides have been made to the present day. The fingerprint is one of the most widely used biometrics as it has uniqueness, permanence, and measurability among its properties. Current electronic devices incorporating fingerprint recognition are de facto standards; in fact, they can be found in laptops or smartphones. Therefore, beyond having a mature theory to support these systems, we also have the technology that is ready to build a recognition system.

Is it possible to consider a fully "secure" fingerprint recognition system?

There have been at least two famous cases where a fingerprint recognition system has been fooled. The first one dates back to 2008 when a woman tried to enter another state using false fingerprints to impersonate some other persons [1]. The second happened most recently in 2013 in a hospital in Sao Paulo where some workers used fingers of colleagues to credit the presence of absentees [1].

However, the research in this field started on 1998 when D. Willis and M. Lee conducted an experiment in which six different biometric fingerprint scanners were tested against artificial fingers. Four out of those six scanners were encountered to be potentially prone to spoof attacks [2]. In the following years, many attempted to replicate this experiment such as Putte et al. [3] in 2001 and Matsumoto et al. [4] in 2002. Later on in 2002, Schuckers [5] deepened into fingerprint scanners that are protected by PAD software, which she claimed to be less vulnerable to spoofing via fake fingers.

After these events, methods proposed to solve the problem were usually classified in the hardware or software type. Hardware solutions are more expensive than the second category as these systems require hardware additions. The software solution is a special module that measures the liveness of a fingerprint placed on the scanner. Such techniques take the name of fingerprint liveness detection (FDL) or fingerprint presentation attack detection (FPAD).

In recent years, to understand the advancement of research by universities and industries, two research centers (Clarkson and Cagliari University) created an international competition on FPAD, called LivDet.[1] Competition datasets are used as benchmarks by everyone, and, in the experimental part of this chapter we use in the LivDet 2011 [6] and LivDet 2015 [7] datasets. The liveness is a property of the fingertip placed in the device surface; thus, it does not depend on the scanner. An ideal scenario is illustrated in the Fig. 4.1, where we have two sensors and a unique features extractor and a pretrained classifier. If we were to change the scanner we would like to keep using the same system, trained with images from a different scanner. This would reduce the maintenance costs of the entire system since the main problem is to collect and reproduce the fake fingerprint. In order to collect a dataset, like those coming from the LivDet competitions, many volunteers are needed, molds and spoof fingerprints have to be created and this involves spending money. Moreover, a

---

[1] http://livdet.org/index.php.

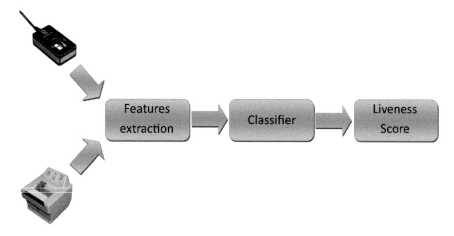

Fig. 4.1 The figure illustrates the ideal solution where a system does not depend on the scanner. We can change the scanners in a system without drop the performances

remarkable manual skill is required in order to replicate the fingerprint. If we reduce this process, we have a cheaper system with respect to the others on the market.

The FPAD systems have so far been built using fingerprint images from a single scanner; thus, the system is customized on just one type of image. These systems cannot generalize the problem; in other words, a system trained on a scanner cannot discriminate images acquired with a different scanner. This lack of interoperability is the subject of this chapter. Figure 4.2 explains our proposed solution: given an old sensor with a trained classifier and a new untrained sensor, the feature extracted from images of the new sensor are moved by the black box from their feature space to the one of the old sensors. This way we can use the old trained classifier on the new system.

The next section proposes a small review of the state-of-the-art FPAD algorithms. Section 4.3 tries to make the reader understand why these systems are not interoperable with each other through experimental evidence. In Sect. 4.4, we propose the model of an interoperable system between scanners. The experimental results described in Sects. 4.5 and 4.6 conclude the chapter.

4.2 Review of Fingerprint Presentation Attacks Detection Methods

In real-world scenarios, one can attempt to circumvent a biometric sensor by using a copy of certain required biometry. The artifact that is used as a counterfeit biometric is also called "spoof" or "fake". Presentation attacks detection (PAD) is the method which distinguishes genuine living biometric from fake ones [8]. An acquired biometric sample may be biased or damaged, but thanks to an enhancing process it can

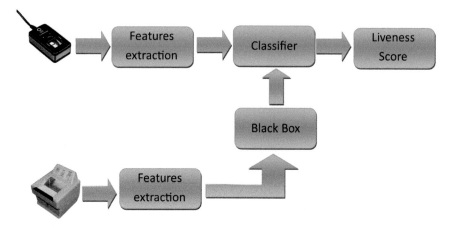

Fig. 4.2 The figure illustrates our solution, we use a black box that allows us to change the scanner

be more accurately classified as "live" or "fake". Such possibility to gather additional information and to make one step beyond a standard verification process is in what PAD is based on. Related to fingerprints, an impostor could use either a dismembered or an artificial finger as a spoof in order to attack a specific system.

4.2.1   Fingerprint Reproduction Process

Generally speaking, two methods are known in order to fabricate a replica of certain finger—cooperative and noncooperative method:

- Cooperative method: The finger of the target individual must be placed into certain ductile material. Several materials are valid for this purpose. The process consists in creating a mold being a negative impression of a fingerprint with either plastic, dental impression material, or even silicone gum. Once the negative mold is ready, it is filled with materials like silicone, gelatine, or PlayDoh.
- Noncooperative method: This is the process to be performed when the subject left a latent fingerprint on a surface and it needs to be enhanced. It can be done by using a photograph to digitize the fingerprint so that the negative image is printed on a transparency sheet. As before, fakes are created by pouring materials like silicone, gelatine, or PlayDoh over the sheet.

4.2.2   Liveness Detection Methods

A fingerprint recognition system must be able to perform the challenging task of distinguishing if a presented fingerprint comes from an artificial finger or a live person. In this sense, several systems are in continuous development in order to be able to detect differences between live fingers and spoofs [9, 10]. We would introduce the reader to the FPAD through an overview of the SOA in this field. It is not and cannot be exhaustive, but it will help to understand the domain adaptation which we talk about in this chapter. We will focus on the software solutions because, as we will explain later, it is cheaper with respect to the hardware modules. We will investigate the FPAD SOA through the Coli's taxonomy [11]. Analyzing the technical evolution of the algorithms it can be noticed that in the first attempts the feature vector contained physical measures. Afterward, textural algorithms were introduced in the field and they outperformed the performance of the first solution category. The textural algorithms analyze the pixel intensity in the fingerprint grayscale image. The grayscale images study conducted by these methods captures the liveness fingerprint characteristics but also information about the source device used for the acquisition. Hence, the feature space distribution depends on the used scanner. We can claim through the next sections that the features space is constrained by the class ω ("live", "fake") and by the device θ. We can write that $X(\omega, \theta)$, where X is the feature space. If we have two different scanners, we have the same feature space but a different location of ω based on the θ. In other words, the samples of two scanners lie in the same hyperspace, but the distributions of "live" and "fake" of one scanner are different with respect to those of the other scanner. Hence, there is no overlapping among the "live" and "fake" samples of the two scanners. In order to understand that we need to introduce the textural algorithms in detail, we try to resume the FPAD SOA.

As mentioned before, attempting to propose a taxonomy for the present field, Coli et al. [11] distinguished two main categories for detection methods: hardware- and software-based:

- Hardware-based: Detection of liveness traits in the fingerprint can be made by using temperature, electrocardiograms, blood pressure or other methods. All of those are additional features that can be incorporated into the system through the use of hardware components.
- Software-based: Instead of using hardware-based solutions, liveness detection capabilities can also be added to the system by using algorithms. Since this approach is totally programming-based, it does not require any additional hardware, consequently being it cheaper. The procedure consists of extracting features from the fingerprint images acquired by the sensors. Those features are then used to determine the liveness condition of the target fingerprint. Depending on the number of images examined, they are known as static or dynamic features. If the feature extraction process is performed against one single fingerprint impression or by comparing different impressions of the same fingerprint, they are called static features. Instead, if the extracted features derive from an analysis of multiple impressions of the same fingerprint, they are called dynamic features. Still, one more subdivision can be defined within the software-based methods according to which physical principle they exploit: the elastic distortion, the morphology, and the perspiration of a fingerprint.

4.2.3 Software-Based Methods State of the Art

As discussed above, software-based liveness detection approaches do not require additional hardware. Then, the state of the art of dynamic and static methods will be analyzed.

4.2.3.1 Dynamic Methods

While the finger makes contact with the surface of the sensor the skin becomes more humid as a consequence of the increasing amount of sweat. Derakhshani et al. in [12] examined the difference among the pores of the fingertip surface within sequential frames. Given that the pores constitute the source of the perspiration process, the authors approached to analyze how they change during a fixed interval of few seconds for either live and fake samples.

Some variations were introduced by Parthasaradhi et al. [13] regarding saturated signals caused by excessive wetness. Two new features were added by them: wet saturation percentage change and dry saturation percentage change. Additionally, two dynamic features were proposed by Coli et al. [14]: the L1-distance of its gray-level histogram and the time variation of the gray-level mean value of the whole image.

Furthermore, by taking human skin elasticity as basis, some new feature extraction methods were introduced by Jia et al. [15]. At the instant, the fingertip is on the scanner surface starts the image capture process. Thereby, a sequence of fingerprint samples describes how the finger is deformed, thus representing the skin elasticity.

With the aim of performing an elastic deformation-based liveness detection Antonelli et al. [16] followed a dynamic procedure. The user must rotate the fingertip after placing the finger on the sensor surface. As a result of such required movement, an elastic tension is caused which correspondingly generates a deformation. It is assumed that live and artificial fingers have a different level of elasticity of the skin. Attempting to use measures observed only in live people, Abhyankar e Schuckers in [17, 18] used the perspiration phenomenon to distinguish live samples from not live samples. Zhang et al. [19] proposed a method based on fingerprint deformation analysis. The subject must first place a finger on the scanner surface. Afterward, some pressure must be applied in four different directions.

4.2.3.2 Static Methods

A novel method for quantifying the perspiration phenomenon in a single image was developed by Tan e Schuckers in [20]. This process follows two principal steps. First, the ridge signal representing the gray-level values along the ridge mask is extracted. After this step, this signal is decomposed into multi-scales by using wavelet transform. Additionally, those authors proposed another liveness detection method in other work [21]. The basis of this methods lies on noise analysis along the valleys in the ridge–valley structure of fingerprint images.

Nikam and Agarwal proposed various methods based on the analysis of a single image. A novel approach based on the ridgelet transform was proposed by them in [22]. Another approach which consists of using curvelet transform for liveness detection is proposed by the two authors [23, 24].

The authors in [25] proposed an integration between local binary pattern (LBP) and wavelet transform. On the one hand, LBP histograms are used to capture details of the texture. On the other hand, ridge frequency and orientation information are determined by wavelet energy features. Other work [26] also consists of using textural measures based on wavelet energy signatures and gray-level co-occurrence matrix (GLCM) features. Within this approach, the authors introduced some statistical measures defined by Haralick [27] with the aim of extracting textural characteristics.

One static feature based on the fast Fourier transform (FFT) of the fingerprint skeleton converted into a mono-dimensional signal was used by Derakhshani et al. [12]. Tan and Schuckers merged their previous works [12, 20, 21] into [28] by defining a measure of the image quality. By observing the finger surface with a high-resolution camera, Moon et al. [29] realized that the surface of a live finger is much less coarse in comparison to an artificial finger. Whereas current sensors present 500 dpi on average, the authors opted to use a 1000 dpi sensor. Based on elastic deformation features Chen et al. [30] proposed a static method by using multiple impressions. Other authors analyzed the frequency domain by using a two-dimensional Fourier transform. Owing to ridgeline discontinuity or the roughness of the skin fingerprint traits at microlevel are sometimes less defined in a fake fingerprint image. As a result, high-frequency details can be either strongly reduced or removed. Coli et al. [11] computed the modulus of the Fourier transform which is typically known as power spectrum for the purpose of measuring such details reduction.

Based on a single fingerprint image, H. Choi et al. [31] proposed another liveness detection method by using multiple static features. With the aim of minimizing the energy associated with phase and orientation maps, Abhyankar et al. [32] introduced a multiresolution texture analysis technique. Cross ridge frequency analysis of fingerprint images was performed by means of statistical measures. Two measures were proposed by Tidu et al. [33] in order to discriminate live and artificial: the use of the number of pores and the mean distance between them.

A novel approach was proposed by Marasco and Sansone [34], relying on static features derived from visual textures of the image. The measures proposed in this work are obtained through first-order statistics, intensity-based features, and the use of signal processing methods. Galbally et al. [35] were able to gather a different set of features by using various sources of information: angle information obtained from the direction field, Gabor filters representing a different method of the direction angle, power spectrum, and pixel intensity of the grayscale image. A novel invariant descriptor of fingerprint ridge texture called histograms of invariant gradients (HIG) is proposed by Gottschlich et al. [36]. Scale invariant feature transform (SIFT) and histograms of oriented gradients (HOG) were invariant feature descriptors on which the authors were based.

4.2.3.3 Textural Algorithms in Fingerprint Presentation Attacks Detection

In this work, we focus our attention on textural algorithms since their performances in the FPAD field proved to be at the SOA in the last few years. We do not claim that the lack of interoperability is typical of those algorithms, on the contrary, it is a more general problem as proved by other works [37, 38] in which the convolutional neural networks were used.

The LBP operator was originally employed for two-dimensional textures analysis. Its excellent performances were improved by the version invariant with respect to gray level, orientation, and rotation [39, 40]. Uniform patterns, corresponding to micro-features in the image, are extracted and the image is characterized by the histogram of these uniform patterns occurrence that combines structural (identification of structures like lines and borders) and statistical (microstructures distribution) approaches.

In a grayscale image, we define the texture T in the circular neighborhood of each pixel as

$$T = t(g_c, g_0, ..., g_{P-1}) \tag{4.1}$$

This represents the distribution of the P surrounding pixels. The grayscale value of the selected pixel is g_c and, given the radius $R > 0$, g_p are the pixels in the circular neighborhood, with $p = 0, ..., P - 1$. Given the origin as g_c position, then the P g_p points are in $(-Rsin(2\pi p/P); Rcos(2\pi p/P))$.

If we subtract the central value from the circular neighborhood values we obtain

$$T = t(g_c, g_0 - g_c, g_1 - g_c, ..., g_{P-1} - g_c) \tag{4.2}$$

Assuming that $g_p - g_c$ values are independent from g_c:

$$T \approx t(g_c)t(g_0 - g_c, g_1 - g_c, ..., g_{P-1} - g_c) \tag{4.3}$$

The overall luminance of the image, unrelated to the local texture, is described in (4.3) by $t(g_c)$. Hence, much of the information is contained in

$$T \approx t(g_0 - g_c, g_1 - g_c, ..., g_{P-1} - g_c) \tag{4.4}$$

By considering the signs of the differences and not their exact values then invariance with respect to the gray-level scaling is achieved

$$T \approx t(s(g_0 - g_c), s(g_1 - g_c), ..., s(g_{P-1} - g_c)) \tag{4.5}$$

with

$$s(x) = \begin{cases} 1, & x \geq 0 \\ 0, & x < 0 \end{cases} \tag{4.6}$$

By assigning the factor 2^p for each sign $s(g_p - g_c)$ it can be obtained a unique $LBP_{P,R}$ value. The possible results are 2^p different binary patterns:

$$LBP_{P,R} = \sum_{p=0}^{P-1} s(g_p - g_c)2^p \tag{4.7}$$

Rotation invariance, namely, to assign a unique identifier to each rotation invariant local binary pattern, is achieved as

$$LBP_{P,R}^{ri} = min\{ROR(LBP_{P,R,i}) \mid i = 0, 1, ..., P - 1\} \tag{4.8}$$

where $ROR(x, i)$ is a function that rotates the neighbor set clockwise thus many times that a maximal number of the most significant bits, starting from g_{P-1}, is 0. A measure of uniformity is the number U of spatial transitions (bitwise 0/1 changes) in the neighborhood pixels sequence. Patterns are defined "uniform" if they have a U value of two at most and the following operator is used:

$$LBP_{P,R}^{riu2} = \begin{cases} \sum_{p=0}^{P-1} s(g_p - g_c) & if \quad U(LBP_{P,R}) \leq 2 \\ P + 1 & otherwise \end{cases} \tag{4.9}$$

were

$$U(LBP_{P,R}) = |s(g_{P-1} - g_c) - s(g_0 - g_c)| + \sum_{p=1}^{P-1} |s(g_p - g_c) - s(g_{p-1} - g_c)| \tag{4.10}$$

The 2^p original values obtained with the $LBP_{P,R}$ are $P + 2$ in the $LBP_{P,R}^{riu2}$. These values, extracted for each pixel of the image, are inserted in a histogram that is used as a feature vector.

The experiments were performed using the rotation invariant version with three different (P, R) values combination: (8, 1), (16, 2) and (24, 3). The three obtained histograms were then united in a single feature vector of $10 + 18 + 26 = 54$ values.

After Nikam and Agarwal published their work [25] the capabilities of the LBP method to capture the different characteristics of live and spoof fingerprints became evident. These capabilities are due to the fact that many different primitive textures are detected by the LBP (Fig. 4.3). The final histogram points out the number textures as spots, line ends, edges, corners, and so on. This ability to filter the image and extract similar primitives is common to textural algorithms. For this reason, from that point on, other textural algorithms started being introduced in the FPAD field.

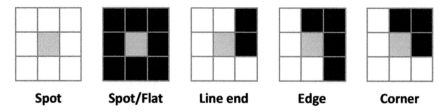

Spot   **Spot/Flat**   **Line end**   **Edge**   **Corner**

Fig. 4.3 Different texture primitives detected by the LBP [40]

A rotation invariant extension of the local phase quantization (LPQ) [41] was proposed by Ghiani et al. [42]. This blur insensitive texture classification technique relies on working at low-frequency values in the frequency domain. For every pixel of the image, the local spectra is computed using a short-term Fourier transform (STFT) in the local neighborhood and four low-frequency components are extracted. Presumably, the key to the capacity of this method for distinguishing a live finger from a spoof one is its blur invariance. Despite the fact that the results achieved were already competitive with the state of the art, they were further improved by using a feature level merge with the LBP.

Another approach related to the Weber local descriptor (WLD) [43] was used by Gragnaniello et al. [44]. This method is based on the differential excitation and the orientation. It relies on the original intensity of certain stimuli such as lighting, more than only on their change. Best results were achieved by combining LPQ and WLD methods.

In 2013, Ghiani et al. [45] proposed to use the binarized statistical image features (BSIF) [46]. It consists of a local image descriptor built through the binarization of the responses to linear filters. However, as opposed to other binary descriptors, these filters are learned using natural images by means of independent component analysis (ICA) method. The local fingerprint texture is then very effectively encoded into a feature vector by a quantization-based representation.

Jia et al. [47] introduced an innovative fingerprint vitality detection approach which lies on multi-scale block local ternary patterns (MBLTP). The local ternary patterns (LTP) are an LBP extension. Instead of the two values, 0 and 1 in Eq. 4.6, a constant threshold t is used to obtain three possible values:

$$s(x) = \begin{cases} 1, & x \geq t \\ 0, & |x| < t \\ -1, & x \leq t \end{cases} \tag{4.11}$$

The computation focuses on the average value of pixel blocks rather than on a single pixel. In order to reflect the differences between a selected threshold and the pixels, the ternary pattern is established.

In a subsequent work, another fake fingerprint detection method was presented by Jia et al. [48] which lies on using two different types of multi-scale local binary pattern (MSLBP). On the one hand, with the first type, the authors just increased the radius of the LBP operator. On the other hand, with the second MSLBP type, they first applied a set of filters to the image, and finally, they applied the LBP operator in the fixed radius. Many of the fake fingerprint detection methods were overtaken by both MSLBP types by properly selecting scales.

4.3   The Interoperability Problem in FPAD Systems

In the previous section, we reviewed the state of the art of FPAD. While the first algorithms used features based on fingerprint physical processes (perspiration, pores detector), lately the extracted features are usually of textural type. One of the first textural feature extractors was the LBP that paved the way to new approaches. Basically, an image is decomposed into a series of fundamental patterns as shown in Fig. 4.3. The frequencies of these patterns compose our histogram, that is, the feature vector. The patterns are identified using the principle of locality of a central pixel. In other words, a pixel is classified in a pattern based on the closest pixels around it. Thus, a textural algorithm works through a statistic of image pixels. The features extraction depends on how the scanner codified the fingerprint ridges and valleys in the image. The image coding depends on the hardware and software differences among scanners.

Some works already raised the FPAD interoperability issue, but they did not dwell into details and the authors did not seem fully aware that it could be a general problem in this field. As a matter of fact, to our knowledge, the works that attempted to make a cross-database experiment noticed a drop in the performance with respect to the use of a unique dataset, without concluding that the problem was related to the lack of interoperability of the extracted features.

For instance, in [48], the authors conducted experiments by training the classifiers with images acquired by a fingerprint scanner and testing with images acquired by another one. They asserted that there are huge differences among the acquired images from different scanners. In the next paragraph, thanks to this claim we will analyze the differences between the images of the different sensors in order to understand this phenomenon.

The interoperability arose even by "skipping" the feature extraction step someway. Marasco et al. [37] avoided the problem of features extraction by adopting three different convolutional neural networks: CaffeNet [49], GoogLeNet [50], and Siamese Network [51]. They used pretrained models, it is a common technique of Transfer Learning. This technique is used when in the new task there are less patterns. These models are further trained with the FPAD dataset. They performed the cross-dataset experiment, that expected train with a dataset coming from a scanner and test with a different dataset coming from another sensor. However, their experiments pointed out that these networks are less able to adapt to the scanner changes. In the authors' opinion, the main problem was the limited number of training images.

CNNs were also used in [38]; in this work, they used pretrained CNNs models. The used CNNs are CNN-VGG [52], CNN-Alexnet [49], and CNN-Random [53], and the authors took pretrained models using different object recognition dataset. These models are customized with an additional training step with FPAD datasets. These models are created for each FPAD dataset and they tested the scanner interoperability with high error rates. Thus, these solutions were not robust to sensor change. In order to estimate the CNNs performance, the authors used the LBP algorithm as a comparison term. In the paper, the authors did not give a scientific explanation of the phenomenon, but they stopped to observe it. However, the paper is not focused on interoperability problem among scanners.

Finally, preliminary experiments reported in our previous paper [54] concurred to include the interoperability problem in the FPAD domain. In that paper, we also presented a possible solution which is detailed in Sect. 4.4.3.

4.3.1 The Origin of the Interoperability Problem

All methods previously reported works on the pixel intensities, with specific regard to CNNs (the filters-based layers) and all the textural features. We do not exclude that this problem may arise with other features (for example, perspiration-based ones), but, for sake of clarity, we limited this chapter scope to the study on textural-based features. Further investigations will be conducted to eventually extend the impact of this phenomenon in a next publication.

Due to the focusing on the processing of each pixel intensity, extracted features depend from each fingerprint scanner characteristics. Each model differs from others in terms of both hardware and software. The principal difference in hardware is the sensor type, which can be optical, solid state or ultrasound [55]. The different physical phenomena codify in grayscale the valleys and ridges fingerprints. DPI (dot per inch), scanning area or geometric accuracy are some of the image characteristics that serve to classify the scanners. Specifically, the DPI is a key point for matcher and liveness detector; it represents the maximum resolution between two points. Some details such as pores are usually highlighted by high-resolution scanners, which is very useful for FPAD. DPI is also extremely important for interdistance measurement of minutiae regarding the identification/verification task. The Bozorth matcher is a good example since it only works at about 500 DPI [56]. The portion of captured fingerprint is defined by the scanning area. For example, due to the small size of their scanners, smartphones do not acquire the entire fingerprint. With the aim of highlighting ridges and valleys, many preprocessing steps can be performed and dynamic of gray levels changes at every step. Contrary to the performance of a presentation attack detector, which lies on a high-frequency analysis, the fingerprint matcher or comparator lies on the minutiae position and it is not dependent on these operations.

Figure 4.4 displays the differences of geometric distortion between two optical fingerprint scanners. They are GreenBit DactyScan 26 and Digital Persona U.are.U 5160 sensors, respectively, which are employed in the fourth edition of the International

Fig. 4.4 Example of the same fingerprint acquired with both a GreenBit DactyScan 26 (left) and a Digital Persona U.are.U 5160 (right) scanner

Competition on Fingerprint Liveness Detection. The Digital Persona scanner generates a more rounded image than the GreenBit one (both images belong to the same fingerprint of the same person).

The histogram representing the gray-level values is another particularity. In Fig. 4.5a, we can observe the mean grayscale histogram for the GreenBit sensor, which displays both training and test sets for LivDet2015.

With the aim of eliminating the background effect, first the center of each fingerprint image is found and then they are segmented into an ROI of 200×200 pixel. All four plots present a quite similar trend. Furthermore, in Fig. 4.5b we can observe on the right side the average histogram of gray levels for all images. Some differences can be appreciated among the four fingerprint sensors (Hi Scan, GreenBit, Digital Persona, Crossmatch) and presumably the feature vectors which were calculated by using methods such as textural algorithms are different as well. The different way of representing the ridges and valleys of a fingerprint affects the frequencies of the histogram. Thus, an image from a scanner is composed of different frequencies of fundamental patterns with respect to another one (Fig. 4.3). Hence, there is no interoperability among scanners and this is a huge problem.

This problem is relevant because we did not succeed in building an interoperable system. The problem phenomenology is due to the different grayscale from different scanners. The problem is of both economical and technological type. The economical side is due to the time and money required in order to replicate the fingerprint. As a matter of fact to collect a spoof dataset we need volunteers that allow us to create a replica of their fingertips and this process takes time for each finger and each used material. Furthermore, each created spoof fingertip has to be acquired with the scanner. All these phases request time and money. Moreover, not all materials are cheap. Related to the technological side we cannot change the scanner in a built system. We cannot use a legacy system with a new scanner; thus, it requires to build a new system. Each different scanner has different characteristics. Moreover, a software company cannot make economies of scale about FPAD tools. In other words, a software company cannot sell a unique FPAD software for all scanners. Hence, it cannot reduce costs, if the tool is custom for the scanner.

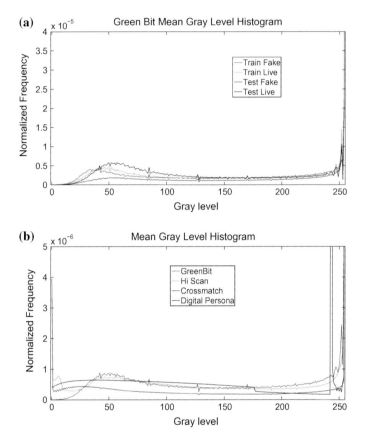

Fig. 4.5 Mean histogram of grayscale GreenBit for all the four subsets (**a**) and differences between the mean histogram of the four LivDet 2015 datasets (**b**)

4.4   Domain Adaptation for the FPAD Interoperability Problem

4.4.1   Problem Definition

The main idea is to project the feature distribution of the novel sensor into that of the "old" sensor. This process allows to keep separate "live" and "fake" classes without need of training a novel classifier, under the assumption of having a small set of novel samples available. Our solution is therefore temporary until a large set of samples from the novel device is available. In the rest of the chapter, we will use the **source** term when we will refer to the old scanner, that is what we would like to change in favor of the new sensor. In the same way, the new scanner will be called **target**. This nomenclature is used when you want to use source knowledge in the target problem. In our case, we want to use the trained model classifier (old) of the source scanner with the target sensor (new). Let $X \in \Re^N$ be a possible feature space, which in our case was built using the algorithms LBP, LPQ, and BSIF, previously described. We call X_S and X_T, respectively, the source and target features space. The source and target features spaces are derived from different scanners; in other words, the fingerprint images are collected using different scanners. Let be $P(X)$ as the marginal probability. We call $P_S(X)$ and $P_T(X)$ as the marginal probability of the source and target features space. The marginal probability is the number of occurrences for each feature. Using the LBP fundamental patterns pictured in Fig. 4.3, the $P_S(X)$ has a different number of occurrences in terms of fundamental patterns with respect to $P_T(X)$.

The Domain Adaptation (DA) is a subfield of the transfer learning theory, where we may have the same features configurations for the same pattern (also called sample, or observation), namely, X_S and X_T, that is, $X_S = X_T$, but different marginal probabilities, that is, $P_S(X) \neq P_T(X)$. This may happen when the same feature extraction steps are applied to an information coming from different acquisition scanners. In our specific case, the same fingerprint is acquired by two different capture devices.

In Sect. 4.2.2, we wrote $X(\omega, \theta)$ where ω represents "live" and "fake" classes and θ is a scanner parameter. The parameter θ may model the different liveness fingerprint measures among scanners. In other words, $X_T(\omega, \theta_T) \neq X_S(\omega, \theta_S)$. For sake of brevity, $X_T(\omega, \theta_T) = X_T$ and $X_S(\omega, \theta_S) = X_S$. In the next section, we will prove through experiments that the interoperability problem is a DA problem. Moreover, the black box in Fig. 4.2 in the Sect. 4.1 is our solution and we take advantage of the DA theory in order to explain it. We would like the marginal probability $P_T(X)$ to be similar to $P_S(X)$; thus, we can use the source classifier. This is the focus of Sect. 4.4.3.

4.4.2 Experimental Evidences ($P_S(X) \neq P_T(X)$)

In this section, we simply show that it is not possible to use the source information in order to classify the target features space. First of all, the ω parameter falls in the {"*live*", "*fake*"} set. In other words, we have the standard problem of FPAD.

The feature extractor (one of the textural descriptors described in Sect. 4.2.3.3) is represented by the function f_e. This function extracts a feature vector from every image, that is, $f_e : I \longrightarrow \Re^N$, where N is the vector size.

Independent of the fingerprint capture device, the features space is always the same, as a result of the function f_e. Thus, we may have for one device the measurement X_S and for another one X_T. In principle, $X_S = X_T$, but, being $P_S(X_S) \neq P_T(X)$ due to the capture device, this is unrealistic and lead to a completely different spread of samples coming from different capture devices.

This is shown by experiments in this section. First, we tried to classify the target features space using a linear SVM [57] trained with source features space. We used the train set of each datasets to train the classifier and we test with the test set of

Table 4.1 LivDet 2015—LBP results: The accuracies are obtained training a classifier with the dataset in the first column and testing on the dataset in the first row. The accuracies in the main diagonal are obtained training and testing with the same scanner. Outside of the main diagonal, the accuracies are obtained training with a scanner and testing with a different one

	Hi Scan test (%)	GreenBit test (%)	Digital P. test (%)	Crossmatch test (%)
Hi Scan train	**85.00**	43.52	40.04	51.66
GreenBit train	59.96	**89.28**	69.36	63.13
Digital P. train	58.12	62.84	**87.68**	71.54
Crossmatch train	58.24	44.40	55.56	**91.76**

Table 4.2 LivDet 2015—LPQ results: The accuracies are obtained training a classifier with the dataset in the first column and testing on the dataset in the first row. The accuracies in the main diagonal are obtained training and testing with the same scanner. Outside of the main diagonal, the accuracies are obtained training with a scanner and testing with a different one

	Hi Scan test (%)	GreenBit test (%)	Digital P. test (%)	Crossmatch test (%)
Hi Scan train	**94.76**	75.16	74.16	57.67
GreenBit train	60.00	**94.40**	86.52	81.11
Digital P. train	50.72	51.88	**89.68**	74.83
Crossmatch train	54.76	48.20	45.80	**94.81**

Table 4.3 LivDet 2015—BISF results: The accuracies are obtained training a classifier with the dataset in the first column and testing on the dataset in the first row. The accuracies in the main diagonal are obtained training and testing with the same scanner. Outside of the main diagonal, the accuracies are obtained training with a scanner and testing with a different one

	Hi Scan test (%)	GreenBit test (%)	Digital P. test (%)	Crossmatch test (%)
Hi Scan train	**91.08**	49.32	41.36	59.36
GreenBit train	60.00	**93.68**	83.60	74.97
Digital P. train	53.72	51.36	**91.16**	76.32
Crossmatch train	47.88	45.56	51.36	**94.95**

each datasets. Tables 4.1, 4.2, 4.3, 4.4, 4.5 and 4.6 report the liveness detection accuracy and cross-accuracy among fingerprint capture devices. The value in row 3 and column 2 of Table 4.1, for example, is the liveness detection accuracy calculated by means of the features extracted from Hi Scan images. Note that those images are submitted to the classifier trained by using features extracted from GreenBit scanner images. Therefore, these values imply that the distributions for features space of live and spoof images are different for all sensors.

Tables 4.1, 4.2, 4.3, 4.4, 4.5 and 4.6 report the accuracy of these experiments. We can notice the good performances in the main diagonal; this is due to train and test images that came from the same scanner. Outside the main diagonal the accuracy varies from 32.07 to 86.52%, thus the range is too big in order to find an explanation.

Let $acc(i, j)$ be an element of the accuracy table t, and i, j be the indices that indicate the position in t. The index t indicates all tables from 4.1, 4.2, 4.3, 4.4, 4.5

Table 4.4 LivDet 2011—LBP results: The accuracies are obtained training a classifier with the dataset in the first column and testing on the dataset in the first row. The accuracies in the main diagonal are obtained training and testing with the same scanner. Outside of the main diagonal, the accuracies are obtained training with a scanner and testing with a different one

	Biometrika test (%)	Italdata test (%)	Digital P. test (%)	Sagem test (%)
Biometrika train	**88.85**	55.65	64.35	48.92
Italdata train	51.20	**81.35**	65.65	32.07
Digital P. train	47.65	50.55	**89.40**	50.59
Sagem train	51.85	48.20	49.90	**91.65**

Table 4.5 LivDet 2011—LPQ results: The accuracies are obtained training a classifier with the dataset in the first column and testing on the dataset in the first row. The accuracies in the main diagonal are obtained training and testing with the same scanner. Outside of the main diagonal the accuracies are obtained training with a scanner and testing with a different one

	Biometrika test (%)	Italdata test (%)	Digital P. test (%)	Sagem test (%)
Biometrika train	**85.20**	50.00	57.75	54.27
Italdata train	55.90	**86.40**	59.00	53.44
Digital P. train	55.55	50.00	**88.55**	54.66
Sagem train	60.15	50.00	54.80	**92.78**

Table 4.6 LivDet 2011—BISF results: The accuracies are obtained training a classifier with the dataset in the first column and testing on the dataset in the first row. The accuracies in the main diagonal are obtained training and testing with the same scanner. Outside of the main diagonal, the accuracies are obtained training with a scanner and testing with a different one

	Biometrika test (%)	Italdata test (%)	Digital P. test (%)	Sagem test (%)
Biometrika train	**91.95**	50.00	50.00	52.50
Italdata train	70.40	**86.15**	53.80	59.72
Digital P. train	50.00	49.80	**95.85**	50.88
Sagem train	53.05	50.00	50.00	**93.76**

and 4.6. Let $\widetilde{acc}_{cross}$ be the accuracy sample mean of all elements outside of the main diagonal of the tables. The $\widetilde{acc}_{cross}$ value is about 0.56, it indicates that on average the results of a system trained with a sensor and tested with another one are equivalent to a random guess. Thus, it is an experimental evidence that there is no interoperability among devices. In the same manner, let $\widetilde{acc}_{md}$ be the sample mean of all elements in the main diagonal of the tables. The $\widetilde{acc}_{md}$ value is about 0.90, since using train and test set coming from the same scanner, we are able to recognize almost all pattern in the test set.

As a confirmation of $P_S(X) \neq P_T(X)$, we changed the learning task. In other words, we did not classify the image based on "live" or "fake", but based on the origin of the image. Let S and T be two datasets coming from different scanners. The textural feature vectors extracted from the images are inserted in the two corresponding matrices D_S and D_T (in which each row contains a feature vector). Following the LivDet [58] protocol, we split both datasets into train set $D_{S,Train}$ and test set $D_{S,Test}$ such that $D_S = D_{S,Train} \cup D_{S,Test}$ and $D_{T,Train}$ and $D_{T,Test}$ such that $D_T = D_{T,Train} \cup D_{T,Test}$. Then, let us determine the set $D_{Train} = D_{S,Train} \cup D_{T,Train}$ and the set $D_{Test} = D_{S,Test} \cup D_{T,Test}$. The aim of this experiment is to train a classifier in order to discriminate images coming from two different scanners. Thus, we repeat the training process by increasing the number of samples. The classifier used in this experiment is the one called linear SVM [57]. Through this method, the scanners classes are separated by a decision hyperplane. Based on the state-of-the-art features for fingerprint attack detection, results related to LBP [25], LPQ [42] and BSIF [45] histograms are reported in this experiment.

Figures 4.6 and 4.7 plot the mean accuracy rate for all pair combinations of scanners when the number of training pattern is increased. The performance grows up in consonance with the number of patterns as it was expected. However, even with a few samples, the accuracy reaches values higher than 85%. Consequently, it suffices just a few patterns to identify the capture device. This fact proves that the set of feature vectors obtained from different sensors are almost completely separable as well as the distributions of those feature vectors. Therefore, we proved that from two different scanners, we have the same feature space but a different location of ω based on the θ as stated in Sect. 4.2.2. Hence $P_S(X) \neq P_T(X)$. If S and T are our source and target datasets, we would like to classify the target feature vectors using the source information. If we were able to move the target "live" and "fake" features distribution over the corresponding source "live" and "fake" feature distribution we could easily use a classifier trained with the source features on the moved target features.

4.4.3 Proposed Method

In the previous sections, we assert that the features space depends on the class ω ("live" or "fake") and the parameter θ. It is the cause of the noninteroperability between capture devices. Using the DA theory, we assert through the experimental evidences that $X_S = X_T$ but $P_S(X) \neq P_T(X)$. We would like to transform the distribution P_T in order to be similar to P_S, this is the section focus. We do not use the P_T and P_S probabilities, instead the observations of these, that we called D_S and D_T. Let D_S and $D_T \in \Re^{M \times N}$ with $M \geq N$ be the matrices containing the feature vectors (one for each row) that represents the patterns extracted from images of two different scanners. Similarly, we can associate the Y_S and Y_T binary labels.

We assume the existence of a matrix M that allows to goes from target domain to source one; hence, we can write

$$D_S = D_T * M \tag{4.12}$$

Fig. 4.6 LivDet 2011
datasets classification

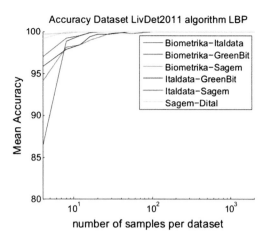

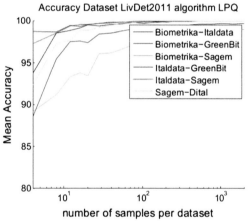

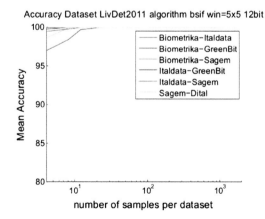

Fig. 4.7 LivDet 2015
datasets classification

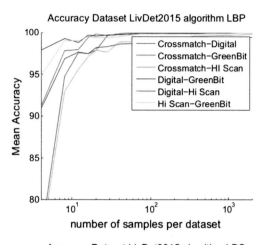

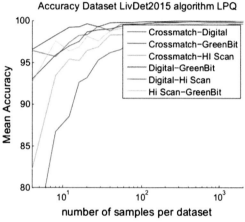

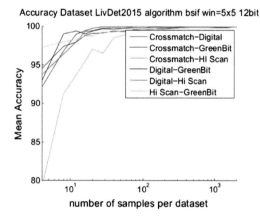

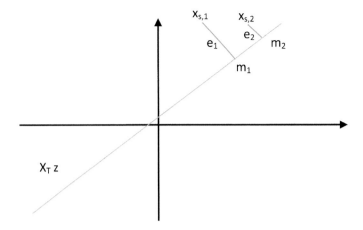

Fig. 4.8 Geometric interpretation of our solution, the span D_S is represented by a straight line. We represent two solutions m_1 and m_2 of the M matrix, with respective errors. The projection of $x_{t,i}$ on m_i is e_i, and e_i is orthogonal to $D_s z$. The smallest segment that passes for a point and intersects a straight line is orthogonal to the latter

Moreover, we hypothesize D_T is invertible thus it is full rank. We can rewrite Eq. 4.12:

$$M = D_T^{-1} D_S \tag{4.13}$$

Given a generic column vector of M that we called m_i, and a generic column vector of D_S called $x_{S,i}$ we can rewrite

$$D_T m_i = x_{S,i} \Rightarrow m_i = D_T^{-1} x_{S,i} \tag{4.14}$$

The solution is not unique because we have $M \geq N$; hence, we have more equations than variables. In order to resolve the equation, we introduce the error, also called residue.

$$e_i = x_{S,i} - D_T m_i \tag{4.15}$$

If $e_i = 0$ we have an exact solution, instead if $e_i \sim 0$ we have a least square (approximation) solution. Given the gradient of norm e_i we can write

$$\nabla_m \|e_i\|^2 = D_T^T D_T m_i - D_T^T x_{S,i} = 0 \tag{4.16}$$

Hence, a generic vector of M is $m_i = (D_T^T D_T)^{-1} D_T^T x_{S,i}$. So we can write m_i as the $x_{S,i}$ projection in X_T space and we can state that vector e_i is orthogonal to X_T.

As can be seen from Fig. 4.8 $X_T z$ is a straight line from multiplying the matrix for all possible values of the feature space z. The minimum distance between $x_{S,i}$ and $X_T z$ is e_i, that is orthogonal to $X_T z$ by definition. We can also say that the error being small is never null and differs for each M vector. We divide $D_S(D_T)$ in two set

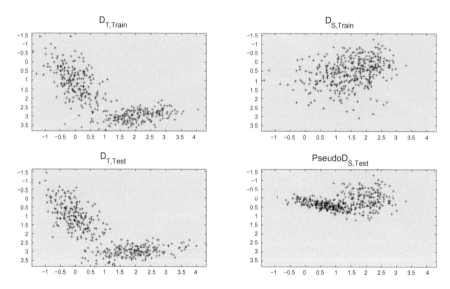

Fig. 4.9 Example of how the algorithm works. We can see how the proposed method changes the features distribution and how the $D_{T,Test}$ is transformed in $PseudoD_{S,Test}$. The distribution "live" and "fake" of $D_{T,Test}$ are roto-translated in order to be comparable to the distributions of $D_{S,Train}$

trains and test using the LivDet protocol [58], and we have $D_T = \{D_{T,train} \cup D_{T,test}\}$ ($D_S = \{D_{S,train} \cup D_{S,test}\}$). By using this method we are able to move every feature vector of D_S toward the feature vectors in D_T by taking into account that we must maintain the distinction between live and spoof fingerprints. Thus, we want the live features in D_S moving to the live features in D_T and the same for the spoof features. In order to understand how the algorithm changes the features space, we can see Fig. 4.9. We generate two synthetic dataset D_T and D_S. Each dataset is composed of two Gaussian distributions, one in order to simulate the "live" samples and one for "fake" patterns and being both datasets are divided in train and test. We can notice that the two trains ($D_{T,Train}$ and $D_{S,Train}$) are almost orthogonal. The two distributions ("live" and "fake") have different shapes in the same space but when the $D_{T,Test}$ is transformed in the $PseudoD_{S,Test}$ its "live" and "fake" distributions become similar to those of D_S. Hence, we have a roto-translation from D_T to D_S in order to have an almost complete overlapping of the feature spaces extracted from the two datasets. As we can see in Fig. 4.9, the patterns locations of $D_{T,Test}$ and $PseudoD_{S,Test}$ are different. This leads to introducing a new marginal probability P_T' that can be estimated using the $PseudoD_{S,Test}$ observations. We can affirm with this method that we try to obtain $P_S(X) \simeq P_T'(X)$.

Let us imagine a hypothetical scenario where we have a large dataset of images collected with a source scanner. We already trained a classifier with it and now we want to move to a newer target scanner, and potentially we will need to capture a whole new set of images. If the number of images captured with the novel scanner is limited could we benefit from the source classifier by transferring the observations in D_T to be similar to the observations in D_S? Would we need live fingerprints only, or spoof samples would be also required? Which is the minimum amount of observations so that we obtain significant results?

4.5 Experiments

In order to answer the questions at the end of the previous section, we performed several experiments through the analysis of the LivDet 2011 [6] and LivDet 2015 [7] datasets. Three different textural algorithms have been used in order to extract the features: LBP [25], LPQ [42] and BSIF [45].

In the first experiment, all possible pair combinations of LivDet 2011 and of LivDet 2015 datasets are used. Given $D_{S,Train}$ and $D_{T,Train}$ matrices coming from each couple of datasets D_S and D_T, the transformation matrix M is calculated. This permits us to move the target features toward the source ones. Once having both M and $D_{T,Test}$ matrices, moving the latter toward the source features is quite an easy task. Thus, we compute $PseudoD_{S,Test}$:

$$PseudoD_{S,Test} = D_{T,Test} \times M \qquad (4.17)$$

The obtained results are presented in Tables 4.7, 4.8, 4.9, 4.10, 4.11 and 4.12, in terms of accuracy. In the diagonal, since D_S and D_T represent the same dataset, the accuracies values are calculated through the classical procedure and making no transformations: a classifier is trained using $D_{S,Train}$ and, thanks to it, the feature vectors in $D_{S,Test}$ are classified. Instead, for non-diagonal values, we first trained the classifier with $D_{S,Train}$ and then we calculate the $PseudoD_{S,Test}$ by Eq. 4.17, so that we can finally classify the feature vectors in it.

Let compare each diagonal value with the other values in the same column. As may be observed, even though almost every accuracy is lower, the values obtained through the transformation of $D_{T,Test}$ into $PseudoD_{S,Test}$ (same column outside the diagonal) are similar compared to those calculated by using the original $D_{S,Test}$ dataset (corresponding diagonal value). Therefore, by using this transformation of

Table 4.7 LivDet 2015—LBP results: The accuracies are obtained by training a classifier with the datasets in the first column and testing on the datasets in the first row where the Pseudo-tests are calculated using both live and fake samples. Using the least square, the trained classifier has about same accuracies testing both with the same scanner and with a different scanner

	Hi Scan Pseudo-test (%)	GreenBit Pseudo-test (%)	Digital P. Pseudo-test (%)	Crossmatch Pseudo-test (%)
Hi Scan train	85.00	87.20	88.12	87.79
GreenBit train	85.80	89.28	87.68	89.96
Digital P. train	85.76	91.80	87.68	89.52
Crossmatch train	80.44	90.16	84.52	91.76

Table 4.8 LivDet 2015—LPQ results: The accuracies are obtained by training a classifier with the datasets in the first column and testing on the datasets in the first row where the Pseudo-tests are calculated using both live and fake samples. Using the least square, the trained classifier has about same accuracies testing both with the same scanner and with a different scanner

	Hi Scan Pseudo-test (%)	GreenBit Pseudo-test (%)	Digital P. Pseudo-test (%)	Crossmatch Pseudo-test (%)
Hi Scan train	94.76	90.68	86.52	86.94
GreenBit train	91.76	94.40	89.12	89.52
Digital P. train	89.60	93.92	89.68	89.96
Crossmatch train	86.88	88.96	82.28	94.81

Table 4.9 LivDet 2015—BSIF results: The accuracies are obtained by training a classifier with the datasets in the first column and testing on the datasets in the first row where the Pseudo-tests are calculated using both live and fake samples. Using the least square, the trained classifier has about same accuracies testing both with the same scanner and with a different scanner

	Hi Scan Pseudo-test (%)	GreenBit Pseudo-test (%)	Digital P. Pseudo-test (%)	Crossmatch Pseudo-test (%)
Hi Scan train	91.08	93.04	88.56	89.01
GreenBit train	86.60	93.68	86.88	88.57
Digital P. train	88.72	93.64	91.16	89.25
Crossmatch train	84.04	91.32	81.96	94.95

Table 4.10 LivDet 2011—LBP results: The accuracies are obtained by training a classifier with the datasets in the first column and testing on the datasets in the first row where the Pseudo-tests are calculated using both live and fake samples. Using the least square, the trained classifier has about same accuracies testing both with the same scanner and with a different scanner

	Biometrika Pseudo-test (%)	Italdata Pseudo-test (%)	Digital P. Pseudo-test (%)	Sagem Pseudo-test (%)
Biometrika train	88.85	80.10	84.15	86.25
Italdata train	86.50	81.35	85.60	88.80
Digital P. train	86.55	79.65	89.40	85.07
Sagem train	83.90	77.15	89.35	91.65

the feature space from one scanner to another, we reduce the impact of moving between two fingerprint devices.

4.5.1 Transformation Using Only Live Samples

In the previous experimental environment, both "live" and "fake" feature vectors are used when calculating the transformation matrix. Unfortunately, acquiring new

Table 4.11 LivDet 2011—LPQ results: The accuracies are obtained by training a classifier with the datasets in the first column and testing on the datasets in the first row where the Pseudo-tests are calculated using both live and fake samples. Using the least square, the trained classifier has about same accuracies testing both with the same scanner and with a different scanner

	Biometrika Pseudo-test (%)	Italdata Pseudo-test (%)	Digital P. Pseudo-test (%)	Sagem Pseudo-test (%)
Biometrika train	85.20	77.65	81.50	88.02
Italdata train	86.45	86.40	81.30	90.52
Digital P. train	79.30	71.20	88.50	87.13
Sagem train	83.70	76.70	84.55	92.78

Table 4.12 LivDet 2011—BSIF results: The accuracies are obtained by training a classifier with the datasets in the first column and testing on the datasets in the first row where the Pseudo-tests are calculated using both live and fake samples. Using the least square, the trained classifier has about same accuracies testing both with the same scanner and with a different scanner

	Biometrika Pseudo-test (%)	Italdata Pseudo-test (%)	Digital P. Pseudo-test (%)	Sagem Pseudo-test (%)
Biometrika train	90.20	77.40	89.85	86.69
Italdata train	87.10	78.70	89.10	84.09
Digital P. train	88.55	74.75	91.15	83.64
Sagem train	88.55	77.80	89.55	89.59

spoofs is usually not as easy and fast as collecting live samples. For this reason, the subsequent experiments have been attempted by using only live samples in order to calculate the transformation matrix.

Regrettably, the values obtained in Tables 4.13, 4.14, 4.15, 4.16, 4.17 and 4.18 show much worse results. This occurs due to the fact that we deliberately avoided using spoof samples; thus, we reduced the available information for the transformation process. This issue clarifies that it is impossible avoiding this information to design the domain adaptation function. It also indirectly evidences the ability of textural algorithms for extracting features from live and fake fingerprints. As an example, most of the artificial fingerprints were misclassified as live within the experiments.

In order to explain the results, we try to reason using Bayesian theory. Our algorithm estimates the $P_S(X)$ ($P_T(X)$) through the observations of D_S (D_T). Furthermore, we introduced P_T' that is the marginal probability estimated using $Pseudo D_S$. And we can write $P_S(X) \simeq P_T'(X)$.

If we expand $P_S(X)$ and $P_T'(X)$ we can write

$$P_S(X) = \sum_{\omega} P_S(X|\omega)P(\omega) \simeq \sum_{\omega} P_T'(X|\omega)P(\omega) = P_T'(X) \qquad (4.18)$$

Table 4.13 LivDet 2015—LBP results: The accuracies are obtained training a classifier with the datasets in the first column and testing on the datasets in the first row where the Pseudo-tests are calculated using only live samples. The information derived from only live samples are not enough in order to reach the interoperability among scanners

	Hi Scan Pseudo-test (%)	GreenBit Pseudo-test (%)	Digital P. Pseudo-test (%)	Crossmatch Pseudo-test (%)
Hi Scan train	85.00	55.76	42.20	58.75
GreenBit train	40.96	89.28	40.80	56.11
Digital P. train	44.36	47.56	87.68	54.92
Crossmatch train	40.32	41.32	39.76	91.76

Table 4.14 LivDet 2015—LPQ results: The accuracies are obtained training a classifier with the datasets in the first column and testing on the datasets in the first row where the Pseudo-tests are calculated using only live samples. The information derived from only live samples are not enough in order to reach the interoperability among scanners

	Hi Scan Pseudo-test (%)	GreenBit Pseudo-test (%)	Digital P. Pseudo-test (%)	Crossmatch Pseudo-test (%)
Hi Scan train	94.76	45.96	45.56	53.43
GreenBit train	48.08	94.40	54.76	50.54
Digital P. train	51.36	46.44	89.68	52.75
Crossmatch train	39.68	39.04	39.24	94.81

Table 4.15 LivDet 2015—BSIF results: The accuracies are obtained training a classifier with the datasets in the first column and testing on the datasets in the first row where the Pseudo-tests are calculated using only live samples. The information derived from only live samples are not enough in order to reach the interoperability among scanners

	Hi Scan Pseudo-test (%)	GreenBit Pseudo-test (%)	Digital P. Pseudo-test (%)	Crossmatch Pseudo-test (%)
Hi Scan train	91.08	42.20	42.56	53.56
GreenBit train	40.20	93.68	47.80	51.70
Digital P. train	42.72	54.72	91.16	59.77
Crossmatch train	39.36	42.12	39.88	94.95

where ω is the class "live" or "fake". Under the hypothesis that the two $P(\omega)$ are equal, we can rewrite Eq. 4.18:

$$P_S(X) \propto \sum_{\omega} P_S(X|\omega) \simeq \sum_{\omega} P'_T(X|\omega) \propto P'_T(X) \qquad (4.19)$$

In this experiment, we try to match the two quantities $P_S(X|\text{"live"}) \simeq P'_T(X|\text{"live"})$, but we do not verify $P_S(X|\text{"fake"}) \simeq P'_T(X|\text{"live"} \text{or} \text{"fake"})$. We do not know where the target "fake" distribution is with respect to the source one. In other words,

Table 4.16 LivDet 2011—LBP results: The accuracies are obtained training a classifier with the datasets in the first column and testing on the datasets in the first row where the Pseudo-tests are calculated using only live samples. The information derived from only live samples are not enough in order to reach the interoperability among scanners

	Biometrika Pseudo-test (%)	Italdata Pseudo-test (%)	Digital P. Pseudo-test (%)	Sagem Pseudo-test (%)
Biometrika train	88.85	57.75	50.00	49.41
Italdata train	72.30	81.35	50.00	50.88
Digital train	49.95	50.00	89.40	49.17
Sagem train	50.00	50.00	50.00	91.65

Table 4.17 LivDet 2011—LPQ results: The accuracies are obtained training a classifier with the datasets in the first column and testing on the datasets in the first row where the Pseudo-tests are calculated using only live samples. The information derived from only live samples are not enough in order to reach the interoperability among scanners

	Biometrika Pseudo-test (%)	Italdata Pseudo-test (%)	Digital P. Pseudo-test (%)	Sagem Pseudo-test (%)
Biometrika train	85.20	52.45	49.75	57.32
Italdata train	56.25	86.40	51.95	52.75
Digital P. train	49.90	49.50	88.50	50.10
Sagem train	50.50	49.90	51.05	92.78

Table 4.18 LivDet 2011—BSIF results: The accuracies are obtained training a classifier with the datasets in the first column and testing on the datasets in the first row where the Pseudo-tests are calculated using only live samples. The information derived from only live samples are not enough in order to reach the interoperability among scanners

	Biometrika Pseudo-test (%)	Italdata Pseudo-test (%)	Digital P. Pseudo-test (%)	Sagem Pseudo-test (%)
Biometrika train	90.20	58.30	50.25	50.25
Italdata train	62.90	78.70	50.75	49.95
Digital P. train	50.85	50.55	91.15	49.51
Sagem train	49.80	49.95	50.45	89.59

using Eq. 4.15 we do not calculate e_i for the "fake" observations, thus we do not move this pattern in the correct hyperplane part.

4.5.2 Number of Feature Vectors

The transformation matrices in the previous subsections were calculated by using all the feature vectors extracted from the train parts of the LivDet 2011 and 2015

Table 4.19 LivDet 2011—LBP results: The accuracies are obtained training the classifiers in the first row and testing on the datasets in the second row obtained using a transformation matrix calculated with the train percentages in the first column. For lack of space, in the second row, the following acronyms are used: BM (Biometrika), IT (Italdata), DP (Digital Persona), and SG (Sagem). By increasing sample numbers, the performances are improved. In other words, adding information helps to solve the problem of interoperability between scanners

Train	Biometrika			Italdata			Digital P.			Sagem		
Pseudo-test (%)	IT (%)	DP (%)	SG (%)	BM (%)	DP (%)	SG (%)	BM (%)	IT (%)	SG (%)	BM (%)	IT (%)	DP (%)
20	73.28	86.06	84.13	77.90	83.81	84.42	79.15	74.64	83.04	81.17	74.99	86.31
40	75.97	87.28	86.77	82.26	87.55	86.00	81.34	74.57	87.48	83.42	75.58	88.16
60	76.41	88.27	88.20	84.83	88.11	87.77	82.56	75.75	88.18	85.20	77.33	89.67
80	77.28	89.18	88.73	85.64	89.17	88.16	83.50	76.16	88.52	85.59	77.12	89.48

Table 4.20 LivDet 2011—LPQ results: The accuracies are obtained training the classifiers in the first row and testing on the datasets in the second row obtained using a transformation matrix calculated with the train percentages in the first column. For lack of space, in the second row, the following acronyms are used: BM (Biometrika), IT (Italdata), DP (Digital Persona), and SG (Sagem). By increasing sample numbers, the performances are improved. In other words, adding information helps to solve the problem of interoperability between scanners

Train	Biometrika			Italdata			Digital P.			Sagem		
Pseudo-test (%)	IT (%)	DP (%)	SG (%)	BM (%)	DP (%)	SG (%)	BM (%)	IT (%)	SG (%)	BM (%)	IT (%)	DP (%)
20	62.56	68.63	70.61	67.24	69.54	71.45	64.64	64.89	69.94	66.97	62.52	69.29
40	70.91	79.06	80.53	75.39	79.50	82.50	75.66	70.09	82.09	74.70	70.97	78.53
60	73.67	82.28	85.11	80.13	83.38	86.02	79.91	73.79	85.75	77.97	75.46	82.45
80	74.06	83.29	87.15	82.30	84.90	87.64	80.80	76.61	86.97	81.53	75.83	85.06

Table 4.21 LivDet 2011—BSIF results: The accuracies are obtained training the classifiers in the first row and testing on the datasets in the second row obtained using a transformation matrix calculated with the train percentages in the first column. For lack of space, in the second row, the following acronyms are used: BM (Biometrika), IT (Italdata), DP (Digital Persona), and SG (Sagem). By increasing sample numbers, the performances are improved. In other words, adding information helps to solve the problem of interoperability between scanners

Train	Biometrika			Italdata			Digital P.			Sagem		
Pseudo-test (%)	IT (%)	DP (%)	SG (%)	BM (%)	DP (%)	SG (%)	BM (%)	IT (%)	SG (%)	BM (%)	IT (%)	DP (%)
20	62.33	72.85	72.25	70.13	72.81	71.66	72.73	62.70	74.64	73.00	65.18	75.66
40	68.50	85.38	80.79	81.12	83.31	80.27	83.59	71.26	82.78	83.66	70.95	85.70
60	73.33	88.78	83.69	85.06	88.06	82.45	86.62	73.59	84.67	88.44	74.38	88.35
80	73.34	89.63	85.71	85.86	89.05	84.57	87.94	75.06	86.07	87.94	76.13	89.97

Table 4.22 LivDet 2015—LBP results: The accuracies are obtained training the classifiers in the first row and testing on the datasets in the second row obtained using a transformation matrix calculated with the train percentages in the first column. For lack of space, in the second row, the following acronyms are used: HS (Hi Scan), GB (GreenBit), DP (Digital Persona), and CM (CrossMatch). By increasing sample numbers, the performance is improved. In other words, adding information helps to solve the problem of interoperability between scanners

Train	Hi Scan			GreenBit			Digital P.			CrossMatch		
Pseudo-test (%)	GB (%)	DP (%)	CM (%)	HS (%)	DP (%)	CM (%)	HS (%)	GB (%)	CM (%)	HS (%)	GB (%)	DP (%)
20	86.29	81.32	88.16	84.68	85.39	90.65	84.50	91.28	90.94	81.62	89.60	83.52
40	89.56	83.94	90.08	86.65	86.80	92.76	86.48	92.83	92.43	84.65	91.01	85.63
60	89.52	83.96	91.09	86.80	87.32	93.23	87.07	93.14	93.28	85.68	91.37	85.88
80	90.24	84.20	91.35	87.58	87.62	93.21	87.86	93.42	93.41	86.16	91.82	85.56

Table 4.23 LivDet 2015—LPQ results: The accuracies are obtained training the classifiers in the first row and testing on the datasets in the second row obtained using a transformation matrix calculated with the train percentages in the first column. For lack of space, in the second row, the following acronyms are used: HS (Hi Scan), GB (GreenBit), DP (Digital Persona), and CM (CrossMatch). By increasing sample numbers, the performance is improved. In other words adding information helps to solve the problem of interoperability between scanners

Train	Hi Scan				GreenBit				Digital P.				CrossMatch			
Pseudo-test (%)	GB (%)	DP (%)	CM (%)	HS (%)	DP (%)	CM (%)	HS (%)	GB (%)	HS (%)	GB (%)	CM (%)	HS (%)	CM (%)	HS (%)	GB (%)	DP (%)
20	73.07	69.56	72.03	72.18	70.84	74.35	76.48	77.91	75.25	73.51	75.47	71.09
40	85.57	77.92	80.58	83.65	79.79	82.92	85.84	89.06	84.62	84.10	87.77	80.25
60	88.02	80.44	82.28	87.09	82.24	85.83	88.68	91.36	88.96	87.37	91.17	83.22
80	89.29	81.26	84.03	87.73	83.39	87.08	90.10	92.71	90.01	88.64	91.57	84.50

Table 4.24 LivDet 2015—BSIF results: The accuracies are obtained training the classifiers in the first row and testing on the datasets in the second row obtained using a transformation matrix calculated with the train percentages in the first column. For lack of space, in the second row, the following acronyms are used: HS (Hi Scan), GB (GreenBit), DP (Digital Persona), and CM (CrossMatch). By increasing sample numbers, the performance is improved. In other words, adding information helps to solve the problem of interoperability between scanners

Train	Hi Scan			GreenBit			Digital P.			CrossMatch		
Pseudo-test (%)	GB (%)	DP (%)	CM (%)	HS (%)	DP (%)	CM (%)	HS (%)	GB (%)	CM (%)	HS (%)	GB (%)	DP (%)
20	74.50	71.28	71.97	69.89	70.25	74.54	72.44	77.34	74.53	70.09	75.16	70.42
40	85.60	79.80	82.89%	79.58	80.27	82.83	81.86	88.24	85.86	80.74	86.03	80.78
60	89.01	82.26	86.38	83.07	82.37	86.09	85.18	91.06	89.18	84.40	89.41	83.99
80	90.82	83.78	88.48	83.89	83.17	88.05	86.62	92.19	90.03	86.27	91.42	85.10

datasets [58], which were approximately 2000 images (1000 lives and 1000 fakes) or more. Live and fakes were used in Tables 4.7, 4.8, 4.9, 4.10, 4.11 and 4.12 and just lives in Table 4.13, 4.14, 4.15, 4.16, 4.17 and 4.18. Since we wanted to simulate a scenario with a limited number of acquisitions, we randomly selected a subset of feature vectors when replicating the experiments. Those subsets had an equal number of live and fake samples. Thus, from the $D_{T,Train}$ and $D_{S,Train}$ matrices we extracted both $SubSet D_{T,Train}$ and $SubSet D_{S,Train}$ subsets and then we calculated the new M matrix, such that

$$PseudoSubSet D_{S,Test} = SubSet D_{T,Test} \times M \qquad (4.20)$$

Tables 4.19, 4.20, 4.21, 4.22, 4.23 and 4.24 present the experiments where transformation matrices were calculated by using subsets containing, respectively, the 20, 40, 60, and 80% of the feature vectors coming from the original train sets.

Reported results show that the bigger is the train percentage, the higher the correct classification rate. However, in the majority of the cases, a subsetting of 40% from the original train set appears to be enough in order to obtain significant results. Consequently, the performance loss is not very significant, and ours can be considered a temporary solution until a proper new dataset of live and fakes has been collected.

Worth noting, the projected features vectors are such that (1) the classification boundary can be held, (2) eventual additional features captured by the novel sensor are also embedded and this could explain the unexpected performance improvement when moving from a "low" performance device to a better one. This appears as true even by a transformation matrix computed on a small set of samples (Tables 4.19, 4.20, 4.21, 4.22, 4.23 and 4.24). As written previously, we estimate $P_S(X)$ and $P_T(X)$ through their respective observations; thus, the number of patterns influences the performances. The estimate of marginal probabilities becomes more and more accurate according to the number of observations.

4.6   Conclusions

In this chapter, we presented an analysis of the interoperability level among different sensors related to the fingerprint presentation attacks detection scope. The existence of this problem, previously pointed out in several scientific works, was further confirmed. We have proven the existence of this problem as well as we stated that there is no trivial solution for it. As a matter of fact, the unique characteristics of each sensor drastically influence the image properties and their corresponding feature space. By the use of textural algorithms, we have proven that there exists a strong "sensor-specific" effect within those features, which results on the related images having a nonoverlapping localization in different regions of the features space itself.

Given this scenario, we proposed a solution based on domain adaptation. Our proposal consists in the shift of sensor-specific feature distributions based on the least squared algorithm. It made possible to reach a significant level of interoperability among fingerprint capture devices. Unfortunately, in order to be effective, the training phase required both live and spoof samples. A transformation based on features extracted only from live images resulted in a substantial drop in performance. Lastly, we reduce the number of feature vectors used in order to calculate the transformation matrix. Experiments proved that, in the majority of the cases, even a 40% of the original train set appears to be enough to obtain a significant performance.

Further improvements are still needed regarding the performance as well as to reduce the required number of feature vectors in order to calculate the transformation matrix. In particular, being able to calculate the domain transformation equation by only using live samples would be a huge improvement. It would reduce efforts with respect to the fingerprint acquisition, thus contributing to develop an efficient algorithm for interoperability among fingerprint liveness detection devices.

References

1. Erdoğmuş N, Marcel S (2014) Introduction, pp 1–11. Springer, London. https://doi.org/10. 1007/978-1-4471-6524-8_1
2. Willis D, Lee M (1998) Six biometric devices point the finger at security. Netw Comput 9(10):84–96 (1998). URL http://dl.acm.org/citation.cfm?id=296195.296211
3. van der Putte T, Keuning J (2001) Biometrical fingerprint recognition: don't get your fingers burned. In: Proceedings of the fourth working conference on smart card research and advanced applications on smart card research and advanced applications. Kluwer Academic Publishers, Norwell, pp 289–303. http://dl.acm.org/citation.cfm?id=366214.366298
4. Matsumoto T, Matsumoto H, Yamada K, Hoshino S (2002) Impact of artificial "gummy" fingers on fingerprint systems, vol 26
5. Schuckers SA (2002) Spoofing and anti-spoofing measures. Inf Sec Tech Rep 7(4):56–62. https://doi.org/10.1016/S1363-4127(02)00407-7
6. Yambay D, Ghiani L, Denti P, Marcialis GL, Roli F, Schuckers S (2012) Livdet 2011 - fingerprint liveness detection competition 2011. In: 2012 5th IAPR international conference on biometrics (ICB). IEEE, pp 208–215
7. Mura V, Ghiani L, Marcialis GL, Roli F, Yambay DA, Schuckers SA (2015) Livdet 2015 fingerprint liveness detection competition 2015. In: 2015 IEEE 7th international conference on biometrics theory, applications and systems (BTAS). IEEE, pp 1–6
8. Jain AK, Flynn P, Ross AA (eds) Handbook of biometrics. Springer (2008). https://doi.org/10. 1007/978-0-387-71041-9
9. Marasco E, Ross A (2014) A survey on antispoofing schemes for fingerprint recognition systems. ACM Comput Surv 47(2):28:1–28:36. https://doi.org/10.1145/2617756
10. Sousedik C, Busch C (2014) Presentation attack detection methods for fingerprint recognition systems: a survey. IET Biometr 3(4):219–233. https://doi.org/10.1049/iet-bmt.2013.0020
11. Coli P, Marcialis GL, Roli F (2007) Vitality detection from fingerprint images: a critical survey. In: Advances in biometrics: international conference, ICB 2007, Seoul, Korea, August 27–29, 2007. Proceedings. Springer, Berlin, pp 722–731. https://doi.org/10.1007/978-3-540-74549-5-76
12. Derakhshani R, Schuckers S, Hornak LA, O'Gorman L (2003) Determination of vitality from a non-invasive biomedical measurement for use in fingerprint scanners. Pattern Recogn 36:383–396
13. Parthasaradhi STV, Derakhshani R, Hornak LA, Schuckers SAC (2005) Time-series detection of perspiration as a liveness test in fingerprint devices. IEEE Trans Syst Man Cybern Part C (Appl Rev) 35(3):335–343. https://doi.org/10.1109/TSMCC.2005.848192

14. Coli P, Marcialis GL, Roli F (2006) Analysis and selection of features for the fingerprint vitality detection. Springer, Berlin, pp 907–915. https://doi.org/10.1007/11815921-100
15. Jia J, Cai L, Zhang K, Chen D (2007) A new approach to fake finger detection based on skin elasticity analysis. Springer, Berlin, pp 309–318. https://doi.org/10.1007/978-3-540-74549-5_33
16. Antonelli A, Cappelli R, Maio D, Maltoni D (2006) Fake finger detection by skin distortion analysis. IEEE Trans Inf Forensics Sec 1(3):360–373. https://doi.org/10.1109/TIFS.2006.879289
17. Aditya Shankar Abhyankar SCS (2004) A wavelet-based approach to detecting liveness in fingerprint scanners. pp 5404 – 5404 – 9 (2004). https://doi.org/10.1117/12.542939
18. Schuckers S, Abhyankar A (2004) Detecting liveness in fingerprint scanners using wavelets: results of the test dataset. Springer, Berlin, pp 100–110. https://doi.org/10.1007/978-3-540-25976-3_10
19. Zhang Y, Tian J, Chen X, Yang X, Shi P (2007) Fake finger detection based on thin-plate spline distortion model. Springer, Berlin, pp 742–749. https://doi.org/10.1007/978-3-540-74549-5_78
20. Tan B, Schuckers S (2006) Liveness detection for fingerprint scanners based on the statistics of wavelet signal processing. In: 2006 conference on computer vision and pattern recognition workshop (CVPRW'06), pp 26–26 (2006). https://doi.org/10.1109/CVPRW.2006.120
21. Tan B, Schuckers SAC (2008) New approach for liveness detection in fingerprint scanners based on valley noise analysis 17(011):009
22. Nikam SB, Agarwal S (2008) Fingerprint anti-spoofing using ridgelet transform. In: 2008 IEEE second international conference on biometrics: theory, applications and systems, pp 1–6. https://doi.org/10.1109/BTAS.2008.4699347
23. Nikam SB, Agarwal S (2008) Fingerprint liveness detection using curvelet energy and co-occurrence signatures. 2008 fifth international conference on computer graphics, imaging and visualisation, pp 217–222
24. Nikam SB, Agarwal S (2010) Curvelet-based fingerprint anti-spoofing. Signal, Image Video Process 4(1):75–87. https://doi.org/10.1007/s11760-008-0098-8
25. Nikam SB, Agarwal S (2008) Texture and wavelet-based spoof fingerprint detection for fingerprint biometric systems. In: 2008 first international conference on emerging trends in engineering and technology, pp 675–680. https://doi.org/10.1109/ICETET.2008.134
26. Nikam SB, Agarwal S (2008) Wavelet energy signature and glcm features-based fingerprint anti-spoofing. In: 2008 international conference on wavelet analysis and pattern recognition, vol 2, pp 717–723. https://doi.org/10.1109/ICWAPR.2008.4635872
27. Haralick RM, Shanmugam K, Dinstein I (1973) Textural features for image classification. IEEE Trans Syst Man Cybern SMC-3(6):610–621 (1973). https://doi.org/10.1109/TSMC.1973.4309314
28. Tan B, Schuckers S (2010) Spoofing protection for fingerprint scanner by fusing ridge signal and valley noise. Pattern Recogn 43(8):2845–2857. https://doi.org/10.1016/j.patcog.2010.01.023
29. Moon YS, Chen JS, Chan KC, So K, Woo KC (2005) Wavelet based fingerprint liveness detection. Electron Lett 41(20):1112–1113. https://doi.org/10.1049/el:20052577
30. Chen Y, Jain A, Dass S (2005) Fingerprint deformation for spoof detection. In: Biometric symposium
31. Choi H, Kang R, Choi K, Kim J (2007) Aliveness detection of fingerprints using multiple static features. World academy of science, engineering and technology, vol 2
32. Abhyankar A, Schuckers S (2006) Fingerprint liveness detection using local ridge frequencies and multiresolution texture analysis techniques. In: 2006 international conference on image processing, pp 321–324. https://doi.org/10.1109/ICIP.2006.313158
33. Marcialis GL, Roli F, Tidu A (2010) Analysis of fingerprint pores for vitality detection. In: 2010 20th international conference on pattern recognition, pp 1289–1292. https://doi.org/10.1109/ICPR.2010.321

34. Marasco E, Sansone C (2010) An anti-spoofing technique using multiple textural features in fingerprint scanners. In: 2010 IEEE workshop on biometric measurements and systems for security and medical applications, pp 8–14. https://doi.org/10.1109/BIOMS.2010.5610440

35. Galbally J, Alonso-Fernandez F, Fierrez J, Ortega-Garcia J (2012) A high performance fingerprint liveness detection method based on quality related features. Future Gener Comput Syst 28(1):311–321. http://dx.doi.org/10.1016/j.future.2010.11.024

36. Gottschlich C, Marasco E, Yang AY, Cukic B (2014) Fingerprint liveness detection based on histograms of invariant gradients. In: IEEE international joint conference on biometrics, pp 1–7. https://doi.org/10.1109/BTAS.2014.6996224

37. Marasco E, Wild P, Cukic B (2016) Robust and interoperable fingerprint spoof detection via convolutional neural networks. In: 2016 IEEE symposium on technologies for homeland security (HST), pp 1–6. https://doi.org/10.1109/THS.2016.7568925

38. Frassetto Nogueira R, Lotufo R, Machado R (2016) Fingerprint liveness detection using convolutional neural networks. IEEE Trans Inf Forensics Sec 11:1–1

39. Ojala T, Pietikainen M, Maenpaa T (2002) Multiresolution gray-scale and rotation invariant texture classification with local binary patterns. IEEE Trans Pattern Anal Mach Intell 24(7):971–987. https://doi.org/10.1109/TPAMI.2002.1017623

40. Mäenpää T, Pietikäinen M (2005) Texture analysis with local binary patterns. Handbook of pattern recognition and computer vision 3:197–216

41. Heikkila J, Ojansivu V (2009) Methods for local phase quantization in blur-insensitive image analysis. In: International workshop on local and non-local approximation in image processing, 2009. LNLA 2009. IEEE, pp 104–111

42. Ghiani L, Marcialis GL, Roli F (2012) Fingerprint liveness detection by local phase quantization. In: 2012 21st international conference on pattern recognition (ICPR), pp 537–540

43. Chen J, Shan S, He C, Zhao G, Pietikainen M, Chen X, Gao W (2010) Wld: a robust local image descriptor. IEEE Trans Pattern Anal Mach Intell 32(9):1705–1720

44. Gragnaniello D, Poggi G, Sansone C, Verdoliva L (2013) Fingerprint liveness detection based on weber local image descriptor. In: 2013 IEEE workshop on biometric measurements and systems for security and medical applications, pp 46–50. https://doi.org/10.1109/BIOMS.2013.6656148

45. Ghiani L, Hadid A, Marcialis GL, Roli F (2013) Fingerprint liveness detection using binarized statistical image features. In: 2013 IEEE sixth international conference on biometrics: theory, applications and systems (BTAS), pp 1–6. https://doi.org/10.1109/BTAS.2013.6712708

46. Kannala J, Rahtu E (2012) Bsif: Binarized statistical image features. In: 21st international conference on pattern recognition (ICPR) 2012. IEEE, pp 1363–1366

47. Jia X, Yang X, Zang Y, Zhang N, Dai R, Tian J, Zhao J (2013) Multi-scale block local ternary patterns for fingerprints vitality detection. In: 2013 international conference on biometrics (ICB), pp 1–6. https://doi.org/10.1109/ICB.2013.6612964

48. Jia X, Yang X, Cao K, Zang Y, Zhang N, Dai R, Zhu X, Tian J (2014) Multi-scale local binary pattern with filters for spoof fingerprint detection. Inf Sci 268:91–102. https://doi.org/10.1016/j.ins.2013.06.041. http://www.sciencedirect.com/science/article/pii/S0020025513004787. (New sensing and processing technologies for hand-based biometrics authentication)

49. Krizhevsky A, Sutskever I, Hinton G.E (2012) Imagenet classification with deep convolutional neural networks. In: Pereira F, Burges CJC, Bottou L, Weinberger KQ (eds) Advances in neural information processing systems, vol 25. Curran Associates, Inc., pp 1097–1105. http://papers.nips.cc/paper/4824-imagenet-classification-with-deep-convolutional-neural-networks.pdf

50. Szegedy C, Liu W, Jia Y, Sermanet P, Reed S, Anguelov D, Erhan D, Vanhoucke V, Rabinovich A (2015) Going deeper with convolutions. In: 2015 IEEE conference on computer vision and pattern recognition (CVPR), pp 1–9. https://doi.org/10.1109/CVPR.2015.7298594

51. Chopra S, Hadsell R, LeCun Y (2005) Learning a similarity metric discriminatively, with application to face verification. In: 2005 IEEE computer society conference on computer vision and pattern recognition (CVPR'05), vol 1, pp 539–546. https://doi.org/10.1109/CVPR.2005.202

52. Simonyan K, Zisserman A (2014) Very deep convolutional networks for large-scale image recognition. CoRR (2014). http://arxiv.org/abs/1409.1556
53. Nogueira RF, de Alencar Lotufo R, Machado RC (2014) Evaluating software-based fingerprint liveness detection using convolutional networks and local binary patterns. In: 2014 IEEE workshop on biometric measurements and systems for security and medical applications (BIOMS) Proceedings, pp 22–29. https://doi.org/10.1109/BIOMS.2014.6951531
54. Ghiani L, Mura V, Tuveri P, Marcialis GL (2017) On the interoperability of capture devices in fingerprint presentation attacks detection
55. Maltoni D, Maio D, Jain AK, Prabhakar S (2009) Handbook of fingerprint recognition, 2nd edn. Springer Publishing Company, Incorporated, Berlin
56. Watson CI, Garris MD, Tabassi E, Wilson CL, Mccabe RM, Janet S, Ko K, User's guide to nist biometric image software (nbis)
57. Chang CC, Lin CJ (2011) LIBSVM: A library for support vector machines. ACM Trans Intell Syst Technol 2:27:1–27:27
58. Ghiani L, Yambay DA, Mura V, Marcialis GL, Roli, F, Schuckers SAC (2017) Review of the fingerprint liveness detection (livdet) competition series. Image Vis Comput 58(C):110–128 (2017). https://doi.org/10.1016/j.imavis.2016.07.002

Chapter 5
Review of Fingerprint Presentation Attack Detection Competitions

**David Yambay, Luca Ghiani, Gian Luca Marcialis, Fabio Roli
and Stephanie Schuckers**

Abstract A spoof or artifact is a counterfeit biometric that is used in an attempt
to circumvent a biometric sensor. Presentation attacks using an artifact have proven
to still be effective against fingerprint recognition systems. Liveness detection aims
to distinguish between live and fake biometric traits. Liveness detection is based on
the principle that additional information can be garnered above and beyond the data
procured by a standard authentication system, and this additional data can be used to
determine if a biometric measure is authentic. The Fingerprint Liveness Detection
Competition (LivDet) goal is to compare both software-based and hardware-based
fingerprint liveness detection methodologies. The competition is open to all academic
and industrial institutions. The number of competitors grows at every LivDet edition
demonstrating a growing interest in the area.

5.1 Introduction

Among biometrics, fingerprints are probably the best known and widespread because
of the its properties: universality, durability, and individuality. Unfortunately as introduced earlier, fingerprint systems have still been shown to be vulnerable to presentation attacks. Numerous competitions have been held in the past to address matching

D. Yambay (✉) · S. Schuckers
Clarkson University, 10 Clarkson Avenue, Potsdam, NY 13699, USA
e-mail: yambayda@clarkson.edu

S. Schuckers
e-mail: sschucke@clarkson.edu

L. Ghiani · G. L. Marcialis · F. Roli
Department of Electrical and Electronic Engineering, University of Cagliari,
Cagliari, Italy
e-mail: luca.ghiani@diee.unica.it

G. L. Marcialis
e-mail: marcialis@diee.unica.it

F. Roli
e-mail: roli@diee.unica.it

© Springer Nature Switzerland AG 2019   109
S. Marcel et al. (eds.), *Handbook of Biometric Anti-Spoofing*,
Advances in Computer Vision and Pattern Recognition,
https://doi.org/10.1007/978-3-319-92627-8_5

in biometrics, such as the Fingerprint Verification Competition held in 2000, 2002, 2004, and 2006 [1] and the ICB Competition on Iris Recognition (ICIR2013) [2]. However, these competitions did not consider presentation attacks.

Since 2009, in order to assess the main achievements of the state of the art in fingerprint liveness detection, University of Cagliari and Clarkson University organized the first Fingerprint Liveness Detection Competition.

The First International Fingerprint Liveness Detection Competition (LivDet) 2009 [3] provided an initial assessment of software systems based on the fingerprint image only. The second, third, and fourth Liveness Detection Competitions (LivDet 2011 [4], 2013 [5] and 2015 [6]) were created in order to ascertain the progressing state of the art in liveness detection and also included integrated system testing.

This chapter reviews the previous LivDet competitions and the evolution of the competitions over the years. The following sections will describe the methods used in testing for each of the LivDet competitions as well as descriptions of the datasets that have been generated from each competition. Also discussed are the trends across the different competitions that reflect changes to the art of presentation attacks as well as advances in the state of the art in presentation attack detection. Further, conclusions from previous LivDet competitions and the future of LivDet are discussed.

5.2 Background

The Liveness Detection Competition series started in 2009 and created a benchmark for measuring liveness detection algorithms, similar to matching performance, through the use of open competitions and publically released datasets for future testing of presentation attack detection systems. At that time, there had been no other public competitions held that have examined the concept of liveness detection as part of a biometric modality in deterring spoof attacks. In order to understand the motivation of organizing such a competition, we observed that the first trials to face with this topic were often carried out with home-made datasets that were not publicly available, the experimental protocols were not unique, and the same reported results were obtained on very small datasets. We pointed out these issues in [7].

Therefore, the basic goal of LivDet has been, since its birth, to allow researchers testing their own algorithms and systems on publicly available datasets, obtained and collected with the most updated techniques to replicate fingerprints enabled by the experience of Clarkson and Cagliari laboratories, both active on this problem since 2000 and 2003, respectively. At the same time, using a "competition" instead of simply releasing datasets could be an assurance of a free-of-charge, third-party testing using a sequestered test set. (Clarkson and Cagliari have never took part in LivDet as competitors due to conflict of interest.)

LivDet 2009 provided results which demonstrated the state of the art at that time [3] for fingerprint systems. LivDet continued in 2011, 2013, and 2015 [4–6] and contained two parts: evaluation of software-based systems in Part 1: Algorithms, and evaluation of integrated systems in Part 2: Systems.

Table 5.1 Number of LivDet
citations on Google Scholar

	Citations
LivDet 2009	119
LivDet 2011	84
LivDet 2013	62
LivDet 2015	13

Since 2009, the evaluation of spoof detection for facial systems was performed in the Competition on Counter Measures to 2-D Facial Spoofing Attacks, first held in 2011 and then held a second time in 2013. The purpose of this competition is to address different methods of detection for 2-D facial spoofing [8]. A subset was released for training and then another subset of the dataset was used for testing purposes.

During these years, many works cited the publications related to the first three LivDet competitions, 2009 [3], 2011 [4] and 2013 [5]. A quick Google Scholar research produced 119 results for 2009, 84 for 2011, 62 for 2013, and 13 for 2015. These values are shown in Table 5.1. In Tables 5.2, 5.3, 5.4, and 5.5, a partial list of these publications is presented.

5.3 Methods and Datasets

Each LivDet competition is composed of two distinct parts: Part 1: Algorithms which feature strictly software-based approaches to presentation attack detection, and Part 2: Systems which feature software or hardware-based approaches in a fully packaged device. The protocols of each are described further in this section along with descriptions of each dataset created through this competition.

Table 5.2 Publications that cite the LivDet 2009 paper

Authors	Algorithm type	Performance (average classification error) (%)
J. Galbally et al. [9]	Quality related features	6.6
E. Marasco and C. Sansone [10]	Perspiration and morphology-based static features	12.5
J. Galbally et al. [11]	Image quality assessment	8.2
E. Marasco and C. Sansone [12]	Multiple textural features	12.5
L. Ghiani et al. [13]	Comparison of algorithms	N.A.
D.Gragnaniello et al. [14]	Wavelet-Markov local	2.8
R. Nogueira et al. [15]	Convolutional networks	3.9
Y. Jiang and L. Xin [16]	Co-occurrence matrix	6.8

Table 5.3 Publications that cite the LivDet 2011 paper

Authors	Algorithm type	Performance (average classification error) (%)
L. Ghiani et al. [13]	Comparison of algorithms	N.A.
X. Jia et al. [17]	Multi-scale local binary pattern	7.5 and 8.9
D. Gragnaniello et al. [18]	Local contrast phase descriptor	5.7
N. Poh et al. [19]	Likelihood ratio computation	N.A.
A.F. Sequeira and J.S. Cardoso [20]	Modeling the live samples distribution	N.A.
L. Ghiani et al.	Binarized statistical image features	7.2
X. Jia et al. [21]	Multi-scale local ternary patterns	9.8
G.L. Marcialis et al. [21]	Comparison of algorithms	N.A.
R. Nogueira et al. [15]	Convolutional networks	6.5
Y. Zhang et al. [22]	Wavelet analysis and local binary pattern	12.5
A. Rattani et al. [23]	Textural algorithms	N.A.
P. Johnson and S. Schuckers [24]	Pore characteristics	12.0
X. Jia et al. [25]	One-class SVM	N.A.
Y. Jiang and L. Xin [16]	Co-occurrence matrix	11.0

Table 5.4 Publications that cite the LivDet 2013 paper

Authors	Algorithm type	Performance (average classification error) (%)
C. Gottschlich et al. [26]	Histograms of invariant gradients	6.7
R. Nogueira et al. [15]	Convolutional networks	3.6
Y. Zhang et al. [22]	Wavelet analysis and local binary pattern	2.1
P. Johnson and S. Schuckers [24]	Pore characteristics	N.A.

Table 5.5 Publications that cite the LivDet 2015 paper

Authors	Algorithm type	Performance (average classification error) (%)
T. Chugh et al. [27]	CNN-Inception v3 + Minutiae-based local patches	1.4

5.3.1 Performance Evaluation

The parameters adopted for the performance evaluation are the following:

- *FerrLive*: Rate of misclassified live fingerprints.
- *FerrFake*: Rate of misclassified fake fingerprints.
- *Average Classification Error (ACE)*: $ACE = (\frac{FerrLive+FerrFake}{2})$.
- *Equal Error Rate (EER)*: Rate at which FerrLive and FerrFake are equal.
- *Accuracy*: Rate of correctly classified live and fake fingerprints at a 0.5 threshold.

Original terminology is presented although the nomenclature for error rates has been changed in recent competitions to reflect the current standards. FerrLive is equivalent to the Bona Fide Presentation Classification Error Rate (BPCER). FerrFake is equivalent to the Attack Presentation Classification Error Rate (APCER). Each of the algorithms returned a value representing a percentage of posterior probability of the live class (or a degree of "liveness"), given the image normalized in the range 0–100 (100 is the maximum degree of liveness and 0 means that the image is fake). The threshold value for determining liveness was set at 50. This threshold is used to calculate Attack Presentation Classification Error Rate (APCER) and Bona Fide Presentation Classification Error Rate (BPCER) error estimators. Being able to see a range of values for a system is beneficial for understanding how the system is performing against different types of data, however, each competitor will have a different method of determining whether an image is live or spoof. A standardized method was created with which all competitors would have to normalize their outputs to fit. This 0–100 range and a threshold of 50 are arbitrary values provided to competitors but they provide a common range for all competitors to normalize their scores within. The competitors choose how they wish to adjust their system to work within the confines of the competition.

To select a winner, the average of APCER and BPCER was calculated for each participant across datasets and the competitor with the lowest average classification error rate is declared the winner. In terms of the competition, both APCER and BPCER are considered equal cost for failure to keep a balance between systems used for convenience and systems used for high security.

5.3.2 Part 1: Algorithm Datasets

Each iteration of Part 1: Algorithms feature its own set of datasets. Each competition features three to four different sets of fingerprint data. Individual sensors have been used in different competitions, although new data is created for each individual competition. Error rates are calculated for each algorithm on each dataset separately and then the BPCER and APCER for each dataset are averaged to create an overall BPCER and APCER. This overall is averaged to the average classification error rate described in the previous section.

(a)   **(b)**   **(c)**

(d)   **(e)**   **(f)**

(g)   **(h)**   **(i)**

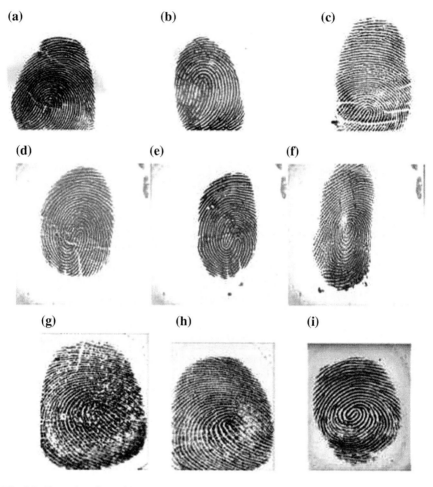

Fig. 5.1 Examples of spoof images of the LivDet 2009 datasets. Crossmatch (top): **a** Playdoh, **b** gelatin, **c** silicone; Identix (middle): **d** playdoh, **e** gelatin, **f** silicone; Biometrika (top): **g** playdoh, **h** gelatin, **i** silicone

LivDet 2009 consisted of data from three optical sensors; Crossmatch, Identix, and Biometrika. The fingerprint images were collected using the consensual approach from three different spoof material types: gelatin, silicone, and playdoh, and the numbers of images available can be found in [3]. Figure 5.1 shows example images from the datasets.

The dataset for LivDet 2011 consisted of images from four different optical devices, Biometrika, Digital Persona, ItalData, and Sagem. The spoof materials were gelatin, latex, ecoflex, playdoh, silicone, and wood glue. More information can be found in [4]. Figure 5.2 shows the images used in the database.

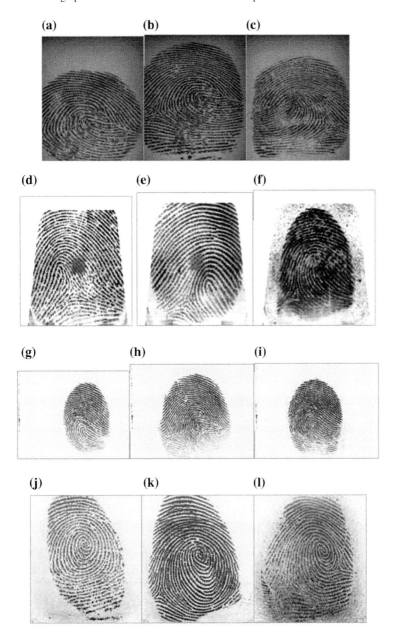

Fig. 5.2 Examples of fake fingerprint images of the LivDet 2011 datasets, from Biometrika **a** latex, **b** gelatin, **c** silicone; from Digital Persona: **d** latex, **e** gelatin, **f** silicone; from Italdata: **g** latex **h** gelatin **i** silicone; from Sagem: **j** latex **k** gelatin **l** silicone

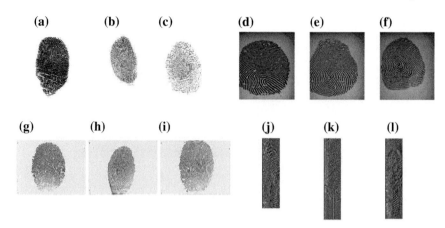

Fig. 5.3 Examples of fake fingerprint images of the LivDet 2013. From Crossmatch **a** body double, **b** latex, **c** wood glue, from Biometrika **d** gelatine, **e** latex, **f** wood glue, from Italdata **g** gelatine, **h** latex, **i** wood glue, from Swipe **j** body double, **k** latex, **l** wood glue

The dataset for LivDet 2013 consisted of images from four different devices; Biometrika, Crossmatch, ItalData, and Swipe. Spoofs were made from gelatin, body double, latex, playdoh, ecoflex, modasil, and wood glue. LivDet 2013 featured the first use of the non-cooperative method for creating spoof images and was used for Biometrika and ItalData while Crossmatch and Swipe continued to use spoofs generated from consensual molds. More information can be found in [5]. Figure 5.3 gives example images from the databases.

The dataset for LivDet 2015 consists of images from four different optical devices; Green Bit, Biometrika, Digital Persona, and Crossmatch. The spoof materials were Ecoflex, gelatin, latex, wood glue, a liquid Ecoflex and RTV (a two-component silicone rubber) for the Green Bit, the Biometrika and the Digital Persona datasets, and playdoh, Body Double, Ecoflex, OOMOO (a silicone rubber) and a novel form of gelatin for Crossmatch dataset. The Crossmatch dataset continued to use strictly the consensual method for obtaining fingerprint molds. More information can be found in [6].

5.3.3 Part 2: Systems Submissions

Data is collected on the submitted systems, however, the datasets generated from the systems submissions are not made public. Unlike in Part 1: Algorithms where data was pre-generated before the competition and distributed into a training and a testing set, Part 2: Systems data comes from systematic testing of the submitted hardware. LivDet 2011 consisted of 500 live attempts from 50 people (totaling 5 images for each of the R1 and R2 fingers) as well as 750 attempts with spoofs of five materials (playdoh, gelatin, silicone, body double, and latex). For LivDet 2013,

1000 live attempts were created from 50 subjects as well as 1000 spoof attempts from 20 subjects from the materials; Playdoh, gelatin, Ecoflex, Modasil, and latex. In 2015, the system was tested using the three known spoof recipes that were previously given to the competitors. Two unknown spoof recipes were also tested to examine the flexibility of the sensor toward novel spoof methods. The known and unknown spoof materials were kept the same from the algorithms portion of the competition. The known recipes were playdoh, Body Double, and Ecoflex and the two unknown recipes used were OOMOO and a novel form of gelatin. 2011 attempts were completed with 1010 live attempts from 51 subjects (2 images each of all 10 fingers) and 1001 spoof attempts across the five different materials giving approximately 200 images per spoof type. 500 spoofs were created from each of 5 fingers of 20 subjects for each of the five spoof materials. Two attempts were performed with each spoof.

The submitted system needs to be able to output a file with the collected image as well as a liveness score in the range of 0–100 with 100 being the maximum degree of liveness and 50 being the threshold value to determine if an image is live or spoof. If the system is not able to process a live subject, it is counted as a failure to enroll and counted against the performance of the system (as part of FerrLive). However, if the system is unable to process a spoof finger, it is considered as a fake nonresponse and counted as a positive in terms of system effectiveness for spoof detection and is classified as the Attack Presentation Nonresponse Rate (APNRR).

5.3.4 Image Quality

Fingerprint image quality has a powerful effect on the performance of a matcher. Many commercial fingerprint systems contain algorithms to ensure that only higher quality images are accepted to the matcher. This rejects low quality images where low quality images have been shown to degrade the performance of a matcher [28]. The algorithms and systems submitted for this competition did not use a quality check to determine what images would proceed to the liveness detection protocols. Through taking into account the quality of the images before applying liveness detection, a more realistic level of error can be shown.

Our methodology uses the NIST Fingerprint Image Quality (NFIQ) software to examine the quality of all fingerprints used for the competition and examine the effects of removing lower quality fingerprint images on the liveness detection protocols submitted. NFIQ computes a feature vector from a quality image map and minutiae quality statistics as an input to a multilayer perceptron neural network classifier [28]. The quality of the fingerprint is determined from the neural network output. The quality for each image is assigned on a scale from 1 (highest quality) to 5 (lowest quality).

5.3.5 Specific Challenges

In the last two editions of the competition, specific challenges were introduced. Two of the 2013 datasets, unlike all the other cases, contain spoofs that were collected using latent fingerprints. The 2015 edition had two new components: (1) the testing set included images from two kinds of spoof materials which were not present in the training set in order to test the robustness of the algorithms with regard to unknown attacks, and (2) one of the datasets was collected using a 1000 dpi sensor.

5.3.5.1 LivDet 2013 Consensual Versus Semi-consensual

In the consensual method, the subject pushed his finger into a malleable material such as silicon gum creating a negative impression of the fingerprint as a mold. The mold was then filled with a material, such as gelatin. The "semi-consensual method" consisted of enhancing a latent fingermark pressed on a surface, and digitizing it through the use of a common scanner.[1] Then, through a binarization process and with an appropriate threshold choice, the binarized image of the fingerprint was obtained. The thinning stage allowed the line thickness to be reduced to one pixel obtaining the skeleton of the fingerprint negative. This image was printed on a transparency sheet in order to have the mold. A gelatin or silicone material was dripped over this image, and, after solidification, separated and used as a fake fingerprint.

The consensual method leads to an almost perfect copy of a live finger, whose mark on a surface is difficult to recognize as a fake unless through an expert dactyloscopist. On the other hand, the spoof created by semi- or nonconsensual method is much less similar. In a latent fingerprint, many details are lost and the skeletonization process further deteriorates the spoof quality making it easier to distinguish a live from a fake. However, in a real-world scenario, obtaining a consensual mold of a subject's fingers is an arduous task to achieve, however acquiring latent prints could potentially prove easier. The spoof images in the Biometrika and Italdata 2013 datasets were created by printing the negative image on a transparency sheet. As we will see in the next section, the error rates, as would be expected, are lower than those of the other datasets.

5.3.5.2 LivDet 2015 Hidden Materials and 500 Versus 1000 dpi

As already stated, the testing sets of LivDet 2015 included spoof images of materials not present in the training sets provided to the competitors. These materials were liquid Ecoflex and RTV for Green Bit, Biometrika, and Digital Persona datasets, as well as OOMOO and a novel gelatin for Crossmatch dataset. Our aim was to assess the reliability of algorithms. As a matter of fact, in a realistic scenario, the material

---

[1]Obviously all subjects were fully aware of this process, and gave the full consent to replicate their fingerprints from their latent marks.

used to attack a biometric system could be considered unknown as a liveness detector should be able to deal with any kind of spoof material.

Another peculiarity of the 2015 edition was the presence of the Biometrika HiScan-PRO, a sensor with a resolution of 1000 dpi instead of ~500 dpi resolution for most of the datasets used so far in the competition. It is reasonable to hypothesize that by doubling the image resolution, the feature extraction phase and final performance of the system should benefit. The results that we will show in the next section do not confirm this hypothesis however.

5.4 Examination of Results

In this section, we analyze the experimental results for the four LivDet editions. Results show the growth and improvement across the four competitions.

5.4.1 Trends of Competitors and Results for Fingerprint Part 1: Algorithms

The number of competitors for Fingerprint Part 1: Algorithms have increased during the last years. LivDet 2009 contained a total of 4 algorithm submissions. LivDet 2011 saw a slight decrease in competitors with only 3 organizations submitting algorithms, however LivDet 2013 and 2015 gave rise to the largest of the competitions with 11 submitted algorithms in 2013 followed by 12 in LivDet 2015. Submissions for each LivDet are detailed in Table 5.6.

This increase of participants has shown the grown of interest in the topic, which has been coupled with the general decrease of the error rates.

First of all, the two best algorithms for each competition, in terms of performance, are detailed in Table 5.7 based on the average error rate across the datasets where "Minimum Average" error rates are the best results and "Second Average" are the second best results.

There is a stark difference between the results seen from LivDet 2009 to LivDet 2015. LivDet 2009 to LivDet 2011 did not see much decrease in error, whereas LivDet 2013 and LivDet 2015 each decreased in error from the previous competition.

The mean values of the ACE (Average Classification Error) over all the participants calculated for each dataset confirm this trend. Mean and standard deviation are shown in Table 5.8.

The standard deviation values range between 5 and 18% depending on the dataset and competition editions. Mean ACE values confirm the error increase in 2011 due to the high quality of cast and fake materials. The low values in 2013 for the Biometrika and Italdata are due, as stated before, to the use of latent fingerprints in the spoof creation process, creating lower quality spoofs that are easier to detect. In order to

Table 5.6 Participants for part 1: algorithms

Participants LivDet 2009	Algorithm name
Dermalog Identification Systems GmbH	Dermalog
Universidad Autonoma de Madrid	ATVS
Anonymous	Anonymous
Anonymous2	Anonymous2
Participants LivDet 2011	Algorithm name
Dermalog Identification Systems GmbH	Dermalog
University of Naples Federico II	Federico
Chinese Academy of Sciences	CASIA
Participants LivDet 2013	Algorithm name
Dermalog Identification Systems GmbH	Dermalog
Universidad Autonoma de Madrid	ATVS
HangZhou JLW Technology Co Ltd	HZ-JLW
Federal University of Pernambuco	Itautec
Chinese Academy of Sciences	CAoS
University of Naples Federico II (algorithm 1)	UniNap1
University of Naples Federico II (algorithm 2)	UniNap2
University of Naples Federico II (algorithm 3)	UniNap3
First Anonymous Participant	Anonym1
Second Anonymous Participant	Anonym2
Third Anonymous Participant	Anonym3
Participants LivDet 2015	Algorithm name
Instituto de Biociencias, Letras e Ciencias Exatas	COPILHA
Institute for Infocomm Research (I2R)	CSI
Institute for Infocomm Research (I2R)	CSI_MM
Dermalog	Hbirkholz
Universidade Federal de Pernambuco	Hectorn
Anonymous Participant	Anonym
Hangzhou Jinglianwen Technology Co., Ltd	Jinglian
Universidade Federal Rural de Pernambuco	UFPE I
Universidade Federal Rural de Pernambuco	UFPE II
University of Naples Federico II	Unina
New York University	Nogueira
Zhejiang University of Technology	Titanz

Table 5.7 Two best error rates for each competition. A positive trend in terms of both FerrLive and FerrFake parameters can be noticed. In particular, 2011 and 2015 exhibited very difficult tasks due to the high quality of fingerprint images thus they should be taken into account as a reference of current liveness detector performance against the "worst scenario", that is, the high quality reproduction of a subject's fingerprint

Minimum Avg FerrLive (%)	Minimum Avg FerrFake (%)	Second Avg FerrLive (%)	Second Avg FerrFake (%)
2009			
13.2	5.4	20.1	9.0
2011			
11.8	24.8	24.5	24.8
2013			
11.96	1.07	17.64	1.10
2015			
5.13	2.79	6.45	4.26

Table 5.8 Mean and standard deviation ACE values for each dataset of the competition

	Mean (%)	Std. Dev. (%)
2009		
Identix	8.27	4.65
Crossmatch	15.59	5.60
Biometrika	32.59	9.64
2011		
Biometrika	31.30	10.25
ItalData	29.50	9.42
Sagem	16.70	5.33
Digital persona	23.47	13.70
2013		
Biometrika	7.32	8.80
Italdata	12.25	18.08
Swipe	16.67	15.30
2015		
GreenBit	11.47	7.10
Biometrika	15.81	7.12
DigitalPersona	14.89	13.72
Crossmatch	14.65	10.28

confirm that, we compared these values with those obtained for the same sensors in 2011 (see Table 5.9). Last two rows of Table 5.9 report average classification error and related standard deviation over above sets. Obviously, further and independent experiments are needed because participants of 2011 and 2013 were different so

Table 5.9 Comparison between mean ACE values for Biometrika and Italdata datasets from LivDet 2011 and 2013

	Consensual (LivDet 2011) (%)	Semi-consensual (LivDet 2013) (%)
Biometrika	31.30	7.32
ItalData	29.50	12.25
Mean	30.40	9.78
Standard deviation	1.27	3.49

Table 5.10 Crossmatch 2013 error rates across 1000 tests

	Average error rate (%)	Standard deviation (%)
FerrLive	7.57	2.21
FerrFake	13.4	1.95
Equal error rate	9.92	1.42

that different algorithms are likely also. However, results highlight the performance virtually achievable over two scenarios: a sort of "worst case", namely, the one represented by LivDet 2011, where quality of spoofs is very high, and a sort of "realistic case" (LivDet 2013), where spoofs are created from latent marks as one may expect. The fact that even in this case the average error is 10%, while the standard deviation does not differ with regard to LivDet 2011, should not be underestimated. The improvement could be likely due to the different ways of creating the fakes.

The abnormally high values for the LivDet 2013 Crossmatch dataset occurred due to an occurrence in the live data. The Live images were difficult for the algorithms to recognize. All data was collected in the same time frame and data in training and testing sets was determined randomly among the data collected. A follow-up test was conducted using benchmark algorithms at University of Cagliari and Clarkson University which revealed similar scores on the benchmark algorithms as the submitted algorithms with initial results had an EER of 41.28%. The data was further tested with 1000 iterations of train/test dataset generation using a random selection of images for the live training and test sets (with no common subjects between the sets which is a requirement for all LivDet competitions). This provided new error rates shown (see Table 5.10 and Figs. 5.4 and 5.5).

Examining these results allows us to draw the conclusion that the original selection of subjects was an anomaly that caused improper error rates because each other iteration, even only changing a single subject, dropped FerrLive error rates to 15% and below. Data is being more closely examined in future LivDet competitions in order to counteract this problem with data being processed on a benchmark algorithm before being given to participants. The solution to this for the LivDet 2013 data going forward is to rearrange the training and test sets for future studies using this data. Researchers will need to be clear which split of training/test they used in their study. For this reason, we removed from the experimental results of those obtained with the Crossmatch 2013 dataset.

Fig. 5.4 FerrLive Rates across 1000 tests for Crossmatch 2013

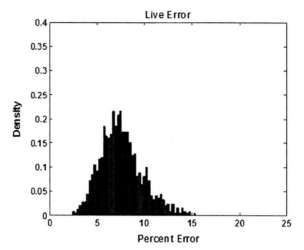

Fig. 5.5 FerrFake Rates across 1000 tests for Crossmatch 2013

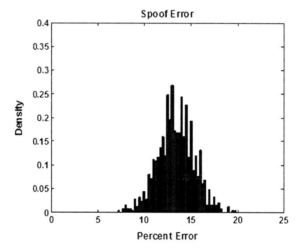

LivDet 2015 error rates confirm a decreasing trend with respect to attacks made up of high quality spoofs, with all the mean values between 11 and 16%. These results are summarized in Fig. 5.6.

Reported error rates suggest a slow but steady advancement in the art of liveness and artifact detection. This gives supporting evidence that the technology is evolving and learning to adapt and overcome the presented challenges.

Comparing the performance of the LivDet 2015 datasets (as shown in Table 5.8 and in more details in [6]), two other important remarks can be made: the higher resolution for Biometrika sensor did not necessarily achieve the best classification performance, while the small size of the images for the Digital Persona device generally degrades the accuracy of all algorithms.

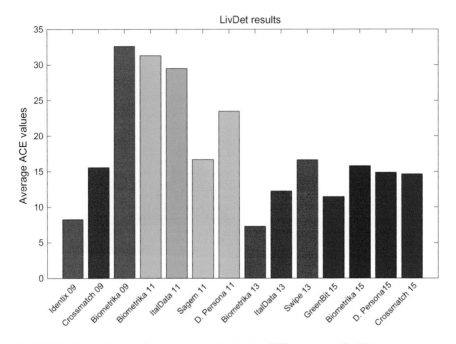

Fig. 5.6 LivDet results over the years: near red colors for 2009, near green for 2011, near magenta for 2013, and near blue for 2015

Table 5.11 FerrFake values of the best algorithms calculated when $FerrLive = 1\%$ for the Crossmatch dataset

	All materials (%)	Known materials (%)	Unknown materials (%)
Unina	7.42	2.47	14.49
Nogueira	2.66	1.94	3.69
Anonym	18.61	10.75	29.82
Average	9.56	5.05	16.00
Std. dev.	8.19	4.94	13.13

Another important indicator of an algorithm validity is the FerrFake value calculated when $FerrLive = 1\%$. This value represents the percentage of spoofs able to hack into the system when the rate of legitimate users that are rejected is no more than 1%. As a matter of fact, by varying the threshold value, different FerrFake and FerrLive values are obtained and, as the threshold grows from 0 to 100, FerrFake decreases and FerrLive increases. Obviously FerrLive value must be kept low to minimize the inconvenience to authorized users but, just as important, the low FerrFake value limits the number of unauthorized users able to enter into the system.

Results in Tables 5.11, 5.12, 5.13, and 5.14 show that even the best performing algorithm (nogueira) is not yet good enough since when $FerrLive = 1\%$, the FerrFake values (testing on all materials) range from 2.66 to 19.10%. These are the

Table 5.12 FerrFake values of the best algorithms calculated when $FerrLive = 1\%$ for the digital persona dataset

	All materials (%)	Known materials (%)	Unknown materials (%)
Unina	51.30	50.85	52.20
Nogueira	19.10	16.00	25.30
Anonym	80.83	79.25	84.00
Average	50.41	48.70	53.83
Std. dev.	30.87	31.68	29.38

Table 5.13 FerrFake values of the best algorithms calculated when $FerrLive = 1\%$ for the green bit dataset

	All materials (%)	Known materials (%)	Unknown materials (%)
Unina	41.80	36.10	53.20
Nogueira	17.90	15.15	23.40
Anonym	75.47	75.25	75.90
Average	45.06	42.17	50.83
Std. dev.	28.92	30.51	26.33

Table 5.14 FerrFake values of the best algorithms calculated when $FerrLive = 1\%$ for the biometrika dataset

	All materials (%)	Known materials (%)	Unknown materials (%)
Unina	11.60	7.50	19.80
Nogueira	15.20	12.60	20.40
Anonym	48.40	44.05	57.10
Average	25.07	21.38	32.43
Std. dev.	20.29	19.79	21.36

percentage of unauthorized users that the system is unable to correctly classify. If we consider only the unknown materials, the results are even worse, ranging from 3.69 to 25.30%.

On the basis of such results, we can say that there is no specific algorithm, among the analyzed ones, able to generalize against never-seen-before spoofing attacks. We observed a performance drop and also found that the amount of the drop is unpredictable as it depends on the material. This should be a matter of future discussions.

5.4.2   Trends of Competitors and Results for Fingerprint Part 2: Systems

Fingerprint Part 2: Systems, over the course of three competitions, have not shown growth in the number of participants. It is not unexpected as the systems' competition requires the submission of a fully packaged fingerprint system with presentation attack detection module built-in. There has also been a general lack of interest in companies shipping full systems for testing and it appears that there is more comfort in submitting an algorithm as it only requires the submission of a software package. Both Livdet 2011 and LivDet 2013 had two submissions while LivDet 2015 had only one. Information about competitors is shown in Table 5.15.

This portion of the LivDet competitions has distinct recognition for the rapid decrease in error rates. In the span of 2 years, the best results from LivDet 2011 were worse than the worst results of LivDet 2013. This shows that system tests have displayed a quicker decrease in error rates as well as the one systems submission in 2013 had lower error rates than any submitted algorithms in LivDet.

In 2011, Dermalog performed at a FerrLive of 42.5% and a FerrFake of 0.8%. GreenBit performed at a FerrLive of 38.8% and a FerrFake of 39.47%. Both systems had high FerrLive scores and can be seen in Table 5.16.

The 2013 edition produced much better results since Dermalog performed at a FerrLive of 11.8% and a FerrFake of 0.6%. Anonymous1 performed at a FerrLive of 1.4% and a FerrFake of 0.0%. Both systems had low FerrFake rates. Anonymous1 received a perfect score of 0.0% error, successfully determining every spoof finger presented as a spoof and can be seen in Table 5.17.

Anonymous2, in 2015, scored a FerrLive of 14.95% and a FerrFake of 6.29% at the (given) threshold of 50 Table 5.6 showing an improvement over the general results seen in LivDet 2011, however, the anonymous system did not perform as well as what was seen in LivDet 2013. There is an 11.09% FerrFake for known recipes and 1% for unknown recipes and seen in Table 5.18. This result is opposite to what has been seen in previous LivDet competitions where known spoof types typically have a better performance than unknown spoof types. The error rate for spoof materials was

Table 5.15 Participants for part 2: systems

Participants LivDet 2011	Algorithm name
Dermalog identification systems GmbH	Dermalog
GreenBit	GreenBit
Participants LivDet 2013	Algorithm name
Dermalog identification systems GmbH	Dermalog
Anonymous	Anonymous1
Participants LivDet 2015	Algorithm name
Anonymous	Anonymous2

Table 5.16 FerrLive and FerrFake for submitted systems in LivDet 2011

Submitted system	FerrLive (%)	FerrFake (%)
Dermalog	42.5	0.8
Greenbit	39.5	38.8
Submitted system	FerrFake known (%)	FerrFake unknown (%)
Dermalog	0.4	1.3
Greenbit	19.1	70

Table 5.17 FerrLive and FerrFake for submitted systems in LivDet 2013

Submitted system	FerrLive (%)	FerrFake (%)
Dermalog	11.8	0.6
Morpho	1.4	0
Submitted System	FerrFake known (%)	FerrFake unknown (%)
Dermalog	0.3	1
Morpho	0	0

Table 5.18 FerrLive and FerrFake for submitted systems in LivDet 2015

Submitted system	FerrLive (%)	FerrFake (%)
Anonymous	14.95	6.29
Submitted system	FerrFake known (%)	FerrFake unknown (%)
Anonymous	11.09	1.00

primarily due to impact on color differences error for the playdoh. Testing across six different colors of playdoh found that certain colors behaved in different ways. For yellow and white playdoh, the system detected spoofs as fake with high accuracy. For brown and black playdoh, the system would not collect an image. Therefore, it was recorded as a fake nonresponse and not an error in detection of spoofs. For pink and lime green playdoh, the system incorrectly accepted spoofs as live for almost 100% of images collected. The fact that almost all pink and lime green playdoh images were accepted as live images resulted in a 28% total error rate for playdoh. The system had a 6.9% Fake Nonresponse Rate primarily due to brown and black playdoh. This is the first LivDet competition where color of playdoh has been examined in terms of error rates and provides vital information for future testing of presentation attack detection systems.

Examining the trends of results over the three competitions has shown that since 2011, there has been a downward trend in error rates for the systems. FerrLive in 2015 while higher than 2013, is drastically lower than 2011. FerrFake has had similar error rates over the years. While the 2015 competition showed a 6.29% FerrFake, the majority of that error stems from playdoh materials, particularly pink and lime colors. If you discount the errors seen in playdoh, the FerrFake is below 2% for

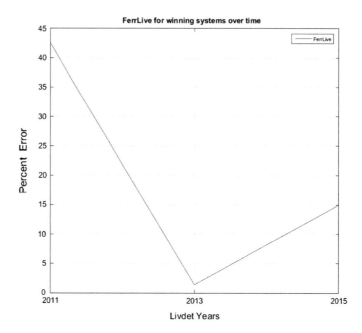

Fig. 5.7 FerrLive for winning systems by year

the 2015 system. The error rates for winning systems over the years are shown in Figs. 5.7 and 5.8.

5.5 Future of LivDet

LivDet-Fingerprint has continued in 2017. The two parts of the competition are being held separately. LivDet-Fingerprint 2017 Fingerprint Systems Liveness Detection Competition is being held with an additional focus toward systems geared for mobile devices. The competition will include not only spoofs generated from the consensual methods but also will include nonconsensual methods of spoof mold generation. This competition also includes a new challenge for competitors in the addition of a verification mode testing. Submitted systems will need to be submitted not only with a presentation attack detection module as in previous systems competitions but will need to have an enrollment feature where subjects will enroll their fingers and both live and spoof fingers will be tested against in the system in verification mode.

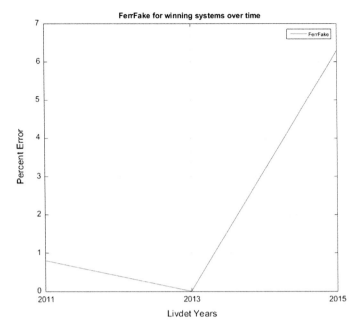

Fig. 5.8 FerrFake for winning systems by year

5.6 Conclusions

Since its first edition in 2009, the Fingerprint Liveness Detection Competition was aimed to allow research centers and companies a fair and independent assessment of their anti-spoofing algorithms and systems.

We have seen over time an increasing interest for this event, and the general recognition for the enormous amount of data made publicly available. The number of citations that LivDet competitions have collected is one of the tangible signs of such interest, with over 100 citations, and further demonstrates the benefits that the scientific community has received from LivDet events.

The competition results show that liveness detection algorithms and systems strongly improved their performance: from about 70% classification accuracy achieved in LivDet 2011 to 90% classification accuracy in LivDet 2015. This result, obtained under very difficult conditions like the ones of the consensual methodology of fingerprints replication, is comparable with that obtained in LivDet 2013 (first two datasets), where the algorithms' performance was tested under the easier task of fingerprints replication from latent marks. Moreover, the two challenges characterizing the last edition, namely, the presence of 1000 dpi capture device and the evaluation against "unknown" spoofing materials, further contributed to show the great improvement that researchers achieved on these issues: submitted algorithms performed very well on both 500 and 1000 dpi capture devices, and some of them also exhibited a good robustness degree against never-seen-before attacks. Results reported on fusion also show that the liveness detection could further benefit from the combination of multiple features and approaches. A specific section on algorithms and systems fusion might be explicitly added to a future LivDet edition.

It is evident that, despite the remarkable results reported from the LivDet competitions, further improvements in system performance are needed. Current performance levels for most submissions are not yet accurate enough for embedding a presentation attack detection algorithm into fingerprint verification system due to the error rate being still too high for many real applications where security is of paramount importance. In the authors' opinion, discovering and explaining benefits and limitations of the currently used features is still an issue whose solution should be encouraged, because only the full understanding of the physical process which leads to the finger's replica and what features extraction process exactly does will shed light on the characteristics most useful for classification. This task is daunting and it may require many years before concrete results can be shown. However, we believe this could be the next challenge for a future edition of LivDet, the Fingerprint Liveness Detection Competition.

Acknowledgements The first and second author had equal contributions to the research. This work has been supported by the Center for Identification Technology Research and the National Science Foundation under Grant No. 1068055, and by the project "Computational quantum structures at the service of pattern recognition: modeling uncertainty" [CRP-59872] funded by Regione Autonoma della Sardegna, L.R. 7/2007, Bando 2012.

References

1. Cappelli R (2007) Fingerprint verification competition 2006. Biom Technol Today 15(7):7–9. https://doi.org/10.1016/S0969-4765(07)70140-6
2. Alonso-Fernandez F, Bigun J (2013) Halmstad university submission to the first ICB competition on IRIS recognition (ICIR2013)
3. Marcialis GL et al (2009) First international fingerprint liveness detection competition–livdet 2009. In: International conference on image analysis and processing. Springer, Berlin. https://doi.org/10.1007/978-3-642-04146-4_4
4. Yambay D et al (2012) Livdet 2011 - fingerprint liveness detection competition 2011. In: 5th IAPR/IEEE International conference on biometrics (ICB 2012)
5. Marcialis G et al (2013) Livdet 2013 - fingerprint liveness detection competition 2013. In: 6th IAPR/IEEE International conference on biometrics (ICB2013)
6. Marcialis G et al (2014) LivDet 2015 - fingerprint liveness detection competition 2015. In: 7th IEEE international conference on biometrics: theory, applications and systems (BTAS 2015)
7. Coli P, Marcialis G, Roli F (2007) Vitality detection from fingerprint images: a critical survey
8. Chakka MM et al (2011) Competition on counter measures to 2-D facial spoofing attacks. https://doi.org/10.1109/IJCB.2011.6117509
9. Galbally J et al (2012) A high performance fingerprint liveness detection method based on quality related features. Future Gener Comput Syst. https://doi.org/10.1016/j.future.2010.11.024
10. Marasco E, Sansone C (2012) Combining perspiration- and morphology-based static features for fingerprint liveness detection. Pattern Recognit Lett 33(9):1148–1156 (1 July 2012)

11. Galbally J et al (2014) Image quality assessment for fake biometric detection: application to iris, fingerprint, and face recognition. IEEE Trans Image Process 23(2):710–724. https://doi. org/10.1109/TIP.2013.2292332

12. Marasco E, Sansone C (2010) An anti-spoofing technique using multiple textural features in fingerprint scanners. In: 2010 IEEE workshop on biometric measurements and systems for security and medical applications (BIOMS), 9 September 2010, pp 8–14. https://doi.org/10. 1109/BIOMS.2010.5610440

13. Ghiani L et al (2014) Experimental results on fingerprint liveness detection. In: Proceedings of the 7th international conference on articulated motion and deformable objects (AMDO 2012). Springer, Berlin, pp 439–441. https://doi.org/10.1049/el.2013.4044

14. Gragnaniello D et al (2014) Wavelet-Markov local descriptor for detecting fake fingerprints. Electron Lett 50(6):439–441

15. Nogueira R et al (2014) Evaluating software-based fingerprint liveness detection using convolutional networks and local binary patterns. In: 2014 Proceedings of the IEEE workshop on biometric measurements and systems for security and medical applications (BIOMS), 17 October 2014, pp 22–29. https://doi.org/10.1109/BIOMS.2014.6951531

16. Jiang Y, Xin L (2015) Spoof fingerprint detection based on co-occurrence matrix. Int J Signal Process Image Process Pattern Recognit 8(8):373–384

17. Jia X et al (2014) Multi-scale local binary pattern with filters for spoof fingerprint detection. Inf Sc 268:91–102 (1 June 2014). ISSN 0020-0255

18. Gragnaniello D et al (2015) Local contrast phase descriptor for fingerprint liveness detection. Pattern Recognit 48(4):1050–1058

19. Poh N et al (2014) Anti-forensic resistant likelihood ratio computation: a case study using fingerprint biometrics. In: Signal processing conference (EUSIPCO), pp 1377–1381

20. Sequeira AF, Cardoso JS (2015) Fingerprint liveness detection in the presence of capable intruders. Sensors (Basel) 15(6):14615–14638. https://doi.org/10.3390/s150614615

21. Jia X et al (2013) Multi-scale block local ternary patterns for fingerprints vitality detection. In: 2013 International conference on biometrics (ICB), 4–7 June 2013, pp 1–6. https://doi.org/10. 1109/ICB.2013.6612964

22. Zhang Y et al (2014) Fake fingerprint detection based on wavelet analysis and local binary pattern. In: 2014 Biometric recognition (CCBR 2014), vol 8833, pp 191–198

23. Rattani A et al (2015) Open set fingerprint spoof detection across novel fabrication materials. IEEE Trans Inf Forensics Secur 10(11):2447–2460. https://doi.org/10.1109/TIFS.2015. 2464772

24. Johnson P, Schuckers S (2014) Fingerprint pore characteristics for liveness detection. In: Proceedings of the international conference of the biometrics special interest group (BIOSIG), pp 1–8

25. Jia X et al (2014) One-class SVM with negative examples for fingerprint liveness detection. Biometric recognition. Springer International Publishing, New York

26. Gottschlich C et al (2014) Fingerprint liveness detection based on histograms of invariant gradients. In: 2014 IEEE international joint conference on biometrics (IJCB). IEEE

27. Chugh T, Cao K, Jain A (2017) Fingerprint spoof detection using minutiae-based local patches. In: 2017 IEEE international joint conference on biometrics (IJCB). IEEE. https://doi.org/10. 1109/BTAS.2017.8272745

28. Watson C et al Users guide to NIST biometric image software

Part II
Iris Biometrics

Chapter 6
Introduction to Iris Presentation Attack Detection

Aythami Morales, Julian Fierrez, Javier Galbally and Marta Gomez-Barrero

Abstract Iris recognition technology has attracted an increasing interest since more than two decades in which we have witnessed a migration from laboratories to real-world applications. The deployment of this technology in real applications raises questions about the main vulnerabilities and security threats related to these systems. Presentation attacks can be defined as presentation of human characteristics or artifacts directly to the input of a biometric system trying to interfere with its normal operation. These attacks include the use of real irises as well as artifacts with different levels of sophistication. This chapter introduces iris presentation attack detection methods and its main challenges. First, we summarize the most popular types of attacks including the main challenges to address. Second, we present a taxonomy of presentation attack detection methods to serve as a brief introduction on this very active research area. Finally, we discuss the integration of these methods into iris recognition systems according to the most important scenarios of practical application.

A. Morales (✉)
School of Engineering, Universidad Autonoma de Madrid, Madrid, Spain
e-mail: aythami.morales@uam.es

J. Fierrez
Universidad Autonoma de Madrid, Madrid, Spain
e-mail: julian.fierrez@uam.es

J. Galbally
European Commission - DG Joint Research Centre, Ispra, Italy
e-mail: javier.galbally@ec.europa.eu

M. Gomez-Barrero
da/sec - Biometrics and Internet Security Research Group, Hochschule Darmstadt, Darmstadt, Germany
e-mail: marta.gomez-barrero@h-da.de

© Springer Nature Switzerland AG 2019
S. Marcel et al. (eds.), *Handbook of Biometric Anti-Spoofing*,
Advances in Computer Vision and Pattern Recognition,
https://doi.org/10.1007/978-3-319-92627-8_6

6.1 Introduction

The iris is one of the most popular biometric modes inside the biometric person recognition technologies. Since the earliest Daugman publications proposing the iris as a biometric characteristic [1] to most recent approaches based on latest machine learning and computer vision techniques [2–4], iris recognition has evolved improving performance, ease of use, and security. Such advances have attracted the interest of researchers and companies boosting the number of products, publications, and applications. The first iris recognition devices were developed to work as stand-alone systems [5]. However, today iris recognition technology is included as an authentication service in some of the most important operating systems (e.g., Android, Microsoft Windows) and devices (e.g., laptop or desktop computers, smartphones). One-seventh of the world population (1.14 billion people) has been enrolled in the Aadhaar India national biometric ID program [6] and iris is one on three biometric modes (in addition to fingerprint and face) employed for authentication in this program. The main advantages of iris can be summarized as follows:

- The iris is generated during the prenatal gestation and presents highly random patterns. Such patterns are composed of complex and interrelated shapes and colors. The highly discriminant characteristics of the iris make possible that recognition algorithms obtain performances comparable to the most accurate biometric modes [2].
- The genetic prevalence on iris is limited and therefore irises from people with shared genes are different. Both irises of a person are considered as different instances, which do not match each other.
- The iris is an internal organ of the eye that is externally visible. The iris can be acquired at a distance and the advances on acquisition sensors allow to easily integrate iris recognition into portable devices [7].

The fast deployment of iris recognition technology in real applications has increased the concerns about its security. The applications of iris biometrics include a variety of different scenarios and security levels (e.g., banking, smartphone user authentication, and governmental ID programs). Among all threats associated to biometric systems, the resilience against attacks emerges as one of the most active research areas in the recent iris biometrics literature. The security of commercial iris systems is questioned by users. In 2017, the Chaos Computer Club reported their successful attack to the Samsung Galaxy S8 iris scanner using a simple photograph and a contact lens [8]. In the context of biometric systems, presentation attacks are defined as presentation of human characteristics or artifacts directly to the input of a biometric system trying to interfere with its normal operation [9]. This definition includes spoofing attacks, evasion attacks, and the so-called zero-effort attacks. Most of the literature on iris Presentation Attack Detection (PAD) is focused on spoofing attacks detection. The term liveness detection is also employed in the literature to propose systems capable of classifying between bona fide samples and artifacts used to attack biometric systems. Depending on the motivations of the attacker, we can distinguish two types of attacks:

- Impostor: The attacker tries to impersonate the identity of other subjects by using his own iris (e.g., zero-effort attacks) or an artifact mimicking the iris of the spoofed identity (e.g., photo, video or synthetic iris). This type of attack requires certain level of knowledge about the iris of the impersonated user and the characteristics of the iris sensor in order to increase the success of the attack (see Sect. 6.2).
- Identity concealer: The attacker tries to evade the iris recognition. Examples in this case include the enrollment of users with fake irises (e.g., synthetically generated) or modified irises (e.g., textured contact lens). These examples represent a way to masquerade the real identities.

The first PAD approaches proposed in the literature were just theoretical exercises based on potential vulnerabilities [10]. In recent years, the number of publications focused on this topic has increased significantly. Some of the PAD methods discussed in the recent literature have been inspired by methods proposed for other biometric modes such as face [11–13]. However, the iris has various particularities which can be exploited for PAD, such as the dynamic, fast, and involuntary responses of the pupil and the heterogeneous characteristics of the eyes tissue. The eye reacts according to the amount and nature of the light received. Another large group of PAD methods exploits these dynamic responses and involuntary signals produced by the eye.

This chapter presents a description of the most important types of attacks from zero-effort attacks to the most sophisticated synthetic eyes. We introduce iris Presentation Attacks Detection methods and its main challenges. The PAD methods are organized according to the nature of features employed with a taxonomy divided into three main groups: hardware-based, software-based, and challenge–response approaches. Please note that the material presented in this chapter tries to be up to date, but keeping an introductory nature. A more comprehensive survey of iris PAD can be found in [4].

The rest of the chapter is organized as follows: Sect. 6.2 presents the main vulnerabilities of iris recognition systems with special attention to different types of presentation attacks. Section 6.3 summarizes the presentation attacks detection methods while Sect. 6.4 presents the integration with iris recognition systems. Finally, conclusions are presented in Sect. 6.5.

6.2   Vulnerabilities in Iris Biometrics

Traditional block diagrams of Iris Recognition Systems (IRS) are similar to block diagrams of other biometric modes. As any other biometric recognition technology, iris recognition is vulnerable to attacks. Figure 6.1 includes a typical block diagram of an IRS and its vulnerable points. The vulnerabilities depend on the characteristics of each module and cover communication protocols, data storage or resilience to artifact presentations, among others. Several subsystems and not just one will define the security of an IRS [14]:

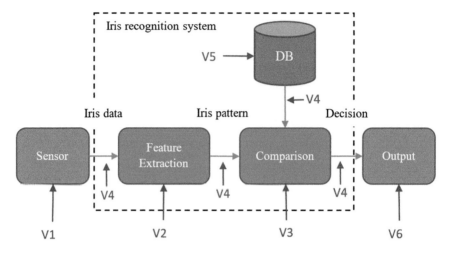

Fig. 6.1 Block diagram of traditional iris recognition systems and main vulnerabilities [14]

- Sensor (V1): CCD/CMOS are the most popular sensors including visible and near-infrared imaging. The iris pattern is usually captured in form of image or video. The most important vulnerability is related to the presentation of artifacts (e.g., photos, videos, synthetic eyes) that mimic the characteristics of real irises.
- Feature extraction and matcher modules (V2-V3): These software modules are composed of the algorithms in charge of preprocessing, segmentation, generation of templates, and comparison. Attacks to these modules include the alteration of algorithms to carry out not legitimate operations (e.g., modified templates, altered comparison).
- Database (V5): The database is composed of structured data associated to the subject information, devices, and iris templates. Any alteration on this information can affect the final response of the system. The security level of the database storage differs depending on the applications. The use of encrypted templates is crucial to ensure the unlinkability between systems and attacks based on weak links [15].
- Communication channel and actuators (V4 and V6): Including internal (e.g., communication between software modules) and external communications (e.g., communication with mechanical actuators or cloud services). The most important vulnerabilities rely on alterations of the information sent and received by the different modules of the IRS.

In this work, we will focus on presentation attacks on the sensor (V1 vulnerabilities). Key properties of these attacks are the high attack success ratio of spoofed irises (if the system is not properly protected) and the low amount of information about the system needed to perform the attack. Other vulnerabilities not covered by this work include attacks to the database (V5), to the software modules (V2-V3), the communication channels (V4) or actuators at the output (V6). This second group of vulnerabilities requires access to the system, and countermeasures to these attacks are more related to general system security protocols. These attacks are beyond the scope of this chapter but should not be underestimated.

Regarding the nature of the Presentation Attack Instrument (PAI) employed to spoof the system, the most popular presentation attacks can be divided into the following categories:

- Zero-effort attacks,
- photo and video attacks,
- contact lens attacks, and
- synthetic eye attacks.

6.2.1 Zero-Effort Attacks

The attack is performed using the iris from the attacker. The impostor does not use any artifact or information about the identity under attack. The iris pattern from the impostor does not match the legitimate pattern and the success of the attack is exclusively related to the False Match Rate (FMR) of the system [16, 17]. Systems with high FMR will be more vulnerable to this type of attack. Note that the FMR is related to the False Non-Match Rate (FNMR) and both are related to the operational point of the system. An operational setup to obtain a low FMR can produce an increment of the FNMR and therefore a higher number of false negatives (legitime users are rejected).

- Information needed to perform the attack: No information needed.
- Generation and acquisition of the PAIs: No need to generate a fake iris. The system is attacked using real irises of the attacker.
- Expected impact of the attack: Most iris recognition systems present very low false acceptance rates. The success rate of these attacks can be considered low.

6.2.2 Photo and Video Attacks

The attack is performed displaying a printed photo, digital image or video from the spoofed iris directly to the sensor of the IRS. Photo attacks are the ones most studied in the literature [12, 18–22] because of two main aspects.

First, with the advent of digital photography and social image sharing (e.g., Flickr, Facebook, Picasa Web, and others), headshots of attacked clients from which the iris can be extracted are becoming increasingly easy to obtain. The face or the voice are biometric modes more exposed than iris. However, iris patterns can be obtained from high-resolution face images (e.g., 200 dpi resolution).

Second, it is relatively easy to print high-quality iris photographs using commercial cameras (up to 12 Megapixels sensors in most of the nowadays smartphones) and ink printers (1200 dpi in most of the commercial ink printers). Alternatively, most of the mobile devices (smartphones and tablets) are equipped with high-resolution

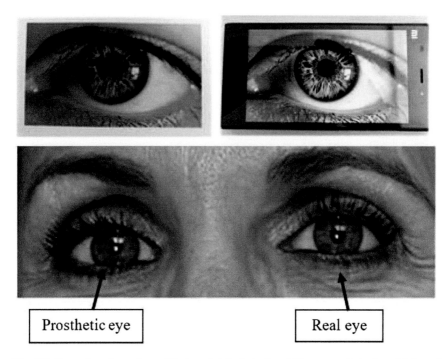

Fig. 6.2 Examples of spoofing artifacts: printed photo (top-left), screen photo (top-right), and prosthetic eye (down). Adapted from Soper Brothers and Associates [http://www.soperbrothers. com]

screens capable of reproducing very natural images and videos in the visible spectrum (see Fig. 6.2).

Video attacks are a sophistication of photo attacks that allows to mimic both the static patterns and the dynamic information of the eye [23–26].

- Information needed to perform the attack: Image or video of the iris of the subject to be impersonated.
- Generation and acquisition of the PAIs: It is relatively easy to obtain high-resolution face photographs from social media and Internet profiles. Other options include the capture using a concealed/hidden camera. Once a photo is obtained, if it is of sufficient quality, one can print the iris region and then present it in front of the iris camera. A screen could also be used for presenting the photograph to the camera. Another way to obtain the iris would be to steal the raw iris image acquired by an existing iris recognition system in which the subject being spoofed was already enrolled.
- Expected impact of the attack: The literature offers a large number of approaches with good detection rates of printed photo attacks [12, 18–21, 27]. However, most of these methods exploit the lack of realism/quality of printed images in comparison with bona fide samples. The superior quality of new screens capable of reproducing digital images and video attacks with a high quality represents a

difficult challenge for PAD approaches based on visible spectrum imaging but not for Near-infrared sensors of commercial systems.

6.2.3 Contact Lens Attacks

This type of attack uses contact lenses created to mimic the pattern of other users (impostor attack) or contact lenses created to masquerade the identity (identity concealer attack). Although the impression of a real iris pattern into contact lenses is theoretically possible, it implies practical difficulties that mitigate the likelihood of this attack. The second scenario is particularly worrying because nowadays, more and more people wear contact lenses (e.g., approximately 125 million people worldwide wear contact lens). We can differentiate between transparent contact lenses and textured contact lenses (also known as printed). Textured contact lenses change the original iris information by the superposition of synthetic patterns (e.g., cosmetic lens to change the color). Although these contact lenses are mostly intended for cosmetics, the same technology can be potentially used to print iris patterns from real users. Once users have been enrolled into the IRS without taking off the textured contact lenses, the IRS can be fooled. Note that asking to remove the contact lenses before the recognition is a non-desirable solution as it clearly decreases the user comfort and usability.

- Information needed to perform the attack: Image of the iris of the client to be attacked for impostor attacks. No information needed for masquerade attacks.
- Generation and acquisition of the fakes: In comparison with photo or video attacks, the generation of textured contact lenses requires a more sophisticated method based on optometrist devices and protocols.
- Expected impact of the attack: These types of attacks represent a great challenge for either automatic PAD systems or visual inspection by humans. It has been reported by several researchers that it is actually possible to spoof iris recognition systems with well-made contact lens [23, 26, 28–31].

6.2.4 Synthetic Eye Attacks

This type of attack is the most sophisticated. The attack uses synthetic eyes generated to mimic the characteristics of real ones. Prosthetic eyes have been used since the beginning of twentieth century to reduce the esthetic impact related to the absence of eyes (e.g., blindness, amputations, etc.). Current technologies for prosthetic manufacturing allow mimicking the most important attributes of the eye with very realistic results. The similarity goes beyond the visual appearance including manufacturing materials with similar physical properties (e.g., elasticity, density).

Table 6.1 Literature on presentation attack detection methods. Summary of key literature about iris PAD methods depending on the type of attack

Type of attack	References	Public databases	Detection errors %
Photo and Video	[11, 12, 33–40]	[20, 21, 24, 27, 35, 41, 42]	0–6
Contact lens	[30, 31, 33, 39, 40]	[27, 43]	0.2–10
Synthetic	[33, 39]	none	0.2–0.3

The number of studies including attacks to iris biometric systems using synthetic eyes is still low [32].

- Information needed to perform the attack: Image of the eye of the client to be attacked.
- Generation and acquisition of the PAIs: This is probably the most sophisticated attack method as it involves the generation of both 2D images and 3D structures. Manually made in the past, 3D-printers and their application to the prosthetic field have revolutionized the generation of synthetic body parts.
- Expected impact of the attack: Although the number of studies is low, the detection of prosthetic eyes represents a big challenge. The detection of these attacks by techniques based on image features is difficult. On the other hand, PAD methods based on dynamic features can be useful to detect the unnatural dynamics of synthetic eyes.

Table 6.1 lists key literature on iris PAD including the most popular public databases available for research purposes.

6.3 Presentation Attack Detection Approaches

These methods are also known in the literature as liveness detection, anti-spoofing, or artifact detection among others. The term Presentation Attack Detection (PAD) was adopted in the ISO/IEC 30107-1:2016 [9], and it is now largely accepted by the research community.

The different PAD methods can be categorized according to several characteristics. Some authors propose a taxonomy of PAD methods based on the nature of both methods and attacks: passive or active methods employed to detect static or dynamic attacks [34]. Passive methods include those capable of extracting features from samples obtained by traditional iris recognition systems (e.g., image from the iris sensor). Active methods modify the recognition system in order to obtain features for the PAD method (e.g., dynamic illumination, challenge–response). Static attacks refer to those based on individual samples (e.g., image) while dynamic attacks include artifacts capable to change with time (e.g., video or lens attacks).

In this chapter, we introduce the most popular PAD methods according to the nature of the features used to detect the forged iris: hardware-based, software-based,

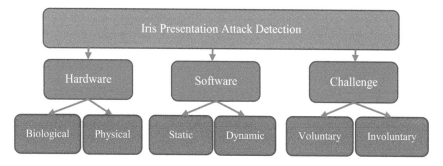

Fig. 6.3 Taxonomy of iris presentation attack detection methods

and challenge–response. The challenge-response approach and most of the hardware methods can be considered active approaches, as they need additional sensors or collaboration from the subject. On the other hand, most of the software methods employ passive approaches in which PAD features are directly obtained from the biometric sample acquired by the iris sensor. Figure 6.3 presents a taxonomy of the iris PAD methods introduced in this chapter.

6.3.1   Hardware-Based Approaches

Also known as sensor-based approaches in the literature. These methods employ specific sensors (in addition to the standard iris sensor) to measure biological and physical characteristics of the eye. These characteristics include optical properties related with the reflectance (e.g., light absorption of the different eye layers), color or composition (e.g., melanin or blood vessel structures in the eye), electrical properties (e.g., electrooculography), or physical properties (e.g., density of the eye tissues). These methods include:

- Multispectral imaging [44–47]: The eye includes complex anatomical structures enclosed in three layers. These layers are made of organic tissue with different spectrographic properties. The idea underlying these methods is to use the spectroscopic print of the eye tissues for PAD. Nonliving tissue (e.g., paper, crystal from the screens or synthetic materials including contact lenses) will present reflectance characteristics different to those obtained from a real eye. These approaches exploit illuminations with different wavelengths that vary according to the method proposed and the characteristic involved (e.g., hemoglobin presents an absorption peak in near-infrared bands).
- 3D imaging [21, 48]: The curvature and 3D nature of the eye have been exploited by researchers to develop PAD methods. The 3D profile of the iris is captured in [48] by using two Near-Infrared light sources and a simple 2D sensor. The idea underlying the method is to detect the shadows on real irises produced by nonuni-

form illumination provided from two different directions. Light Field Cameras (LFC) are used in [21] to acquire multiple depth images and detect the lack of volumetric profiles of photo attacks.

- Electrooculography [49]: The standing potential between the cornea and retina can be measured and the resulting signal is known as electrooculogram. This potential can be used as a liveness indicator but the acquisition of these signals is invasive and includes the placement of at least two electrodes in the eye region. Advances on nonintrusive new methods to acquire the electrooculogram can boost the interest on these approaches.

6.3.2   Software-Based Approaches

Software-based PAD methods use features directly extracted from the samples obtained by the standard iris sensor. These methods exploit pattern recognition techniques in order to detect fake samples. Techniques can be divided into static or dynamic depending on the nature of the information used. While static approaches search for patterns obtained from a single sample (e.g., one image), dynamic approaches exploit time sequences or multiple samples (e.g., a video sequence).

Some authors propose methods to detect the clues or imperfections introduced by printing devices used during manufacturing of PAIs (e.g., printing process for photo attacks). These imperfections can be detected by Fourier image decomposition [18, 19, 36], wavelet analysis [23], or Laplacian transform [24]. All these methods employ features obtained from the frequency domain in order to detect artificial patterns in fake PAIs. Other authors have explored iris quality measures for PAD. The quality of biometric samples has a direct impact on the performance of biometric systems. The literature includes several approaches to measure the quality of image-based biometric samples. The application of quality measures as PAD features for iris biometrics has been studied in [11, 50]. These techniques exploit iris and image quality in order to detect photo attacks.

Advances in image processing techniques have also allowed to develop new PAD methods based on the analysis of features obtained at pixel level. These approaches include features obtained from gray level values [51], edges [30], or color [52]. The idea underlying these methods is that the texture of manufacturing materials shows different patterns due to the nonliving properties of materials (e.g., density, viscosity). In this line, the method proposed in [52] analyzes image features obtained from near-infrared and visible spectrums. Local descriptors have been also used for iris PAD: local binary patterns [31, 35, 53, 54], binary statistical image features [13, 53], scale invariant feature transform [26, 53, 55], and local phase quantization [53].

Finally, in [12], researchers evaluate the performance of deep learning techniques for iris photo attack detection with encouraging results. How to use these networks in more challenging attacks requires a deeper study and novel approaches [12].

6.3.3 Challenge–Response Approaches

These methods analyze voluntary and involuntary responses of the human eye. The involuntary responses are part of the processes associated to the neuromotor activities of the eye while the voluntary behavior is response to specific challenges. Both voluntary and involuntary responses can be driven by external stimuli produced by the PAD system (e.g., changes in the intensity of the light, blink instructions, gaze tracking during dedicated challenges, etc.). The eye reacts to such external stimuli and these reactions can be used as a proof of life to detect attacks based on photos or videos. In addition, there are eye reactions inherent to a living body that can be measured in terms of signals (e.g., permanent oscillation of the eye pupil called hippus, microsaccades, etc.). These reactions can be considered as involuntary challenges noncontrolled by the subject. The occurrence of these signals can be also considered as a proof of life. As a main drawback, these methods increase the level of collaboration demanded from the subjects.

The pupil reactions in presence of uniform light or lighting events were early proposed in [28] for PAD applications and more deeply studied in [34]. As mentioned above, the hippus are permanent oscillations of the pupil that are visible even with uniform illumination. These oscillations range from 0.3 to 0.7 Hz and decline with age. The PAD methods based on hippus have been explored to detect photo attacks and prosthetic eye attacks [19, 56]. However, the difficulties to perform a reliable detection reduce the performance of these methods. Based on similar principles related to eye dynamics, the use of biomechanical models to serve as PAD methods was evaluated in [57].

The eye is a complex organ that includes different types of surfaces. The reflection of the light in the lens and cornea produces a well-known effect named Purkinje reflections. This effect is an involuntary reflection of the eye to external illumination. At least, four Purkinje reflections are usually visible. The reflections change depending on the light source and these changes can be used for liveness detection [39, 45]. Simple photo and video attacks can be detected by these PAD methods. However, their performance against contact lens or synthetic eye attacks is not clear due to the natural reflections on real pupils (contact lens or photo attacks with pupil holes) or sophisticated fabrication methods (synthetic eyes).

6.4 Integration with Iris Recognition Systems

PAD approaches should be integrated into iris recognition systems granting a correct and normal workflow. Software-based PAD methods are usually included as modules in the feature extraction algorithms. A potential problem associated to the inclusion of PAD software is a delay in the recognition time. However, most PAD approaches based on software methods report a low computational complexity that mitigates this concern.

The automatic detection of contact lenses plays an important role in software-based approaches. The effects of wearing contact lenses can be critical in case of textured lenses. In [58], authors reported that textured lenses can cause the FNMR to exceed 90%. The detection of contact lenses represents a first step in IRS, and specific algorithms have been developed and integrated as a preprocessing module [12, 43, 58]. The final goal of these algorithms is to detect and to filter the images to remove the image patterns generated by contact lenses.

Hardware-based PAD approaches are usually integrated before the iris sensor or as an independent parallel module (see Fig. 6.4). In addition to the execution time concerns, hardware-based approaches increase the complexity of the system and the authentication process. Therefore, the main aspects to be analyzed during the integration of those approaches come from the necessity of dedicated sensors and its specific restrictions related to size, time, and cost. These are barriers that difficult the integration of hardware-based approaches into mobile devices (e.g., smartphones).

The main drawback of challenge–response approaches is the increased level of collaboration needed from the user. This collaboration usually introduces delays in the recognition process and some users can perceive it as an unfriendly process.

There are two basic integration schemes:

- Parallel integration: The outputs of the IRS and PAD systems are combined before the decision module. The combination method depends on the nature of the output to be combined (e.g., score level or decision level fusion) [3, 59].
- Series integration: The sample is first analyzed by the PAD system. In case of a legitimate user, the IRS processes the sample. Otherwise, the detection of an attack will avoid unnecessary recognition and the sample will be directly discarded.

6.5 Conclusions

Iris recognition systems have been improved during the last decade achieving better performance, more convenient acquisition at a distance, and full integration with mobile devices. However, the robustness against attacks is still a challenge for the research community and industrial applications. Researchers have shown the vulnerability of iris recognition systems, and there is a consensus about the necessity of finding new methods to improve the security of iris biometrics. Among the different types of attacks, presentation attacks represent a key concern because of its simplicity and high attack success rates. The acquisition at a distance achieved by recent advances on new sensors and the public exposure of the face, and therefore the iris, make relatively easy to obtain iris patterns and use them for malicious purposes. The literature on PAD methods is large including a broad variety of methods, databases, and protocols. In the next years, it will be desirable to unify the research community into common benchmarks and protocols. Even if the current technology shows high detection rates for the simplest attacks (e.g., zero-effort and photo attacks), there are

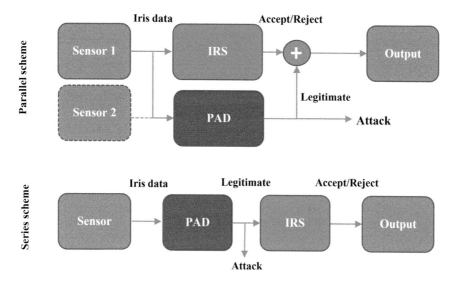

Fig. 6.4 Integration of Presentation Attack Detection (PAD) with Iris Recognition Systems (IRS) in parallel (top) and serial (down) schemes

still challenges associated to most sophisticated attacks such as those using textured contact lenses and synthetic eyes.

Acknowledgements This work was done in the context of the TABULA RASA and BEAT projects funded under the 7th Framework Programme of EU. This work was supported in part by the CogniMetrics Project under Grant TEC2015-70627-R from MINECO/FEDER.

References

1. Daugman J (1993) High confidence visual recognition of persons by a test of statistical independence. IEEE Trans Pattern Anal Mach Intell 15:1148–1161
2. Burge MJ, Bowyer KW (eds) (2013) Handbook of iris recognition. Springer, Berlin
3. Fierrez J, Morales A, Vera-Rodriguez R, Camacho D (2018) Multiple classifiers in biometrics. part 1: fundamentals and review. Inf Fusion 44:57–64
4. Galbally J, Gomez-Barrero M (2017) chapter. In: Rathgeb C, Busch C (eds) Iris and periocular biometric recognition. Presentation attack detection in iris recognition, IET Digital Library, pp 235–263
5. Flom L, Safir A (1987) Iris recognition system. US Patent US4641349 A
6. Abraham R, Bennett ES, Sen N, Shah NB (2017) State of aadhaar report 2016–17. Tech Rep, IDinsight
7. Nguyen K, Fookes C, Jillela R, Sridharan S, Ross A (2017) Long range iris recognition: a survey. Pattern Recognit 72:123–143
8. Chaos Computer Club Berlin: chaos computer clubs breaks iris recognition system of the samsung galaxy s8 (2017). https://www.ccc.de/en/updates/2017/iriden
9. ISO/IEC CD 30107-1. Information technology - biometrics - presentation attack detection - Part 1: framework (2016)

10. Daugman J (1999) Biometrics. Personal identification in a networked society. In: Chapter, Recognizing persons by their iris patterns. Kluwer Academic Publishers, Dordrecht, pp 103–121
11. Galbally J, Marcel S, Fierrez J (2014) Image quality assessment for fake biometric detection: application to iris, fingerprint and face recognition. IEEE Trans Image Process 23:710–724
12. Menotti D, Chiachia G, Pinto A, Schwartz WR, Pedrini H, Falcao AX, Rocha A (2015) Deep representations for iris, face, and fingerprint spoofing detection. IEEE Trans Inf Forensics Secur 10:864–878
13. Raghavendra R, Busch C (2014) Presentation attack detection algorithm for face and iris biometrics. In: Proceedings of the IEEE European signal processing conference (EUSIPCO), pp 1387–1391
14. Ratha NK, Connell JH, Bolle RM (2001) Enhancing security and privacy in biometrics-based authentication systems. IBM Syst J 40(3):614–634
15. Gomez-Barrero M, Maiorana E, Galbally J, Campisi P, Fierrez J (2017) Multi-biometric template protection based on homomorphic encryption. Pattern Recogn 67:149–163
16. Hadid A, Evans N, Marcel S, Fierrez J (2015) Biometrics systems under spoofing attack. IEEE Signal Process Mag 32:20–30
17. Johnson P, Lazarick R, Marasco E, Newton E, Ross A, Schuckers S (2012) Biometric liveness detection: framework and metrics. In: Proceedings of the NIST international biometric performance conference (IBPC)
18. Czajka A (2013) Database of iris printouts and its application: development of liveness detection method for iris recognition. In: Proceedings of the international conference on methods and models in automation and robotics (MMAR), pp 28–33
19. Pacut A, Czajka A (2006) Aliveness detection for iris biometrics. In: Proceedings of the IEEE international Carnahan conference on security technology (ICCST), pp 122–129
20. Ruiz-Albacete V, Tome-Gonzalez P, Alonso-Fernandez F, Galbally J, Fierrez J, Ortega-Garcia J (2008) Direct attacks using fake images in iris verification. In: Proceedings of the COST 2101 workshop on biometrics and identity management (BioID). LNCS, vol 5372. Springer, Berlin, pp 181–190
21. Raghavendra R, Busch C (2014) Presentation attack detection on visible spectrum iris recognition by exploring inherent characteristics of light field camera. In: Proceedings of the IEEE international joint conference on biometrics (IJCB) (2014)
22. Thalheim L, Krissler J (2002) Body check: biometric access protection devices and their programs put to the test. ct magazine, pp 114–121
23. He X, Lu Y, Shi P (2009) A new fake iris detection method. In: Proceedings of the IAPR/IEEE international conference on biometrics (ICB). LNCS, vol 5558. Springer, Berlin, pp 1132–1139
24. Raja KB, Raghavendra R, Busch C (2015) Presentation attack detection using laplacian decomposed frequency response for visible spectrum and near-infra-red iris systems. In: Proceedings of the of IEEE international conference on biometrics: theory and applications (BTAS)
25. Raja KB, Raghavendra R, Busch C (2015) Video presentation attack detection in visible spectrum iris recognition using magnified phase information. IEEE Trans Inf Forensics Secur 10:2048–2056
26. Zhang H, Sun Z, Tan T, Wang J (2011) Learning hierarchical visual codebook for iris liveness detection. In: Proceedings of the IEEE international joint conference on biometrics (IJCB)
27. Yambay D, Doyle JS, Boyer KW, Czajka A, Schuckers S (2014) Livdet-iris 2013 - iris liveness detection competition 2013. In: Proceedings of the IEEE international joint conference on biometrics (IJCB)
28. Daugman J (2004) Iris recognition and anti-spoofing countermeasures. In: Proceedings of the international biometrics conference (IBC)
29. von Seelen UC (2005) Countermeasures against iris spoofing with contact lenses. In: Proceedings of the biometrics consortium conference (BCC)
30. Wei Z, Qiu X, Sun Z, Tan T (2008) Counterfeit iris detection based on texture analysis. In: Proceedings of the IAPR international conference on pattern recognition (ICPR)

31. Zhang H, Sun Z, Tan T (2010) Contact lense detection based on weighted LBP. In: Proceedings of the IEEE international conference on pattern recognition (ICPR), pp 4279–4282
32. Lefohn A, Budge B, Shirley P, Caruso R, Reinhard E (2003) An ocularist's approach to human iris synthesis. IEEE Trans Comput Graph Appl 23:70–75
33. Chen R, Lin X, Ding T (2012) Liveness detection for iris recognition using multispectral images. Pattern Recogn Lett 33:1513–1519
34. Czajka A (2015) Pupil dynamics for iris liveness detection. IEEE Trans Inf Forensics Secur 10:726–735
35. Gupta P, Behera S, Singh MVV (2014) On iris spoofing using print attack. In: IEEE international conference on pattern recognition (ICPR)
36. He X, Lu Y, Shi P (2008) A fake iris detection method based on FFT and quality assessment. In: Proceedings of the IEEE Chinese conference on pattern recognition (CCPR)
37. Huang X, Ti C, zhen Hou Q, Tokuta A, Yang R (2013) An experimental study of pupil constriction for liveness detection. In: Proceedings of the IEEE workshop on applications of computer vision (WACV), pp 252–258
38. Kanematsu M, Takano H, Nakamura K (2007) Highly reliable liveness detection method for iris recognition. In: Proceedings of the SICE annual conference, international conference on instrumentation, control and information technology (ICICIT), pp 361–364
39. Lee EC, Yo YJ, Park KR (2008) Fake iris detection method using Purkinje images based on gaze position. Opt Eng 47(067):204
40. Yambay D, Becker B, Kohli N, Yadav, D, Czajka, A, Bowyer KW, Schuckers S, Singh R, Vatsa M, Noore A, Gragnaniello D, Sansone C, Verdoliva L, He L, Ru Y, Li H, Liu N, Sun Z, Tan T (2017) Livdet iris 2017, iris liveness detection competition 2017. In: Proceedings of the IEEE international joint conference on biometrics (IJCB), pp 1–6
41. Raghavendra R, Busch C (2015) Robust scheme for iris presentation attack detection using multiscale binarized statistical image features. IEEE Trans Inf Forensics Secur 10:703–715
42. Sequeira AF, Oliveira HP, Monteiro JC, Monteiro JP, Cardoso JS (2014) MobILive 2014 - mobile iris liveness detection competition. In: Proceedings of the IEEE international joint conference on biometrics (IJCB)
43. Yadav D, Kohli N, Doyle JS, Singh R, Vatsa M, Bowyer KW (2014) Unraveling the effect of textured contact lenses on iris recognition. IEEE Trans Inf Forensics Secur 9:851–862
44. He Y, Hou Y, Li Y, Wang Y (2010) Liveness iris detection method based on the eye's optical features. In: Proceedings of the SPIE optics and photonics for counterterrorism and crime fighting VI, p 78380R
45. Lee EC, Park KR, Kim J (2006) Fake iris detection by using Purkinje image. In: Proceedings of the IAPR international conference on biometrics (ICB), pp 397–403
46. Lee SJ, Park KR, Lee YJ, Bae K, Kim J (2007) Multifeature-based fake iris detection method. Opt Eng 46(127):204
47. Park JH, Kang MG (2005) Iris recognition against counterfeit attack using gradient based fusion of multi-spectral images. In: Proceedings of the of international workshop on biometric recognition systems (IWBRS). LNCS, vol 3781. Springer, Berlin, pp 150–156
48. Lee EC, Park KR (2010) Fake iris detection based on 3D structure of the iris pattern. Int J Imaging Syst Technol 20:162–166
49. Krupiski R, Mazurek P (2012) Estimation of electrooculography and blinking signals based on filter banks. In: Proceedings of the of the 2012 international conference on computer vision and graphics, pp 156–163
50. Galbally J, Ortiz-Lopez J, Fierrez J, Ortega-Garcia J (2012) Iris liveness detection based on quality related features. In: Proceedings of the IAPR international conference on biometrics (ICB), pp 271–276
51. He X, An S, Shi P (2007) Statistical texture analysis-based approach for fake iris detection using support vector machines. In: Proceedings of the IAPR international conference on biometrics (ICB), LNCS, vol 4642. Springer, Berlin, pp 540–546
52. Alonso-Fernandez F, Bigun J (2014) Fake iris detection: a comparison between near-infrared and visible images. In: Proceedings of the IEEE international conference on signal-image technology and internet-based systems (SITIS), pp 546–553

53. Gragnaniello D, Poggi G, Sansone C, Verdoliva L (2015) An investigation of local descriptors for biometric spoofing detection. IEEE Trans Inf Forensics Secur 10:849–863
54. He Z, Sun Z, Tan T, Wei Z (2009) Efficient iris spoof detection via boosted local binary patterns. In: Proceedings of the IEEE international conference on biometrics (ICB)
55. Sun Z, Zhang H, Tan T, Wang J (2014) Iris image classification based on hierarchical visual codebook. IEEE Trans Pattern Anal Mach Intell 36:1120–1133
56. Park KR (2006) Robust fake iris detection. In: Proceedings of the of articulated motion and deformable objects (AMDO). LNCS, vol 4069. Springer, Berlin, pp 10–18
57. Komogortsev O, Karpov A (2013) Liveness detection via oculomotor plant characteristics: attack of mechanical replicas. In: Proceedings of the international conference of biometrics (ICB) (2013)
58. Bowyer KW, Doyle JS (2014) Cosmetic contact lenses and iris recognition spoofing. IEEE Comput 47:96–98
59. Biggio B, Fumera G, Marcialis G, Roli F (2017) Statistical meta-analysis of presentation attacks for secure multibiometric systems. IEEE Trans Pattern Anal Mach Intell 39(3):561–575

Chapter 7
Application of Dynamic Features of the Pupil for Iris Presentation Attack Detection

Adam Czajka and Benedict Becker

Abstract This chapter presents a comprehensive study on the application of stimulated pupillary light reflex to presentation attack detection (PAD) that can be used in iris recognition systems. A pupil, when stimulated by visible light in a pre-defined manner, may offer sophisticated dynamic liveness features that cannot be acquired from dead eyes or other static objects such as printed contact lenses, paper printouts, or prosthetic eyes. Modeling of pupil dynamics requires a few seconds of observation under varying light conditions that can be supplied by a visible light source in addition to the existing near-infrared illuminants used in iris image acquisition. The central element of the presented approach is an accurate modeling and classification of pupil dynamics that makes mimicking an actual eye reaction difficult. This chapter discusses new data-driven models of pupil dynamics based on recurrent neural networks and compares their PAD performance to solutions based on the parametric Clynes–Kohn model and various classification techniques. Experiments with 166 distinct eyes of 84 subjects show that the best data-driven solution, one based on long short-term memory, was able to correctly recognize 99.97% of attack presentations and 98.62% of normal pupil reactions. In the approach using the Clynes–Kohn parametric model of pupil dynamics, we were able to perfectly recognize abnormalities and correctly recognize 99.97% of normal pupil reactions on the same dataset with the same evaluation protocol as the data-driven approach. This means that the data-driven solutions favorably compare to the parametric approaches, which require model identification in exchange for a slightly better performance. We also show that observation times may be as short as 3 s when using the parametric model, and as short as 2 s when applying the recurrent neural network without substantial loss in accuracy. Along with this chapter we also offer: (a) all time series representing pupil dynamics for 166 distinct eyes used in this study, (b) weights of the trained recurrent neural network offering the best performance, (c) source codes

A. Czajka (✉)
Research and Academic Computer Network (NASK), ul. Kolska 12, 01045 Warsaw, Poland
e-mail: aczajka@nd.edu

A. Czajka · B. Becker
University of Notre Dame, 384 Fitzpatrick Hall, Notre Dame, IN 46556, USA
e-mail: bbecker5@nd.edu

© Springer Nature Switzerland AG 2019
S. Marcel et al. (eds.), *Handbook of Biometric Anti-Spoofing*,
Advances in Computer Vision and Pattern Recognition,
https://doi.org/10.1007/978-3-319-92627-8_7

151

of the reference PAD implementation based on Clynes–Kohn parametric model, and (d) all PAD scores that allow the reproduction of the plots presented in this chapter. To our best knowledge, this chapter proposes the first database of pupil measurements dedicated to presentation attack detection and the first evaluation of recurrent neural network-based modeling of pupil dynamics and PAD.

7.1 Introduction

Presentation attack detection (PAD) is a key aspect of biometric system's security. PAD refers to an automated detection of presentations to the biometric sensor that has the goal to interfere with an intended operation of the biometric system [1]. Presentation attacks may be realized in various ways, and using various *presentation attack instruments (PAI)*, such as presentation of fake objects or cadavers, non-conformant presentation, or even coerced use of biometric characteristics. This chapter focuses on detection of liveness features of an iris, i.e., changes in pupil dilation under varying light conditions that indicate the authenticity of the eye. This liveness test can prove useful in both iris and ocular recognition systems.

Iris PAD has a significant representation in scientific literature. Most of the methods are based on static properties of the eye and iris to detect iris paper printouts and cosmetic contact lenses. Examples of such methods include the use of image quality metrics (Galbally et al. [2], Wei et al. [3]) and texture descriptors such as local binary patterns (Doyle et al. [4], Ojala et al. [5]), local phase quantization (Ojansivu and Heikkilä [6]), binary Gabor pattern (Zhang et al. [7]), hierarchical visual codebook (Sun et al. [8]), histogram of oriented gradients (Dalal and Triggs [9]), and binarized statistical image features (Doyle and Bowyer [10], Raghavendra and Busch [11]). The recent advent of deep learning, especially convolutional neural networks, has caused a dynamic increase in solutions based on neural networks [12]. However, the results of the last LivDet-Iris 2017 competition [13] suggest that these static artifacts are still challenging when the algorithms are tested on data unknown to developers: the winning algorithms were able to achieve an average of APCER $= 14.71\%$ and an average of BCPER $= 3.36\%$ on a combined dataset (paper printouts and textured contact lenses collected at five different universities—organizers of LivDet). Also, the results achieved in an open-set scenario (unknown PAI species) were often worse by an order of magnitude than those observed in a closed-set scenario (known PAI species).

The application of dynamic features of the eye is less popular. An interesting solution proposed by Raja et al. [14] was based on Eulerian Video Magnification that detects micro phase deformations of iris region. Also, Komogortsev et al. proposed an application of eyeball dynamics to detect mechanical eye replicas [15]. Pupil dynamics are rarely used in biometric presentation attack detection. Most existing approaches are based on methods proposed by Pacut and Czajka [16–18], which deploy a parametric Clynes–Kohn model of the pupil reaction. Readers interested in the most recent summary of the iris PAD topic are encouraged to look at the survey by Czajka and Bowyer [19].

To our knowledge, there are no publicly available databases of iris videos acquired under visible light stimuli. We have thus collected a dataset of pupil reactions measured for 166 distinct eyes when stimulating the eye with positive (a sudden increase of lightness) and negative (a sudden decrease of lightness) stimuli. Bona fide samples are represented by time series presenting pupil reaction to either positive, negative or both stimuli. Attack samples are represented by time series acquired when the eye was not stimulated by the light. In many attack samples, one may still observe spontaneous pupil oscillation and noise caused by blinking, eye off-axis movements, or imprecise iris segmentation. These are good complications making the classification task more challenging. The measurements of pupil size (in a time series) are made available along with this chapter to interested research groups.

In this chapter, we present and compare two approaches to classify the acquired time series. The state-of-the-art approach is based on the parametric Clynes–Kohn model of pupil dynamics and various classifiers such as support vector machines, logistic regression, bagged trees, and k nearest neighbors. Solutions employing the parametric model require model identification for each presentation attack detection, which incorporates minimization procedures that often results in nondeterministic behavior. However, once trained this approach presents good generalization capabilities. The second approach presented in this chapter is based on data-driven models realized by four variants of recurrent neural networks, including long short-term memory. These solutions present slightly worse results than those based on parametric model, but they deliver more deterministic outputs. It is also sometimes convenient to use an end-to-end classification algorithm; hence, data-driven models may serve as an interesting PAD alternative to algorithms employing parametric models.

In Sect. 7.2, we provide technical details of the employed dataset. Section 7.3 briefly summarizes the application of Clynes–Kohn model of pupil dynamics. In Sect. 7.4, we explain how the recurrent neural networks were applied to build data-driven models of pupil reflex. Results are presented in Sect. 7.5, and in Sect. 7.6, we summarize this approach along with some limitations.

7.2 Database

7.2.1 Acquisition

The acquisition scenario followed the one applied in [18]. All images were captured in near-infrared light ($\lambda = 850$ nm) and the recommendations for iris image quality given in ISO/IEC 19794-6 and ISO/IEC 29794-6 standards were easily met. The sensor acquired 25 images per second. Volunteers presented their eyes in a shaded box in which the sensor and four visible light-emitting diodes were placed to guarantee a stable acquisition environment throughout the experiment. A single acquisition

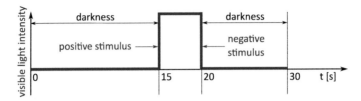

Fig. 7.1 Visible light intensity profile in the data collection for a single attempt

attempt lasted 30 s: spontaneous pupil movement was recorded during the first 15 s, then the pupil reaction to a **positive light stimulus** (dark→light) was recorded during the next 5 s, and during the last 10 s, a pupil reaction to a **negative light stimulus** (light→dark) was measured. Figure 7.1 presents a visible light profile applied in each attempt.

84 subjects presented their left and right eyes, except for two subjects who presented only a single eye. Thus, the database is comprised of measurements from 166 distinct eyes, with up to 6 attempts per eye. All attempts for a given subject were organized on the same day, except for four persons (eight eyes) who had their pupil dynamics measured 12 years before and then repeated (using the same hardware and software setup). The total number of attempts for all subjects is 435, and in each attempt, we collected 750 iris images (30 s × 25 frames per second).

7.2.2 Estimation of Pupil Size

All samples were segmented independently using Open Source IRIS [20] which implements a Viterbi algorithm to find a set of points on the iris boundaries. Least squares curve fitting was then used to find two circles approximating the inner and outer iris boundaries. We tried to keep the acquisition setup identical during all attempts. However, small adjustments to the camera before an attempt may have introduced different optical magnifications for each participant. Also, iris size may slightly differ across the population of subjects. Hence, despite a stable distance between a subject and a camera, we normalized the pupil radius by the iris radius to get the *pupil size* ranging from 0.2 to 0.8 in our experiments. Instead of using the iris radius calculated in each frame, we used its averaged value calculated for all frames considered in a given experiment. This approach nicely compensates for possible changes in the absolute iris size and does not introduce additional noise related to fluctuations in outer boundary detection.

7.2.3 Noise and Missing Data

Figure 7.2 presents example iris images acquired immediately after positive light stimulus (increase of the light intensity at $t = 0$) along with the segmentation results.

Fig. 7.2 Example iris images and segmentation results 0s (**left**), 1s (**middle**), and 5s (**right**) after positive light stimuli. Circular approximations of iris boundaries are shown in green, and irregular occlusions are shown in red

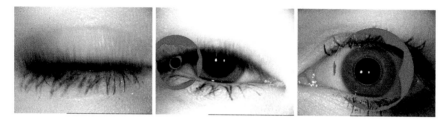

Fig. 7.3 Closed eyes prevent the segmentation from providing meaningful data (**left**). Significant eyelid coverage results in inaccurate segmentation (**middle**). Occasionally, the segmentation may also fail for a good-quality sample (**right**). All these cases introduce a natural noise into pupil dynamics time series. As in Fig. 7.2, circular approximations of iris boundaries are shown in green, and irregular occlusions are shown in red

Correct segmentation delivers a valid pupil size for each video frame. However, there are at least two types of processing errors that introduce a natural noise into the data:

- *missing points*: the segmentation method may not be able to provide an estimate of pupil and/or iris position; this happens typically when the eye is closed completely (Fig. 7.3 left);
- *inaccurate segmentation*: caused mainly by low quality data, due to factors such as blinking, off-axis gaze, or significant occlusion (Fig. 7.3 middle), or by occasional segmentation algorithm errors (Fig. 7.3 right).

These noisy measurements can be observed in each plot presented in Fig. 7.4 as points departing from the expected pupil size. Hence, we applied a median filtering within a one-second window to smooth out most of the segmentation errors and get *denoised pupil size* (cf. black dots in Fig. 7.4). Denoised pupil size was used in all parametric model experiments presented in this chapter. For data-driven models, we simply filled the gaps by taking the previous value of the pupil size as a predictor of a missing point.

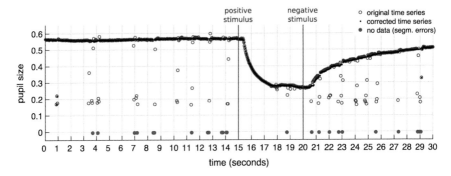

Fig. 7.4 Original (blue circles) and denoised (black dots) pupil size depending on the stimulus. Original data points departing from the expected pupil size, including missing points marked as red dots, correspond to blinks and iris segmentation errors

7.2.4 Division of Data and Recognition Scenarios

To train the classifiers, both bona fide and attack examples were generated for each time series. The first 15 s and the last 5 s of each time series represent no pupil reaction, and thus were used to generate attack presentations. The time series starting from the 15th second and ending at the 25th second represent authentic reactions to light stimuli, and thus were used to generate bona fide presentations. We consider three scenarios of pupil stimulation:

- **s1**: presentation of only positive stimulus; in this case we observe the eye up to 5 s after the visible light is switched on,
- **s2**: presentation of only negative stimulus; in this case we observe the eye up to 5 s after the visible light is switched off, and
- **s3**: presentation of both stimuli sequentially; in this case, we observe the eye for 10 s after the visible light is switched on and sequentially switched off.

Since the pupil reaction is different for each type of stimulus, we evaluate separate data-driven models. Also, when only one stimulus is used (positive or negative), the observation can be shorter than 5 s. Shorter observations are also investigated in this chapter.

7.3 Parametric Model of Pupil Dynamics

The parametric model-based method for pupil dynamics recognition follows past works by Czajka [18] and uses the Clynes and Kohn pupil reaction model [21]. For completeness, we briefly characterize this approach in this subsection.

The Clynes and Kohn model, illustrated as a two-channel transfer function of a complex argument s in Fig. 7.5, accounts for asymmetry in pupil reaction y depending

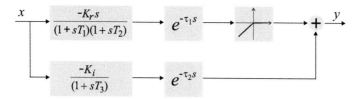

Fig. 7.5 Pupil dynamics model deployed in this work and derived from an original proposal of Kohn and Clynes [21]. Graph adapted from [17]

on the polarity of the stimulus x. Positive stimuli (darkness → lightness) engage two channels of the model, while for negative stimuli (lightness → darkness), the upper channel is cut by a nonlinear component, and only lower channel is used to predict the pupil size. That is, for positive stimuli, each reaction is represented by a point in seven-dimensional space: three time constants (T_1, T_2 and T_3), two lag elements (τ_1 and τ_2), and two gains (K_r and K_i). For negative stimuli, we end up with three-dimensional feature space corresponding to lower channel parameters: K_i, T_3, and τ_2.

One can easily find the model response $y(t; \phi)$ in time domain by calculating the inverse Laplace transform, given a model shown in Fig. 7.5. Assuming that the light stimuli occur at $t = 0$, the upper channel response

$$y_{\text{upper}}(t; \phi_1) = \begin{cases} -\frac{K_r}{T_1^2}(t - \tau_1)e^{-\frac{t-\tau_1}{T_1}} & \text{if } T_1 = T_2 \\ \frac{K_r}{T_2-T_1}\left(e^{-\frac{t-\tau_1}{T_1}} - e^{-\frac{t-\tau_1}{T_2}}\right) & \text{otherwise,} \end{cases} \tag{7.1}$$

where

$$\phi_1 = [K_r, T_1, T_2, \tau_1].$$

The lower channel response

$$y_{\text{lower}}(t; \phi_2) = -K_i\left(1 - e^{-\frac{t-\tau_2}{T_3}}\right), \tag{7.2}$$

where

$$\phi_2 = [K_i, T_3, \tau_2],$$

and the model output $y(t; \phi)$ is simply a sum of both responses

$$y(t; \phi) = y_{\text{upper}}(t; \phi_1) + y_{\text{lower}}(t; \phi_2), \tag{7.3}$$

where

$$\phi = [\phi_1, \phi_2] = [K_r, T_1, T_2, \tau_1, K_i, T_3, \tau_2].$$

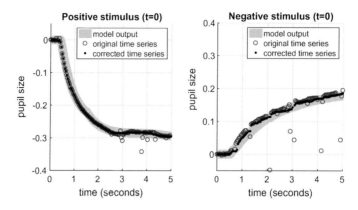

Fig. 7.6 Original (blue circles) and denoised (black dots) pupil size depending on the stimulus (positive on the **left**, negative on the **right**) along with the Clynes–Kohn model output (gray thick line)

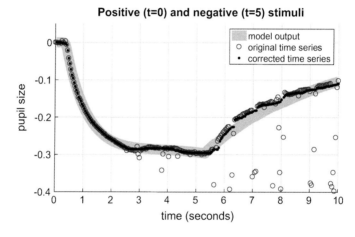

Fig. 7.7 Same as in Fig. 7.6, except that the results for a long reaction (10 s) after both stimuli (positive, then negative) are illustrated

In case of negative light stimuli, the model output is based on the lower channel response since $y_{upper}(t; \phi_1) = 0$. The vector ϕ constitutes the *liveness features* and is found as a result of the optimization process used to solve the model fitting problem. Figures 7.6 and 7.7 present model outputs obtained, for example, time series and the three recognition scenarios listed in Sect. 7.2.4. Figure 7.8 illustrates a correct model behavior in case of no stimuli.

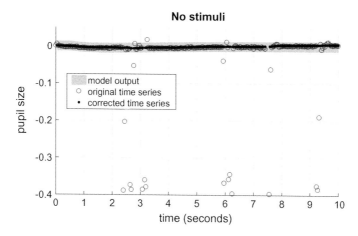

Fig. 7.8   Same as in Fig. 7.6, except that the pupil actual size and model output are shown when the eye is not stimulated by the light

7.4   Data-Driven Models of Pupil Dynamics

7.4.1   Variants of Recurrent Neural Networks

Recurrent neural networks (RNN) belong to a family of networks used for processing sequential data. RNNs can scale to much longer sequences than networks having no sequence-based specialization, such as convolutional neural networks (CNN). Regardless of the sequence length, the learned RNN-based model always has the same input size. These basic properties make the RNN a well-suited candidate to model the time series that represent pupil dynamics.

Graphs with cycles are used to model and visualize recurrent networks, and the cycles introduce dynamics by allowing the predictions made at the current time step to be influenced by past predictions. Parameter sharing in an RNN differs from the parameter sharing applied in a CNN: instead of having the same convolution kernel at each time step (CNN), the same transition function (from hidden state $h^{(t-1)}$ to hidden state $h^{(t)}$) with the same parameters is applied at every time step t. More specifically, a conventional RNN maps an input sequence $\left(\mathbf{x}^{(0)}, \mathbf{x}^{(1)}, \ldots, \mathbf{x}^{(T)}\right)$ into an output sequence $\left(\mathbf{o}^{(0)}, \mathbf{o}^{(1)}, \ldots, \mathbf{o}^{(T)}\right)$ in the following way:

$$\mathbf{o}^{(t)} = \mathbf{c} + \mathbf{V}\mathbf{h}^{(t)} \tag{7.4}$$

$$\mathbf{h}^{(t)} = g(\underbrace{\mathbf{b} + \mathbf{W}\mathbf{h}^{(t-1)} + \mathbf{U}\mathbf{x}^{(t)}}_{\mathbf{z}^{(t)}}), \tag{7.5}$$

where $\mathbf{W}$, $\mathbf{U}$, and $\mathbf{V}$ are weight matrices for hidden-to-hidden (recurrent), input-to-hidden, and hidden-to-output connections, respectively; $\mathbf{b}$ and $\mathbf{c}$ denote bias vectors;

Fig. 7.9 Repeating element in the recurrent neural network (*basic RNN cell*)

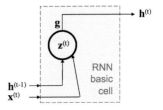

and g is an activation function. The repeating element, *basic RNN cell*, is shown in Fig. 7.9. Conventional RNN is the **first data-driven model of pupil dynamics evaluated in this chapter**.

Learning long dependencies by conventional RNNs is theoretically possible, but can be difficult in practice due to problems with gradient flow [22]. Among different variants proposed to date, the long short-term memory (LSTM) [23] was designed to avoid the long-term dependency problem and allows the error derivatives to flow unimpeded. The LSTM repeating cell is composed of four nonlinear layers interacting in a unique way (in contrast to a conventional RNN, which uses just a single nonlinear layer in the repeating cell), Fig. 7.10. The LSTM hidden state $\mathbf{h}^{(t)}$ in time moment t can be expressed as

$$\mathbf{h}^{(t)} = \gamma^{(t)} \circ g\left(\mathbf{m}^{(t)}\right), \tag{7.6}$$

where

$$\gamma^{(t)} = \sigma\Big(\underbrace{\mathbf{W}_{h\gamma}\mathbf{h}^{(t-1)} + \mathbf{U}_{x\gamma}\mathbf{x}^{(t)} + \mathbf{W}_{m\gamma}\mathbf{m}^{(t)} + \mathbf{b}_{\gamma}}_{\mathbf{z}_{\gamma}^{(t)}}\Big)$$

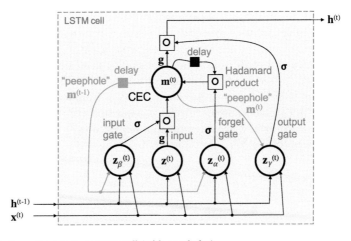

Fig. 7.10 Long short-term memory cell (with peepholes)

is the *output gate*, ∘ denotes the Hadamard product, and

$$\mathbf{m}^{(t)} = \alpha^{(t)} \circ \mathbf{m}^{(t-1)} + \beta^{(t)} \circ g\left(\mathbf{x}^{(t)}, \mathbf{h}^{(t-1)}\right)$$

is the LSTM cell state (the Constant Error Carousel, or CEC). $\alpha^{(t)}$ and $\beta^{(t)}$ are the so-called *forget gate* and *input gate*, namely:

$$\alpha^{(t)} = \sigma\left(\underbrace{\mathbf{W}_{h\alpha}\mathbf{h}^{(t-1)} + \mathbf{U}_{x\alpha}\mathbf{x}^{(t)} + \mathbf{W}_{m\alpha}\mathbf{m}^{(t-1)} + \mathbf{b}_\alpha}_{\mathbf{z}_\alpha^{(t)}}\right)$$

$$\beta^{(t)} = \sigma\left(\underbrace{\mathbf{W}_{h\beta}\mathbf{h}^{(t-1)} + \mathbf{U}_{x\beta}\mathbf{x}^{(t)} + \mathbf{W}_{m\beta}\mathbf{m}^{(t-1)} + \mathbf{b}_\beta}_{\mathbf{z}_\beta^{(t)}}\right),$$

where σ is a sigmoid activation function. The closer the gate's output is to 1, the more information the model will retain and the closer the output is to 0, the more the model will forget. Additional $\mathbf{W}_{m\alpha}$, $\mathbf{W}_{m\beta}$ and $\mathbf{W}_{m\gamma}$ matrices (compared to a conventional RNN) represent "peephole" connections in the gates to the CEC of the same LSTM. These connections were proposed by Gers and Schmidhuber [24] to overcome a problem of closed gates that can prevent the CEC from getting useful information from the past. LSTMs (with and without peepholes) are the **next two data-driven models of pupil dynamics evaluated in this work**.

Greff et al. [25] evaluated various types of recurrent networks, including variants of the LSTM, and found that coupling input and forget gates may end up with a similar or better performance as a regular LSTM with fewer parameters. The combination of the forget and input gates of the LSTM was proposed by Cho et al. [26] and subsequently named the gated recurrent unit (GRU). The GRU merges the cell state and hidden state, that is

$$\mathbf{h}^{(t)} = \alpha^{(t)} \circ \mathbf{h}^{(t-1)} + (1 - \alpha^{(t)}) \circ g\left(\mathbf{x}^{(t)}, \gamma^{(t)} \circ \mathbf{h}^{(t-1)}\right), \tag{7.7}$$

where

$$\alpha^{(t)} = \sigma\left(\underbrace{\mathbf{W}_{h\alpha}\mathbf{h}^{(t-1)} + \mathbf{U}_{x\alpha}\mathbf{x}^{(t)} + \mathbf{b}_\alpha}_{\mathbf{z}_\alpha^{(t)}}\right)$$

is the *update gate*, and

$$\gamma^{(t)} = \sigma\left(\underbrace{\mathbf{W}_{h\gamma}\mathbf{h}^{(t-1)} + \mathbf{U}_{x\gamma}\mathbf{x}^{(t)} + \mathbf{b}_\gamma}_{\mathbf{z}_\gamma^{(t)}}\right)$$

is the *reset gate*. GRU uses no peephole connections and no output activation function. A repeating GRU cell is shown in Fig. 7.11. GRU will be **the fourth data-driven model of pupil dynamics deployed in this work**.

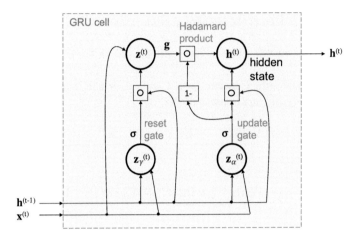

Fig. 7.11 Gated recurrent unit cell

7.4.2 Implementation and Hyperparameters

In all data-driven models, a neural network with two layers was used: a type of recurrent neural network composed of 24 hidden neurons, and a perceptron layer composed of two neurons with a softmax output classifying samples as bona fide or attack classes. For training, a static learning rate of 0.0001 was used, and the networks were trained using a batch size of 16 for a total of 200,000 iterations, or almost 87 epochs. In all cases, the optimizer used for training was RMSProp proposed by Hinton [27], the loss function used was categorical cross entropy and the network's weight initializer was "Xavier" initialization [28].

Each network took in time series data of the form 5×1 s intervals. The period of one second was chosen so that each time window would have a significant amount of data. Thus, the network input $\mathbf{x}^{(t)}$ is comprised of 25 data points that correspond to 25 iris images acquired per second. Shorter time windows would not provide enough context for the network and longer time windows would defeat the purpose of using recurrent networks. The length of training sequences for both bona fide and attack examples was kept the same.

7.5 Results

The parametric Clynes and Kohn model transforms each time series into a multidimensional point in the model parameter space. A binary classifier makes a decision whether the presented time series corresponds to an authentic reaction of the pupil, or to a reaction that is odd or noisy. Several classifiers were tested, namely, linear and nonlinear support vector machines (SVM), logistic regression, bagged trees, and *k* nearest neighbors (kNN). In turn, the data-driven model based on recurrent neural network makes the classification all the way from the estimated pupil size to the decision. In all experiments (both for parametric and data-driven models), the same leave-one-out cross validation was applied. That is, to make validation subject-disjoint, all the data corresponding to a single subject was left for validation, while the remaining data was used to train the classifier. This train-validation split could thus be repeated 84 times (equal to the number of subjects). The presented results are averages over all 84 of these validations.

We follow ISO/IEC 30107-1:2016 and use the following PAD-specific error metrics:

- **Attack Presentation Classification Error Rate (APCER)**: proportion of *attack presentations* incorrectly classified as *bona fide (genuine) presentations*, and
- **Bona Fide Presentation Classification Error Rate (BPCER)**: proportion of *bona fide (genuine) presentations* incorrectly classified as *presentation attacks*.

Table 7.1 presents the results obtained from both parametric and data-driven models and for all three scenarios of eye stimulation (only positive, only negative, and both stimuli). The best solution in the **positive stimuli scenario** was based on the **parametric model** (Clynes–Kohn + SVM). It recognized 99.77% of the normal pupil reactions (BPCER = 0.23%) and 99.54% of the noise (APCER = 0.46%). The **data-driven solution** based on LSTM with no peephole connections achieved similar accuracy recognizing 98.62% of the normal pupil reactions (BPCER = 1.38%) and 99.77% of the noisy time series (APCER = 0.23%).

The **negative stimulus** was harder to classify by a **data-driven model**, as the best accuracy, obtained with a conventional recurrent neural network, recognized 96.09% of normal pupil reactions (BPCER = 3.91%) and 98.74% of noisy time series (APCER = 1.26%). In turn, **parametric model** with bagged trees or kNN classification was perfect in recognizing spontaneous pupil oscillations (APCER = 0) and correctly classified 99.77% of bona fide pupil reactions (BPCER = 0.23%).

Increasing the observation time to 10 s and applying **both positive and negative stimuli** (Table 7.1, last column) do not result in a more accurate solution. Therefore, we have investigated the winning solutions applied to positive or negative stimuli for **shorter times**, starting from 2 s. One should expect to get both APCER and BPCER below 1% when the parametric model is applied to time series of approximately 3 s, Fig. 7.12. For data-driven models, we do not observe a substantial decrease in performance even when the eye is observed for approximately 2 s, Fig. 7.13. These results increase a potential for practical implementations.

7.6 Discussion

In this chapter, we have presented how recurrent neural networks can be applied to serve as models of pupil dynamics for use in presentation attack detection, and compared this approach with methods based on a parametric model. The results show

Table 7.1 Error rates (in %) for all modeling and classification approaches considered in this paper. Best approaches (in terms of average of APCER and BPCER) for each combination of model type and stimulus type are shown in bold font

Model	Variant		Positive stimulus (5 s)		Negative stimulus (5 s)		Both stimuli (10 s)	
			APCER	BPCER	APCER	BPCER	APCER	BPCER
Parametric	SVM [18]	Linear	**0.46**	**0.23**	1.15	0.23	1.61	0.23
		Polynomial	**0.46**	**0.23**	0.46	0.00	1.15	0.92
		Radial basis	**0.46**	**0.23**	0.00	0.46	0.69	0.69
	Logistic regression		0.92	0.92	0.92	0.23	1.61	0.23
	Bagged trees		0.69	0.46	0.23	0.23	**0.23**	**0.92**
	kNN	k = 1	0.92	0.23	**0.00**	**0.23**	1.15	0.46
		k = 10	0.92	0.23	0.69	0.00	1.38	0.23
Data-driven	Basic RNN		0.46	1.84	**1.26**	**3.91**	1.15	1.15
	GRU		0.69	2.99	2.30	6.67	2.07	0.69
	LSTM	No peepholes	**0.23**	**1.38**	1.38	3.91	1.61	1.38
		With peepholes	0.46	1.84	1.95	3.45	**0.69**	**1.38**

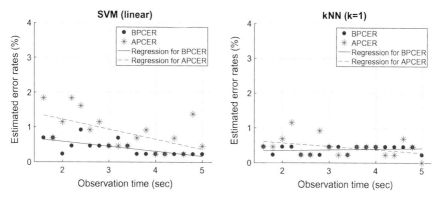

Fig. 7.12 The best configurations found for the entire positive and negative stimuli (5 s) analyzed for shorter horizons with a **parametric model**. Results for a positive stimulus (**left**) and a negative stimulus (**right**)

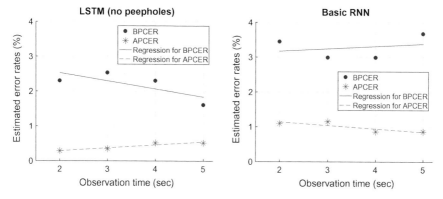

Fig. 7.13 Same as in Fig. 7.12, except that a **data-driven model** was used. Results for a positive stimulus (**left**) and a negative stimulus (**right**)

that data-driven models can be effective alternatives to parametric ones. To make a fair evaluation of our findings, we discuss shortly both limitations and advantages of the proposed iris PAD solution in this concluding section.

Merits. The results of distinguishing the true pupil reaction from a noisy input representing incorrect pupil reactions (or no reaction) are astonishing. Pupil dynamics as a behavioral presentation attack detection has a potential to detect sophisticated attacks, including coercion, that would be difficult to detect by methods designed to detect static artifacts such as iris paper printouts of cosmetic contact lenses. This chapter shows that pupil size can be effectively modeled and classified by simple recurrent neural networks, and the recent popularity of various deep learning software tools facilitates such implementations. The minimum observation time of 2–3 s required to achieve the best accuracy presented in this study makes this PAD close to practical implementations since a typical iris acquisition requires a few seconds even when only a still image is to be taken.

Limitations. It has been shown in the literature that pupil reaction depends on many factors such as health conditions, emotional state, drug induction, and fatigue. It also depends on the level of ambient light. Therefore, the models developed in this work may not work correctly in adverse scenarios, unless we have knowledge or data that could be used to adapt our classifiers. Generating attack presentations for pupil dynamics is difficult; hence, the next necessary step is the reformulation of this problem into one-class classification. Also, the application of negative or positive stimuli depends on the implementation environment. That is, it will be probably easier to stimulate the eye by increasing level of light than evoke pupil dilation that requires a sudden decrease in the level of ambient light. An interesting observation is that BPCER (recognition of authentic eyes) for the data-driven solution is significantly higher (3–4%) than for the parametric model-based approach (up to 1%). ACPER (recognition of no/odd reactions) remains similar in both approaches. This may suggest that more flexible neural network-based classification, when compared to a model specified by only seven parameters, is more sensitive to subject-specific fluctuations in pupil size and presents lower generalization capabilities than methods based on a parametric model.

Acknowledgements The authors would like to thank Mr. Rafal Brize and Mr. Mateusz Trokielewicz, who collected the iris images in varying light conditions under the supervision of the first author. The application of Kohn and Clynes model was inspired by research of Dr. Marcin Chochowski, who used parameters of this model as individual features in biometric recognition. This author, together with Prof. Pacut and Dr. Chochowski, has been granted a US patent No. 8,061,842 which partially covers the ideas related to parametric model-based PAD and presented in this work.

References

1. ISO/IEC: Information technology – Biometric presentation attack detection – Part 1: Framework, 30107-3 (2016)
2. Galbally J, Marcel S, Fierrez J (2014) Image quality assessment for fake biometric detection: application to iris, fingerprint, and face recognition. IEEE Trans Image Process (TIP) 23(2):710–724. https://doi.org/10.1109/TIP.2013.2292332
3. Wei Z, Qiu X, Sun Z, Tan T (2008) Counterfeit iris detection based on texture analysis. In: International conference on pattern recognition, pp 1–4. https://doi.org/10.1109/ICPR.2008.4761673
4. Doyle JS, Bowyer KW, Flynn PJ (2013) Variation in accuracy of textured contact lens detection based on sensor and lens pattern. In: IEEE international conference on biometrics: theory applications and systems (BTAS), pp 1–7. https://doi.org/10.1109/BTAS.2013.6712745
5. Ojala T, Pietikainen M, Maenpaa T (2002) Multiresolution gray-scale and rotation invariant texture classification with local binary patterns. IEEE Trans Pattern Anal Mach Intell (TPAMI) 24(7):971–987. https://doi.org/10.1109/TPAMI.2002.1017623
6. Ojansivu V, Heikkilä J (2008) Blur insensitive texture classification using local phase quantization. Springer, Berlin, pp 236–243. https://doi.org/10.1007/978-3-540-69905-7_27
7. Zhang L, Zhou Z, Li H (2012) Binary Gabor pattern: an efficient and robust descriptor for texture classification. In: IEEE international conference on image processing (ICIP), pp 81–84. https://doi.org/10.1109/ICIP.2012.6466800

8. Sun Z, Zhang H, Tan T, Wang J (2014) Iris image classification based on hierarchical visual codebook. IEEE Trans Pattern Anal Mach Intell (TPAMI) 36(6):1120–1133. https://doi.org/10.1109/TPAMI.2013.234
9. Dalal N, Triggs B (2005) Histograms of oriented gradients for human detection. In: IEEE international conference on computer vision and pattern recognition (CVPR), vol 1, pp 886–893. https://doi.org/10.1109/CVPR.2005.177
10. Doyle JS, Bowyer KW (2015) Robust detection of textured contact lenses in iris recognition using BSIF. IEEE Access 3:1672–1683. https://doi.org/10.1109/ACCESS.2015.2477470
11. Raghavendra R, Busch C (2015) Robust scheme for iris presentation attack detection using multiscale binarized statistical image features. IEEE Trans Inf Forensics Secur (TIFS) 10(4):703–715. https://doi.org/10.1109/TIFS.2015.2400393
12. Menotti D, Chiachia G, Pinto A, Schwartz W, Pedrini H, Falcao A, Rocha A (2015) Deep representations for iris, face, and fingerprint spoofing detection. IEEE Trans Inf Forensics Secur (TIFS) 10(4):864–879. https://doi.org/10.1109/TIFS.2015.2398817
13. Yambay D, Becker B, Kohli N, Yadav D, Czajka A, Bowyer KW, Schuckers S, Singh R, Vatsa M, Noore A, Gragnaniello D, Sansone C, Verdoliva L, He L, Ru Y, Li H, Liu N, Sun Z, Tan T (2017) LivDet iris 2017 – iris liveness detection competition 2017. In: IEEE international joint conference on biometrics (IJCB), pp 1–6
14. Raja K, Raghavendra R, Busch C (2015) Video presentation attack detection in visible spectrum iris recognition using magnified phase information. IEEE Trans Inf Forensics Secur (TIFS) 10(10):2048–2056. https://doi.org/10.1109/TIFS.2015.2440188
15. Komogortsev OV, Karpov A, Holland CD (2015) Attack of mechanical replicas: liveness detection with eye movements. IEEE Trans Inf Forensics Secur (TIFS) 10(4):716–725. https://doi.org/10.1109/TIFS.2015.2405345
16. Pacut A, Czajka A (2006) Aliveness detection for iris biometrics. In: IEEE international Carnahan conferences security technology (ICCST), pp 122–129. https://doi.org/10.1109/CCST.2006.313440
17. Czajka A, Pacut A, Chochowski M (2011) Method of eye aliveness testing and device for eye aliveness testing, United States Patent, US 8,061,842
18. Czajka A (2015) Pupil dynamics for iris liveness detection. IEEE Trans Inf Forensics Secur (TIFS) 10(4):726–735. https://doi.org/10.1109/TIFS.2015.2398815
19. Czajka A, Bowyer KW (2018) Presentation attack detection for iris recognition: an assessment of the state of the art. ACM Comput Surv (CSUR) 1(1):1–35. https://doi.org/10.1145/3232849, arXiv:1804.00194
20. Sutra G, Dorizzi B, Garcia-Salitcetti S, Othman N (2017) A biometric reference system for iris. OSIRIS version 4.1. http://svnext.it-sudparis.eu/svnview2-eph/ref_syst/iris_osiris_v4.1/. Accessed 1 Aug 2017
21. Kohn M, Clynes M (1969) Color dynamics of the pupil. Ann N Y Acad Sci 156(2):931 950. Available online at Wiley Online Library (2006)
22. Bengio Y, Simard P, Frasconi P (1994) Learning long-term dependencies with gradient descent is difficult. IEEE Trans Neural Netw 5(2):157–166. https://doi.org/10.1109/72.279181
23. Hochreiter S, Schmidhuber J (1997) Long short-term memory. Neural Comput 9(8):1735–1780. https://doi.org/10.1162/neco.1997.9.8.1735
24. Gers FA, Schmidhuber J (2000) Recurrent nets that time and count. In: International joint conference on neural network (IJCNN), vol 3, pp 189–194. https://doi.org/10.1109/IJCNN.2000.861302
25. Greff K, Srivastava RK, Koutnk J, Steunebrink BR, Schmidhuber J (2016) LSTM: a search space Odyssey. IEEE Trans Neural Netw Learn Syst (99):1–11. https://doi.org/10.1109/TNNLS.2016.2582924
26. Cho K, van Merriënboer B, Bahdanau D, Bengio Y (2014) On the properties of neural machine translation: encoder–decoder approaches. In: Workshop on syntax, semantics and structure in statistical translation (SSST), pp 1–6
27. Hinton G, Srivastava N, Swersky K (2017) Neural networks for machine learning. Lecture 6a: overview of mini-batch gradient descent. http://www.cs.toronto.edu/~tijmen/csc321. Accessed 28 April 2017

28. Glorot X, Bengio Y (2010) Understanding the difficulty of training deep feedforward neural networks. In: Teh YW, Titterington M (eds) International conference on artificial intelligence and statistics (AISTATS), Proceedings of machine learning research, vol 9. PMLR, Chia Laguna Resort, Sardinia, Italy, pp 249–256

Chapter 8
Review of Iris Presentation Attack Detection Competitions

David Yambay, Adam Czajka, Kevin Bowyer, Mayank Vatsa, Richa Singh, Afzel Noore, Naman Kohli, Daksha Yadav and Stephanie Schuckers

Abstract Biometric recognition systems have been shown to be susceptible to presentation attacks, the use of an artificial biometric in place of a live biometric sample from a genuine user. Presentation Attack Detection (PAD) is suggested as a solution to this vulnerability. The LivDet-Iris—Iris Liveness Detection Competition started in 2013 strives to showcase the state-of-the- art in presentation attack detection by assessing the software-based iris PAD methods (Part 1), as well as hardware-based iris PAD methods (Part 2) against multiple datasets of spoof and live fingerprint images. These competitions have been open to all institutions, industrial and academic, and competitors which can enter as either anonymous or using

D. Yambay (✉) · S. Schuckers
Clarkson University, 10 Clarkson Avenue, Potsdam, NY 13699, USA
e-mail: yambayda@clarkson.edu

S. Schuckers
e-mail: sschucke@clarkson.edu

A. Czajka
Warsaw University of Technology, Nowowiejska 15/19 Str., 00-665 Warsaw, Poland
e-mail: aczajka@nd.edu

K. Bowyer
Notre Dame University, 46556 Notre Dame, IN, France
e-mail: kwb@nd.edu

M. Vatsa · R. Singh
IIIT-Delhi Okhla Industrial Estate, Phase III, New Delhi 110020, India
e-mail: mayank@iiitd.ac.in

R. Singh
e-mail: rsingh@iiitd.ac.in

A. Noore · N. Kohli · D. Yadav
West Virginia University, Morgantown, WV 26506, USA
e-mail: afzel.noore@mail.wvu.edu

N. Kohli
e-mail: nakohli@mix.wvu.edu

D. Yadav
e-mail: dayadav@mix.wvu.edu

© Springer Nature Switzerland AG 2019
S. Marcel et al. (eds.), *Handbook of Biometric Anti-Spoofing*,
Advances in Computer Vision and Pattern Recognition,
https://doi.org/10.1007/978-3-319-92627-8_8

the name of their institution. There have been two previous fingerprint competitions through LivDet; 2013 and 2015. LivDet-Iris 2017 is being conducted during 2017. LivDet-Iris has maintained a consistent level of competitors for Part 1: Algorithms throughout the two previous competitions and 2017 competition has begun to garner further interest.

8.1 Introduction

The iris biometric is becomingly an increasingly popular biometric modality for access control and commercial products due to its properties such as uniqueness, stability, and performance. However, these systems have been shown to be vulnerable to the presentation attacks in the form of patterned contact lenses and printed iris images. Numerous competitions have been held in the past to address matching in biometrics such as the Fingerprint Verification Competition held in 2000, 2002, 2004, and 2006 [1] and the ICB Competition on Iris Recognition (ICIR2013) [2]. However, these competitions did not consider presentation attacks.

Since 2009, in order to assess the main achievements of the state-of- the-art in fingerprint liveness detection, University of Cagliari and Clarkson University organized the first Fingerprint Liveness Detection Competition. This expanded to include the iris biometric in 2013.

The First International Iris Liveness Detection Competition (LivDet) 2013 [3], provided the first assessment on iris presentation attack detection. LivDet-Iris has continued in 2015 and 2017. LivDet 2015 [4] and 2017 were created in order to examine the progressing state-of-the-art in presentation attack detection.

LivDet-Fingerprint has been hosted in 2009 [5], 2011 [6], 2013 [7], and 2015 [8] and are discussed in the previous chapter.

This chapter reviews the previous LivDet-Iris competitions and the evolution of the competitions over the years. The following sections will describe the methods used in testing for each of the LivDet competitions as well as descriptions of the datasets that have been generated from each competition. Also discussed are the trends across the different competitions that reflects change to the art of presentation attacks as well as advances in the state-of-the-art in presentation attack detection. Further, conclusions from previous LivDet competitions and the future of LivDet is discussed.

8.2 Background

The Liveness Detection Competition series was started in 2009 and created a benchmark for measuring fingerprint presentation attack detection algorithms, similar to matching performance. At that time, there had been no other public competitions held that has examined the concept of liveness detection as part of a biometric modality in deterring spoof attack and in 2013, the competition was expanded to include the iris modality.

The goal of LivDet since its conception has been to allow the researcher to test their own algorithms and systems on publicly available data sets, with results published to establish baselines for future research and to examine the state-of-the-art as it develops.

Since the start of LivDet, evaluation of spoof detection for facial systems were performed in the Competition on Counter Measures to 2D Facial Spoofing Attacks, first held in 2011 and then, held a second time in 2013. The purpose of this competition is to address different methods of detection for 2D facial spoofing [9]. The competition dataset consisted of 400 video sequences, 200 of them real attempts, and 200 attack attempts [9]. A subset was released for training and then, another subset of the dataset was used for testing purposes.

8.3 Methods and Datasets

Each LivDet competition is composed of two distinct parts: Part 1: Algorithms which features strictly software-based approaches to presentation attack detection and Part 2: Systems which features software or hardware-based approaches in a fully packaged device. Although LivDet-Iris has not yet held a systems competition. The protocol of Part 1: Algorithms are described further in this section along with descriptions of each dataset created through this competition. There are two main presentation attack types during the competition: printed iris images and patterned contact lenses. Printed iris images are used to impersonate another person while patterned contact lenses are used to obfuscate the attacker's natural iris pattern.

8.3.1 Part 1: Algorithm Datasets

Each iteration of the competition features at least three different datasets of iris data. Individual sensors have been used in different competitions, although new data is created for each individual competition. Due to the fact that LivDet data is made public and some sensors were used in multiple competitions, competitors were able to use previous competition data in their algorithm training.

LivDet 2013 consisted of data from three iris sensors: DALSA (Clarkson), LG4000 (Notre Dame), and IrisGuard AD100 (Warsaw). Presentation attacks were created for each dataset using printed iris images for IrisGuard, as well as using patterned contact lenses for DALSA and LG4000. Number of images in each dataset is shown in Fig. 8.1 and further information can be found in [3] (Table 8.1).

Textured contact lenses for the Notre Dame dataset came from Johnson & Johnson [10], CIBA Vision [11], and Cooper Vision [12]. Contacts were purchased of varying colors from each manufacturer and some contacts are "toric" lenses which means

Table 8.1 Dataset description LivDet 2013

	Training		Testing	
	Live	Spoof	Live	Spoof
Notre Dame:LG4000	2000	1000	800	400
Warsaw: IrisGuard AD100	228	203	624	612
Clarkson: DALSA	270	400	246	440

that they are designed to maintain a preferred orientation around the optical axis. The training and testing sets were split equally into three classes: (1) no contact lenses, (2) non-textured contact lenses, and (3) textured contact lenses. The sample images can be seen in Fig. 8.1.

The Warsaw dataset was generated using printouts of iris images using a laser printer on matte paper and making a hole where the pupil was in order to have a live eye behind the spoof iris image. Two different printers were used to build Warsaw subset: (a) HP LaserJet 1320 and (b) Lexmark c534dn. The HP LaserJet was used to print "low resolution" iris images of approximately 600 dpi, whereas the Lexmark was used to print "high resolution" images of 1200 dpi. Only high-resolution images were present in the training set, whereas both low- and high resolution were used in the testing set. The sample images can be seen in Fig. 8.2.

The Clarkson dataset was created using patterned contact lenses. The images were collecting using an NIR video camera, DALSA. The camera is modified to capture in the NIR spectrum similar to commercial iris cameras. It captures a section of the face of each subject that includes both eyes. The eyes are then cropped out of the images to create the subset. A total of 64 eyes were used in the live dataset with varying illuminations as well as varying levels of blur. The training set contained 5 images per illumination, but the higher blur level was too strong and was removed giving 3 images per illumination in the testing set. The spoof set contained 6 subjects wearing 19 patterned contacts each. The samples images can be seen in Fig. 8.3. A list of patterned contacts used in the Clarkson Dataset is shown in Table 8.2.

The dataset for LivDet 2015 consisted of three datasets: DALSA (Clarkson), LG IrisAccess EOU2200 (Clarkson), and IrisGuard AD100 (Warsaw). Presentation attacks consisted of printed iris images for all three datasets and patterned contact lenses for both of Clarkson's datasets. Number of images in each dataset is shown in Fig. 8.3 and further information can be found in [4].

The Clarkson dataset used the same spoof attacks for both the two datasets. The printed images use a variety of configurations including 1200 dpi versus 2400 dpi printouts, contrast adjustment versus raw images, pupil hole versus no pupil hole, and glossy paper versus matte paper. The printouts came from images collected from both the LG iris camera and from the Dalsa camera. The LG camera is similar to the DALSA camera, and in that it captures video sequences, however, the LG camera is designed as an iris camera. The sample images for both LG and DALSA can be seen in Figs. 8.4 and 8.5, respectively. The patterned contact lenses used by the Clarkson datasets is shown in Table 8.4.

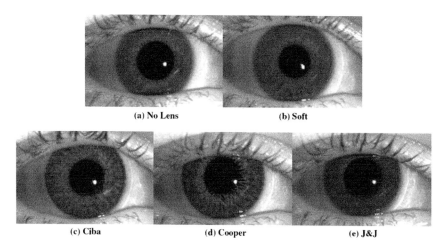

(a) No Lens **(b) Soft**

(c) Ciba **(d) Cooper** **(e) J&J**

Fig. 8.1 Sample images of University of Notre Dame subset showing the variety of cosmetic lens textures that are available from different manufacturers. All images are from the same subject eye

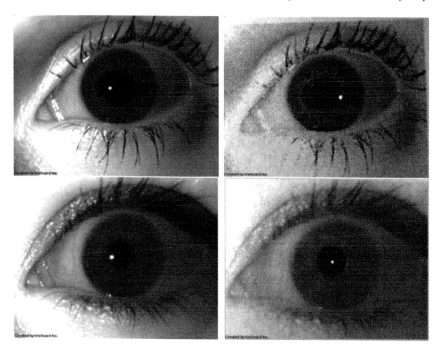

Fig. 8.2 Sample images of Warsaw subset. Images of the authentic eyes are shown in the left column, and their fake counterparts are shown in the right column. Low- and high-resolution printouts are presented in the upper and lower rows, respectively

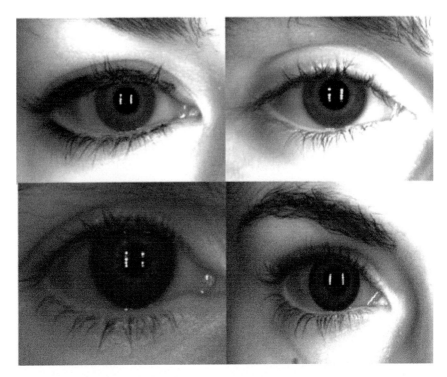

Fig. 8.3 Sample images from Clarkson dataset. Live images (top) and spoof images (bottom)

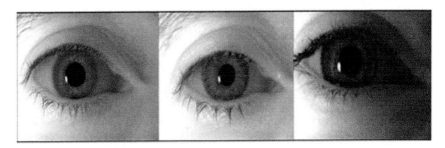

Fig. 8.4 Images from LG dataset. Left to right; live, patterned, printed

The Warsaw dataset is created using the same Lexmark 534dn used previously in LivDet 2013 and only used 1200 dpi images. Printouts were completed using both color printing and black and white printing. In addition, pupil holes are added in order to have a live user presented behind the printouts. This is to counter a camera that searches for specular reflection of a live cornea. The sample images for the Warsaw dataset can be seen in Fig. 8.6 (Table 8.3).

LivDet 2017 was comprised of four different datasets: LG IrisAccess EOU2200 (Clarkson), IrisGuard AD100 (Warsaw), LG 4000 and AD100 (Notre Dame), and IriShield MK2120U (IIITD-WVU). Warsaw and Notre Dame had datasets with a

Table 8.2   Clarkson patterned contact types 2013

Number	Contact type	Color
1	Freshlook dimensions	Pacific blue
2	Freshlook dimensions	Sea green
3	Freshlook colorblends	Green
4	Freshlook colorblends	Blue
5	Freshlook colorblends	Brown
6	Freshlook colors	Hazel
7	Freshlook colors	Green
8	Freshlook colors	Blue
9	Phantasee natural	Turquoise
10	Phantasee natural	Green
11	Phantasee vivid	Green
12	Phantasee vivid	Blue
13	Phantasee diva	Black
14	Phantasee diva	Brown
15	ColorVue biggerEyes	Cool blue
16	ColorVue biggerEyes	Sweet honey
17	ColorVue 3 Tone	Green
18	ColorVue elegance	Aqua
19	ColorVue elegance	Brown

Table 8.3   Dataset description LivDet 2015

	Training			Testing		
	Live	Patterned	Printed	Live	Patterned	Printed
Warsaw	852	N/A	815	2002	N/A	3890
Clarkson LG	400	576	846	378	576	900
Clarkson DALSA	700	873	846	378	558	900

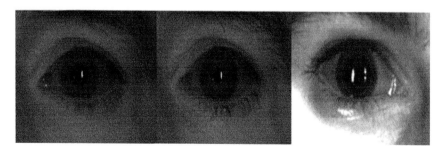

Fig. 8.5   Images from Dalsa dataset. Left to right; live, patterned, printed

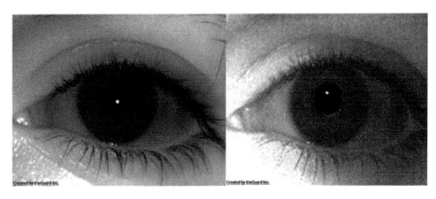

Fig. 8.6 Images from Warsaw dataset. Left to right; live, printed

Table 8.4 Patterned contact types from LG and Dalsa datasets. Unknown (test set only) patterns in bold

Number	Contact type	Color
1	Expressions colors	Brown
2	Expressions colors	Jade
3	Expressions colors	Blue
4	Expressions colors	Hazel
5	Air optix colors	Brown
6	Air optix colors	Green
7	**Air optix colors**	**Blue**
8	Freshlook colorblends	Brilliant blue
9	Freshlook colorblends	Brown
10	Freshlook colorblends	Honey
11	Freshlook colorblends	Green
12	Freshlook colorblends	Sterling gray
13	Freshlook one-day	Green
14	Freshlook one-day	Pure hazel
15	Freshlook one-day	Gray
16	Freshlook one-day	Blue
17	**Air optix colors**	**Gray**
18	**Air optix colors**	**Honey**
19	**Expressions colors**	**Blue topaz**
20	**Expressions colors**	**Green**

single presentation attack type, printed iris images, and patterned contact lenses, respectively. Clarkson and IIITD-WVU used both printed iris images and patterned contacts in their datasets. Number of images in each dataset is shown in Fig. 8.5 (Table 8.5).

Table 8.5 Dataset description LivDet 2017

	Training			Testing		
	Live	Patterned	Printed	Live	Patterned	Printed
Warsaw	1844	N/A	2669	974	N/A	2016
Clarkson	2469	1122	1346	1485	765	908
Notre dame	600	600	N/A	900	1800	N/A
IIITD-WVU	2250	1000	3000	702	701	2806

8.3.2 Specific Challenges

LivDet 2013 and 2015 were organized as similar competitions, however, LivDet 2017 was used as a chance to propose more difficult challenges to participants. Each dataset contained some form of unknown presentation attack in the testing dataset, which was not always the case in previous iterations of the competition. Beyond this, the Clarkson dataset included two additional challenges beyond the unknown presentation attacks. The first of which is that included in the testing set were three patterned contact types that only obscure part of the iris. Acuvue Define is designed to accentuate the wearer's eyes and only cover a portion of the natural iris pattern. Beyond this, the printed iris image testing set contained iris images that were captured by an iPhone 5 in the visible spectrum. The red channel was extracted and converted to grayscale and presented back to the sensor. The Warsaw and IIITD-WVU testing datasets both had unknown presentation attacks that were captured on a different camera than the training data, although the resolution was kept consistent with the training sets. The patterned contacts used by Clarkson can be seen in Table 8.6.

The LivDet 2017 competition also included a cross-sensor challenge. In the standard competition, each algorithm is trained to a specific training set and only data from that testing set is compared to that version of the competitor's algorithm. In the cross-sensor challenge, each competitor had a single algorithm that was tested against each of the testing datasets to compute a singular error rate.

Due to all of these additional challenges, LivDet 2017 is considered by the competition organizers as vastly more difficult to correctly classify the images.

8.3.3 Performance Evaluation

Each of the algorithms returned a value representing a percentage of posterior probability of the live class (or a degree of "liveness") given the image normalized in the range 0–100 (100 is the maximum degree of liveness, 0 means that the image is fake). The threshold value for determining liveness was set at 50. This threshold is used to calculate Attack Presentation Classification Error Rate (APCER) and Bona Fide Presentation Classification Error Rate (BPCER) error estimators, where

Table 8.6 Patterned contact types from LG dataset. Unknown (test set only) patterns in bold

Number	Contact type	Color
1	**Acuvue define**	**Natural shimmer**
2	**Acuvue define**	**Natural shine**
3	**Acuvue define**	**Natural sparkle**
4	Air optix	Green
5	Air optix	Brilliant blue
6	Air optix	Brown
7	Air optix	Honey
8	Air optix	Hazel
9	Air optix	Sterling gray
10	Expressions colors	Aqua
11	Expressions colors	Hazel
12	Expressions colors	Brown
13	Expressions colors	Green
14	Expressions colors	Jade
15	Freshlook colorblends	Amethyst
16	Freshlook colorblends	Brown
17	Freshlook colorblends	Green
18	Freshlook colorblends	Turquoise
19	**Freshlook colors**	**Blue**
20	**Freshlook colors**	**Green**

- APCER is the rate of misclassified spoof images (spoof called live) and
- BPCER is the rate of misclassified live images (live called spoof).

Both APCER and BPCER are calculated for each dataset separately, as well as the average values across all datasets. Being able to see a range of values for a system is beneficial for understanding how the system is performing against different types of data, however, each competitor will have a different method of determining whether an image is live or spoof. A standardized method was created with which all competitors would have to normalize their outputs to fit. This 0–100 range and a threshold of 50 are arbitrary values provided to competitors, but they provide a common range for all competitors to normalize their scores within. The competitors choose how they wish to adjust their system to work within the confines of the competition. To select a winner, the average of APCER and BPCER was calculated for each participant across datasets. The weight of importance between APCER to BPCER will change based on use case scenario. In particular, low BPCER is more important for low- security implementations such as unlocking phones, however, low APCER is more important for high-security implementations. Due to this, APCER and BPCER are given equal weight in the LivDet competition series.

Error rate curves demonstrating a changing threshold are shown in the individual competition papers.

Processing time per image is also considered as long processing times can cause throughput issues in systems.

This performance evaluation is examined a second time for the cross-sensor challenge in LivDet 2017 which has results that are separate from the main competition.

In LivDet competitions before 2017, a different terminology was used to determine error rates. FerrLive is equivalent to the BPCER. FerrFake is equivalent to the APCER.

8.4 Examination of Results

In this section, we analyze the experimental results for the three LivDet editions. The results show the growth and improvement across the three competitions. The overall competition results are shown in this chapter. More in-depth analysis of the results as well error rate curves can be seen in the individual competition reports for each LivDet.

8.4.1 Participants

The competition is open to all academic and industrial institutions. Upon registration, each participant was required to sign a database release agreement detailing the proper usage of data made available through the competition. Participants were then given a database access letter with a username and password to access the server to download the training data.

Participants have the ability to register as either an anonymous submission or as their organization. After the results are tallied, each competitor is sent their organization's personal results and given the option to be anonymous or as their organization before publication. Placement in the competition and results of other submissions are not provided to competitors at that time.

8.4.2 Trends of Competitors and Results for Fingerprint Part 1: Algorithms

Unlike LivDet-Fingerprint, the number of competitors for LivDet-Iris has remained relatively constant from year to year. LivDet 2013 was comprised of three competitors and these competitors are listed in Table 8.7. LivDet 2015 had four total submissions and these competitors can be seen in Table 8.8. In LivDet 2017, the competition had three participants and their information is shown in Table 8.9. Participation in LivDet-Iris has not fallen over time, however, unfortunately unlike the fingerprint competitions, it has not seen the same level of entrants.

Table 8.7 Participants and acronyms for LivDet 2013

Participant name	Algorithm name
ATVS—Biometric Recognition Group, Universidad Autonoma de Madrid	ATVS
University of Naples Federico II	Federico
Faculdade de engenharia de Universidade do Porto	Porto

Table 8.8 Participants and acronyms for LivDet 2015

Participant name	Algorithm name
Anonymous 0	Anon0
Anonymous 1	Anon1
Anonymous 2	Anon2
Anonymous University of Naples Federico II	Federico

Table 8.9 Participants and acronyms for LivDet 2017

Participant name	Algorithm name
Anonymous	Anon1
Universita' deli Studi di Napoli	UNINA
Chinese Academy of Sciences	CASIA

Table 8.10 Best error rate for each competition.

2013	
Minimum avg BPCER	Minimum avg APCER
28.56%	5.72%
2015	
Minimum avg BPCER	Minimum avg APCER
7.53%	1.34%
2017	
Minimum avg BPCER	Minimum avg APCER
3.36%	14.71%

The best algorithm for each competition, in terms of performance, are detailed in Table 8.10. These values represent the best competitor algorithm taking the average error rates across all testing datasets. Average values are computed by taking the mean of the error rates with no adjustment for dataset size. The lack of adjustment for dataset size is due to the fact that each dataset is constructed with entirely different properties and have different attacks and quality levels. Also since each partner creates their own dataset, the sizes of each dataset have the potential to be wildly different sizes. Due to this, adjusting error rates by dataset size when calculating the overall error rate

Table 8.11 Mean and standard deviation values for each dataset of the competitions

2013

	Mean BPCER	Std. dev. BPCER	Mean APCER	Std. dev. APCER
Clarkson	29.68	18.69	26.8	30.57
Warsaw	17.55	10.98	6.75	5.70

2015

	Mean BPCER	Std. dev. BPCER	Mean APCER	Std. dev. APCER
Clarkson LG	9.39	5.12	6.12	7.81
Clarkson Dalsa	9.79	4.51	11.99	14.28
Warsaw	1.6	1.47	2.39	4.45

2017

	Mean BPCER	Std. dev. BPCER	Mean APCER	Std. dev. APCER
Clarkson	3.10	2.86	21.17	15.19
Warsaw	4.74	4.29	5.91	2.41
Notre Dame	2.61	4.28	35	44.11
IIITD-WVU	6.70	8.38	40.61	25.01

would heavily skew the error rates towards an abnormally large dataset compared to other datasets, such as Warsaw from LivDet 2015.

LivDet 2013 saw error rates of above 10% for BPCER for all submissions, however, in both LivDet 2015 and 2017, BPCER dropped below 10% for all competitors, which is a substantial decrease in error rates and a consistent decrease over the course of the competitions.

Examining APCER across the three competitions, it can be seen that there has been a pendulum effect in the error rates. APCER dropped considerably from 5.72 to 1.34% from 2013 to 2015, however, LivDet 2017 saw APCER spike higher. This is not unexpected as LivDet 2017 included a number of new challenges from a presentation attack perspective. APCER rates are as expected given the increased difficulty of the challenge.

For each competition, the mean and standard deviation of the submissions across each dataset can be calculated. Although LivDet 2013 has three datasets, only one competitor completed algorithms against all three datasets and thus, mean and standard deviation cannot be calculated for the Notre Dame dataset for LivDet 2013. These values are shown in Table 8.11. The high standard deviations seen are due to some datasets where one or more algorithms have low error rates while others have extraordinarily high error rates. With this, it can be seen that especially with LivDet 2013 and 2017 the situation of highly varied error rates by algorithms for a specific dataset is prevalent. Notre Dame in 2017 shows this the most with two error rates hovering around 10% for APCER but one algorithm having an error rate of 85.89%.

The results do suggest that there has been an advancement in the state-of-the-art for presentation attack detection, however, systems were not fully prepared for the sudden increase in presentation attack difficulty. However, there is still evidence that systems are advancing over time.

8.5   Future of LivDet

LivDet has continued in 2017 with the fingerprint competition taking place in the fall of 2017. The new challenges posed in the iris competition in 2017 show that more advancement needs to be made in iris presentation attack detection algorithms. With the 2017 dataset becoming public, it is hoped that algorithms will be more well equipped to handle the new challenges in the next LivDet competition.

The cross-sensor challenge especially is an interesting challenge being examined for future competitions as it shows how an algorithm can be trained to handle data from multiple different sensors.

8.6   Conclusions

Since the inception of the LivDet-Iris competition, LivDet has been aimed to allow research centers and industries the ability to have an independent assessment of their presentation attack detection algorithms. While error rates have improved over time, it is evident that more improvements can be made in the area of iris presentation attack detection and the authors look forward to what future researchers can accomplish with the datasets. All datasets are publicly available and can be requested at livdet.org.

Acknowledgements The first and second author had equal contributions to the research. This work has been supported by the Center for Identification Technology Research and the National Science Foundation under Grant No. 1068055.

References

1. Cappelli R (2007) Fingerprint verification competition 2006. Biom Technol Today 15(7):7–9. https://doi.org/10.1016/S0969-4765(07)70140-6
2. Alonso-Fernandez F, Bigun J (2013) Halmstad university submission to the first ICB competition on iris recognition (icir2013)
3. Yambay D et al (2014) Livdet-iris 2013-iris liveness detection competition 2013. https://doi.org/10.1109/BTAS.2014.6996283
4. Yambay D, Walczak B, Schuckers S, Czajka A (2017) LivDet-iris 2015–Iris liveness detection. In: IEEE international conference on identity, security and behavior analysis (ISBA), pp 1–6
5. Marcialis GL et al (2009) First international fingerprint liveness detection competition—LivDet 2009. In: International conference on image analysis and processing. Springer, Berlin. https://doi.org/10.1007/978-3-642-04146-4_4
6. Yambay D et al (2011) Livdet 2011 - fingerprint liveness detection competition 2011. In: 5th IAPR/IEEE international conference on biometrics (ICB 2012)
7. Marcialis G et al (2013) Livdet 2013 - fingerprint liveness detection competition 2013. In: 6th IAPR/IEEE international conference on biometrics (ICB2013)
8. Marcialis G et al (2014) LivDet 2015–fingerprint liveness detection competition 2015. In: 7th IEEE international conference on biometrics: theory, applications and systems (BTAS 2015)

9. Chakka MM et al (2011) Competition on counter measures to 2-d facial spoofing attacks. https://doi.org/10.1109/IJCB.2011.6117509
10. JohnsonJohnson (2013). http://www.acuvue.com/products-acuvue-2-colours
11. CibaVision (2013). http://www.freshlookcontacts.com
12. Vision C (2013). http://coopervision.com/contact-lenses/expressions-color-contacts

Part III
Face Biometrics

Chapter 9
Introduction to Face Presentation Attack Detection

Javier Hernandez-Ortega, Julian Fierrez, Aythami Morales
and Javier Galbally

Abstract The main scope of this chapter is to serve as a brief introduction to face presentation attack detection. The next pages present the different presentation attacks that a face recognition system can confront, in which an attacker presents to the sensor, mainly a camera, an artifact (generally a photograph, a video, or a mask) to try to impersonate a genuine user. First, we make an introduction of the current status of face recognition, its level of deployment, and the challenges it faces. In addition, we present the vulnerabilities and the possible attacks that a biometric system may be exposed to, showing that way the high importance of presentation attack detection methods. We review different types of presentation attack methods, from simpler to more complex ones, and in which cases they could be effective. Later, we summarize the most popular presentation attack detection methods to deal with these attacks. Finally, we introduce public datasets used by the research community for exploring the vulnerabilities of face biometrics and developing effective countermeasures against known spoofs.

9.1 Introduction

Over the last decades, there have been numerous technological advances that helped to bring new possibilities to people in the form of new devices and services. Some

J. Hernandez-Ortega (✉)
Biometrics and Data Pattern Analytics - BiDA Lab, Universidad Autonoma de Madrid,
Madrid, Spain
e-mail: javier.hernandezo@uam.es

J. Fierrez
Universidad Autonoma de Madrid, Madrid, Spain
e-mail: julian.fierrez@uam.es

A. Morales
School of Engineering, Universidad Autonoma de Madrid, Madrid, Spain
e-mail: aythami.morales@uam.es

J. Galbally
European Commission - DG Joint Research Centre, Ispra, Italy
e-mail: javier.galbally@ec.europa.eu

© Springer Nature Switzerland AG 2019
S. Marcel et al. (eds.), *Handbook of Biometric Anti-Spoofing*,
Advances in Computer Vision and Pattern Recognition,
https://doi.org/10.1007/978-3-319-92627-8_9

years ago, it would have been almost impossible to imagine having in the market devices like current smartphones and laptops, at affordable prices that allow a high percentage of the population to have their own piece of top-level technology at home, a privilege that historically has been restricted to big companies and research groups.

Thanks to this quick advance in technology, specially in computer science and electronics, it has been possible to broadly deploy biometric systems for the first time. Nowadays, they are present in a high number of scenarios like border access control, surveillance, smartphone authentication, forensics, and online services like e-learning and e-commerce.

Among all the existing biometric traits, face recognition is currently one of the most extended. The face has been studied as a mean of recognition since the 60s, acquiring special relevance in the 90s following the evolution of computer vision [1]. Some interesting properties of the human faces for biometrics are acquisition at a distance, nonintrusively, and the good discriminant characteristics of the face to perform identity recognition.

At present, face is one of the biometric traits with the highest economic and social impact due to several reasons:

- Face is the second most largely deployed biometric at world level in terms of market quota right after fingerprints [2]. Each day more and more manufacturers are including face recognition in their products, like Apple with its Face ID technology.
- Face is adopted in most identification documents such as the ICAO-compliant biometric passport [3] or national ID cards [4].

Given their high level of deployment, attacks having a face recognition system as their target is not restricted anymore to theoretical scenarios, becoming a real threat. There exist all kinds of applications and sensitive information that can be menaced by attackers. Giving to each face recognition application an appropriate level of security, as it is being done with other biometric traits, like iris or fingerprint, should be a top priority.

Historically, the main focus of research in face recognition has been given to the improvement of the performance at the verification and identification tasks, i.e., distinguishing better between subjects using the available information of their faces. To achieve that goal, a face recognition system should be able to optimize the differences between the facial features of each user [5], and also the similarities among samples of the same user. Within the variability factors that can affect the performance of face recognition systems there are occlusions, low-resolution, different viewpoints, lighting, etc. Improving the performance of recognition systems in the presence of these variability factors is currently an active area in face recognition research.

Contrary to the optimization of their performance, the security vulnerabilities of face recognition systems have been much less studied in the past, and only over the recent few years some attention has been given to detecting different types of attacks [6]. Regarding these security vulnerabilities, Presentation Attack Detection (PAD) consists on detecting whether a biometric trait comes from a living person or it is a fake.

The rest of this chapter is organized as follows: Sect. 9.2 overviews the main vulnerabilities of face recognition systems, making a description of several presentation attack approaches. Section 9.3 introduces presentation attack detection techniques. Section 9.4 presents some available public databases for research and evaluation of face presentation attack detection. Sections 9.5 and 9.6 discuss about architectures and applications of face PAD. Finally, concluding remarks are drawn in Sect. 9.7.

9.2 Vulnerabilities in Face Biometrics

In the present chapter, we concentrate on Presentation Attacks, i.e., attacks against the sensor of a face recognition system [7] (see point V1 in Fig. 9.1). An overview of indirect attacks to face systems can be found elsewhere [8]. Indirect attacks (points V2–V7 in Fig. 9.1) can be prevented by securing certain points of the face recognition system, i.e., the communication channels, the equipment and the infrastructure involved. The techniques needed for improving those modules are more related to "classical" cybersecurity than to biometrics, so they will not be covered in this chapter.

On the other hand, presentation attacks are a purely biometric vulnerability that is not shared with other IT security solutions and that needs specific countermeasures. In these attacks, intruders use some type of artifact, typically artificial (e.g., a face photo, a mask, a synthetic fingerprint or a printed iris image), or try to mimic the aspect of genuine users (e.g., gait, signature) to fraudulently access the biometric system.

A high amount of biometric data are exposed, (e.g., photographs and videos at social media sites) showing the face, eyes, voice, and behavior of people. Presentation attackers are aware of this reality and take advantage of those sources of information to try to circumvent face recognition systems [9]. This is one of the well-known

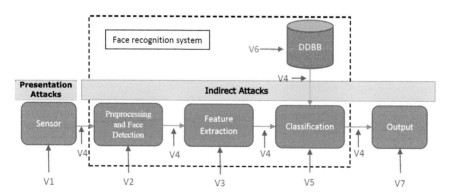

Fig. 9.1 Scheme of a generic biometric system. In this type of system, there exist several modules and points that can be the target of an attack (V1–V7). Presentation attacks are performed at sensor level (V1), without the need of having access to the interior of the system. Indirect attacks (V2–V7) can be performed at the database, the matcher, the communication channels, etc. In this type of attack the attacker needs access to the interior of the system

drawbacks of biometrics: "biometric traits are not secrets" [10]. In this context, it is worth noting that a factor that makes face an interesting trait for person recognition, i.e., easiness to capture, makes face biometrics also specially vulnerable to attackers, who may easily find example faces of the identities to attack.

In addition to being fairly easy to obtain a face image of the real users under attack, face recognition systems are known to respond weakly to presentation attacks, for example, using one of these three categories of attacks:

1. Using a photograph of the user to be impersonated [11].
2. Using a video of the user to be impersonated [12].
3. Building and using a 3D model of the attacked face, for example, an hyperrealistic mask [13].

The success probability of an attack may vary considerably depending on the characteristics of the face recognition system, for example, if it uses visible light or works in another range of the spectrum [14], if it has one or several sensors, the resolution, the lighting, and also depending on the characteristics of the artifact: quality of the texture, the appearance, the resolution of the presentation device, the type of support used to present the fake, or the background conditions.

Without implementing presentation attack detection measures, most of the state-of-the-art facial biometric systems are vulnerable to simple attacks that a regular person would detect easily. This is the case, for example, of trying to impersonate a subject using a photograph of his face. Therefore, in order to design a secure face recognition system in a real scenario, for instance for replacing password-based authentication, Presentation Attack Detection (PAD) techniques should be a top priority from the initial planning of the system.

Given the discussion above, it could be stated that face recognition systems without PAD techniques are at clear risk, so a question often rises: What technique(s) should be adopted to secure them? The fact is that counterfeiting this type of threats is not a straightforward problem, as new specific countermeasures need to be developed and adopted whenever a new attack appears.

With the scope of encouraging and boosting the research in presentation attack detection techniques in face biometrics, there are numerous and very diverse initiatives in the form of dedicated tracks, sessions, and workshops in biometric-specific and general signal processing conferences [15, 16]; organization of competitions [17]; and acquisition of benchmark datasets [13, 18] that have resulted in the proposal of new presentation attack detection methods [7]; standards in the area [19, 20]; and patented PAD mechanisms for face recognition systems [21].

9.2.1 Attacking Methods

Typically, face recognition systems can be spoofed by presenting to the sensor (e.g., a camera) a photograph, a video, or a 3D mask of a targeted person (see Fig. 9.2). There are other possibilities in order to circumvent a face recognition system, such as

GENUINE

ATTACKS

PHOTO   VIDEO   3D MASK   OTHERS

Fig. 9.2 Examples of face presentation attacks: The upper image shows an example of a genuine user, and below it there are some examples of presentation attacks, depending on the artifact shown to the sensor: a photo, a video, a 3D mask, and others

using makeup [22] or plastic surgery. However, using photographs and videos are the most common type of attacks due to the high exposition of face (e.g., social media, video surveillance), and the low cost of high-resolution digital cameras, printers, or digital screens.

Regarding the attack types, a general classification can be done taking into account the nature and the level of complexity of the artifact used to attack: photo-based, video-based, and mask-based (as can be seen in Fig. 9.2). It must be remarked that this is only a classification of the most common types of attacks, but there could exist more complex and newer attacks that may not fall into in any of these categories, or that may belong to several categories at the same time.

9.2.1.1 Photo Attacks

A photo attack consists in displaying a photograph of the attacked identity to the sensor of the face recognition system [23] (see example in Fig. 9.2).

Photo attacks are the most critical type of attack because of several factors. For example, printing color images from the face of the genuine user is really cheap and easy to do. These are usually called print attacks in the literature [24]. Alternatively, the photos can be displayed in the high-resolution screen of a device (e.g., a smartphone, a tablet, or a laptop). It is also easy to obtain samples of genuine faces thanks to the recent growth of social media sites like Facebook, Twitter, Flickr, etc. [9]. With the price reduction that digital cameras have experimented in recent years, it is also possible to obtain photos of a legitimate user simply by using a hidden camera.

Among the photo attack techniques, there are also more complex ones like photographic masks. This technique consists in printing a photograph of the subject's face and then making holes for the eyes and the mouth [18]. This is a good way to avoid presentation attack detection techniques based on blinking and mouth movements detection.

Even if these attacks seem too simple to work in a real scenario, some studies performed by private security firms indicate that many commercial systems are vulnerable to them [25]. Due to the easiness of carrying out this type of attack, implementing robust countermeasures that perform well against them should be a must for any facial recognition system.

9.2.1.2 Video Attacks

Similarly to the case of photo attacks, video acquisition of people intended to be impersonated is also becoming increasingly easier with the growth of public video sharing sites and social networks, or even using a hidden camera. Another reason to use this type of attack is that it increases the probability of success by introducing liveness appearance to the displayed fake biometric sample [26].

Once a video of the legitimate user is obtained, one attacker could play it in any device that reproduces video (smartphone, tablet, laptop, etc.) and then present it to the sensor/camera [27], (see Fig. 9.2). This type of attacks is often referred to in the literature as replay attacks, a more sophisticated version of photo attacks.

Replay attacks are more difficult to detect, compared to the photo spoofs, as not only the face texture and shape is emulated but also its dynamics, like eye blinking, mouth and/or facial movements [12]. Due to their higher sophistication, it is reasonable to assume that systems that are vulnerable to photo attacks will perform even worse with respect to video attacks, and also that being robust against photo attacks does not mean to be equally strong against video attacks [18]. Therefore, specific countermeasures need to be developed and implemented.

9.2.1.3 Mask Attacks

In this type of attack, the presented artifact is a 3D mask of the user's face. The attacker builds a 3D reconstruction of the face and presents it to the sensor/camera. Mask attacks require more skills to be well executed than the previous attacks, and also access to extra information in order to construct a realistic mask of the genuine user [28].

There are different types of masks depending on the complexity of the manufacturing process and the amount of data that is required. Some examples, ordered from simpler to more complex are:

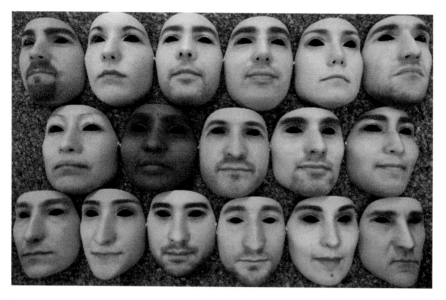

Fig. 9.3 Example of 3D masks. These are the 17 hard-resin facial masks used to create the 3DMAD dataset, from [13]

- The simplest method is to print a 2D photograph of the user's face and then stick it to a deformable structure. Examples of this type of structures could be a t-shirt or a plastic bag. Finally, the attacker can put the bag on his face and present it to the biometric sensor. This attack can mimic some deformable patterns of the human face, allowing to spoof some low-level 3D face recognition systems.
- Image reconstruction techniques can generate 3D models from two or more pictures of the genuine user's face, e.g., one frontal photo and a profile photo. Using these photographs, the attacker could be able to extrapolate a 3D reconstruction of the real face (see Fig. 9.2). This method is unlikely to spoof top-level 3D face recognition systems, but it can be an easy and cheap option to spoof a high number of standard systems.
- A more sophisticated method consists in making directly a 3D capture of a genuine user's face [29] (see Fig. 9.3). This method entails a higher level of difficulty than the previous ones since a 3D acquisition can be done only with dedicated equipment and it is complex to obtain without the cooperation of the end user. However, this is becoming more feasible and easier with the new generation of affordable 3D acquisition sensors [30].

When using any of the two last methods, the attacker would be able to build a 3D mask with the model he has computed. Even though the price of 3D printing devices is decreasing, 3D printers with sufficient quality and definition are still expensive. See reference [29] for a recent work evaluating face attacks with 3D-printed masks. There are some companies where such 3D face models may be obtained for a reasonable price.[1]

This type of attack may be more likely to succeed due to the high realism of the spoofs. As the complete structure of the face is imitated, it becomes difficult to find effective countermeasures. For example, the use of depth information becomes inefficient against this particular threat.

These attacks are far less common than the previous two categories because of the difficulties mentioned above to generate the spoofs. Despite the technical complexity, mask attacks have started to be systematically studied thanks to the acquisition of the first specific databases which include masks of different materials and sizes [13, 28, 29, 31].

9.3 Presentation Attack Detection

Face recognition systems try to differentiate between genuine users, not to determine if the biometric sample presented to the sensor is real or a fake. A presentation attack detection method is usually accepted to be any technique that is able to automatically distinguish between real biometric traits presented to the sensor and synthetically produced artifacts.

This can be done in four different ways [6]: (i) with available sensors to detect in the signal any pattern characteristic of live traits, (ii) with dedicated hardware to detect an evidence of liveness, which is not always possible to deploy, (iii) with a challenge response method where a presentation attack can be detected by requesting the user to interact with the system in a specific way, or (iv) employing recognition algorithms intrinsically robust against attacks.

Due to its easiness of deployment, the most common countermeasures are based on employing the already existing hardware and running software PAD algorithms over it. A selection of relevant PAD works based on software techniques are shown in Table 9.1. A high number of the software-based PAD techniques are based on liveness detection without needing any special help of the user. This type of approach is really interesting as it allows to upgrade the countermeasures in existing systems without the requirement of new pieces of hardware, and permitting authentication to be done in real time as it does not need user interaction. These presentation attack detection techniques aim to detect physiological signs of life (such as eye blinking, facial expression changes, mouth movements, etc.), or any other differences between presentation attack artifacts and real biometric traits (e.g., texture and deformation).

There are works in the literature that use special sensors such as 3D scanners to verify that the captured faces are not 2D (i.e., flat objects) [32], or thermal sensors to detect the temperature distribution associated with real living faces [33]. However, these approaches are not popular, even though they tend to achieve higher presentation

---

[1]http://real-f.jp, http://www.thatsmyface.com, https://shapify.me, and http://www.sculpteo.com.

Table 9.1 Selection of relevant works in software-based face PAD

Method	Year	Type of images	Database used	Type of features
[34]	2009	Visible and IR photo	Private	Color (reflectance)
[24]	2011	RGB video	PRINT-ATTACK	Face background motion
[12]	2012	RGB video	REPLAY-ATTACK	Texture based
[35]	2013	RGB photo and video	NUAA PI, PRINT-ATTACK and CASIA FAS	Texture based
[36]	2013	RGB photo and video	PRINT-ATTACK and REPLAY ATTACK	Texture based
[23]	2013	RGB video	PHOTO ATTACK	Motion correlation analysis
[37]	2014	RGB video	REPLAY-ATTACK	Image quality based
[38]	2015	RGB video	Private	Color (challenge reflections)
[39]	2016	RGB video	3DMAD and private	rPPG (color based)
[40]	2017	RGB video	OULU-NPU	Texture based
[41]	2018	RGB and NIR video	3DMAD and private	rPPG (color based)

detection rates, because in most systems the required hardware is expensive and not broadly available.

9.3.1 PAD Methods

The software-based PAD methods can be divided into two main categories depending on whether they take into account temporal information or not: static and dynamic analysis.

9.3.1.1 Static Analysis

This subsection refers to the development of techniques that analyze static features like the facial texture to discover unnatural characteristics that may be related to presentation attacks.

The key idea of the texture-based approach is to learn and detect the structure of facial micro-textures that characterize real faces but not fake ones. Micro-texture analysis has been effectively used in detecting photo attacks from single face images: extraction of texture descriptions such as Local Binary Patterns (LBP) [12] or Gray-Level Co-occurrence Matrices (GLCM) followed by a learning stage to perform discrimination between textures.

Another group of methods exploits the fact that the printing of an image to create a spoof usually introduces quality degradation in the sample, making it possible to distinguish between a genuine access attempt and an attack, by analyzing their textures [37].

The major drawback of texture-based presentation attack detection is that high-resolution images are required in order to extract the fine details from the faces that are needed for discriminating genuine faces from presentation attacks. These countermeasures will not work properly with bad illumination conditions that make the captured images to have bad quality in general.

Most of the time, the differences between genuine faces and artificial materials can be seen in images acquired in the visual spectrum with or without a preprocessing stage. However, sometimes, a translation to a more proper feature space [42], or working with images from outside the visible spectrum [43] is needed in order to distinguish between real faces and spoof attack images.

Additionally to the texture, there are other properties of the human face and skin that can be exploited to differentiate between real and fake samples. Some of these properties are absorption, reflection, scattering, and refraction [34].

This type of approaches may be useful to detect photo attacks, video attacks, and also mask attacks, since all kinds of spoofs may present texture or optical properties different than real faces.

9.3.1.2 Dynamic Analysis

These techniques have the target of distinguishing presentation attacks from genuine access attempts based on the analysis of motion. The analysis may consist of detecting any physiological sign of life, for example, pulse, eye blinking, facial expression changes, or mouth movements. This objective is achieved using knowledge of the human anatomy and physiology.

As stated in Sect. 9.2, photo attacks are not able to reproduce all signs of life because of their static nature. However, video attacks and mask attacks can emulate blinking, mouth movements, etc. Related to these types of presentation attacks, it can be assumed that the movement of the presented artifacts differs from the movement of real human faces which are complex nonrigid 3D objects with deformations.

One simple approximation to this type of countermeasures consists in trying to find correlations between the movement of the face and the movement of the background respect to the camera [23, 27]. If the fake face presented contains also a piece of fake background, the correlation between the movement of both regions should be high. This could be the case of a replay attack, in which the face is shown on the screen of some device. This correlation in the movements allows to evaluate the degree of synchronization within the scene during a defined period of time. If there is no movement, as in the case of a fixed support attack, or too much movement, as in a hand-based attack, the input data is likely to come from a presentation attack. Genuine authentication will usually have uncorrelated movement between the face and the background, since the user's head generally moves independently from the background.

Some works on dynamic analysis for face liveness detection are [44] or [35], which exploit the fact that humans blink on average three times per minute and analyzed videos to develop an eye blink-based presentation attack detection scheme.

Other works like [36] provide more evidence of liveness using Eulerian video magnification [45] applying it to enhance small changes in face regions, that often go unnoticed. Some changes that are amplified thanks to this technique are, for example, small color and motion changes on the face caused by the human blood flow, by finding peaks in the frequency region that corresponds to the human heartbeat rate.

As mentioned above, motion analysis approaches usually require some level of motion between different head parts or between the head and the background. Sometimes this can be achieved through user cooperation [38]. Therefore, some of these techniques can only be used in scenarios without time requirements as they may need time for analyzing a piece of video and/or for recording the user's response to a command. Due to the nature of these approaches, some videos and well-performed mask attacks may deceive the countermeasures.

9.4 Face Presentation Attacks Databases

In this section, we overview some publicly available databases for research in face PAD. The information contained in these datasets can be used for the development and evaluation of new face PAD techniques against presentation attacks.

As it has been mentioned in the past sections, with the recent spread of biometric applications, the threat of presentation attacks has grown, and the biometric community is starting to acquire large and complete databases to make recognition systems more robust to presentation attacks.

International competitions have played a key role to promote the development of PAD measures. These competitions include the IJCB 2017 Competition on Generalized Face Presentation Attack Detection in Mobile Authentication Scenarios [46], and the 2011 and 2013 2D Face Anti-Spoofing contests [17, 47].

Despite the increasing interest of the community in studying the vulnerabilities of face recognition systems, the availability of PAD databases is still scarce. The acquisition of new datasets is highly difficult because of two main reasons:

- Technical aspects: the acquisition of presentation attack data offers additional challenges to the usual difficulties encountered in the acquisition of standard biometric databases [48] in order to correctly capture similar fake data than the present in real attacks (e.g., generation of multiple types of artifacts).
- Legal aspects: as in the face recognition field in general, data protection legislation limits the sharing of biometric databases among research groups. These legal restrictions have forced most laboratories or companies working in the field of presentation attacks to acquire their own datasets usually small and limited.

In the area of face recognition PAD, we can find the following public databases:

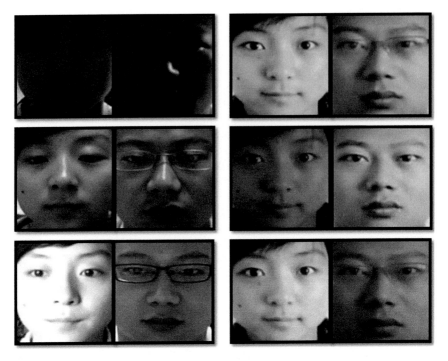

Fig. 9.4 Samples from the NUAA Photo Imposter Database [11]. Samples from two different users are shown. Each row corresponds to a different session. In each row, the left pair are from a live human and the right pair from a photo fake. Images have been taken from [11]

- The NUAA Photo Imposter Database (NUAA PI DB) [11] was one of the first efforts to generate a large public face PAD dataset. It contains images of real access attempts and print attacks of 15 users. The images contain frontal faces with a neutral expression captured using a webcam. Users were also told to avoid eye blinks. The attacks are performed using printed photographs on photographic paper. Examples from this database can be seen in Fig. 9.4. The NUAA PI DB is property of the Nanjing University of Aeronautics and Astronautics, and it can be obtained at http://parnec.nuaa.edu.cn/xtan/data/nuaaimposterdb.html.

- The YALE-RECAPTURED DB [49] appeared shortly after, adding to the attacks of the NUAA PI DB also the difficulty of varying illumination conditions as well as considering LCD spoofs, not only printed photo attacks. The dataset consists of 640 static images of real access attempts and 1920 attack samples, acquired from 10 different users. The YALE-RECAPTURED DB is a compilation of images from the NUAA PI DB and the Yale Face Database B made by the University of Campinas.

- The PRINT-ATTACK DB [24] represents another step in the evolution of face PAD databases, both in terms of the size (50 different users were captured) and of the types of data acquired (it contains video sequences instead of still images). It only

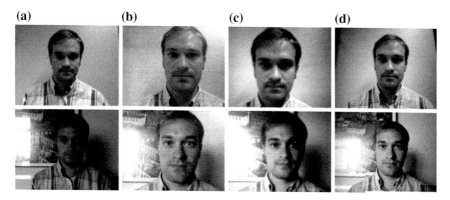

Fig. 9.5   Examples of real and fake samples from the REPLAY-ATTACK DB [12]. The images come from videos acquired in two illumination and background scenarios (controlled and adverse). The first row belongs to the controlled scenario while the second row represents the adverse condition. **a** Shows real samples, **b** shows samples of a printed photo attack, **c** corresponds to a LCD photo attack, and **d** to a high-definition photo attack

considers the case of photo attacks. It consists of 200 videos of real accesses and 200 videos of print attack attempts from 50 different users. Videos were recorded under two different background and illumination conditions. Attacks were carried out with hard copies of high-resolution photographs of the 50 users, printed on plain A4 paper. The PRINT-ATTACK DB is property of the Idiap Research Institute, and it can be obtained at https://www.idiap.ch/dataset/printattack.

– The PHOTO ATTACK database [23] is an extension of the PRINT-ATTACK database. It also provides photo attacks with the difference that the attack photographs are presented to the camera using different devices such as mobile phones and tablets. It can be obtained at https://www.idiap.ch/dataset/photoattack.

– The REPLAY-ATTACK database [12], is also an extension of the PRINT-ATTACK database. It contains short videos of both real access and presentation attack attempts of 50 different subjects. The attack attempts present in the database are video attacks using mobile phones and tablets. The attack attempts are also distinguished depending on how the attack device is held: hand-based and fixed support. Examples from this database can be seen in Fig. 9.5. It can be obtained at https://www.idiap.ch/dataset/replayattack.

• The CASIA FAS DB [18], similarly to the REPLAY-ATTACK database contains photo attacks with different supports (paper, phones, and tablets) and also replay video attacks. The main difference with the REPLAY-ATTACK database is that while in the REPLAY DB only one acquisition sensor was used with different attacking devices and illumination conditions, the CASIA FAS DB was captured using sensors of different quality under uniform acquisition conditions. The CASIA FAS DB is property of the Institute of Automation, Chinese Academy of Sciences (CASIA), and it can be obtained at http://www.cbsr.ia.ac.cn/english/Databases.asp.

- The 3D MASK-ATTACK DB [13], as its name indicates, contains information related to mask attacks. As described above, all previous databases contain attacks performed with 2D artifacts (i.e., photo or video) that are very rarely effective against systems capturing 3D face data. The attacks in this case were performed with real-size 3D masks manufactured by ThatsMyFace.com[2] for 17 different subjects. For each access attempt, a video was captured using the Microsoft Kinect for Xbox 360, that provides RGB data and also depth information. That allows to evaluate both 2D and 3D PAD techniques, and also their fusion [29]. Example masks from this database can be seen in Fig. 9.3. The 3D MASK-ATTACK DB is property of the Idiap Research Institute, and it can be obtained at https://www.idiap.ch/dataset/3dmad.

- The OULU-NPU DB [40], is a recent dataset that contains information of PAD attacks acquired with mobile devices. Nowadays mobile authentication is one of the most relevant scenarios due to the widespread use of smartphones. However, in most datasets, the images are acquired in constrained conditions. This type of data may present motion, blur, and changing illumination conditions, backgrounds and head poses. The database consists of 5940 videos of 55 subjects recorded in 3 distinct illumination conditions, with 6 different smartphone models. The resolution of all videos is 1920×1080 including print and video replay attacks. The OULU-NPU DB is property of the University of Oulu, it has been used in the IJCB 2017 Competition on Generalized Face Presentation Attack Detection [46], and it can be obtained at https://sites.google.com/site/oulunpudatabase/.

In Table 9.2 we show a comparison of the most relevant features of the above-mentioned databases.

9.5 Integration with Face Recognition Systems

In order to create a face recognition system resistant to presentation attacks, the proper PAD techniques have to be selected. After that, the integration of the PAD countermeasures with the face recognition system can be done at different levels, namely score level or decision-level fusion [50].

The first possibility consists of using score level fusion as shown in Fig. 9.6. This is a popular approach due to its simplicity and the good results given in fusion of multimodal biometric systems [51–53]. In this case, the biometric data enter at the same time to both the face recognition system and the PAD system, and each one computes their own scores. Then, the scores from each system are combined into a new final score that is used to determine if the sample comes from a genuine user or not. The main advantage of this approach is its speed, as both modules, i.e., the PAD

---

[2]http://www.thatsmyface.com/.

Table 9.2 Features of the main public databases for research in face PAD. Comparison of the most relevant features of each of the databases described in this chapter

Database	Users # (real/fakes)	Samples # (real/fakes)	Attack types	Support	Attack illumination
NUAA PI [11]	15/15	5,105/7,509	Photo	Held	Uncont.
YALE-RECAPTURED [49]	10/10	640/1,920	Photo	Held	Uncont.
PRINT-ATTACK[a] [12, 23, 24]	50/50	200/1,000	Photo and video	Held and fixed	Cont. and Uncont.
CASIA FAS [18]	50/50	150/450	Photo and video	Held	Uncont.
3D MASK-ATTACK [13]	17/17	170/85	Mask	Held	Cont.
OULU-NPU [40]	55/55	1,980/3,960	Photo and video	Mobile	Uncont.

[a]Containing also PHOTO-ATTACK DB and REPLAY-ATTACK DB

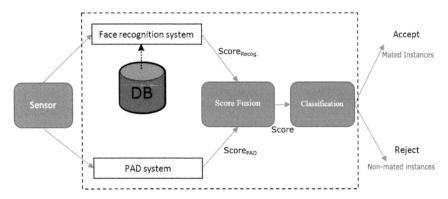

Fig. 9.6 Scheme of a parallel score level fusion between a PAD and a face recognition system. In this type of scheme, the input biometric data is sent at the same time to both the face recognition system and the PAD system, and each one generates an independent score, then the two scores are fused to take one unique decision

and face recognition modules, perform their operations at the same time. This fact can be exploited in systems with good parallel computation specifications, such as those with multicore/multithread processors.

Another common way to combine PAD and face recognition systems is a serial scheme, as in Fig. 9.7, in which the PAD system makes its decision first, and only if the samples are determined to come from a living person, then they are processed by the face recognition system. Thanks to this decision-level fusion, the face recognition system will search for the identity that corresponds to the biometric sample knowing previously that the sample does not come from a presentation attack. On the other hand, in the serial scheme the average time for an access attempt will be longer due

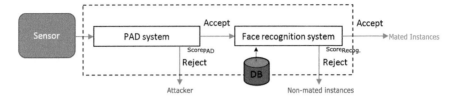

Fig. 9.7 Scheme of a serial fusion between a PAD and a face recognition system. In this type of scheme, the PAD system makes its decision first, and only if the samples are determined to come from a living person, then they are processed by the face recognition system

to the consecutive delays of the PAD and the face recognition modules. However, this approach avoids the extra work of the face recognition system in the case of a PAD attack, since the computation will end at an early stage.

9.6 Discussion

Attackers can use a great number of spoofs with no constraints, each one of different nature. Therefore, it is important to collect new databases with new scenarios in order to develop more effective PAD methods. Otherwise, it will be difficult to grant an acceptable level of security of face recognition systems. However, it is especially challenging to recreate real attacking conditions in a laboratory evaluation. Under controlled conditions, systems are tested against a restricted number of typical presentation artifacts. These restrictions make it unfeasible to collect a database with all the different fake spoofs that may be found in the real world.

Normally, PAD techniques are developed to fight against one concrete type of attack (e.g., printed photos), retrieved from a specific dataset. The countermeasures are thus designed to achieve high presentation attack detection against that particular spoof technique. However, when testing these same techniques against other types of fake artifacts (e.g., video replay attacks), usually the system is unable to efficiently detect them. There is one important lesson to be learned from this fact: there is not a superior PAD technique that outperforms all the others in all conditions; so knowing which technique to use against each type of attack is a key element. It would be interesting to use different countermeasures that have proved to be robust against particular types of artifacts, in order to develop fusion schemes that combine their results, achieving that way a high performance against a variety of presentation attacks [6, 51].

On the other hand, as technology progresses constantly, new hardware devices and software techniques continue to appear. It is important to keep track of this quick technological progress since some of the advances can be the key to develop novel and efficient presentation attack techniques. For example, focusing the research on the biological nature of biometric traits (e.g., thermogram, blood flow, etc.) should be considered [39], as the standard techniques based on texture and movement seem to be inefficient against some spoof artifacts.

9.7 Conclusions

Face recognition systems are increasingly being deployed in a diversity of scenarios and applications. Due to this widespread use, they have to withstand a high variety of attacks. Among all these threats, one with high impact is presentation attacks [6].

In this chapter, a review of the strengths and vulnerabilities of face as a biometric trait has been presented. We have described the main presentation attacks, differentiating between multiple approaches, the corresponding PAD countermeasures, and the public databases that can be used to evaluate new protection techniques [7]. The weak points of the existing countermeasures have been discussed, and also some possible future directions to deal with those weaknesses have been commented.

Due to the nature of face recognition systems, without the correct PAD countermeasures, most of the state-of-the-art systems are vulnerable to attacks. Existing databases are useful resources to study presentation attacks, but the PAD techniques developed using them might not be robust in all possible attack scenarios. The combination of countermeasures with fusion schemes [52], and the acquisition of new challenging databases could be a key asset to counterfeit the new types of attacks that could appear [29].

To conclude this introductory chapter, it could be said that even though a great amount of work has been done to fight against face presentation attacks, there are still big challenges to be faced in this topic, due to the evolving nature of the attacks, and the critical applications in which these systems are deployed in the real world.

Acknowledgements This work was done in the context of the TABULA RASA and BEAT projects funded under the 7th Framework Programme of EU, the project CogniMetrics (TEC2015-70627-R), the COST Action CA16101 (Multi-Foresee), and project Bio-Guard (Ayudas Fundación BBVA a Equipos de Investigación Científica 2017). Author J. H.-O. is supported by a FPI Fellowship from Universidad Autonoma de Madrid.

References

1. Turk MA, Pentland AP (1991) Face recognition using eigenfaces. In: Computer society conference on computer vision and pattern recognition (CVPR), pp 586–591
2. Biometrics: Market Shares, Strategies, and Forecasts, Worldwide, 2015–2021 (2015). Wintergreen Research, Inc
3. Gipp B, Beel J, Rössling I (2007) ePassport: The Worlds New Electronic Passport. Risks and its Security. CreateSpace, A Report about the ePassports Benefits
4. Garcia C (2004) Utilización de la firma electrónica en la Administración española iv: Identidad y firma digital. El DNI electrónico, Administración electrónica y procedimiento administrativo
5. Jain AK, Li SZ (2011) Handbook of face recognition. Springer, Berlin
6. Hadid A, Evans N, Marcel S, Fierrez J (2015) Biometrics systems under spoofing attack: an evaluation methodology and lessons learned. IEEE Signal Process Mag 32(5):20–30
7. Galbally J, Marcel S, Fierrez J (2014) Biometric antispoofing methods: a survey in face recognition. IEEE Access 2:1530–1552

8. Gomez-Barrero M, Galbally J, Fierrez J, Ortega-Garcia J (2013) Multimodal biometric fusion: a study on vulnerabilities to indirect attacks. In: Iberoamerican congress on pattern recognition, Springer, Berlin, pp 358–365

9. Newman LH (2016). https://www.wired.com/2016/08/hackers-trick-facial-recognition-logins-photos-facebook-thanks-zuck/

10. Goodin D (2008) Get your german interior ministers fingerprint here. Register 30

11. Tan X, Li Y, Liu J, Jiang L (2010) Face liveness detection from a single image with sparse low rank bilinear discriminative model. Comput Vis–ECCV 504–517

12. Chingovska I, Anjos A, Marcel S (2012) On the effectiveness of local binary patterns in face anti-spoofing. In: IEEE BIOSIG

13. Erdogmus N, Marcel S (2014) Spoofing face recognition with 3D masks. IEEE Trans Inf Forensics Secur 9(7):1084–1097

14. Gonzalez Sosa E, Vera-Rodriguez R, Fierrez J, Patel V (2018) Person recognition beyond the visible spectrum: combining body shape and texture from mmW images. In: International conference on biometrics (ICB)

15. Proceedings of the IEEE international conference acoust. speech signal process. (ICASSP) (2017)

16. Proceedings of the IEEE/IAPR international joint conference biometrics (IJCB) (2017)

17. Chingovska I, Yang J, Lei Z, Yi D, Li SZ, Kahm O, Glaser C, Damer N, Kuijper A, Nouak A et al (2013) The 2nd competition on counter measures to 2D face spoofing attacks. In: International conference on biometrics (ICB)

18. Zhang Z, Yan J, Liu S, Lei Z, Yi D, Li SZ (2012) A face antispoofing database with diverse attacks. In: International conference on biometrics (ICB), pp 26–31

19. ISO: Information technology security techniques security evaluation of biometrics, ISO/IEC Standard ISO/IEC 19792:2009, 2009. International organization for standardization (2009). https://www.iso.org/standard/51521.html

20. ISO: Information technology – biometric presentation attack detection – Part 1: Framework. international organization for standardization (2016). https://www.iso.org/standard/53227.html

21. Kim J, Choi H, Lee W (2011) Spoof detection method for touchless fingerprint acquisition apparatus. Korea Patent 1(054):314

22. Dantcheva A, Chen C, Ross A (2012) Can facial cosmetics affect the matching accuracy of face recognition systems? In: 2012 IEEE Fifth international conference on biometrics: theory, applications and systems (BTAS), IEEE, pp 391–398

23. Anjos A, Chakka MM, Marcel S (2013) Motion-based counter-measures to photo attacks in face recognition. IET Biom 3(3):147–158

24. Anjos A, Marcel S (2011) Counter-measures to photo attacks in face recognition: a public database and a baseline. In: International joint conference on biometrics (IJCB), pp 1–7

25. Nguyen D, Bui Q (2009) Your face is NOT your password. BlackHat DC

26. da Silva Pinto A, Pedrini H, Schwartz W, Rocha A (2012) Video-based face spoofing detection through visual rhythm analysis. In: SIBGRAPI conference on graphics, patterns and images, pp 221–228

27. Kim Y, Yoo JH, Choi K (2011) A motion and similarity-based fake detection method for biometric face recognition systems. IEEE Trans Consum Electron 57(2):756–762

28. Liu S, Yang B, Yuen PC, Zhao G (2016) A 3D Mask Face Anti-spoofing Database with Real World Variations. In: Proceedings of the IEEE conference on computer vision and pattern recognition workshops, pp 100–106

29. Galbally J, Satta R (2016) Three-dimensional and two-and-a-half-dimensional face recognition spoofing using three-dimensional printed models. IET Biom 5(2):83–91

30. Intel: (2017). https://software.intel.com/realsense

31. Kose N, Dugelay JL (2013) On the vulnerability of face recognition systems to spoofing mask attacks. In: (ICASSP) International conference on acoustics, speech and signal processing, IEEE, pp 2357–2361

32. Lagorio A, Tistarelli M, Cadoni M, Fookes C, Sridharan S (2013) Liveness detection based on 3D face shape analysis. In: International workshop on biometrics and forensics (IWBF), IEEE
33. Sun L, Huang W, Wu M (2011) TIR/VIS correlation for liveness detection in face recognition. In: International conference on computer analysis of images and patterns, Springer, Berlin, pp 114–121
34. Kim Y, Na J, Yoon S, Yi J (2009) Masked fake face detection using radiance measurements. JOSA A 26(4):760–766
35. Yang J, Lei Z, Liao S, Li SZ (2013) Face liveness detection with component dependent descriptor. In: International conference on biometrics (ICB), pp 1–6
36. Bharadwaj S, Dhamecha TI, Vatsa M, Singh R (2013) Computationally efficient face spoofing detection with motion magnification. In: Proceedings of the IEEE conference on computer vision and pattern recognition workshops, pp 105–110
37. Galbally J, Marcel S, Fierrez J (2014) Image quality assessment for fake biometric detection: Application to iris, fingerprint, and face recognition. IEEE Trans Image Process 23(2):710–724
38. Smith DF, Wiliem A, Lovell BC (2015) Face recognition on consumer devices: reflections on replay attacks. IEEE Trans Inf Forensics Secur 10(4):736–745
39. Li X, Komulainen J, Zhao G, Yuen PC, Pietikäinen M (2016) Generalized face anti-spoofing by detecting pulse from face videos. In: 23rd international conference on pattern recognition (ICPR), IEEE, pp 4244–4249
40. Boulkenafet Z, Komulainen J, Li L, Feng X, Hadid A (2017) OULU-NPU: a mobile face presentation attack database with real-world variations. In: IEEE International conference on automatic face gesture recognition, pp 612–618
41. Hernandez-Ortega J, Fierrez J, Morales A, Tome P (2018) Time analysis of pulse-based face anti-spoofing in visible and NIR. In: IEEE CVPR computer society workshop on biometrics
42. Zhang D, Ding D, Li J, Liu Q (2015) PCA based extracting feature using fast fourier transform for facial expression recognition. In: Transactions on engineering technologies, pp 413–424
43. Gonzalez-Sosa E, Vera-Rodriguez R, Fierrez J, Patel V (2017) Exploring body shape from mmW images for person recognition. IEEE Trans Inf Forensics Secur 12(9):2078–2089
44. Pan G, Wu Z, Sun L (2008) Liveness detection for face recognition. In: Recent advances in face recognition. InTech
45. Wu HY, Rubinstein M, Shih E, Guttag J, Durand F, Freeman W (2012) Eulerian video magnification for revealing subtle changes in the world. ACM Trans Graph 31(4)
46. Boulkenafet Z, Komulainen J, Akhtar Z, Benlamoudi A, Samai D, Bekhouche S, Ouafi A, Dornaika F, Taleb-Ahmed A, Qin L, et al (2017) A competition on generalized software-based face presentation attack detection in mobile scenarios. In: International joint conference on biometrics (IJCB), pp 688–696
47. Chakka MM, Anjos A, Marcel S, Tronci R, Muntoni D, Fadda G, Pili M, Sirena N, Murgia G, Ristori M, Roli F, Yan J, Yi D, Lei Z, Zhang Z, Li SZ, Schwartz WR, Rocha A, Pedrini H, Lorenzo Navarro J, Castrilln-Santana M, Määttä J, Hadid A, Pietikäinen M (2011) Competition on counter measures to 2-D facial spoofing attacks. In: International joint conference on biometrics (IJCB)
48. Ortega-Garcia J, Fierrez J, Alonso-Fernandez F, Galbally J, Freire MR, Gonzalez-Rodriguez J, Garcia-Mateo C, Alba-Castro JL, Gonzalez-Agulla E, Otero-Muras E (2010) The multiscenario multienvironment biosecure multimodal database (BMDB). IEEE Trans Pattern Anal Mach Intell 32(6):1097–1111
49. Peixoto B, Michelassi C, Rocha A (2011) Face liveness detection under bad illumination conditions. In: International conference on image processing (ICIP), pp 3557–3560
50. Chingovska I, Anjos A, Marcel S (2013) Anti-spoofing in action: joint operation with a verification system. In: Proceedings of the IEEE conference on computer vision and pattern recognition workshops, pp 98–104
51. de Freitas Pereira T, Anjos A, De Martino JM, Marcel S (2013) Can face anti-spoofing countermeasures work in a real world scenario? In: International conference on biometrics (ICB), pp 1–8

52. Fierrez J, Morales A, Vera-Rodriguez R, Camacho D (2018) Multiple classifiers in biometrics. Part 1: Fundamentals and review. Inf Fusion 44:57–64
53. Ross AA, Nandakumar K, Jain AK (2006) Handbook of multibiometrics, Springer

Chapter 10
Recent Advances in Face Presentation Attack Detection

Sushil Bhattacharjee, Amir Mohammadi, André Anjos and Sébastien Marcel

Abstract The undeniable convenience of face recognition (FR) based biometrics has made it an attractive tool for access control in various application areas, from airports to remote banking. Widespread adoption of face biometrics, however, depends on the perception of robustness of such systems. One particular vulnerability of FR systems comes from presentation attacks (PA), where a subject **A** attempts to impersonate another subject **B**, by presenting, say, a photograph of **B** to the biometric sensor (i.e., the camera). PAs are the most likely forms of attacks on face biometric systems, as the camera is the only component of the biometric system that is exposed to the outside world. Presentation attack detection (PAD) methods provide an additional layer of security to FR systems. The first edition of the Handbook of Biometric Anti-Spoofing included two chapters on face-PAD. In this chapter we review the significant advances in face-PAD research since the publication of the first edition of this book. In addition to new face-PAD methods designed for color images, we also discuss advances involving other imaging modalities, such as near-infrared (NIR) and thermal imaging. Research on detecting various kinds of attacks, both planar as well as involving three-dimensional masks, is reviewed. The chapter also summarizes a number of recently published datasets for face-PAD experiments.

S. Bhattacharjee (✉) · A. Mohammadi · A. Anjos · S. Marcel
Biometrics Security and Privacy Group, Idiap Research Institute,
Centre du Parc, Rue Marconi 19, PO Box 592, 1920 Martigny, Switzerland
e-mail: sushil.bhattacharjee@idiap.ch

A. Mohammadi
e-mail: amir.mohammadi@idiap.ch

A. Anjos
e-mail: andre.anjos@idiap.ch

S. Marcel
e-mail: sebastien.marcel@idiap.ch

© Springer Nature Switzerland AG 2019
S. Marcel et al. (eds.), *Handbook of Biometric Anti-Spoofing*,
Advances in Computer Vision and Pattern Recognition,
https://doi.org/10.1007/978-3-319-92627-8_10

10.1   Introduction

As pointed out by Ratha et al. [1] and many other researchers, biometrics-based access-control systems can be attacked in several ways. Most kinds of attacks on a biometric system require privileged access to the various components of the system. The biometric sensor in the system is the most susceptible to attacks, as it is the only exposed component in the system. By design, privileged access is not necessary to interact with the sensor. Attacks on the biometric sensor are called *Presentation Attacks* (PA). The ISO standard[1] for biometric Presentation Attack Detection (PAD) defines a PA as "*a presentation to the biometric data capture subsystem with the goal of interfering with the operation of the biometric system.*" An attacker, **A**, mounts a PA on a previously enrolled identity, **B**, using a *Presentation Attack Instrument* (PAI). For FR systems, common PAIs are images, videos, or even three-dimensional (3D) facial masks of the victim **B**. Such attacks fall into the category of *impersonation* attacks. It is important to note that the ISO standard also includes *obfuscation* as a kind of PA. An obfuscation attack is said to occur when the attacker attempts to spoof the biometric sensor in order to avoid being correctly recognized. Classic examples of obfuscation in face biometrics are the use of clothing or facial makeup, or a mask, to avoid identification by a FR system.

PAD is an crucial component in any secure biometric system. The first edition of this handbook included a comprehensive chapter describing the approaches face-PAD. In this chapter we review advances in face-PAD research since the publication of the first edition. Specifically, we review significant works in face-PAD published since the year 2015. Besides discussing the significant face-PAD methods proposed in the past three years, we also describe recently published datasets useful for research on this topic.

10.1.1   Standardization Efforts

One of the most significant developments in PAD has been the formal adoption of ISO standards. Among other things, the standard defines several metrics for reporting experimental results. The metrics relevant to this chapter are listed below:

- IAPMR the *Impostor Attack Presentation Match Rate* quantifies the vulnerability of a biometric system, and is given as the proportion of impostor attack presentations that are incorrectly accepted by the biometric security system,
- APCER: *Attack Presentation Classification Error Rate* gives the proportion of PAs that is accepted by the system in question, and,
- BPCER: *Bona Fide Presentation Classification Error Rate* specifies the proportion of *bona fide* presentations that are incorrectly rejected by the system as PA.

---

[1]ISO/IEC 30107-1: 2016 Part 1.

Note that the IAPMR is computed in the licit scenario (the scenario where PAs are not expected, and every presentation is considered *bona fide*), whereas APCER and BPCER are computed in the PA scenario. There is a further subtlety to be taken into account when computing the APCER in a given experiment, namely, that APCER values should be computed separately for each PAI. In other words, for a FR system, separate APCER values should be determined for print attacks, video-replay attacks, 3D-mask attacks, and so on. If an experiment includes attacks based on different PAIs, that is, if a certain test dataset contains PAs involving different kinds of PAIs, then the APCER corresponding to the PAI that is the most expensive (in terms of cost, as well as manufacturing effort) should be specified as the overall APCER achieved in the experiment. It is often more practical to report the BPCER when the APCER is no greater than a preset value, for example BPCER @ APCER = 10% (sometimes abbreviated as BPCER10).

10.1.2 Structure of the Chapter

The remainder of the chapter is organized in four sections. In Sect. 10.2 we discuss some recent studies on the vulnerability of FR systems to PAs. This section highlights the importance of continuing research and development of face-PAD technology. Following the discussion on vulnerability, a range of recent research publications relevant to face-PAD are summarized in Sect. 10.3. To facilitate comparison with the state of the art, most research publications on face-PAD include results on publicly available datasets. As technology for mounting PAs improves, new datasets are needed to evaluate the performance of face-PAD algorithms. Section 10.4 presents a number of recent public datasets for face-PAD experiments. We end the chapter with concluding remarks in Sect. 10.5.

10.2 Vulnerability of FR Systems to PA

FR systems are explicitly trained to handle session variability, that is, variability due to changes in scale, orientation, illumination, facial expressions, and to some extent even makeup, facial grooming, and so on. This capacity to deal with session variability also opens the door to presentation attacks. In 2016, a wide-ranging European project (TABULA RASA[2]) hypothesized that the higher the efficacy of a FR system in distinguishing between *genuine* and *zero-effort-impostor* (ZEI) presentations, the more vulnerable the system is to PAs. Several studies investigating the vulnerability to PAs of various FR systems, under different scenarios, have provided quantitative evidence that most FR schemes are very vulnerable in this respect.

---

[2]http://www.tabularasa-euproject.org/.

Hadid [2] analyses the vulnerability of a FR system that uses a parts-based Gaussian mixture model (GMM). His experiments show that when the false rejection rate (FRR) is constrained to 0.1%, the presence of spoof attacks causes the false acceptance rate (FAR) of the trained GMM is 80%. In standardized metric terms, for this GMM-FR system, the IAPMR @ FAR = 0.1% is 80%.

Ramachandra et al. [3] report on the vulnerability of a FR system relying on presentations in different spectral ranges. Their study is based on the Sparse Representation based Classifier (SRC) [4]. They capture 2D color-print PAIs (color face images printed on two types of printers: laser, and inkjet) in several wavelength bands, ranging from visible light (RGB) to near-infrared (NIR) (specifically, at the following seven wavelengths: 425, 475, 525, 570, 625, 680, and 930 nm). Evaluating the vulnerability in individual bands separately, they show that in almost all cases the chosen FR system shows very high vulnerability (IAPMR in the range of 95–100%). Only in one case, namely, laser-printed PAIs captured in the 930 nm wavelength, does the IAPMR drop to acceptable levels (IAPMR = 1.25%). This experimental result is consistent with the finding that the reflectance of facial skin dips sharply in a narrow spectral-band around 970 nm [5].

Deep-learning based FR systems are now considered the state of the art. In the current decade convolutional neural networks (CNN) based FR systems have achieved near-perfect FR performance [6–8] on highly unconstrained datasets, such as the well known Labeled Faces in the Wild (LFW) dataset [9]. Mohammadi et al. [10] have studied the vulnerability of several CNN-FR systems. Their study, based on several publicly available PAD datasets, shows that CNN-FR systems are in fact more vulnerable (IAPMR up to 100%) to PAs than older FR methods.

One class of PAs not often considered is the morphed-image attack [11, 12]. Here, face images of two different subjects, say, **A** and **B**, are morphed into a single image. The morphed image is constructed to resemble both subjects sufficiently closely to pass a quick visual inspection. Then, if, say, subject **A** wishes to avoid detection at an international border, he may alter his passport using such a morphed image to impersonate **B**. Ramachandra et al. [13] have shown, using a commercial off-the-shelf (COTS) FR system, that vulnerability of FR systems to morphed-image attacks may be as high as 100%.

10.3 Recent Approaches to Face-PAD

It is not straightforward to impose a neat taxonomy on existing face-PAD approaches. Chingovska et al. [14] group face-PAD methods into three categories: motion based, texture based, and image quality based. Other works [15] have considered image quality based face-PAD methods as a subclass of texture-based methods. In the hierarchical organization of face-PAD methods offered by Ramachandra and Busch [16] the most general (top level) groups are "hardware-based" and "software-based".

Here, we do not propose any specific taxonomy of face-PAD methods. To provide some order to our discussion, however, we have organized our survey of recent face-PAD methods in several sections: methods that operate on visible-light imagery, methods that rely on inputs captured in wavelengths outside the visible-range of light, and a separate category of methods designed to detect 3D-mask-based attacks. In the following discussion, the term *extended-range (ER) imagery* refers to data captured in wavelengths outside the visible-range of light.

10.3.1 Visible-Light Based Approaches

A majority of studies on face-PAD so far have relied exclusively on visible-light imagery (commonly called color imagery) as input. The term *visible light* here refers to the range of the electromagnetic spectrum – approximately from 380 to 750 nm – that is typically perceptible by the human visual system. One reason for the use of color imagery is that the proliferation of high-quality and low-cost color cameras has made digital color imagery widely accessible. Another reason is the need for face-PAD on mobile devices such as laptops, smartphones, and tablet devices. With the sharp increase in the use of mobile devices in sensitive applications such as remote banking and online education, secure identity-verification on such devices has become a critical issue. Although recently some companies have introduced products that include NIR cameras, a large majority of mobile devices still come with only color cameras. It is, therefore, important to continue developing face-PAD methods that can function with only color imagery as input.

Rudd et al. [17] have demonstrated that a low-cost polarization filter (based on twisted nematic liquid crystal (TNLC) in this case) can easily detect common kinds of 2D PAs, such as print attacks and digital-replay attacks. In appropriately polarized images, *bona fide* face presentations are clearly visible as faces, whereas 2D attack presentations are not.

Successful application of histograms of local binary patterns (LBP) coefficients to the problem of face-PAD [14, 18, 19] has made LBP and its various variants a mainstay for face-PAD. Initial LBP based methods for face-PAD relied on gray-level images. Boulkenafet et al. [20, 21] have used LBP features to characterize color-texture. For a given color image in RGB color-space, they first generate the YC_bC_r as well as HSV representations of the image. Uniform LBP histograms are then computed on the Y, C_b, C_r, H, S, and V components and concatenated together to generate the final feature-vector representing the input color image. These color-texture feature-vectors may be classified using support vector machines (SVM). Boulkenafet et al. have shown that color-texture features outperform gray-level LBP features in the face-PAD task [20]. In a separate work [21], they have also shown that this color-texture representation leads to significantly better generalization to unknown attacks, compared to other hand-crafted face-PAD features. Indeed, in a recent face-PAD competition [22], the winning entry also combined motion information with color-texture information using LBP histograms.

Notwithstanding the success of LBP-based methods, in the past 3 years researchers have also explored other approaches for face-PAD. Prominent recent works using color imagery have focused on a variety of features characterizing local motion, local texture, and more generally, image quality. Wen et al. [23] propose several features for image distortion analysis (IDA) to tackle the problem of face-PAD for 2D (print and video-replay) attacks. Their features characterize the color-diversity, image-sharpness and the presence of specular regions in the input images. The IDA features are computed only over the face region (i.e., on the output of the face-detection step), and are classified using a two-class SVM classifier. The authors present results on several public datasets, including a new dataset (MSU-MFSD, see Sect. 10.4) introduced in this paper. In intra-database experiments the IDA features perform competitively to other face-PAD approaches. Cross-dataset experiments [23] indicate that these features show better generalization properties than previous approaches, notably when compared to LBP+SVM (i.e., LBP features classified using a SVM).

The IDA features [23] complement the image quality measures (IQM) proposed earlier by Galbally et al. [24]. The IQM features are all computed on gray-level images. The IDA features provide a way of additionally capturing information relevant to face-PAD available in the color domain.

Costa-Pazo et al. [15] have proposed a face-PAD approach using a set of Gabor features, which characterize the image-texture over the face region. This work represents the first use of Gabor features for face-PAD. Their experiments show that the Gabor features perform better than the IQM features [24] in detecting PAs. Texture information, captured using shearlets, has also been exploited in the method proposed by Li et al. [25].

Certain face-PA cues are not as consistent as others. For example, the set of IDA feature-set includes several features characterizing the amount of specularity in a image. The underlying expectation is that the presence of large specular regions indicates that the input is a PA. There are, however, many instances of PAs that do not include significant specularity. Similarly, although the presence of Moiré patterns is also a strong indicator of PAs [26, 27], the absence of Moiré patterns does not rule of a PA.

Tirunagari et al. [28] exploit motion cues to detect face liveness. Specifically, they detect micro-motions, such as slight head movements, lip movements, and eye-blinks, to identify *bona fide* presentations. Unlike the work of Anjos et al. [29] – where motion information derived from optical flow computation is directly used to identify PAs – here the video is treated a three-dimensional data, and apply dynamic mode decomposition (DMD) to this 3D data. The result of the DMD procedure is an image where regions of high local micro-motion are marked with brighter pixels. The micro-texture information in the resulting image is characterized using LBP histograms, which are subsequently classified using a SVM.

In the past few years several specific research directions have attracted attention in the context of face-PAD. Unsurprisingly, the application of deep-learning methods for face-PAD has become a popular research track. The idea of personalized face-PAD, where client information is incorporated into the PAD process, has also been explored. Several works have been published on the subject of detecting obfuscation attacks. Finally, as the question of detecting previously unseen kinds of PAs becomes important, several researchers have posed face-PAD as an anomaly-detection problem. In the following sections we discuss publications on each of these topics separately.

10.3.1.1 Deep-Learning Approaches To PAD

Following the success of deep-learning based approaches for face recognition, there has been a proliferation in CNN-based approaches for face-PAD. One reason why researchers are looking into the use of deep networks for face-PAD is that as the quality of PAIs improves, it is becoming increasing difficult to design explicit hand-crafted features able to distinguish PAs from *bona fide* presentation. Face-PAD methods based on deep networks have explored both kinds of approaches, the use of network embeddings as well as end-to-end architectures. Here, we highlight a few representative works, to provide readers with a general idea about current research activities on this topic.

In one of the first works in this area, Yang et al. [30][3] have proposed a CNN with the same architecture as ImageNet [31], but with the output layer configured for only two outputs: *bona fide* or PA. In this work the authors augment the training data by using input images at multiple scales and also multiple frames of video. The trained CNN is used to extract a feature-vector (from the penultimate fully connected layer, fc7, of the network) for each input test image. The feature-vector is then classified using a two-class SVM.

More recent works on the use of CNNs for face-PAD have focused on newer CNN architectures. Lucena et al. have proposed FASNet[4] [32], a deep network for face-anti-spoofing. They start with the VGGNet16 (16-layer VGGNet [33]) and modify only the top fully connected section of the network by removing one fc-layer, and changing the sizes of the subsequent two fc-layers to 256 units and 1 unit, respectively. FASNet shows a small improvement over SpoofNet [34] on the two datasets, 3DMAD and REPLAY-ATTACK, used in both works.

Nagpal and Dubey [35] compare the performances of three different CNN architectures: the Inception-v3 [36] and two versions of ResNet [37], namely ResNet50 (a 50-layer ResNet) and ResNet152 (the 152-layer version). For each architecture, they have conducted six experiments, by training the networks with different parameter-settings. Their study is based on the MSU-MSFD dataset (see Sect. 10.4), which is a relatively small dataset. The authors augment their training data by using flipped versions of each frame in the training set as well. The best result achieved in this work is an accuracy of 97.52%, produced by the ResNet152 initialized with weights taken from the ImageNet, and where only the final densely connected layers have been re-trained using the MSU-MSFD data. Their experiments also seem to indicate that using lower learning-rates may lead to better discrimination in face-PAD tasks.

---

[3]Open source implementation available on https://github.com/mnikitin/Learn-Convolutional-Neural-Network-for-Face-Anti-Spoofing.

[4]Open-source implementation of FASNet is available on https://github.com/OeslleLucena/FASNet.

Li et al. have used a hybrid CNN [38] to model *bona fide* and attack presentations in a parts-based fashion. The face region is divided into rectangular sub-regions, and a separate two-class CNN (VGG-Face network [6]) is trained for each sub-region. Given a test image, a feature-vector is constructed by concatenating the output vectors from the last fully connected layer of each CNN. This feature-vector is then classified using a SVM.

Nguyen et al. [39] have explored the idea of combining hand-crafted features with deep-learning-based features. They train a 19-layer VGGNet [33] (with only two output classes), and take the output of the fc7 layer as a descriptor for the input test image. The descriptors from the CNN are concatenated with a multi-level LBP (MLBP) histogram, a set of hand-crafted features, to construct a combined feature-vector. Principal Component Analysis (PCA) is used as a dimensionality-reduction step to reduce the combined feature-vector to a much shorter feature-vector (reduced from 7828-D to between 90-D and 530-D depending on the dataset). Finally, the reduced feature-vectors are classified using a two-class SVM classifier.

Xu et al. [40] combine a long short-term memory (LSTM) network with a CNN to extract features that encode both temporal as well as spatial information. The input to the LSTM-CNN network is a short video, instead of individual frames. The LSTM is plugged on top of the CNN, to model the temporal information in the video. The authors show that this network can outperform straightforward CNNs, as well as various hand-crafted features.

Liu et al. [41] combine a CNN and a LSTM network for face-PAD. In this architecture, the CNN is trained on individual video-frames (images) to extract image-feature-maps as well as depth-maps of the face region. The LSTM network takes the feature-map produced by the CNN, and is trained to extract a remote photo-plethysmography (rPPG) signal from the video. They present results on the OULU-NPU dataset (see Sect. 10.4. A new dataset, named Spoof in the Wild (SiW, discussed in Sect. 10.4) is also introduced in this paper.

In general, current datasets for face-PAD are too small to train CNNs from scratch. Most works involving CNNs for face-PAD so far have adapted existing FR CNNs for face-PAD applications, using transfer-learning.

10.3.1.2 Client-Specific Face-PAD

In real-world applications PAD systems are not expected to function in isolation – a PAD system is usually deployed in conjunction with a biometric-verification system. The client-identity information available to the verification system may also be incorporated into the PAD process to improve the PAD performance. This approach to PAD has been explored in various other biometric modalities (such as for fingerprint PAD).

Chingovska and Anjos [42] have proposed client-specific face-PAD methods using both discriminative as well as generative approaches. In both cases, essentially, a separate classifier is constructed for each enrolled client. In the discriminative scheme, for each client, they train a two-class SVM in a one-versus-all configuration.

In the generative approach, GMMs are trained for each client using a cohorts-based approach to compensate for the lack of adequate numbers of PAs for each client.

Although the idea of a client-specific approach to face-PAD sounds attractive, one severely limiting factor is the cost of constructing a sufficient variety and number of PAs for every enrolled client. Indeed, the cost may quickly become prohibitive when PAs based on custom silicone 3D-masks are considered. Yang et al. [43] have also proposed a face-PAD method that incorporates client-specific information. Again, they train a separate classifier for each enrolled client. They propose an innovative solution to the problem of lack of sufficient PA samples to train classifiers for newly enrolled clients. Their solution is to use domain-adaptation to generate virtual PA samples to train the client-specific classifiers. The domain-adaptation model learns the relationship between the *bona fide* and attack presentations from the training partition of a dataset. Thereafter, the trained adaptation model is used to generate PA samples for clients in the test partition.

10.3.1.3 Obfuscation Attacks

An obfuscation attack is said to occur if the attacker actively attempts to alter one's appearance to the extent that FR systems may fail to recognize the subject. Obfuscation attacks may take the form of the use of extreme facial makeup, the use of clothing, or simple medical masks, to occlude significant portions of the face, or even the use of facial masks (mask that resemble faces) made of various materials.

In case of severe occlusion, even localizing the face region in the image (face detection) is a significant challenge. Ge et al. [44] have proposed a LLE-CNN – combining CNN-based feature-extraction with locally linear embedding (LLE) – to detect the face region even in the presence of extensive occlusion. For subjects wearing makeup, Wang and Fu [45] have proposed a method for reconstructing makeup-free face images, using local low-rank dictionary learning. Kose et al. [46] use a combination of LGBP (LBP histograms computed over a set of Gabor-filtered images) and histogram of gradients (HOG) to classify face images as containing makeup or not. Agarwal et al. [47] tackle the problem of detecting obfuscation using 3D flexible masks, that is, detecting whether the subject in the presentation is wearing a mask, using multispectral imagery. Specifically, they capture images in visible, NIR and thermal wavelength-ranges of the spectrum. Their experiments, based on a variety of local texture descriptors, show that thermal imagery is the best suited for detecting masks reliably. (The use of multispectral data for face-PAD is discussed in more detail in Sect. 10.3.2.)

The morphed-image attacks mentioned in Sect. 10.2 may be seen as a kind of obfuscation attack. Ramachandra et al. [13] have demonstrated the superiority of binarized statistical image features (BSIF) over LBP histograms in detecting morphed-image attacks.

10.3.1.4 One-Class Classification for PAD

Most researchers approach PAD as a two-class problem. That is, data is collected for both *bona fide* and attack presentations, and, using suitable feature-descriptors, a two-class classifier is trained to discriminate between *bona fide* presentations and attacks. The greatest disadvantage of this general scheme is poor generalization to unknown attacks. A recent face-PAD competition [48] showed that the performance of all entries deteriorated in the test-protocol involving unknown attacks, relative to their respective performances in test-protocols involving known attacks. Most published face-PAD methods have performed relatively poorly in cross-dataset tests (see, for example [23, 24]). The reason is that different datasets include attacks of different kinds (different PAIs, or even just different devices used for performing the attacks). Consequently, the attacks in a given dataset are very likely to be unknown to the classifier that has been trained on a different dataset. This issue – generalization to unknown attacks – has emerged as the most significant challenge in face-PAD.

Indeed, when implementing countermeasures to PAs, the goal is simply to detect PAs, and not necessarily to identify the class of the PA. The problem of PAD may therefore be formulated as one of anomaly detection, where only the *bona fide* class is modeled using a one-class classifier (OCC). In general OOCs may be grouped under two categories: generative and non-generative. A GMM modeling only the *bona fide* class is an example of a generative OCC. A one-class SVM, on the other hand, is a non-generative OCC. Arashloo and Kittler [49] have investigated the use of both kinds of OCCs for the purpose of face-PAD. They report results using a SVM as the non-generative classifier, and a SRC [4] as the generative classifier. The authors compare the performances of two-class GMM and two-class SVM with one-class GMM and one-class SVM respectively, for face-PAD. In total they have considered 20 different scenarios, that is 20 different combinations of classifiers and features. From their experiments, performed with three publicly available datasets, the authors conclude that the OCC based outlier-detection approach can perform comparably to a two-class system. More importantly, the OCC results are better than their two-class counterparts in tests involving unknown PAs (i.e., tests where certain PAs are not represented in the training dataset).

Nikisins et al. [50] have also studied the use of OCCs for face-PAD. They base their work on an aggregate dataset composed of three publicly available datasets: REPLAY-ATTACK, REPLAY-MOBILE, and MSU-MFSD (discussed in Sect. 10.4). The difference between this work and that of Arashloo and Kittler [49] is that Nikisins et al. [50] train their classifiers using the *bona fide* presentations from all three component datasets at once, where as Arashloo and Kittler use *bona fide* presentations of only one dataset at a time in a given experiment. Nikisins et al. [50] use a one-class GMM (a generative OCC) to model the distribution of *bona fide* presentations in the aggregated dataset, using a set of image-quality features [23, 24]. Their experiments also show that although two-class classifiers perform better than their one-class counterparts for known attacks (i.e., the case where samples of the attack-types have been included in the training set), their performance deteriorates sharply when presented with unknown attacks, that is PAIs that were not included in the training set.

By contrast, the one-class GMM appears to generalize better to unknown classes of PAs [50].

The advantage of using a one-class system is that only data for *bona fide* presentations is necessary. Although experimental test datasets usually include a variety of attack presentations, in real scenarios it is quite difficult to collect sufficient data for all the various possible kinds of attacks.

10.3.2   Approaches Based on Extended-Range Imagery

Broadly speaking, visible-light based approaches rely on identifying subtle qualitative differences between *bona fide* and attack presentations. As the quality (color-fidelity, resolution, and so on) of PA devices improves, distinctions between the two kinds of presentations are becoming increasing narrower. That is, progress in PAI quality impacts the performance of existing face-PAD methods. This phenomenon is concretely illustrated by Costa-Pazo et al. [15]. They apply the same face-PAD method – SVM classification using a set of image-quality measures – to two datasets. Their experiment shows that the performance of the chosen face-PAD method is significantly worse on the newer dataset (REPLAY-MOBILE [15]) than on the older (REPLAY-ATTACK [14]) dataset. The reason is that as technology (cameras, electronic screens, printers, etc.) improves, the quality of PAs in visible-light is also approaching that of *bona fide* presentations, and therefore it is becoming increasingly difficult to separate the two classes.

A new approach to face-PAD involves the use of ER imagery. Both active- as well as passive-sensing approaches have been considered in recent works. In active ER imagery, the subject is illuminated under a chosen wavelength band, for example, with NIR and Short-wave IR (SWIR) illumination, and the biometric sensor (camera) is equipped with appropriate filters, to be able to capture data only in the chosen wavelength band. In passive sensing no specific illumination is used, and the camera is designed to capture radiation in a given wavelength band. One example of passive sensing is the use of thermal cameras to capture the heat radiated by human subjects.

When using active ER imagery for face-PAD, the general idea is to model the reflectance properties of human skin at different wavelengths. Steiner et al. [51] have proposed the design of a multispectral SWIR camera for face-PAD applications. The camera captures images at four narrow wavelength bands, namely, 935, 1060, 1300, and 1550 nm. The image-sensor is sensitive in the range 900–1700 nm. The camera is equipped with a ring-illuminator consisting of LEDs emitting SWIR in the four wavelength bands of interest. During image-acquisition the camera cycles through the illumination in the different bands one by one, and synchronizes the image-capture to the duration of illumination at a given wavelength. Thus, the camera captures at multispectral stack of four images at each time interval. This camera can capture 20 stacks, or frames per second (FPS) – a significant improvement on a previous design of a SWIR camera proposed by Bourlai [52], which was able to capture image at an average rate of 8.3 FPS. Using this camera, human skin can be reliably distinguished from other materials. Steiner et al. show results demonstrating the efficacy of face-PAD using data acquired with this camera.

Ramachandra et al. [53] have used seven-band multispectral imagery for face-PAD, captured using a SpectroCam™multispectral camera. This device captures presentations in narrow bands centered at the following wavelengths: 425, 475, 525, 570, 625, 680, and 930 nm. The authors propose two face-PAD approaches based on:

- image fusion, where the seven images in a given multispectral stack are fused into a single image, and a PAD algorithm processes the fused image, and
- score fusion, where the individual images in the multispectral stack are classified separately, and the 7 scores are then fused to generate the final classification score.

Quantitative results [53] show that the score-fusion approach performs significantly better than the image-fusion approach.

Bhattacharjee and Marcel [54] have also investigated the use of ER imagery for face-PAD. They demonstrate that a large class of 2D attacks, specifically, video-replay attacks, can be easily detected using NIR imagery. In live presentations under NIR illumination the human face is clearly discernible. However, electronic display monitors appear almost uniformly dark under NIR illumination. Therefore, using NIR imagery, it is possible to design simple statistical measures to distinguish between *bona fide* presentations and attacks. This approach may also be applied to detect print-based attacks. It may fail, however, if the PAIs are printed using metallic inks. The authors also demonstrate that NIR imagery is not particularly useful in detecting 3D mask based attacks. They go on to show that thermal or imagery can be used to easily distinguish *bona fide* presentations from mask-based attacks. This is because, in a *bona fide* presentation, the heat emanating from the subject's face renders it very brightly in the thermal image. In contrast, in a mask attack, the mask appears very dark in the image, because it has a much lower temperature than the subject's body.

This direction of research is still in its infancy. One reason why research in ER imagery has not yet been widely explored is the high cost of IR and thermal cameras. In recent years, however, low-cost options such as the Microsoft , Intel's RealSense range of sensors, and inexpensive thermal cameras such as from and have become widely available. Availability of affordable hardware will be a key factor in advancing research in this direction.

10.3.3   Detection of 3D Mask Attacks

Good quality 3D masks present clear threats in both impersonation as well as obfuscation categories. As custom 3D masks become increasingly affordable, research on PAD for 3D masks is also gaining critical importance. Bhattacharjee et al. [55] have recently demonstrated empirically, that several state-of-the-art FR CNNs are significantly vulnerable to attacks based on custom silicone 3D masks (IAPMR is at least 10 times greater than the false non-match rate (FNMR)).

Initial research was directed towards detecting custom rigid masks, typically made of sandstone powder and resin, with hand-painted facial features. Publicly available datasets 3DMAD [56] and HKBU-MARs [57] contain data pertaining to custom rigid masks. More recent face-PAD research has focused on detecting attacks based on hyper-realistic flexible custom masks, usually made of silicone. Although custom silicone masks are still fairly expensive to manufacture, in the coming years the cost of creating such masks is expected to drop to affordable levels.

Another strand of research involving 3D masks is to detect obfuscation attacks mounted using readily available, generic latex masks. Agarwal et al. [58] have used texture cues characterized using a set of features computed over co-occurence matrices (so called Haralick-features) to detect rigid-mask attacks in the 3DMAD dataset [56]. Liu et al. [57] have published the more recent HKBU-MARs dataset containing images of 3D-mask based PAs. They have proposed a rPPG based approach to detecting 3D-mask PAs.

Manjani et al. [59] present an observational study into obfuscation attacks using 3D-masks. They describe PAD experiments based on the SMAD dataset (see Sect. 10.4), which consists of public-domain videos collected from the World-wide Web. Although observational studies such as this may indicate association between variables (in this case between the true labels of the test videos and the classifier-score), the influence of other confounding variables here cannot be ruled out. To demonstrate the efficacy of a method for detecting 3D-mask based PAs, it is important to design a controlled experiment to highlight exclusively the causal effect of 3D-masks on the resulting classifier-score.

10.4 New Datasets for Face-PAD Experiments

One significant reason for rapid advances in face-PAD research is the availability of publicly shared datasets, which facilitates comparison of the performance of new PAD algorithms with existing baseline results. As the quality of devices used to mount attacks improves, the older datasets tend to become less relevant. It is, therefore, important for the research community to continually collect new datasets, representing attacks created using state-of-the- art technology.

Table 10.1 lists some recently published face-PA datasets. The MSU-MFSD, UVAD, REPLAY-MOBILE, MSU-USSA, OULU-NPU, and SiW datasets contain 2D attacks captured under the visible-light illumination. The other datasets include data representing 3D attacks (HKBU-MARs and SMAD) or 2D attacks captured under non-standard illumination, such as extended-range (multispectral) imagery (MS-Face, EMSPAD and MLFP), or light-field imagery (GUC-LiFFAD). Brief descriptions of these datasets follow:

- MSU-MFSD: The public version of the MSU-MFSD dataset [23] includes real-access and attack videos for 35 subjects. Real-access videos ($\sim$12 s long) have been captured using two devices: a 13" MacBook Air (using its built-in camera),

Table 10.1 Recently published datasets for face-PAD experiments

Dataset name	Year	PAIs	Comment
MSU-MFSD [23]	2015	2D attacks: print and replay	70 *bona fide* and 210 PA videos representing 35 subjects, collected using laptop and smartphone
GUC-LiFFAD [60]	2015	2D attacks: print and replay	Light-field imagery collected from 80 subjects, using a Lytro camera For each presentation several images are collected, each at a different depth-of-focus
UVAD [61]	2015	2D attacks: video-replay	17,076 videos corresponding to 404 identities
REPLAY-MOBILE [15]	2016	2D attacks: print, replay	1200 *bona fide* and attack videos representing 40 subjects, captured using only smartphone and tablet
MSU-USSA [27]	2016	2D attacks: print, replay	1000 *bona fide* presentations and 8000 PAs representing 1000 subjects, captured using only smartphone and tablet
MS-Face [62]	2016	2D attacks: print (visible and NIR)	Based on 21 subjects print PAIs Data captured using hi-res. CMOS sensor
HKBU-MARs [57]	2016	3D attacks: rigid masks	1008 videos corresponding to 12 subjects and their masks
SMAD [59]	2017	3D attacks: silicone masks	The dataset contains 130 presentations: 65 *bona fide*, and 65 mask attacks
EMSPAD [3]	2017	2D attacks: print (laser and inkjet)	7-band multispectral data for 50 subjects
OULU-NPU [22]	2017	2D attacks: print, video-replay	5,940 videos corresponding to 55 subjects using 6 different smartphones, captured in 3 different environments
MLFP [47]	2017	3D attacks: obfuscation with latex masks	1350 videos based on 10 subjects in visible, NIR and thermal bands, captured in indoor and outdoor environments
SiW [41]	2018	2D attacks: 2 print- and 4 replay-attacks	4620 videos based on 165 subjects, captured in various head-poses and environments Replay-attacks captured using four different PAIs

and a Google Nexus 5 (Android 4.4.2) phone. Videos captured using the laptop camera have a resolution of 640 × 480 pixels, and those captured using the Android camera have a resolution of 720 × 480 pixels. The dataset also includes PA videos representing printed photo attacks, and mobile video replay-attacks where video captured on an iPhone 5s is played back on an iPhone 5s, and high-definition (HD) (1920 × 1080) video-replays (captured on a Canon 550D SLR, and played back on an iPad Air).

- GUC-LiFFAD: The GUC Light Field Face Artefact Database (GUC-LiFFAD) has been created for face-PAD experiments based on light-field imagery. Specifically, the biometric sensor used in this dataset is a Lytro[5] camera, which, for every presentation, captures several images, each at a different depth-of-focus. Data corresponding to 80 subjects is included in this dataset. Only print attacks, based on high-quality photographs (captured using a Canon EOS 550D DSLR camera, at 18 megapixel resolution, and printed on both laser and inkjet printers) are represented in this dataset.

- UVAD: The Unicamp Visual Attack Database (UVAD) consists of 17,076 *bona fide* and attack presentation videos corresponding to 404 identities. All videos have been recorded at full-HD resolution, but subsequently cropped to a size of 1366 × 768. The dataset includes *bona fide* videos collected using six different cameras. Two videos have been captured for each subject, both using the same camera but under different ambient conditions. PA videos corresponding a given subject have also been captured using the same camera as that used for the *bona fide* videos of the subject in question. The PAs have been generated using seven different electronic monitors, and all PA videos have also been cropped to the same shape as the *bona fide* videos.

- REPLAY-MOBILE: This dataset contains short (∼10 s long) full-HD resolution (720 × 1280) videos corresponding to 40 identities, recorded using two mobile devices: an iPad Mini 2 tablet and a LG-G4 smartphone. The videos have been collected under six different lighting conditions, involving artificial as well as natural illumination. Four kinds of PAs are represented in this database have been constructed using two PAIs: matte-paper for print attacks, and matte-screen monitor for digital-replay attacks. For each PAI, two kinds of attacks have been recorded: one where the user holds the recording device in hand, and the second where the recording device is stably supported on a tripod.

- MSU-USSA: The Unconstrained Smartphone Spoof Attack dataset from MSU (MSU-USSA) aggregates *bona fide* presentations from a variety of Internet-accessible sources. In total 1000 *bona fide* presentations of celebrities have been included in this dataset. Two cameras (front and rear camera of a Google Nexus 5 smartphone) have been used to collect 2D attacks using four different PAIs (laptop, tablet, smartphone, and printed photographs), resulting in a total of 8000 PAs.

- HKBU-MARs: This dataset is designed to test countermeasures for 3D rigid-mask based attacks. The second version (V2) of this dataset contains data corresponding to 12 subjects. Rigid masks created by two different manufacturers have been used

---

[5]www.lytro.com.

to construct this dataset. Presentations have been captured using seven different cameras (including mobile devices), under six different illumination conditions.

- MS-Face: This is the first public dataset to explore the use of NIR imagery for face-PAD. Specifically, data is collected under two kinds of illumination: visible-light and 800 nm (NIR) wavelengths. The dataset contains data captured from 21 subjects. *Bona fide* presentations in this dataset have been collected under five different conditions. Only print attacks have been considered in this dataset. For PAs under visible-light, high-quality color prints have been used, whereas PAs under NIR illumination have been created using gray-level images printed at 600 dpi.

- SMAD: the Silicone Mask Attack Database (SMAD) consists of videos collected from the Internet. The authors [59] have collected 65 videos of celebrities (which form the *bona fide* presentations) as well as 65 videos of actors wearing a variety of flexible masks. Although the authors refer to the masks as silicone masks, some of the masks in the dataset appear to be constructed from latex, instead of silicone. Some of the original videos collected for this dataset may be rather long. For the purposes of experiments, long videos have been trimmed, so that all videos in the dataset are between 3 and 10 s long.

- EMSPAD: the Extended Multispectral Presentation Attack Database (EMSPAD) contains images captured using a Pixelteq SpectroCam™ camera. This camera captures multispectral images using a set of filters mounted on a continuously rotating wheel. The dataset contains seven-band multispectral stacks per time-instant, that is, for each frame, 7 images have been captured in narrow wavelength bands centered at the following values: 425, 475, 525, 570, 625, 680, and 930 nm. *Bona fide* and attack presentations for 50 subjects comprise this dataset. *Bona fide* presentations have been collected in two sessions, and in each session, five frames (i.e., 5 × 7 images) have been collected for each subject. This dataset includes only one kind of PAI, namely, 2D color-print attacks. To construct the attacks, high-quality color photographs of each subject have been printed on two kinds of printers – a color laser printer, and a color inkjet printer – at 600 dpi resolution, and multispectral images of these printed photographs have been captured using the SpectroCam camera.

- OULU-NPU: This dataset includes data corresponding to 55 subjects. Front cameras of six different mobile devices have been used to capture the images included in this dataset. The images have been collected under three separate conditions, each corresponding to a different combination of illumination and background. PAs include print attacks created using two printers, as well as video-replay attacks using two different displays. In total, 4950 *bona fide* and attack videos comprise the dataset.

- MLFP: The Multispectral Latex Mask based Video Face Prepresentation Attack (MLFP) dataset has been prepared for experiments in detecting obfuscation attacks using flexible latex masks. The dataset consists of 150 *bona fide* and 1200 attack videos, corresponding to 10 subjects. In fact the attacks have been performed using seven latex masks and three paper masks. Data has been collected in both indoor and outdoor environments.

- SiW: The Spoof in the Wild dataset consists of 1320 *bona fide* videos captured from 165 subjects, and 3300 attack videos. Liu et al. [41] mention that the dataset encapsulates greater racial diversity than previous datasets. Varying ambient conditions, as well as different facial expressions and head-poses are also represented in the SiW dataset. Two kinds of print attacks and four kinds of video replay-attacks have been included in this dataset. Replay-attacks have been created using four PAIs: two smartphones, a tablet device, and a laptop-monitor screen.

For detailed descriptions of the datasets, such as the experimental protocols as well as how to access the datasets, the reader is referred to the respective references cited in Table 10.1.

10.5 Conclusion

As several studies have quantitatively demonstrated, modern face recognition (FR) methods are highly susceptible to presentation attacks (PA). This vulnerability is a consequence of the desired ability of FR methods to handle inter-session variability. In order to have secure face-verification systems, the underlying FR methods need to be augmented with appropriate presentation attack detection (PAD) methods. Consequently, face-PAD has become a topic of intense research in recent years. In this chapter we have attempted to summarize several prominent research directions in this field.

A large majority of face-PAD methods operate on color imagery. Several new kinds of features characterizing local motion information, image quality, as well as texture information have been proposed in the recent scientific literature. Deep-learning based methods for face-PAD have also been widely explored. Most works involving deep-learning methods have started with a CNN designed for FR, and have adapted the network for face-PAD using transfer-learning. The reason for this approach is that current face-PAD datasets are still too small to train really deep networks from scratch. Given this constraint on the size of available training data, perhaps researchers should investigate the use of relatively smaller networks for face-PAD.

In addition to well studied categories of 2D attacks, namely, print attacks and video-replay attacks, several research groups are now developing methods to detect attacks performed using hyper-realistic custom-made masks. Attacks based on both rigid and flexible masks have been considered. In the past this category of attacks did not receive much attention as constructing custom-masks was prohibitively expensive. Although, even today the cost of manufacturing high-quality custom masks remains high, the costs have come down significantly, and we may expect PAs based on such masks to be highly likely in the near future. The research community would benefit from a concerted effort to produce large and significantly diverse datasets based on a variety of custom-made masks.

Extended-range (ER) imagery, that is, imagery in wavelengths outside the visible-light spectrum, is proving to be a valuable tool in tackling both 2D and 3D PAs. Given the availability of low-cost infrared and thermal cameras, this is a promising direction of research in face-PAD.

Besides impersonation attacks, the recently adopted ISO standard for PAD also considers obfuscation attacks as PAs. Specifically, there is a need to detect presentations where makeup or a mask is used to hide one's identity. This category of PA has not received the same amount of attention as impersonation attacks. The availability of carefully constructed datasets representing obfuscation attacks is key to the progress of research on this topic.

We note, in general, that most recent papers on face-PAD still report results on relatively old datasets, such as CASIA and REPLAY-ATTACK – datasets that are more than 5 years old now. With ever-improving technology for constructing PAs, older datasets become increasingly irrelevant. In order to have the true snapshot of the state of the art, besides publishing new datasets at a steady rate, it is also important that face-PAD researchers report results on recent datasets.

Although most state-of-the-art face-PAD methods seem to perform well in intra-dataset tests, generalization in cross-dataset scenarios remains a significant challenge. Cross-dataset generalization is an important goal, because it indicates the ability of a given PAD method to tackle previously unseen attacks. In this context the use of one-class classifiers (OCC) have been shown to be a step in the right direction.

There is a growing interest in developing face-PAD methods for scenarios involving previously unseen attacks. We expect this trend to grow in the coming years. Another research direction with great potential is the use of ER imagery to tackle various kinds of PAs. So far, deep-learning based methods for face-PAD have been shown to be roughly as accurate as state-of-the-art methods relying on hand-crafted features. As mentioned earlier, current efforts involving deep learning start with well understood deep networks designed for object recognition or FR. Further research is required in this area, perhaps involving[6] bespoke deep architectures for face-PAD.

Acknowledgements This work has been supported by the European H2020-ICT project TeSLA (grant agreement no. 688520), the project on Secure Access Control over Wide Area Networks (SWAN) funded by the Research Council of Norway (grant no. IKTPLUSS 248030/O70), and by the Swiss Center for Biometrics Research and Testing.

References

1. Ratha NK, Connell JH, Bolle RM (2001) An analysis of minutiae matching strength. In: Bigun J, Smeraldi F (eds) Audio- and video-based biometric person authentication. Springer, Heidelberg, pp 223–228
2. Hadid A (2014) Face biometrics under spoofing attacks: vulnerabilities, countermeasures, open issues, and research directions. In: Proceedings of IEEE computer society conference on computer vision and pattern recognition workshops (cVPRW), pp 113 – 118
3. Ramachandra R, Raja KB, Venkatesh S, Cheikh FA, Büsch C (2017) On the vulnerability of extended multispectral face recognition systems towards presentation attacks. In: Proceedings

---

[6]www.tesla-project.eu.

of IEEE international conference on identity, security and behavior analysis (ISBA), pp 1 – 8. https://doi.org/10.1109/ISBA.2017.7947698

4. Wright J, Yang AY, Ganesh A, Sastry SS, Ma Y (2009) Robust face recognition via sparse representation. IEEE Trans Pattern Anal Mach Intell 31(2):210–227. https://doi.org/10.1109/TPAMI.2008.79

5. Kanzawa Y, Kimura Y, Naito T (2011) Human skin detection by visible and near-infrared imaging. In: Proceedings of the 12th IAPR conference on machine vision applications, MVA 2011. Nara, Japan

6. Parkhi OM, Vedaldi A, Zisserman A (2015) Deep face recognition. In: British machine vision conference

7. Wu X, He R, Sun Z (2015) A lightened CNN for deep face representation. CoRR. http://arxiv.org/abs/1511.02683, arXiv:1511.02683

8. Schroff F, Kalenichenko D, Philbin J (2015) FaceNet: a unified embedding for face recognition and clustering. CoRR. http://arxiv.org/abs/1503.03832, arXiv:1503.03832

9. Huang GB, Ramesh M, Berg T, Learned-Miller E (2007) Labeled Faces in the Wild: A Database for Studying Face Recognition in Unconstrained Environments. Technical report 07–49, University of Massachusetts, Amherst (MA), USA

10. Mohammadi A, Bhattacharjee S, Marcel S (2018) Deeply vulnerable: a study of the robustness of face recognition to presentation attacks. IET Biom 7(1):15–26. https://doi.org/10.1049/iet-bmt.2017.0079

11. Ferrara M, Franco A, Maltoni D (2014) The magic passport. In: Proceedings of IEEE international joint conference on biometrics (IJCB). https://doi.org/10.1109/BTAS.2014.6996240

12. Scherhag U, Nautsch A, Rathgeb C, Gomez-Barrero M, Veldhuis RNJ, Spreeuwers L, Schils M, Maltoni D, Grother F, Marcel S, Breithaupt R, Ramachandra R, Büsch C (2017) Biometric systems under morphing attacks: assessment of morphing techniques and vulnerability reporting. In: Proceedings of international conference of the biometrics special interest group (BIOSIG). https://doi.org/10.23919/BIOSIG.2017.8053499

13. Ramachandra R, Raja KB, Büsch C (2016) Detecting morphed face images. In: Proceedings of IEEE 8th international conference on biometrics theory, applications and systems (BTAS), pp 1 – 7. https://doi.org/10.1109/BTAS.2016.7791169

14. Chingovska I, Anjos A, Marcel S (2012) On the effectiveness of local binary patterns in face anti-spoofing. In: Proceedings of the international conference of biometrics special interest group (BIOSIG)

15. Costa-Pazo A, Bhattacharjee S, Vazquez-Fernandez E, Marcel S (2016) The replay-mobile face presentation-attack database. In: Proceedings of international conference of the biometrics special interest group (BIOSIG). https://doi.org/10.1109/BIOSIG.2016.7736936

16. Ramachandra R, Büsch C (2017) Presentation attack detection methods for face recognition systems - a comprehensive survey. ACM Comput Surv 50

17. Rudd EM, Günther M, Boult TE (2016) PARAPH: presentation attack rejection by analyzing polarization hypotheses. In: IEEE conference on computer vision and pattern recognition workshops (CVPRW), pp 171 – 178. https://doi.org/10.1109/CVPRW.2016.28

18. Boulkenafet Z, Komulainen J, Hadid A (2015) Face anti-spoofing based on color texture analysis. In: IEEE international conference on image processing (ICIP), pp 2636–2640

19. Määttä J, Hadid A, Pietikäinen M (2011) Face spoofing detection from single images using micro-texture analysis. In: Proceedings of international joint conference on biometrics (IJCB). https://doi.org/10.1109/IJCB.2011.6117510

20. Boulkenafet Z, Komulainen J, Hadid A (2016) Face spoofing detection using colour texture analysis. IEEE Trans Inf Forensics Secur 11(8):1818–1830. https://doi.org/10.1109/TIFS.2016.2555286

21. Boulkenafet Z, Komulainen J, Hadid A (2018) On the generalization of color texture-based face anti-spoofing. Image Vis Comput. Accepted at the time of writing

22. Boulkenafet Z, Komulainen J, Li L, Feng X, Hadid A (2017) OULU-NPU: a mobile face presentation attack database with real-world variations. In: Proceedings of 12th IEEE international conference on automatic face and gesture recognition (FG 2017)

23. Wen D, Han H, Jain AK (2015) Face spoof detection with image distortion analysis. IEEE Trans Inf Forensics Secur 10(4):746–761

24. Galbally J, Marcel S, Fierrez J (2014) Image quality assessment for fake biometric detection: application to iris, fingerprint, and face recognition. IEEE Trans Image Process 23(2):710–724. https://doi.org/10.1109/TIP.2013.2292332

25. Li Y, Po LM, Xu X, Feng L, Yuan F (2016) Face liveness detection and recognition using shearlet based feature descriptors. In: Proceedings of IEEE international conference on acoustics, speech and signal processing (ICASSP), pp 874 – 877

26. Garcia DC, de Queiroz RL (2015) Face-spoofing 2D-detection based on Moiré-Pattern analysis. IEEE Trans Inf Forensics Secur 10(4):778–786. https://doi.org/10.1109/TIFS.2015.2411394

27. Patel K, Han H, Jain A (2016) Secure face unlock: spoof detection on smartphones. IEEE Trans Inf Forensics Secur 11(10):2268–2283. https://doi.org/10.1109/TIFS.2016.2578288

28. Tirunagari S, Poh N, Windridge D, Iorliam A, Suki N, Ho ATS (2015) Detection of face spoofing using visual dynamics. IEEE Trans Inf Forensics Secur 10(4):762–777. https://doi.org/10.1109/TIFS.2015.2406533

29. Anjos A, Chakka MM, Marcel S (2014) Motion-based counter-measures to photo attacks in face recognition. IET Biom 3(3):147–158. https://doi.org/10.1049/iet-bmt.2012.0071

30. Yang J, Lei Z, Li SZ (2014) Learn convolutional neural network for face anti-spoofing. CoRR. http://arxiv.org/abs/1408.5601, arXiv:1408.5601

31. Krizhevsky A, Sutskever I, Hinton GE (2012) ImageNet classification with deep convolutional neural networks. In: Advances in neural information processing systems, vol 25

32. Lucena O, Junior A, Hugo GMV, Souza R, Valle E, De Alencar Lotufo R (2017) Transfer learning using convolutional neural networks for face anti-spoofing. In: Karray F, Campilho A, Cheriet F (eds.) Proceedings of international conference on image analysis and recognition (ICIAR), Springer International Publishing, Cham, pp 27–34

33. Simonyan K, Zisserman A (2014) Very deep convolutional networks for large-scale image recognition. CoRR. http://arxiv.org/abs/1409.1556, arXiv:1409.1556

34. Menotti D, Chiachia G, Pinto A, Schwartz WR, Pedrini H, Falco A, Rocha A (2015) Deep representations for iris, face, and fingerprint spoofing detection. IEEE Trans Inf Forensics Secur 10(4):864–879. https://doi.org/10.1109/TIFS.2015.2398817

35. Nagpal C, Dubey SR (2018) A performance evaluation of convolutional neural networks for face anti spoofing. CoRR. https://arxiv.org/abs/1805.04176, arXiv:1805.04176

36. Szegedy C, Vanhoucke V, Ioffe S, Shlens J, Wojna Z (2016) Rethinking the inception architecture for computer vision. In: 2016 IEEE conference on computer vision and pattern recognition (CVPR), pp 2818–2826. https://doi.org/10.1109/CVPR.2016.308

37. He K, Zhang X, Ren S, Sun J (2016) Deep residual learning for image recognition. In: Proceedings of IEEE conference on computer vision and pattern recognition (CVPR), pp 770–778. https://doi.org/10.1109/CVPR.2016.90

38. Li L, Xia Z, Li L, Jiang X, Feng X, Roli F (2017) Face anti-spoofing via hybrid convolutional neural network. In: Proceedings of international conference on the frontiers and advances in data science (FADS), pp 120–124. https://doi.org/10.1109/FADS.2017.8253209

39. Nguyen TD, Pham TD, Baek NR, Park KR (2018) Combining deep and handcrafted image features for presentation attack detection in face recognition systems using visible-light camera sensors. J Sens 18(3):699–727. https://doi.org/10.3390/s18030699

40. Xu Z, Li S, Deng W (2015) Learning temporal features using LSTM-CNN architecture for face anti-spoofing. In: Proceedings of 3rd IAPR asian conference on pattern recognition (ACPR), pp 141–145. https://doi.org/10.1109/ACPR.2015.7486482

41. Liu Y, Jourabloo A, Liu X (2018) Learning deep models for face anti-spoofing: binary or auxiliary supervision. In: Proceeding of IEEE computer vision and pattern recognition, Salt Lake City, USA

42. Chingovska I, dos Anjos AR (2015) On the use of client identity information for face anti-spoofing. IEEE Trans Inf Forensics Secur 10(4):787–796. https://doi.org/10.1109/TIFS.2015.2400392

43. Yang J, Lei Z, Yi D, Li SZ (2015) Person-specific face antispoofing with subject domain adaptation. IEEE Trans Inf Forensics Secur 10(4):797–809. https://doi.org/10.1109/TIFS.2015.2403306
44. Ge S, Li J, Ye Q, Luo Z (2017) Detecting masked faces in the wild with LLE-CNNs. In: Proceedings of IEEE conference on computer vision and pattern recognition (CVPR), pp 426–434. https://doi.org/10.1109/CVPR.2017.53
45. Wang S, Yun Fu Y (2016) Face behind makeup. In: Proceedings of the thirtieth conference of the association for the advancement of artificial intelligence, pp 58–64
46. Kose N, Apvrille L, Dugelay JL (2015) Facial makeup detection technique based on texture and shape analysis. In: Proceedings of 11th IEEE international conference on automatic face and gesture recognition, Ljubljana, Slovenia (FG). Ljubljana, SLOVENIA, 4–8 May 2015. https://doi.org/10.1109/FG.2015.7163104. http://www.eurecom.fr/publication/4494
47. Agarwal A, Yadav D, Kohli N, Singh R, Vatsa M, Noore A (2017) Face presentation attack with latex masks in multispectral videos. In: Proceedings of IEEE conference on computer vision and pattern recognition workshops (CVPRW), pp 275–283. https://doi.org/10.1109/CVPRW.2017.40
48. Boulkenafet Z et al (2017) A competition on generalized software-based face presentation attack detection in mobile scenarios. In: Proceedings of IEEE international joint conference on biometrics (IJCB), pp 688–696. https://doi.org/10.1109/BTAS.2017.8272758
49. Arashloo SR, Kittler J (2017) An anomaly detection approach to face spoofing detection: a new formulation and evaluation protocol. In: Proceedings of the IEEE international joint conference on biometrics (IJCB), pp 80–89
50. Nikisins O, Mohammadi A, Anjos A, Marcel S (2018) On effectiveness of anomaly detection approaches against unseen presentation attacks in face anti-spoofing. In: Proceedings of international conference on biometrics (ICB). https://doi.org/10.1109/ICB2018.2018.00022
51. Steiner H, Sporrer S, Kolb A, Jung N (2016) Design of an active multispectral SWIR camera system for skin detection and face verification. J Sens 2016(1):1 – 8. Article ID 9682453, Special Issue on Multispectral, Hyperspectral, and Polarimetric Imaging Technology
52. Bourlai T, Narang N, Cukic B, Hornak L (2012) On designing a SWIR multi-wavelength facial-based acquisition system. In: Proceedings of SPIE: infrared technology and applications, vol 8353
53. Ramachandra R, Raja KB, Venkatesh S, Büsch C (2017) Extended multispectral face presentation attack detection: an approach based on fusing information from individual spectral bands. In: Proceedings of 20th international conference on information fusion (Fusion). https://doi.org/10.23919/ICIF.2017.8009749
54. Bhattacharjee S, Marcel S (2017) What you can't see can help you – extended range imaging for 3d-mask presentation attack detection. In: Proceedings of the 16th international conference of the biometrics special interest group (BIOSIG), Darmstadt, Germany
55. Bhattacharjee S, Mohammadi A, Marcel S (2018) Spoofing deep face recognition with custom silicone masks. In: Proceedings of the IEEE international conference on biometrics: theory, applications and systems (BTAS). Los Angeles, USA)
56. Erdogmus N, Marcel S (2013) Spoofing in 2D Face recognition with 3d masks and anti-spoofing with kinect. In: Proceedings of the IEEE international conference on biometrics: theory, applications and systems (BTAS)
57. Liu S, Yang B, Yuen PC, Zhao G (2016) A 3D mask face anti-spoofing database with real world variations. In: Proceedings of IEEE conference on computer vision and pattern recognition workshops (CVPRW), pp 1551–1557. https://doi.org/10.1109/CVPRW.2016.193
58. Agarwal A, Singh R, Vatsa M (2016) Face anti-spoofing using haralick features. In: Proceedings of the IEEE international conference on biometrics: theory, applications, and systems (BTAS), Niagara Falls, USA, pp 1–6
59. Manjani I, Tariyal S, Vatsa M, Singh R, Majumdar A (2017) Detecting silicone mask-based presentation attack via deep dictionary learning. IEEE Trans Inf Forensics Secur 12(7):1713–1723. https://doi.org/10.1109/TIFS.2017.2676720

60. Ramachandra R, Raja K, Büsch C (2015) Presentation attack detection for face recognition using light field camera. IEEE Trans Image Process 24(3):1–16
61. Pinto A, Schwartz WR, Pedrini H, Rocha ADR (2015) Using visual rhythms for detecting video-based facial spoof attacks. IEEE Trans Inf Forensics Secur 10(5):1025–1038. https://doi.org/10.1109/TIFS.2015.2395139
62. Chingovska I, Erdogmus N, Anjos A, Marcel S (2016) Face recognition systems under spoofing attacks. In: Bourlai T (ed.) Face recognition across the imaging spectrum, Springer, Berlin, pp 165–194 (2016)

Chapter 11
Recent Progress on Face Presentation Attack Detection of 3D Mask Attacks

Si-Qi Liu, Pong C. Yuen, Xiaobai Li and Guoying Zhao

Abstract With the advanced 3D reconstruction and printing technologies, creating a super-real 3D facial mask becomes feasible at an affordable cost. This brings a new challenge to face presentation attack detection (PAD) against 3D facial mask attack. As such, there is an urgent need to solve this problem as many face recognition systems have been deployed in real-world applications. Since this is a relatively new research problem, few studies has been conducted and reported. In order to attract more attentions on 3D mask face PAD, this book chapter summarizes the progress in the past few years, as well as publicly available datasets. Finally, some open problems in 3D mask attack are discussed.

11.1 Background and Motivations

Face presentation attack, a widely used face attack approach where a fake face of an authorized user is present to cheat the face recognition system, is one of the greatest challenges in practice. With the increasing variety of face recognition applications, this security concern has been receiving increasing attentions [1, 2]. Face image or video attacks are the two traditional spoofing methods that can be easily conducted through prints or screens. The face images or videos can also be easily acquired from the Internet with the boosting of social networks. In the last decade, a large number

S.-Q. Liu · P. C. Yuen (✉)
Department of Computer Science, Hong Kong Baptist University,
Kowloon, Hong Kong
e-mail: pcyuen@comp.hkbu.edu.hk

S.-Q. Liu
e-mail: siqiliu@comp.hkbu.edu.hk

X. Li · G. Zhao
Center for Machine Vision and Signal Analysis, University of Oulu, Oulu, Finland
e-mail: xiaobai.li@oulu.fi

G. Zhao
e-mail: guoying.zhao@oulu.fi

© Springer Nature Switzerland AG 2019
S. Marcel et al. (eds.), *Handbook of Biometric Anti-Spoofing*,
Advances in Computer Vision and Pattern Recognition,
https://doi.org/10.1007/978-3-319-92627-8_11

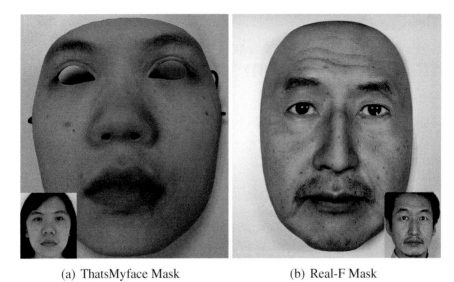

(a) ThatsMyface Mask   (b) Real-F Mask

Fig. 11.1 High-resolution sample images of Thatsmyface mask and REAL-f mask

of efforts have been devoted to face Presentation Attack Detection (PAD) on photo and video attacks [2–16] and encouraging results have been obtained.

Recently, with the rapid development of 3D printing and reconstruction technologies, creating a super-real 3D facial mask at an affordable cost become feasible. For instance, to make a customized Thatsmyface 3D mask as shown in Fig. 11.1a, the user is only required to submit a frontal face image with a few attributed key points. The 3D facial model is reconstructed from it and used to print the 3D mask. Compared with the 2D image or video attacks, 3D masks own 3D structure close to human faces while retaining the appearance in terms of skin texture and facial structure. In this case, the traditional 2D face PAD approaches may not work.

Recent research [17] points out that Thatsmyface[1] masks can successfully spoof many exiting popular face recognition systems. On the 3D Mask Attack Dataset (3DMAD) which is made of Thatsmyface masks, the Inter Session Variability (ISV) modeling method [18] achieves around a Spoofing False Acceptance Rate (SFAR) of around 30%. In addition, the REAL-f mask as shown in Fig. 11.1b using the 3D scan and "Three-Dimension Photo Form (3DPF)" technique to model the 3D structure and print the facial texture, can achieve higher appearance quality and 3D modeling accuracy than the Thatsmyface mask [19]. By observing the detailed textures such as the hair, wrinkle, or even eyes' vessels, without the context information, even a human can hardly identify whether it is a genuine face or not. As such, there is an urgent need to address the 3D mask face PAD.

---

[1]www.thatsmyface.com.

While 3D mask PAD is a new research problem, some work has been developed and reported which can be mainly categorized as the appearance-based approach, motion-based approach, and remote Photoplethysmography-based approach. Similar to the image and video attacks, 3D masks may contain some defects due to the printing quality. Thereby, using the appearance cues such as the texture and color becomes a possible solution [17]. The motion-based approaches detect mask attacks by using the fact that current 3D masks mainly have a hard surface and they cannot retain the subtle facial movements. Motion cues can hardly perform well against 3D mask attacks since 3D masks preserve both the geometric and appearance properties of genuine faces. Moreover, the soft silicone gel mask is able to preserve the subtle movement of the facial skin, which make the motion-based approaches less reliable.

As such, it is necessary to develop a new intrinsic liveness cue that can be independent of the appearance variation and motion patterns of different masks. Recently, studies turned out that the heartbeat signal on a face can be observed through a normal RBG camera by analyzing the color variation of the facial skin. If it is applied to the 3D mask PAD, the periodic heartbeat signal can be detected on a genuine face but not on a masked face since the mask blocks the light transmission [19]. Due to the uniqueness of this new liveness cues, we categorize is as the remote Photoplethysmography-based approach. The book chapter is organized as follows. We have given the background and motivations of this research work. Next, the publicly available 3D mask datasets and evaluation protocols are reviewed in Sect. 11.2. In Sect. 11.3, we discuss the methods developed in three categories for face PAD, namely appearance-based, motion-based, and remote photoplethysmography-based. The performances of the three approaches are evaluated on publicly available datasets in Sect. 11.4. Finally, open challenges in 3D mask attack are discussed in Sect. 11.5.

11.2 Publicly Available Datasets and Experiments Evaluation Protocol

11.2.1 Datasets

As far as we know, there are two rigid and two soft 3D mask attack datasets for 3D mask PAD. The two rigid mask datasets are: **3D M**ask **A**ttack **D**ataset (3DMAD) [20], **H**ong **K**ong **B**aptist **U**niversity 3D **M**ask **A**ttack with **R**eal World Variation**s** (HKBU-MARs) dataset [21]. The two soft mask datasets are: **S**ilicone **M**ask **A**ttack **D**ataset (SMAD) [22] and **M**ulti-spectral **L**atex **M**ask based Video **F**ace **P**resentation Attack (MLFP) dataset [23].

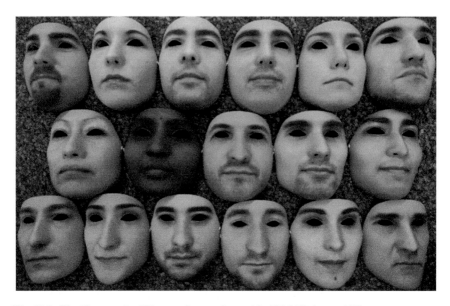

Fig. 11.2 The 17 customized Thatsmyface masks used in 3DMAD dataset [17]

11.2.1.1   3DMAD

The **3D M**ask **A**ttack **D**ataset (3DMAD) [20] is the first publicly available 3D mask attack dataset in which the attackers wear the customized 3D facial masks of a valid user. The Custom Wearable Masks used in 3DMAD, are built from Thatsmyface.com and proved to be good enough to spoof facial recognition system [20]. The dataset contains a total of 255 videos of 17 subjects, as shown in Fig. 11.2. It is noted that the eyes and nose holes are uncovered by the masks for a better wearing experience. The 3DMAD dataset is recorded by Kinect and contain color and depth information of size 640×480 at 30 frames per second. Each subject has 15 videos with ten live faces and five masked faces. This dataset is divided into three sessions that include two real access sessions recorded with a time delay and one attack session captured by a single operator (attacker). 3DMAD is the first public available 3D mask dataset which can be downloaded from https://www.idiap.ch/dataset/3dmad.

11.2.1.2   HKBU-MARs

The **H**ong **K**ong **B**aptist **U**niversity 3D **M**ask **A**ttack with **R**eal World Variations (HKBU-MARs) dataset [21] proposes to simulate the real-world application scenarios by adding variations in terms of type of mask, camera setting, and lighting condition. In particular, the HKBU-MARs dataset contains 12 subjects with 12 masks. To imitate different types of mask attacks in practice, six masks are from Thatsmyface and the other six are from REAL-f. Figure 11.3 shows the masks used in

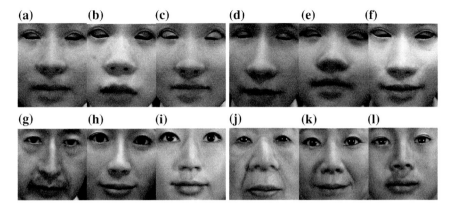

Fig. 11.3 Sample mask images from the HKBU-MARs dataset. **a–f** are ThatsMyFace masks and **g–l** are Real-F masks

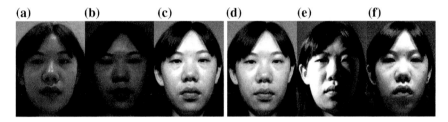

Fig. 11.4 Different lighting conditions in the HKBU-MARs dataset. **a–f** represents the low light, room light, bright light, warm light, side light, and upside light, respectively

HKBU-MARs. Since face recognition systems are widely deployed in different applications, such as the mobile application and the immigration, seven different cameras from the stationary and mobile devices are used to capture around 10 s videos. For the stationary applications, A web camera Logitech C920, an industrial camera and, and a mirrorless camera (Canon EOS M3) are used to represent different types of face acquisition systems. For the mobile devices, three smartphones: Nexus 5, iPhone 6, Samsung S7 and one tablet: Sony Tablet S are used. In addition, the HKBU-MARs dataset considers six lighting conditions to cover the typical scenes of face recognition applications, which includes the Low light, room light, bright light, warm light, side light and upside light as shown in Fig. 11.4. In sum, each subject contains 42 (seven cameras * six lightings) genuine and 42 mask sequences and the total size is 1008 videos. The HKBU-MARs that contains the variations of mask type, camera setting, and lighting condition can be used to evaluate the generalization ability of face PAD systems across different application scenarios in practice. More detailed information can be found at http://rds.comp.hkbu.edu.hk/mars/. A preliminary version of the HKBU-MARs, namely the HKBU-MARs V1 [19] has been publicly available at http://rds.comp.hkbu.edu.hk/mars/. HKBU-MARsV1 contains eight masks including six Thatsmyface masks and two REAL-f masks. It is recorded through a Logitech C920 webcam at a resolution of 1280×720 under natural lighting conditions.

11.2.1.3 SMAD

Different from 3D printed masks, silicon masks with soft surface can retain both the appearance and facial motions. As shown in the Hollywood movie *Mission Impossible*, Ethan Hunt wears silicon masks to impersonate others identities and even human can hardly recognize them. In this case, face recognition systems are highly vulnerable as well. The **S**ilicone **M**ask **A**ttack **D**atabase (SMAD) [22] is proposed to help research in developing 3D mask attack detection methods in such scenarios. The SMAD contains 130 real and attack videos (65 for each) that obtained from online resources under unconstrained settings. The genuine videos are auditioning, interviewing, or hosting shows collected from multiple sources so they contain different application environment in terms of illumination, background and camera settings. The time interval of videos varies from 3 to 15 s. In particular, the silicon masks in attack videos fit the eyes and mouths holes properly and some masks include hair or mustache. Note that the silicon masks in SMAD are not the customized masks with identities in genuine videos due to the very expensive price. The Text file that contains the videos URLs is available at: http://www.iab-rubric.org/resources.html.

11.2.1.4 MLFP

Since the soft 3D mask can be super real while maintaining the facial movement, using appearance or motion-based methods in the visible spectrum may not be effective. The **M**ulti-spectral **L**atex **M**ask based Video **F**ace **P**resentation Attack (MLFP) dataset is proposed to help research in designing multi-spectral based face PAD method [23]. The MLFP contains 1350 videos of 10 subjects with or without seven latex and three paper masks. Note that the masks in MLFP are also not the customized masks with identities in genuine videos. The MLFP is recorded in three different spectrums: visible, near-infrared, and thermal with the environmental variations which include indoor and outdoor with fixed and random backgrounds. The MLFP dataset is not yet publicly available at the writing of the chapter.

11.2.2 Experimental Evaluation Protocol

The performance of a 3D mask face PAD method is mainly evaluated under intra-dataset, cross-dataset, and intra/cross-variation scenarios to test its discriminability and generalizability. The intra-dataset testing is conducted in one dataset by separating the subjects into non-overlapping part as the training, development, and testing sets. The cross-dataset, and intra/cross-variation testing are designed to simulate the scenarios that the training samples are limited and different from the testing samples.

11.2.2.1 Intra-dataset Testing

For intra dataset testing, the classifier is trained on the training set and test with the testing set. It is noted that the development set is used to tune the parameters for real application scenarios (e.g., the testing set). Erdogmus et al. propose to use cross validation to assign different subjects into the training, development and test sets [20]. This protocol is updated to the leave one out cross validation (LOOCV) in [17], which selects one testing subject at each iteration and divides the rest subjects into training and development sets. For instance, the experiments on 3DMAD are done in 17-folds. In each fold, after selecting 1 subject's data as the testing set, for the remaining 16 clients, the first 8 is chosen for training and the remaining for development. Liu et al. updated the LOOCV by randomly assigning the remaining subjects for training and development to avoid the effect caused by the order of subjects in a dataset [19].

11.2.2.2 Cross-Dataset Testing

The cross-dataset protocol uses different datasets for training and testing to simulate the practical scenarios where the training data are limited and may be different from the testing samples. To conduct the cross-dataset testing, one can select one dataset for training and use the other dataset for testing. Due to the limited number of subjects of 3D mask attack datasets, the result may not be representative. We may select part of the subjects from one dataset for training and the final result is summarized by conducting several rounds of cross-dataset testing. For instance, for the HKBU-MARsV1 and 3DMAD cross testing, Liu et al. randomly selected five from the former dataset as the training set and used all data of 3DMAD for testing [19].

11.2.2.3 Intra-variation Testing

For the HKBU-MARs dataset [21] that contains three types of variations: mask type, camera, and lighting condition, the intra variation testing protocol is designed to evaluate the robustness of a 3D mask face PAD method when encountering only one specific variation. Under the selected variation (fixed mask type, camera, and lighting condition), LOOCV is conducted to obtain the final results. Although the intra variation testing may not match the practical scenarios, it is useful to evaluate the robustness of a 3D mask PAD method.

11.2.2.4   Cross-Variation Testing

The cross-variation testing protocol is designed to evaluate the generalization ability across different types of variation in practice. In particular, the leave one variation out cross validation (LOVO) [21] is proposed to evaluate the robustness of one type of variation, the others are fixed. For one specific type of variation, in each iteration, one sub-variation is selected as the training set and the rests are regarded as the testing set. For example, for the LOVO on camera type variations, the data captured by one type of camera is selected as the training set and data captured by the rest types of cameras is selected as the testing set. Note that the other types of variation: mask and lighting condition are fixed. In sum, the LOVO of cameras under different mask types and lighting conditions involves a total of 2×6 (mask types $\times$ lightings) sets of results [21].

11.3   Methods

11.3.1   Appearance-Based Approach

As the appearance of a face in printed photos or videos is different from the real face, several texture-based methods have been used for face PAD and achieve encouraging results [7, 11, 15, 24]. The 3D mask also contains the quality defect that results in the appearance difference from a genuine face, due to the imperfection precision problems of 3D printing technology. For example, the skin texture and detailed facial structures in masks as shown in Fig. 11.1a have perceivable differences compared to those in real faces.

Erdogmus et al. evaluate the LBP-based methods on 3DMAD dataset and show their effectiveness [17]. The Multi-Scale LBP (MS-LBPMS-LBP: Multi-Scale Local Binary Pattern) [7] achieves the best performance under most of the testing protocols. From a normalized face image, MS-LBP extracts $LBP_{16,2}^{u2}$, $LBP_{8,2}^{u2}$ from the entire image and $LBP_{8,1}^{u2}$ from the 3×3 overlapping regions. Therefore, one 243-bin, one 59-bin, and nine 59-bin histograms which contain both the global and local information are generated and then concatenated as the final 833-dimensional feature. It is reported that the MS-LBP achieves 99.4% Area Under Curve (AUC), 5% Equal Error Rate (EER) on the Morpho dataset.[2] Multi-Scale LBP [7] concatenates different LBP settings and achieves promising performance on 3D mask detection [17]. While the results are promising with the above methods, recent studies indicate that they cannot generalize well in a cross-dataset scenario [16, 25]. It is reported that the MS-LBP is less effective (22.6% EER and 86.8% AUC) on HKBU-MARsV1 due to the super-real mask—REAL-f [19].

---

[2]http://www.morpho.com.

Since the differences between 3D masks and real faces are mainly from the textures, analyzing the textures details in the frequency domain can be effective. Agarwal et al. propose the RDWT-Haralick [26] which uses redundant discrete wavelet transform (RDWT) and Haralick descriptors [27] to analyze the input image under different scales. Specifically, the input image is divided into 3×3 blocks and then the Haralick descriptors are extracted from the four sub-bands of the RDWT results and the original image. For video input, the RDWT-Haralick features are extracted from multiple frames and concatenated as the final feature vector. After feature dimension reduction using principal component analysis (PCA), the final result is obtained through Support Vector Machine (SVM) with linear kernel. It is reported that the RDWT-Haralick feature can achieve 100% accuracy and 0% HTER on 3DMAD dataset.

Despite the texture-based methods, the perceivable differences between genuine faces and fraud masks also exist in their 3D geometric appearance. Tang et al. analyze the 3D meshed faces (acquired through a 3D scanner) and highlight the dissimilarities of micro-shape of genuine and masked faces by the principal curvature measures [28] based 3D geometric attribute [29]. Specifically, they design a shape-description-oriented 3D facial feature description scheme which represents the 3D face as a histogram of principle curvature measures (HOC). It is reported that the HOC can achieve 92% true acceptance rate (TAR) when FAR is 0.01 on Morpho dataset.

Recently, with the booming of the deep learning, the community adopts deep networks to extract discriminative appearance features for biometric PAD [30, 31]. Compared to the solutions that rely on domain knowledge, Menotti et al. propose to learn a suitable CNN architecture through the data [30]. In the meantime, the filter weights of the network are learned via back-propagation. The two approaches interact with each other to form the final adapted network, namely the *spoofnet*. The authors report 100% accuracy and 0% HETER on the publicly available 3DMAD dataset. While the performances are significant, the deep learning based features require well designed large-scale training data. Due to the intrinsic data-driven nature [25], the over-fitting problem of the deep learning based methods in cross-dataset scenario remains open.

The hand-crafted features are difficult to be robust across multiple kinds of application scenarios. On the other hand, the deep network based features require significantly large and representative training dataset. Ishan et al. propose to use deep dictionary [32] via greedy learning algorithm (DDGL) for PAD [22]. It is reported that DDGL can achieve impressive performance on both the photo, video, hard mask, and silicon mask attacks. On the other hand, its generalizability under cross-dataset scenarios is less promising since the DDGL is also based on the feature learning framework and the performance, to some extent, still depends on the training data.

11.3.2 Motion-Based Approach

Since most existing 3D masks are made of hard materials, the facial motion, such as eye blinks and mouth movements, and facial expression changes may not be observed on a masked face. As such, the motion-based methods on 2D face PAD, such as the dynamic textures [11] or Histograms of Oriented Optical Flow (HOOF) [33] are effective in detecting these hard 3D masks.

Talha et al. propose the multifeature Videolet by encoding the appearance texture with the motion information [34] of both facial and surrounding regions. The texture feature is based on a configuration of local binary pattern, namely the multi-LBP. The motion feature is encoded by extracting HOOF from different time slots of the input video. The multifeature Videolet method not only achieves 0% EER on the 3DMAD dataset but is also effective in detecting the image and video presentation attacks.

Shao et al. propose to exploit the lower convolutional layer to obtain the dynamic information from fine-grained textures in feature channels [35]. In particular, given a preprocessed input video, fine-grained textures in feature channels of a convolutional layer of every frame are first extracted using a pre-trained VGG [36]. Then the dynamic information of textures in each feature channel (of all frames) are estimated using optical flow [37]. To exploit the most discriminative dynamic information, a channel-discriminability constraint is learned through minimizing intra-class variance and maximizing inter-class variance. The authors report 0.56% and 8.85% EER on 3DMAD and HKBU-MARsV1 dataset, respectively. To evaluate the generalizability, the cross-dataset [19] testing is conducted and yields 11.79% and 19.35% EER for 3DMAD to HKBU-MARsV1, and HKBU-MARsV1 to 3DMAD.

11.3.3 Remote Photoplethysmography-Based Approach

Different from the aforementioned two traditional approaches, a new intrinsic liveness cue based on the remote heartbeat detection is proposed recently [19, 38]. The rationale of the rPPG-based approach, and two state-of-the-art methods are illustrated in the following subsections.

11.3.3.1 What is Remote Photoplethysmography (rPPG)?

The remote heartbeat detection is conducted through the remote photoplethysmography (rPPG). The photoplethysmography (PPG) is an optically obtained plethysmography, which measures the volumetric of an organ, such as the heart. PPG measures the changes in light absorption of a tissue when heart pumps blood in a cardiac cycle [39]. Different from PPG that often uses the pulse oximeter attached to the

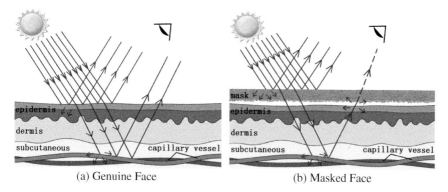

(a) Genuine Face   (b) Masked Face

Fig. 11.5 Effect of rPPG extraction on a genuine face (**a**), and a masked face (**b**) [19]

skin to detect the signal, rPPG capture it remotely through a normal RGB camera (e.g., web camera or mobile phone camera) under ambient lighting conditions.

11.3.3.2   Why rPPG Works for 3D Mask Presentation Attack Detection?

rPPG can be the intrinsic liveness cue for 3D mask face PAD. As shown in Fig. 11.5a, environmental light penetrates skin and illuminates capillary vessel in the subcutaneous layer. When the heart pumps blood in each cardiac cycle, the blood oxygen saturation changes, which results in a periodic color variation, namely the heartbeat signal. The heartbeat signal then transmits back from the vessels and can be observed by a normal RGB camera. Such an intrinsic liveness cue can be very effective on 3D mask face PAD. As shown in Fig. 11.5b for a masked face, the light needs to penetrate the mask and the source heartbeat signal needs to go through the mask again to be observed. Consequently, the rPPG signal extracted from a masked face is too weak to reflect the liveness evidence. In summary, rPPG signals can be detected on genuine faces but not on masked faces, which shows the feasibility of rPPG-based 3D mask face PAD.

11.3.3.3   rPPG-Based 3D Mask Presentation Attack Detection

Liu et al. are the first that exploit rPPG for 3D mask face PAD [19]. First, the input face is divided into local regions based on the facial landmarks and used to extract local rPPG signals. Then they model a correlation pattern from it to enhance the heartbeat signal and weaken the environmental noise. This is because the local rPPG signals share the same heartbeat frequency and the noise does not. Finally, the local rPPG correlation feature is fed into an SVM tuned by the learned confidence map. The experiments shows that this method achieves 16.2% EER and 91.7% AUC on HKBU-MARsV1 and 95.5% AUC and 9.9% EER on a Combined dataset formed of the 3DMAD and HKBU-MARsV1. In addition, since the rPPG signal is independent of the mask's appearance quality, the local rPPG solution yields good generalizability. Under the cross-dataset testing, it achieves 94.9% and 91.2% AUC for 3DMAD to HKBU-MARsV1 and HKBU-MARsV1 to 3DMAD.

Li et al. develop a generalized rPPG-based face PAD which works for both 3D mask and traditional image and video attacks [38]. Given the input face video, the Viola-Jones face detector [40] is used to find the bounding box for landmark detection. A customized region of interest is then defined to extract three raw pulse signals from the RGB channels. Next, they apply temporal filters to remove frequencies not relevant for pulse analysis. Signals are analyzed in the frequency domain and the liveness feature, a vector that consists of the maximum power amplitudes and the signal to noise ratio of the three channels, is extracted from the power density curves. This method achieves 4.71% EER on 3DMAD and 1.58% EER on two REAL-f masks [38]. It is also noted that most of the error cases are false negatives. This is because heart rate is fragile due the factors like darker skin tone and small facial resolution.

11.4 Experiments

Two experiments are conducted to evaluate the performance of the appearance-based, motion-based, and rPPG-based methods. In particular, MS-LBP [17], HOOF [34], RDWT-Haralick [26], convMotion [35], the local rPPG solution (LrPPG) [19], and the global rPPG solution (GrPPG) [38] are implemented and evaluated under both intra and cross-dataset testing protocols. For the appearance based methods, only the first frame of the input video is used. It is also noted that only the HOOF part of the videoLet [34] method is adopted to compare the motion cue with other methods. The final results are obtained through MATLAB SVM with RBF kernel for MS-LBP, HOOF, RDWT-Haralick and LrPPG and linear kernel for GrPPG.

To evaluate the performance, HTER [2, 17], AUC, EER and False Fake Rate (FFR) when False Liveness Rate (FLR) equals 0.1 and 0.01 are used as the evaluation criteria. For the intra-dataset test, HTER is evaluated on the development set and testing set, which are named as HTER_dev and HTER_test, respectively. A ROC curve with FFR and FLR is used for qualitative comparisons.

11.4.1 Intra Dataset Evaluation

To evaluate the discriminability of the 3D mask methods, LOOCV experiments on both 3DMAD and HKBU-MARsV1 datasets are conducted. For the 3DMAD dataset, we randomly choose eight subjects for training and the remaining eight for development. For the HKBU-MARsV1 dataset, we randomly choose three subjects as training set and the remaining four as the development set. We conduct 20 rounds of experiments.

Table 11.1 shows the promising results of the texture-based method on 3DMAD dataset while the performance drop on the HKBU-MARsV1 points out the limitation on the high-quality 3D masks. HOOF achieves similar results on the two datasets while the precisions are below the average. Note that the rPPG-based methods perform better on 3DMAD than on HKBU-MARsV1 (Table 11.2). Since the rPPG signal quality depends on the number of pixels of the region of interests, this circumstances may due to the facial resolution of videos from 3DMAD (around 80×80) are smaller than the videos from HKBU-MARsV1 (around 200×200). Specifically, LrPPG achieves better results due to the robustness of the cross-correlation model and confidence map. It is also noted that, as shown in Fig. 11.6 the major error classifications fall on the false reject due to the weakness of rPPG signals on face. The convMotion that fuses deep learning with motion liveness cue achieves the best performance among the existing methods.

Table 11.1 Comparison of results under intra dataset protocol on the 3DMAD dataset

	HTER_dev (%)	HTER_test (%)	EER (%)	AUC	FFR@ FLR = 0.1	FFR@ FLR = 0.01
MS-LBP [17]	0.15 ± 0.6	1.56 ± 5.5	0.67	100.0	**0.00**	0.42
HOOF [34]	32.9 ± 6.5	33.8 ± 20.7	34.5	71.9	62.4	85.2
RDWT-Haralick [26]	7.88 ± 5.4	10.0 ± 16.2	7.43	94.0	2.78	89.1
convMotion [35]	**0.10 ± 0.1**	**0.95 ± 0.6**	**0.56**	**100.0**	0.00	**0.30**
GrPPG [38]	8.99 ± 3.1	10.7 ± 11.5	9.41	95.3	8.95	70.7
LrPPG [19]	7.12 ± 4.0	7.60 ± 13.0	8.03	96.2	7.36	14.8

Table 11.2 Comparison of results under intra dataset protocol on the HKBU-MARsV1 dataset

	HTER_dev (%)	HTER_test (%)	EER (%)	AUC	FFR@ FLR − 0.1	FFR@ FLR = 0.01
MS-LBP [17]	17.5 ± 10.8	11.9 ± 18.2	21.1	86.8	55.4	93.9
HOOF [34]	25.4 ± 12.8	23.8 ± 23.9	28.9	77.9	75.0	95.6
RDWT-Haralick [26]	16.3 ± 6.8	10.3 ± 16.3	18.1	87.3	31.3	81.7
convMotion [35]	**2.31 ± 1.7**	**6.36 ±3.5**	**8.47**	**98.0**	**6.72**	**27.0**
GrPPG [38]	14.9 ± 6.5	15.8 ± 13.1	16.8	90.2	31.3	73.5
LrPPG [19]	11.5 ± 3.9	13.0 ± 9.9	13.4	93.6	16.2	55.6

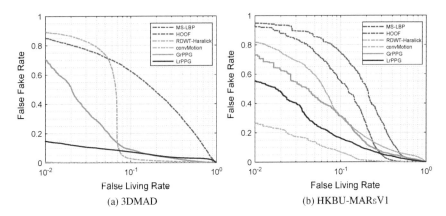

(a) 3DMAD   (b) HKBU-MARsV1

Fig. 11.6 Average ROC curves of two datasets under intra-dataset protocol

11.4.2 Cross-Dataset Evaluation

To evaluate the generalization ability across different datasets, we design cross-dataset experiments where the training and test samples are from two different datasets. In particular, when 3DMAD and HKBU-MARsV1 are used as the training and testing datasets (3DMAD→HKBU-MARsV1 for short), we randomly select eight subjects from 3DMAD for training and use the remaining nine subjects from 3DMAD for development. When HKBU-MARsV1 and 3DMAD are used as the training and testing datasets (HKBU-MARsV1→3DMAD for short), we randomly select four subjects from HKBU-MARsV1 as the training set and the remaining four subjects from HKBU-MARsV1 as the development set. We also conduct 20 rounds of experiments.

As shown in Tables 11.3 and 11.4, the rPPG-based methods and convMotion achieve close results as the intra dataset testing, which shows their encouraging robustness. The main reason behind the rPPG-based method's success is that the rPPG signal for different people under different environment is consistent, so the features of genuine and fake faces can be well separated. On the other hand, the performance of MS-LBP and HOOF drop. Specifically, the appearance based methods, MS-LBP and RDWT-Haralick achieve better results for HKBU-MARsV1→3DMAD (Fig. 11.7), since the HKBU-MARsV1 contains two types of masks while the 3DMAD contains one (the classifier can generalize better when it is trained with larger data variance). It is also noted that the RDWT-Haralick feature achieves better performance than MS-LBP as it analyzes the texture differences from different scales with redundant discrete wavelet transform [26]. The HOOF fails on the cross-dataset as for different dataset, the motion patterns based on optical flow may vary with the different recording settings.

Table 11.3 Cross-dataset evaluation results under 3DMAD→HKBU-MARsV1

	3DMAD→HKBU-MARsV1				
	HTER (%)	EER (%)	AUC (%)	FFR@ FLR = 0.1	FFR@ FLR = 0.01
MS-LBP [17]	43.6 ± 5.9	44.8	57.7	98.8	100.0
HOOF [34]	51.8 ± 12.0	50.6	47.3	84.1	92.1
RDWT-Haralick [26]	23.5 ± 4.7	39.6	68.3	71.6	98.1
convMotion [35]	**10.1 ± 2.1**	11.8	96.2	13.8	**33.9**
GrPPG [38]	29.7 ± 11.9	15.6	90.5	27.4	77.0
LrPPG [19]	10.7 ± 3.7	**11.4**	**96.2**	**12.3**	55.6

Table 11.4 Cross-dataset evaluation results under HKBU-MARsV1→3DMAD

	HKBU-MARsV1→3DMAD				
	HTER (%)	EER (%)	AUC (%)	FFR@ FLR = 0.1	FFR@ FLR = 0.01
MS-LBP [17]	45.5 ± 2.8	25.8	83.4	41.2	69.0
HOOF [34]	42.4 ± 4.1	44.1	57.6	77.6	93.1
RDWT-Haralick [26]	13.8 ± 7.5	21.3	86.7	37.1	90.3
convMotion [35]	17.2 ± 1.3	19.4	89.6	27.4	56.0
GrPPG [38]	29.0 ± 11.6	22.7	83.2	57.3	91.9
LrPPG [19]	**12.8 ± 3.3**	**13.2**	**93.7**	**16.9**	**55.9**

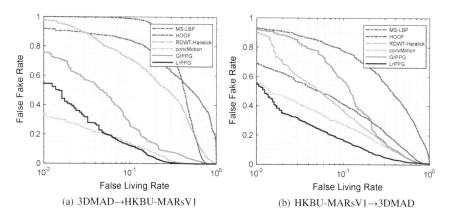

(a) 3DMAD→HKBU-MARsV1   (b) HKBU-MARsV1→3DMAD

Fig. 11.7 Average ROC curves under cross-dataset protocol

11.5 Discussion and Open Challenges

With the development of 3D printing and 3D face reconstruction technology, 3D mask is proved to be able to spoof a face recognition system. As such, this problem has drawn an increasing attention with the boosting numbers of related publications. In this chapter, we revealed the challenges of 3D mask presentation attack, summarized the existing datasets and evaluation protocols, and discussed different approaches that have been proposed. Still, there are issues remain open.

Although the costs of 3D masks are expensive, there are two publicly available datasets address the 3D mask presentation attack challenges. However, the numbers of the subjects and customized 3D masks are quite limited because of the cost. Without sufficient number of data, the evaluation results of toy experiments may not be convincing enough for real-world applications. Additionally, the variations like the recording device and conditions are limited which results in the difficulties of evaluating the generalization capabilities of the methods in practical scenarios. The existing publicly available datasets using customized masks are mainly recorded through a stationary camera under single lighting condition. While in practice, the training data may vary from the testing samples, in terms of the mask types, camera devices, or lighting conditions. For instance, since the mobile applications are getting more and more popular, the scenario of a mobile device with camera motion interferences under unconstrained lighting conditions could be the common situation. Therefore, more data with real-world settings should be designed and collected.

On the other hand, since the 3D mask face PAD is at the beginning stage, current researches mainly focus on the detection under fixed conditions with simple testing protocols, which may not reflect the practical scenarios. The excellent results on single dataset indicate that more challenging evaluation protocols are needed before the 3D mask face PAD can be applied at the practical level. Additionally silicone mask attacks need to be first "collected" and then studied. For the appearance-based methods, adapting different mask qualities, cameras, and lighting conditions are the challenges to be addressed. The motion-based methods may not work on the soft masks that can preserve the facial motion. The rPPG-based methods may not be robust under lower lighting condition or with motion interferences. In sum, larger dataset is still the most critical issue for designing more complicated protocols to evaluate not only the discriminability but also the generalizability.

Acknowledgements This project is partially supported by Hong Kong RGC General Research Fund HKBU 12201215, Academy of Finland and FiDiPro program of Tekes (project number: 1849/31/2015).

References

1. Galbally J, Marcel S, Fierrez J (2014) Biometric antispoofing methods: a survey in face recognition. IEEE Access 2:1530–1552

2. Hadid A, Evans N, Marcel S, Fierrez J (2015) Biometrics systems under spoofing attack: an evaluation methodology and lessons learned. IEEE Signal Process Mag 32(5):20–30
3. Rattani A, Poh N, Ross A (2012) Analysis of user-specific score characteristics for spoof biometric attacks. In: CVPRW
4. Evans NW, Kinnunen T, Yamagishi J (2013) Spoofing and countermeasures for automatic speaker verification. In: Interspeech, pp 925–929
5. Pavlidis I, Symosek P (2000) The imaging issue in an automatic face/disguise detection system. In: Computer vision beyond the visible spectrum: methods and applications
6. Tan X, Li Y, Liu J, Jiang L (2010) Face liveness detection from a single image with sparse low rank bilinear discriminative model. In: Computer vision–ECCV, pp 504–517
7. Määttä J, Hadid A, Pietikäinen M (2011) Face spoofing detection from single images using micro-texture analysis. In: IJCB
8. Anjos A, Marcel S (2011) Counter-measures to photo attacks in face recognition: a public database and a baseline. In: International joint conference on biometrics (IJCB), pp 1–7
9. Zhang Z, Yan J, Liu S, Lei Z, Yi D, Li SZ (2012) A face antispoofing database with diverse attacks. In: International conference on biometrics (ICB), pp 26–31
10. Pan G, Sun L, Wu Z, Lao S (2007) Eyeblink-based anti-spoofing in face recognition from a generic webcamera. In: ICCV
11. de Freitas Pereira T, Komulainen J, Anjos A, De Martino JM, Hadid A, Pietikäinen M, Marcel S (2014) Face liveness detection using dynamic texture. EURASIP J Image Video Process 2014(1):1–15
12. Kose N, Dugelay JL (2014) Mask spoofing in face recognition and countermeasures. Image Vis Comput 32(10):779–789
13. Yi D, Lei Z, Zhang Z, Li SZ (2014) Face anti-spoofing: multi-spectral approach. Handbook of biometric anti-spoofing. Springer, Berlin, pp 83–102
14. Galbally J, Marcel S, Fierrez J (2014) Image quality assessment for fake biometric detection: application to iris, fingerprint, and face recognition. IEEE Trans Image Process 23(2):710–724
15. Kose N, Dugelay JL (2013) Shape and texture based countermeasure to protect face recognition systems against mask attacks. In: CVPRW
16. Wen D, Han H, Jain AK (2015) Face spoof detection with image distortion analysis. IEEE Trans Inf Forensics Secur 10(4):746–761
17. Erdogmus N, Marcel S (2014) Spoofing face recognition with 3D masks. IEEE Trans Inf Forensics Secur 9(7):1084–1097
18. Wallace R, McLaren M, McCool C, Marcel S (2011) Inter-session variability modelling and joint factor analysis for face authentication. In: IJCB
19. Liu S, Yuen PC, Zhang S, Zhao G (2016) 3d mask face anti-spoofing with remote photoplethysmography. In: ECCV
20. Erdogmus N, Marcel S (2013) Spoofing in 2d face recognition with 3d masks and anti-spoofing with kinect. In: BTAS
21. Liu S, Yang B, Yuen PC, Zhao G (2016) A 3D mask face anti-spoofing database with real world variations. In: Proceedings of the IEEE conference on computer vision and pattern recognition workshops, pp 100–106
22. Manjani I, Tariyal S, Vatsa M, Singh R, Majumdar A (2017) Detecting silicone mask based presentation attack via deep dictionary learning. In: TIFS
23. Agarwal A, Yadav D, Kohli N, Singh R, Vatsa M, Noore A (2017) Face presentation attack with latex masks in multispectral videos. SMAD 13:130
24. Boulkenafet Z, Komulainen J, Hadid A (2016) Face spoofing detection using colour texture analysis. IEEE Trans Inf Forensics Secur 11(8):1818–1830
25. de Freitas Pereira T, Anjos A, De Martino JM, Marcel S (2013) Can face anti-spoofing countermeasures work in a real world scenario? In: International conference on biometrics (ICB), pp 1–8
26. Agarwal A, Singh R, Vatsa M (2016) Face anti-spoofing using Haralick features. In: BTAS
27. Haralick RM, Shanmugam K et al (1973) Textural features for image classification. IEEE Trans Syst Man Cybern 1(6):610–621

28. Cohen-Steiner D, Morvan JM (2003) Restricted delaunay triangulations and normal cycle. In: Proceedings of the nineteenth annual symposium on computational geometry, pp 312–321

29. Tang Y, Chen L (2017) 3d facial geometric attributes based anti-spoofing approach against mask attacks. In: FG

30. Menotti D, Chiachia G, Pinto A, Schwartz WR, Pedrini H, Falcão AX, Rocha A (2015) Deep representations for iris, face, and fingerprint spoofing detection. IEEE Trans Inf Forensics Secur 10(4):864–879

31. Yang J, Lei Z, Li SZ (2014) Learn convolutional neural network for face anti-spoofing. arXiv:1408.5601

32. Tariyal S, Majumdar A, Singh R, Vatsa M (2016) Deep dictionary learning. IEEE Access 4:10096–10109

33. Chaudhry R, Ravichandran A, Hager G, Vidal R (2009) Histograms of oriented optical flow and Binet Cauchy kernels on nonlinear dynamical systems for the recognition of human actions. In: CVPR

34. Siddiqui TA, Bharadwaj S, Dhamecha TI, Agarwal A, Vatsa M, Singh R, Ratha N (2016) Face anti-spoofing with multifeature videolet aggregation. In: ICPR

35. Rui Shao XL, Yuen PC (2017) Deep convolutional dynamic texture learning with adaptive channel-discriminability for 3d mask face anti-spoofing. In: IJCB

36. Simonyan K, Zisserman A (2014) Very deep convolutional networks for large-scale image recognition. CoRR. arXiv:1409.1556

37. Barron JL, Fleet DJ, Beauchemin SS (1994) Performance of optical flow techniques. Int J Comput Vis

38. Li X, Komulainen J, Zhao G, Yuen PC, Pietikäinen M (2016) Generalized face anti-spoofing by detecting pulse from face videos. In: 23rd international conference on pattern recognition (ICPR). IEEE, pp 4244–4249

39. Shelley K, Shelley S (2001) Pulse oximeter waveform: photoelectric plethysmography. In: Lake C, Hines R, Blitt C (eds) Clinical monitoring. WB Saunders Company, Philadelphia, pp 420–428

40. Viola P, Jones M (2001) Rapid object detection using a boosted cascade of simple features. In: CVPR

Chapter 12
Challenges of Face Presentation Attack Detection in Real Scenarios

Artur Costa-Pazo, Esteban Vazquez-Fernandez, José Luis Alba-Castro and Daniel González-Jiménez

Abstract In the current context of digital transformation, the increasing trend in the use of personal devices for accessing online services has fostered the necessity of secure cyberphysical solutions. Biometric technologies for mobile devices, and face recognition specifically, have emerged as a secure and convenient approach. However, such a mobile scenario also brings some specific threats, and spoofing attack detection is, without any doubt, one of the most challenging. Although much effort has been devoted in anti-spoofing techniques over the past few years, there are still many challenges to be solved when implementing these systems in real use cases. This chapter analyses some of the gaps between research and real scenario deployments, including generalisation, usability, and performance. More specifically, we will focus on how to select and configure an algorithm for real scenario deployments, paying special attention to use cases involving limited processing capacity devices (e.g., mobile devices), and we will present a publicly available evaluation framework for this purpose.

A. Costa-Pazo (✉) · E. Vazquez-Fernandez · D. González-Jiménez
GRADIANT, CITEXVI, loc. 14., 36310 Vigo, Spain
e-mail: acosta@gradiant.org

E. Vazquez-Fernandez
e-mail: evazquez@gradiant.org

D. González-Jiménez
e-mail: dgonzalez@gradiant.org

J. L. Alba-Castro
Universidade de Vigo, Campus Universitario Lagoas-Marcosende, 36310 Vigo, Spain
e-mail: jalba@gts.uvigo.es

© Springer Nature Switzerland AG 2019
S. Marcel et al. (eds.), *Handbook of Biometric Anti-Spoofing*,
Advances in Computer Vision and Pattern Recognition,
https://doi.org/10.1007/978-3-319-92627-8_12

247

12.1 Introduction

Accessing to personal data using our smartphones has become a part of normal every-day life. Using passwords, unlock patterns, as well as biometric recognition systems is common for a secure access to our social networks, bank apps and so on. Facial recognition is one of the most widespread biometric features on modern applications. However, given the multimedia resources available today on the Internet (i.e. Facebook photos, videos on YouTube), it is easy to obtain audiovisual material from almost every potential user, allowing the creation of tools to perform presentation attacks (PAI, Presentation Attack Instrument).

Over the past few years, the interest in methods to detect this type of attacks in facial recognition systems, the so-called Presentation Attack Detection (PAD) methods, has increased. This has led to the emergence of numerous research works [1, 2], anti-spoofing specific databases (composed of genuine access and attack videos) [3–8], and associated competitions [9, 10] with the aim of evaluating new PAD algorithms.

There are currently several ways to tackle the problem of PAD from an algorithmic point of view. We could simplify them in two approaches: collaborative methods and automatic methods. The first takes advantage of the possibility of interaction with the user, while the latter relies solely on image and video analysis in a non-interactive way.

Nowadays, many of the deployed facial recognition systems have chosen collaborative anti-spoofing systems in order to preserve their security. This approach implies challenge-response strategies such as eye blinking, smiling, looking in different directions or flashing the user with colour lights from the device screen. Some examples of these approaches can be found in.[1,2,3] The main reason for biometrics providers to implement these solutions lies in the robustness against automated attacks. Collaborative anti-spoofing systems cannot be fooled as easily as presenting an image or video in front of the camera, as they require the attack to mimic certain actions such as blinking, smiling, or even moving one's face by following an specific movement pattern. However, advances in computer graphics and the reduction of costs to produce high resolution masks are emerging as a potential risk that threatens even this collaborative countermeasures [11]. This is the case with 3D reconstructions from 2D images using Deep Learning methods like the ones presented in [12]. Using these methods an attacker could reproduce 3D models of a face following the corresponding patterns required for registration (a demo is publicly available in[4]).

Alternatively, automatic face-PAD methods are more convenient and less intrusive for the user. These face-PAD methods are better accepted by the final users than the collaborative ones, since they are more discreet to be used in public, by preventing possible social barriers (i.e. smiling or deliberately blinking in front of the smartphone

[1]https://www.iproov.com.

[2]https://www.zoloz.com/smile.

[3]https://www.bioid.com/liveness-detection/.

[4]http://cvl-demos.cs.nott.ac.uk/vrn/.

in public spaces) and thus are being demanded by the industry. In addition, these methods have also advantages in terms of usability: as the user does not need to cooperate, they do not require a learning process; likewise they do not need any kind of synchronization with the user interface for such interaction. In spite of the advantages, these PAD methods are not as widely used in a real-world scenario due to the difficulty of transferring the performance obtained in databases into real-world environment [13]. We will address this problem in the next section.

Despite the research effort, the deployment of this technology in realistic environments presents many challenges and it is an issue which is still far from being solved. Examples of this are the facial recognition systems included in some of the latest smartphone models. Samsung's latest device, Galaxy S9 line, have been hacked with a simple photograph shortly after their presentation. Despite incorporating specific depth and infrared sensors, iPhone X has also suffered from spoofing attacks. After a week of its release, the first videos of users and companies claiming their capability to get into the device by tricking the system with masks have already appeared. In view of these cases, it is important to consider the challenges of deploying anti-spoofing systems in real-world environments and, therefore, also how we are assessing these systems and whether we can do better.

The main contributions of this chapter are

- An analysis of the weak points of the current face-PAD evaluation procedures, in relation to performance and usability constraints in real environments.
- A publicly available evaluation framework is proposed to help researchers on selecting parameters and tuning algorithms for bringing new face-PAD methods to applications on real scenarios.
- A discussion on the benefits of the presented framework by showing the performance evaluation of two different face-PAD methods.

The chapter is organized as follows. Challenges and related work on face anti-spoofing is reviewed and analysed in Sect. 12.2. Section 12.3 describes the framework for the proposed evaluation. Section 12.4 reports the results obtained. Conclusions are finally presented in Sect. 12.5.

12.2   Challenges of Deploying Face-PAD in Real Scenarios

In this section we provide a brief comment of some relevant aspects for face anti-spoofing that need to be analysed and studied: cross-domain performance, database limitations and usability.

12.2.1   Cross-Domain Performance

As indicated by recent works on cross-dataset evaluations, generalisation is a critical issue in most systems [13]. Systems that have been trained and evaluated in the same dataset obtain very high performance. However when they are evaluated in a different dataset the error rates increase by an order of magnitude [6].

One of the main problems of current systems lies in the difficulty of cross-domain generalisation, losing performance when the system deals with images taken in different conditions than the trained ones. In a recent work [1] Patel et al. studied the influence of the number of training subjects on performance, by using the MSU-USSA database[5] (a 1000-subject dataset). Reported results show that using a larger training set significantly improves the cross-dataset performance. However, in the same work an interesting end-to-end experiment was performed using an Android app for routine smartphone unlock with an integrated face-PAD method previously trained on MSU-USSA. This experiment showed that performance obtained on datasets do not reflect the real performance on actual applications. Despite the fact that they achieved an impressive improvement on intra-dataset performance, it has been demonstrated that end-to-end tests are needed to forecast performance on deployment scenarios. The gap between academic datasets and real scenario accesses and attacks seems clear. The observed drop in performance shows that the type of attacks, use cases and scenarios in real life differ significantly from those covered by academic datasets. Unless databases add more representative variety of attacks and genuine accesses, performance loss in real-world scenarios will continue to be a problem.

12.2.2 Database Limitations

Collecting a dataset for reproducing collaborative face-PAD scenarios is a difficult task that includes synchronization of the user interface and capture, movement and reaction annotations, variety of PAIs, etc. Therefore, there are no publicly available collaborative face-PAD datasets with these characteristics. In this situation, collaborative anti-spoofing countermeasure methods cannot be compared in an open and reproducible framework.

On the other hand, several datasets have been published in recent years trying to represent the automatic authentication scenario on mobile devices. NUAA [3], CASIA FASD [4] and REPLAY-ATTACK [5] helped to evaluate first anti-spoofing countermeasures. Unfortunately, these datasets were captured with cameras that are no longer representative of mobile scenarios. In 2015, Di We et al. presented the MSU-MFSD database [6] which was captured with a *Macbook Air 13* laptop (similar to REPLAY-ATTACK database) and also with a *Google Nexus 5* Android phone, a more representative device of the mobile scenario. Then in 2016, the REPLAY-MOBILE database [7] was captured with a *iPad Mini 2* and a *LG-G4*. This dataset includes different protocols for tuning the system considering both usability and security. This database, besides having the two typical types of attack (print and replay), has five different scenarios available to mimic several actual access conditions, paying special attention to lighting variations. Finally, the *OULU-NPU* [8]

---

[5]http://biometrics.cse.msu.edu/Publications/Databases/MSU_USSA/.

provides a richer representation of current mobile devices (*Samsung Galaxy S6 edge, HTC Desire EYE, MEIZU X5, ASUS Zenfone Selfie, Sony XPERIA C5 Ultra Dual* and *OPPO N3*). Four challenging protocols (*Illumination, PAIs, Leave One Camera Out - LOCO* and a combination of the previous ones) are available in this recent dataset.

As pointed out before, cross-domain performance is an issue in state-of-the-art face-PAD systems. Current databases have several differences such as the type of capturing devices, illumination conditions, camera-holding conditions, user interaction protocols, etc. As we have introduced in previous section, current face-PAD methods seem to overfit these datasets, resulting in poor cross-dataset performance. This leads us to the conclusion that we need larger and more representative of the real problem datasets.

Moreover, some face-PAD systems depend on an specific use case scenario that may not be covered in some datasets. For instance, some motion-based algorithms use background modelling to detect dynamic information from sequences, preventing their use for biometric systems embedded in handheld capturing devices. These mismatching scenarios cause that not all face-PAD methods can be evaluated in all datasets. This leads us to the conclusion that a fair benchmarking of face-PAD systems should be use case dependent.

On top of taking care of cross-domain and use case dependency when measuring face-PAD systems performance, there are at least three other aspects to improve in current benchmarking protocols

- **The type of attacks** available on current datasets are limited and, in some cases, outdated. For example, adversarial models are improving [19] and it is getting easier and easier to manufacture sophisticated attacks. The new datasets must update the type of attacks to cope with evolving picaresque.
- **Number of subjects**: most of the publicly available face spoofing datasets contain no more than 55 subjects. An extra effort is needed to increase this figure in order to capture greater inter-subject variability and increase evaluation fairness. The number of users performing the attacks is also an important element which is overlooked in current databases. If the attacks are performed by a single person or a reduced group of people, the variety and representativeness may suffer.
- **Evaluation and protocols** have a large margin for improvement, even with current publicly available data. On the one hand, as we pointed before, the identities of the available databases are not many, so a union of the databases might improve the quality of the evaluation. Therefore, an analysis of the compatibilities between datasets must be addressed to avoid identity collisions and increase diversity of attacks while maintaining a meaningful protocol for the fused dataset. On the other hand, there are some key parameters (i.e time of acquisition, frame rate, CPU performance, etc.) that are not taken into account by the current evaluation protocols focussed only on the binary misclassification performance. The analysis of these parameters will provide a wider view of real scenario face-PAD performance.

12.2.3   Usability

A critical factor when selecting and implementing an anti-spoofing system is to
ensure that it meets the security requirements. However, we often need to reach a
trade-off between security and usability. For instance, for a bank, a *False Rejection* in
a small wire transfer can imply worse consequences than a *False Acceptance*. Thus,
a trade-off between security and usability is essential to ensure a successful usage
of biometrics systems on real-world applications. On the NIST report Usability &
Biometrics - Ensuring Successful Biometric Systems,[6] some guidelines about usability are defined as a core aspect of design to ensure biometric systems and products became easy to learn, effective to use and enjoyable from the users perspective.
Beyond accept/reject decisions, there are some important aspects that biometric systems have to accomplish, in terms of usability, as (a) effectiveness (effective to use);
(b) efficiency (efficient to use); (c) satisfaction (enjoyable to use); (d) learnability
(easy to learn); and (e) memorability (easy to remember). Adoption of biometric
systems as widespread authenticators is linked to usability improvements, so new
schemes must be effective, allowing an easy capture process to obtain high-quality
samples through an intuitive user interface. Moreover, current systems must improve
its efficiency, in terms of convenience. Otherwise, why would a user start using biometrics instead of a password if it is slower and cumbersome?

In many actual situations an early response is critical for the performance (i.e.
access control, mobile device unlock, payments, etc.). However, current anti-spoofing
evaluation protocols do not take into account time of response, i.e. the time the
PAD method requires for taking a decision. These protocols use entire attempt-to-
access videos, which are generally about 5–10 s long, for testing of the algorithms.
Some studies focused on analysing the usability of biometric access control systems
have measured times from 4 to 12 s for the whole process of facial verification [14,
15]. However, some of the traditional reference guides regarding usability and user
interaction in web and apps indicate that the response time should be no more than
1 s to ensure that the user does not perceive an annoying interruption in the process
[16, 17]. Therefore, we think it is necessary to introduce time constraints in order to
fairly evaluate the performance of anti-spoofing systems. Note we are not speaking
here about computational performance, which an also leads to some delays and must
also be addressed, but to the duration of the user interaction for the system to take a
decision.

Finally, learnability and memorability are very related with user experience,
whereby a good guide of usage (i.e. show user some real examples of correct and
incorrect usages), and appropriate feedback messages are essential for a favourable
performance.

---

[6]https://www.nist.gov/sites/default/files/usability_and_biometrics_final2.pdf.

12.3 Proposed Framework

As presented in the section above, the challenges of face-PAD have been focused on generalisation between databases and unseen attacks, as well as the creation of representative databases which mitigate the constant improvement of the Presentation Attack Instruments (i.e. mobiles devices, printers, monitors and 3D scanners). However, the usability and hardware constraints are passed over in the evaluation process. Taking decisions without considering these constraints does not ensure us to appropriately tune the systems for the best performance. So, how can we evaluate and compare our systems for working on actual applications? In this chapter, we propose an evaluation framework for helping to reduce the existing gap between the research process and real scenario deployment. This framework presents two novel evaluation protocols to analyse the performance of face-PAD considering deployment constraints, as the frame rate and response time. Our proposed framework[7] is publicly available as a Bob [18, 19] package to help researchers to evaluate their face-PAD algorithms over more realistic conditions. Reproducibility principles were considered for designing our software, which is easy to use and extend, and allows the reader to reproduce all the experiments.

Moreover, our software was designed as a python desktop tool to provide researchers with a common, easy-to-use and fair framework for comparing anti-spoofing methods. It is not intended to give an absolute measure about correct parameters for each device, but a framework for comparison between algorithms and algorithm configuration. For example, some systems require a minimum frame rate to perform as expected, e.g., some motion-based algorithms which depend on fast dynamics (as facial micro-expressions). However, the practical implementation of these algorithms may not be feasible in some use cases or scenarios. The proposed framework will provide valuable information to decide which algorithm(s) and algorithm settings are appropriate for each specific use case. Finally, parameters can be tuned on each processor (i.e. ARM, x86, i386) extending our software by fulfilling the proposed interfaces so as to capture the performance statistic.

In order to analyse the different time parameters and delays that come into play in a face recognition system, we must take into account the typical flow of use of a real application. Normally, in real deployment scenarios, face recognition implementations comprises three stages: login screen, biometric procedure and decision screen. Ideally, the user should not experience a noticeable delay at any stage. However, high computational requirements, connection errors (i.e. getting an access token) or difficulties getting high-quality samples on the capture process can affect the usability of a biometric system. In the following, we make a breakdown of some parameters that stand for each step of the biometric procedure. First, we present the adopted terminology; second, we explain the proposed methodology.

---

[7]The evaluation framework may be downloaded using the following URL: https://github.com/Gradiant/bob.chapter.hobpad2.facepadprotocols.

12.3.1 Terminology

Henceforth, T_r (Total time of system response) represents the time between the beginning of the biometric process (i.e. open the camera flow in face-based system) and the time of binary decision (access granted or not). Figure 12.1 shows how T_r is calculated by the addition of two consecutive time intervals: T_a or total time of acquisition of the attempt-to-access video, that includes capturing, feature extraction (sometimes postponed to the next phase) and other concurrent CPU tasks; and T_d or time to take a final decision, that sometimes include feature extraction, algorithmic decision and other concurrent CPU tasks.

$CPUT_{fp}(i)$ is the CPU time dedicated for processing the frame i of the video sequence. The total CPU Time of video processing or $CPUT_{vp}$ is defined by the following formula, as the aggregation of all the time slots of CPU time (or $CPUT_{fp}$) dedicated to the frame acquisition and processing:

$$CPUT_{vp} = \sum_{i=1}^{n} CPUT_{fp}(i) \tag{12.1}$$

where n, or the number of processed frames, is a consequence of the chosen parameters T_a (total time of acquisition) and FR (frame rate of the device in frames per second). T_a helps to evaluate usability and FR helps to evaluate computational suitability in actual deployments. Ideally, $CPUT_{vp}$ has to be shorter than T_a. However, biometrics systems have to coexist and operate on multi-task environments and sometime $CPUT_{vp}$ can be larger. Considering this possibility we defined T_r as:

$$T_r = \max(T_a, CPUT_{vp}) + T_d \tag{12.2}$$

Then, it is quite easy to calculate the use of the processor, during acquisition time, as:

$$CPU_{usage} = \frac{CPUT_{vp}}{T_a} * 100(\%) \tag{12.3}$$

This measure will help us to determine if a system can operate on an actual scenario with specific FR and CPU constraints. Thus, it is very interesting to analyse

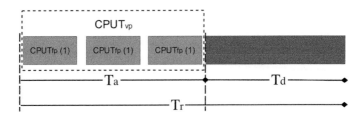

Fig. 12.1 Breakdown of biometric procedure in terms of computational demand

the costs of each module in order to be able to face the deployment stage with more information.

12.3.2   Methodology

From this analysis we have designed a novel evaluation framework to better model the performance of a face-PAD method on a real application. It includes three evaluation procedures that will help us to research some unstudied parameters and their influence in the performance of a deployed system.

- **Algorithmic Unconstrained Evaluation**, or AUE is the given name for the classical (current de facto) algorithmic evaluation. On this stage, every method is evaluated following the defined protocols for each database, without any constraints about on-device implementation (i.e. T_a or FR). This classical evaluation is still fundamental in order to provide a fair performance comparison in terms of error rates, so we propose it to be the starting point for the design, development and parameterization process of a face-PAD method. The calculated error rates and working points on this evaluation are only reliable for an unconstrained comparison, since the algorithms are taking advantage of unrestricted video duration and frame rate. These results and parameters are much less useful on a real implementation, nevertheless they can help on the initial parameterization of the algorithm.
- **Algorithmic Constrained Evaluation**, or ACE, provides information about performance and error rates related to actual deployment constraints. More specifically, FR (frame rate) and T_a (total time of acquisition). This stage consists of evaluating a method cloning each input video but simulating different acquisition settings, obtaining, this way, valuable information to forecast the face-PAD performance. From this evaluation we can determine the best configuration of a face-PAD accompanied by a $Working Point$ (normally represented by a $Threshold$) for a given FR and T_a. The final aim of this protocol is to analyse the real performance of the algorithm under different framerate and timing constraints.
- **End-to-end Simulation**: Once a parameterization laboratory was finished (using both of previous evaluation stages), it is necessary to evaluate the whole system (determined by optimum FR, T_a and a $Working Point$). This protocol simulates the final behaviour of a face-PAD on an actual deployment using a bunch of videos. This end-to-end simulation provides interesting information about the actual conditions on T_r (total time of system response), T_d (time of decision) and CPU_{usage} over a selected subset of videos. Although this evaluation is very useful for an initial decision concerning implementation parameters, we should keep in mind that it does not replace the end-to-end tests running in an actual production device. This evaluation takes the best algorithm configuration selected in the previous two evaluation stages and analyses delays and CPU_{usage}.

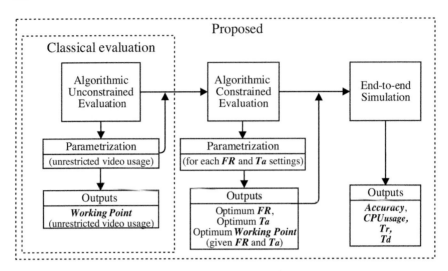

Fig. 12.2 Overview of our proposed evaluation framework

Figure 12.2 shows the proposed database evaluation pipeline. Output boxes represent valuable information for deployment parameterization obtained by each evaluation stage. Our framework allows to answer some crucial questions that arise within the integration process: How long should the anti-spoofing countermeasure last in order to get a performance similar as evaluated? How many images should our face-PAD system compute?, etc. By the selection of an optimum performance related to a T_a and FR parameters we can determine in a more founded way, the answer to this questions from *ACE* stage. Then, on the *End-to-end Simulation* stage we will get helpful information regarding face-PAD suitability to work on real scenarios. A further explanation of *Algorithmic Constrained Evaluation* and *End-to-end Simulation* can be found below.

12.3.2.1   *ACE* - Algorithmic Constrained Evaluation

ACE analyses the performance of the face-PAD method using a number of different configurations. The default parameters are the following, resulting in 20 different configurations:

> list_framerate = [5, 10, 15, 20, 25] (frames per second)
> list_time_video_duration = [500, 1000, 1500, 2000] (milliseconds)

As discussed before, settings as frame rate and acquisition time are crucial on the implementation of a face-PAD, thus performance evaluation needs to consider them.

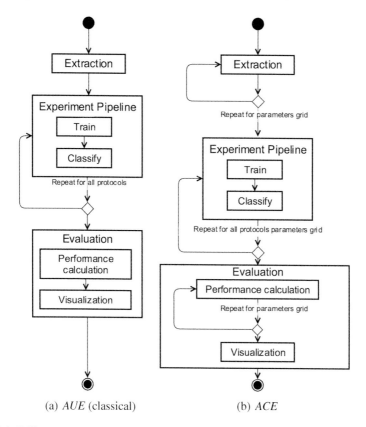

(a) *AUE* (classical) (b) *ACE*

Fig. 12.3 Differences between classical evaluation *AUE* (Algorithmic Unconstrained Evaluation) and proposed *ACE* (Algorithmic Constrained Evaluation)

The evaluation protocol is configurable so these parameters can be changed to adapt it to different use cases.

This evaluation procedure is quite similar to *AUE* (classical). Figure 12.3 shows the workflow of this evaluation, which differs from the classical one in the repetition of the process over several simulated video versions running in the same CPU. These videos are generated from the original ones considering a set of FR and T_a values. Shorter T_a values are implemented by cropping the initial segment of the video up to the T_a value. FR modifications are implemented by sub-sampling the original video. The proposed framework calculates the performance values (i.e. EER, $HTER$ or standard ISO/IEC 30107-3 metrics as $APCER$, $BPCER$ and $ACER$) for all parameter combinations. To represent this information, we have designed a 2-D plot where x-axis stand for the evaluated T_a, and y-axis represents the error rate in percentage. Each configuration of a FR is represented as a curve, where we can observe the evolution of the performance for increasingly higher T_a. Figure 12.4 shows an example of this representation with synthetic data.

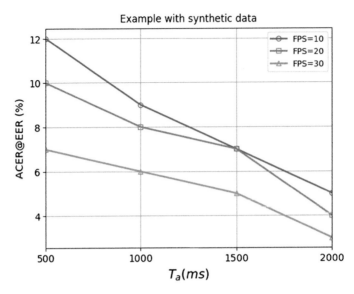

Fig. 12.4 *ACE* representation using synthetic data

This way of conducting experiments allows us to carry out two complementary analysis:

- **Parameter selection**: in this experiment the protocol trains the algorithms on *Train* subset to evaluate them on *Dev* and *Test* subsets. This process is repeated for each version of the dataset (video simulating *FR* and T_a), thus generating several models. As a result, performance data is obtained for all configurations. This information permits to rank the configurations, and select the best "a priori" option for the intended use case. This experiment includes both training and testing of the algorithm
- **Pre-trained performance test**: this method requires a pre-trained algorithm that will be tested on the test set videos but cloned with different settings. Evaluating a system this way allows us to see the "a posteriori" behaviour over the different *ACE* parameters. This experiment is focused only on testing the algorithm under different constraints. Systems with high performance degradation in this evaluation may have problems operating on a wide range of devices with different computational capabilities.

Knowledge about how the system works over different configurations leads us to a better parameterization, and at the same time, gives us valuable clues on face-PAD-decision process. One telling example is the case of decision-making within two approaches with a similar performance using the classical evaluation or *AUE*. If we only consider performance values for the systems working over the whole video and without any constraint regarding *FR*, we will probably make a blind decision. Imagine that when you are evaluating both approaches over different constrained parameters, one system only needs 500 ms to reach its best performance value, and the other needs at least 4000 ms. Now, with this information the decision would be much more founded. In this example, we took some assumptions about time processing considering that, for instance, T_{vp} (time of processing all necessary frames) is less than T_a. This assumptions will be addressed below, at the *end-to-end* evaluation subsection. Moreover, if we carefully observe the information given by *ACE* experiments, it can provide a vision of the suitability of a system for working on a wide range of hardware. That means the less FR-dependent is a face-PAD, the more suitable is to work on any device. For instance, on mobile device scenarios it is very important to determine the range of devices where the system can ensure a demanded security and usability performance.

At the end of the day, the decision process is highly related to use cases and their restrictions (usually given by hardware, software and connectivity). This proposed algorithmic constrained evaluation protocol provides very useful information for orienting face-PAD research and deployment.

12.3.2.2 End-to-End Simulation

The end-to-end simulation replicates the final behaviour of a face-PAD and allows the evaluation of an early prototype with some interesting real-world restrictions. Our proposal is oriented to its use in the last stages of development, providing the researcher with a different perspective of its implementations. The proposed end-to-end simulation forces the researcher to design a real face-PAD implementation considering some parameters not normally evaluated. In order to use this protocol, the face-PAD prototype needs to implement a given interface (See UML diagram on Fig. 12.5.).

The *abstract* class, *FacePad*, requires to be initialized with constant values FR, T_a and $WorkingPoint$, as well as external requirements as information about pre-trained models. Ideally, these values have to be selected on previous evaluation

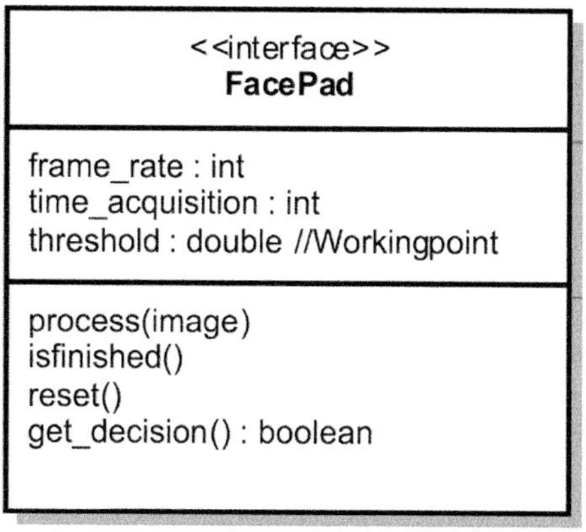

Fig. 12.5 UML diagram of *FacePad* abstract class

```
        <<interface>>
          FacePad

  frame_rate : int
  time_acquisition : int
  threshold : double //Workingpoint

  process(image)
  isfinished()
  reset()
  get_decision() : boolean
```

stages. Besides, researchers must take some decisions regarding the implementation that might have a lot of influence on the outcome. For example, if we implement on *FacePad*'s *process* method a computational demanding process, CPU_{usage} will increase. Otherwise, if computationally expensive operations are implemented on decision step (*get_decision* method), probably the CPU_{usage} will be greatly reduced, but penalizing, that way, the time of response due to the increment of the time of delay (T_d). This can help to structure the processing chain of the proposed method in order to be suitable for real-world applications. Thus, prototype implementation is very useful to find out the bottlenecks on laboratory devices and select the most appropriate processor in the deployment stage later on. In particular, on mobile devices scenario, our proposed protocol can help us to better understand the limitations of a face-PAD using a desktop script. Optimization research would have to be done considering *ARM* architectures. For a fair comparison, results in terms of performance must be reported citing used machine, otherwise results are worthless.

Generally, evaluation software is optimized for parallel running (e.g feature extraction) in order to make the most of the available computational infrastructure. Both *ACE* and AUE were designed and implemented by following this parallelisation approach. However, in order to properly simulate the timing of a real interaction process, the end-to-end simulation runs sequentially for each access. Even so, the proposed framework cannot isolate the evaluation from other processes running on the testing machine, so we recommend to launch the experiments in a dedicated machine in order get a more fair evaluation.

12.4 Experimental Results and Discussion

12.4.1 Evaluated Face-PAD

In order to evaluate the performance in realistic conditions, two different approaches of face-PAD are considered. First, as a baseline, we present the image-quality measures based face-PAD, *IQM* from now on. Further, we present a face-PAD- introduced in the IJCB 2017 competition on generalized face presentation attack detection in mobile authentication scenarios [10], *GRADIANT* from now on.

IQM Based Face-PAD

IQM is an adaptation of the proposed method by Galbally et al. in [20]. This implementation was presented as a baseline of the REPLAY-MOBILE [7], and an implementation is available as a Bob package.[8] This method is based on the extraction of a set of image-quality measures obtaining a 18-length feature vector. For each frame of video, a feature vector is obtained, computed over the entire frame (in this case, rescaling to a maximum size of 320 pixels). With every frame of the training set, a support-vector machine (SVM) with a radial-basis function kernel (RBF) and

---

[8]https://gitlab.idiap.ch/bob/bob.ip.qualitymeasure.

a gamma $= 1.5$ is used for training the system. A final score per video is obtained averaging the scores of the n-frames of a sequence.

GRADIANT Based Face-PAD

This algorithm has an hybrid strategy for features extraction fusing colour, texture and motion information. *GRADIANT* exploits both HSV and YCbCr colour spaces by extracting dynamic information over a given video sequence, mapping temporal variation (for each channel) into a single image. Hence, six images (from six channels) represent motion in a sequence, and for each one, a ROI is cropped based on eye positions over the sequence and rescaled to 160×160. Each ROI is divided into 3×3 and 5×5 rectangular regions from which uniform LBP histogram features are extracted and concatenated.

These two methods are quite different, and will help us to present and analyse how the evaluation framework can contribute in the stage of an anti-spoofing system. *GRADIANT* is a motion-based face-PAD so it is expected that changes on the number of processed frames and differences between them, affect the performance. On the other side, although the final score of *IQM* uses the whole video, it can operate by giving a decision per frame, so it is expected to be more consistent over FR and T_a variations. In this work we perform face spoof detection experiments using the *OULU-NPU* dataset. The larger number of samples, devices and users makes it the most representative current dataset. The sizes of the test sets in the other datasets (i.e. REPLAY-ATTACK, REPLAY-MOBILE or MSU-MFSD) are an order of magnitude lower, so the resolution is not sufficient for the variations to be significant. Nevertheless, the software is prepared for using those publicly datasets as well. Moreover, the existing protocols on *OULU-NPU* in the dataset are designed for the evaluation of algorithms under unseen training conditions. This characteristic is very useful for generalisation evaluation, however, it greatly reduces the number of samples available in the testing subsets, and therefore increases the noise in evaluations. We have extended the dataset by adding a protocol called *Grandtest* incrementing, in that way, the evaluated number of samples on *Dev* and *Test* subsets. This protocol combines all original protocols (Protocol 1, 2, 3 and 4). Future updates of the proposed framework should include the optimal aggregation of datasets to widen representativity of training samples.

12.4.2   AUE - Algorithmic Unconstrained Evaluation

First of all, we present the result for the evaluated systems using classical benchmarking. Table 12.1 shows values for each system. The values in the table show the performance of the system using the whole video without frame rate downsampling.

With a classic evaluation, the *GRADIANT* approach clearly achieves better results with very low error rates in a recent database. However, with this evaluation, we cannot determine if systems are or not suitable to work on actual deployment scenarios. In this stage, as researchers, we are almost blinded regarding some crucial parameters,

Table 12.1 The performance of the proposed methods under *OULU-NPU* (Grandtest) dataset

Method	Dev	Test	
	EER (%)	HTER (%)	ACER (%)
IQM	29.72	29.23	30.69
GRADIANT	1.11	0.76	0.97

and normally, we rank face-PADs systems only with error rates information. This fact leads to the question: Are the classical metrics we are using to evaluate PAD methods the most appropriate to select and adjust systems in real scenarios? Reported results on the following evaluation stages shed some light in order to create a more fairly face-PAD benchmarking.

12.4.3 ACE - Algorithmic Constrained Evaluation

The evaluation carried out in the previous section shows the error rates of the PAD systems without any constraints. However, as indicated in the Guidelines for best practices in biometrics research [21], operational systems need to meet application-specific requirements and constraints in addition to recognition accuracy (or error rate). In those guidelines, the term operational system is used to denote the utilization of a biometric system in a actual application. In this second evaluation protocol we introduce two new operational constraints: time and frame rate. An error rate curve is shown for each frame rate configuration and each curve shows the variation as a function of the video duration (T_a) used in the analysis. It is therefore possible to compare the performance achieved by the different systems with the specific constraints of an operational environment. Performance curves on this work present error rate curves in terms of $ACER$. In order to show the influence of these constraints, we report two different experiments: parameter selection and pre-trained performance test.

Parameter Selection

Figure 12.6 shows the performance of the two evaluated systems trained with different configurations. We can see that the behaviour is similar, and as the T_a is higher and more information can be extracted, the error is reduced. The behaviour for variations of FR is different. On the one hand, in *IQM*, for a greater value of FR the performance improves, while the behaviour in the *GRADIANT* system does not follow a pattern. The uniform decrement of ACER when increasing T_a and FR in *IQM* seems to follow the principle of vanishing Gaussian noise when averaging a larger number of frames. On the other hand, the GRADIANT system, that uses learnt dynamics, also benefits from increasing the total number of frames but it is more sensible to the motion sampling.

From this experiment, it is possible to obtain the best configuration for our system determined by the security restrictions, as well as those of the environment.

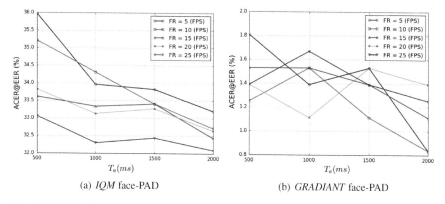

Fig. 12.6 *ACE*: the performance under different configurations (FR and T_a) on *OULU-NPU* (Grandtest) dataset

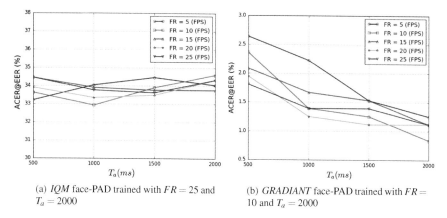

(a) *IQM* face-PAD trained with $FR = 25$ and $T_a = 2000$

(b) *GRADIANT* face-PAD trained with $FR = 10$ and $T_a = 2000$

Fig. 12.7 The performance of a pre-trained face-PAD under different configurations (FR and T_a) on *OULU-NPU* dataset (Grandtest protocol)

Pre-trained Performance Test

Once a candidate system is selected to be the optimal one (it depends on the use case), it is interesting to evaluate it under realistic conditions. *ACE* allows to evaluate a pre-trained model. Two face-PAD methods (one for *IQM* and another for *GRADIANT*) have been chosen for being evaluated here. Performance errors have been evaluated for different configurations in Fig. 12.7.

As we can observe on Fig. 12.7a, the *IQA* system has no significant performance variations depending on test configuration given the best trained system. This is the expected behaviour, as this method does not take full advantage of the temporal information in the video. With this in mind, a configuration with a low analysis time (e.g. 500 ms) and a low frame rate could be selected for a real environment without any significant loss in performance. On the other hand, the *GRADIANT* system (Fig. 12.7b) has variations depending on the configuration, where, as the length of the analysed video sequence increases (from 500 to 2000 ms), the error rate decreases. This behaviour is also expected, as *GRADIANT* method takes advantage of the temporal information of the video sequence.

12.4.4 End-to-End Simulation

Finally, we perform the end-to-end simulation on a Intel®Xeon®CPU X5675 @ 3.07 GHz.[9] The performance of the studied systems are reported in Table 12.2. As we can appreciate, *IQM* implementation exceeds, by far, the 100% of CPU_{usage} making impossible the usage of this configuration on a real-time application. Nevertheless, according to results reported on previous sections, the *IQM* is not greatly affected by variation on FR and T_a. Thus, we can select a less computationally demanding setting, for example $FR = 5$ with the same $T_a = 2000$ ms, for which we obtain a $CPU_{usage} = 136, 8\%$. With this new configuration, we would pass from a total response time (T_r) of 48301.89 to 2736.49 ms, making the system much more usable because its response would be given in less than 3 s. Note that decision time (T_d) is negligible since only a score average is performed in this stage whilst features extraction is made at video processing time.

In contrast to the *IQM*, the *GRADIANT* approach implements features extraction within the decision stage, once the total time of acquisition (T_a) is ended and all dynamic information is analysed. In order to use this system while a facial recognition solution is also running, we need to evaluate the CPU_{usage}. Depending on the computational cost of the verification module, we must have a compromise solution between error rates and computational demand. Based in Fig. 12.6b we can select the system trained with $FR = 5$ and $T_a = 2000$ ms where the performance is equal to the proposed on Table 12.2. With this configuration, the outcome of CPU_{usage} is 63% and the total time of response is 2110 ms making it more suitable to be used on an actual application.

Moreover, these studied systems have a wide margin of improvement in the deployment stage. For instance, most of the processing time of *GRADIANT* face-PAD approach is spent on face and landmark detection using *dlib* library.[10] Replacing these modules with faster solutions would greatly reduce the CPU_{usage}. On the other hand, the *IQM* approach has a large margin of optimization by implementing the quality measures focusing on performance and efficiency.

---

[9] https://ark.intel.com/es-es/products/52577/Intel-Xeon-Processor-X5675-12M-Cache-3_06-GHz-6_40-GTs-Intel-QPI.
[10] http://dlib.net/.

Table 12.2 The end-to-end results for pre-trained *IQM* and *GRADIANT*

Method	CPU usage (%)	T_{vp} (ms)	T_r (ms)	T_a (ms)	T_d (ms)	FR (frames per second)
IQM	2414	48286.89	48301.89	**2000**	0.15	**25**
GRADIANT	155	3104.64	3215.05	**2000**	110.41	**10**

12.5 Conclusion

In this chapter we have reviewed the current challenges of face presentation attack detection, focusing on usability issues and restrictions of real scenarios. We have presented a new evaluation framework, introducing two new protocols, *Algorithmic Constrained Evaluation* and the *End-to-end Simulation*. The first one is focused on analysing the performance of the face-PAD algorithms under different frame rates and timing constraints. This allows us to compare and select the best algorithms and algorithm settings for an specific use case. The second one allows us to take the best algorithm configuration selected in the previous evaluations and analyse the performance in terms of usability taking into account potential delays and CPU usage.

We have made this framework publicly available as a Python package inheriting from bob architecture. Reproducibility principles were considered for designing this software, which should be easy to use and to extend, and should allow the reader to reproduce all the experiments. The proposed protocols and framework are aimed to give us a new perspective of whether or not a face-PAD is suitable to be used in a real scenario. It will allow us to take better decisions when selecting and configuring the systems for the real world, where performance and usability constraints are critical. As an example of this evaluation process, we have analysed two different face-PAD systems in order to show the possibilities and benefits of the proposed protocols and framework.

Acknowledgements We thank the colleagues of the Biometrics Team at Gradient for their assistance in developing the reproducible toolkit.

References

1. Patel K, Han H, Jain A (2016) Secure face unlock: spoof detection on smartphones. IEEE Trans Inf Forensics Secur 11(10), 2268–2283
2. Boulkenafet Z, Komulainen J, Hadid A (2015) Face anti-spoofing based on color texture analysis. In: 2015 IEEE international conference on image processing (ICIP), pp 2636–2640. https://doi.org/10.1109/ICIP.2015.7351280
3. Tan X, Li Y, Liu J, Jiang L (2010) Face liveness detection from a single image with sparse low rank bilinear discriminative model. Springer, Berlin, pp 504–517. https://doi.org/10.1007/978-3-642-15567-3_37
4. Zhang Z, Yan J, Liu S, Lei Z, Yi D, Li SZ (2012) A face antispoofing database with diverse attacks. In: 2012 5th IAPR international conference on biometrics (ICB), pp 26–31. https://doi.org/10.1109/ICB.2012.6199754

5. Chingovska I, Anjos A, Marcel S (2012) On the effectiveness of local binary patterns in face anti-spoofing. In: 2012 BIOSIG - proceedings of the international conference of biometrics special interest group (BIOSIG), pp 1–7
6. Wen D, Han H, Jain AK (2015) Face spoof detection with image distortion analysis. IEEE Trans Inf Forensics Secur 10(4):746–761. https://doi.org/10.1109/TIFS.2015.2400395
7. Costa-Pazo A, Bhattacharjee S, Vazquez-Fernandez E, Marcel S (2016) The replay-mobile face presentation-attack database. In: Proceedings of the international conference on biometrics special interests group (BioSIG)
8. Boulkenafet Z, Komulainen J, Li L, Feng X, Hadid A (2017) OULU-NPU: a mobile face presentation attack database with real-world variations. In: 2017 12th IEEE international conference on automatic face gesture recognition (FG 2017), pp 612–618. https://doi.org/10.1109/FG.2017.77
9. Chingovska I, Yang J, Lei Z, Yi D, Li SZ, Kähm O, Glaser C, Damer N, Kuijper A, Nouak A, Komulainen J, Pereira TF, Gupta S, Khandelwal S, Bansal S, Rai A, Krishna T, Goyal D, Waris M, Zhang H, Ahmad I, Kiranyaz S, Gabbouj M, Tronci R, Pili M, Sirena N, Roli F, Galbally J, Fiérrez J, da Silva Pinto A, Pedrini H, Schwartz WS, Rocha A, Anjos A, Marcel S (2013) The 2nd competition on counter measures to 2d face spoofing attacks. In: International conference on biometrics, ICB 2013, 4–7 June 2013, Madrid, Spain, pp 1–6. https://doi.org/10.1109/ICB.2013.6613026
10. Boulkenafet Z, Komulainen J, Akhtar Z, Benlamoudi A, Samai D, Bekhouche SE, Ouafi A, Dornaika F, Taleb-Ahmed A, Qin L, Peng F, Zhang LB, Long M, Bhilare S, Kanhangad V, Costa-Pazo A, Vazquez-Fernandez E, Pérez-Cabo D, Moreira-Pérez JJ, González-Jiménez D, Mohammadi A, Bhattacharjee S, Marcel S, Volkova S, Tang Y, Abe N, Li L, Feng X, Xia Z, Jiang X, Liu S, Shao R, Yuen PC, Almeida WR, Andaló F, Padilha R, Bertocco G, Dias W, Wainer J, Torres R, Rocha A, Angeloni MA, Folego G, Godoy A, Hadid A (2017) A competition on generalized software-based face presentation attack detection in mobile scenarios. In: IJCB 2017: international joint conference on biometrics (IJCB)
11. Xu Y, Price T, Frahm JM, Monrose F (2016) Virtual u: defeating face liveness detection by building virtual models from your public photos. In: 25th USENIX security symposium (USENIX security 16). USENIX Association, Austin, TX, pp 497–512
12. Jackson AS, Bulat A, Argyriou V, Tzimiropoulos G (2017) Large pose 3d face reconstruction from a single image via direct volumetric CNN regression. CoRR. arXiv:1703.07834
13. de Freitas Pereira T, Anjos A, Martino JMD, Marcel S (2013) Can face anti-spoofing countermeasures work in a real world scenario? In: 2013 international conference on biometrics (ICB), pp 1–8. https://doi.org/10.1109/ICB.2013.6612981
14. Miguel-Hurtado O, Guest R, Lunerti C (2017) Voice and face interaction evaluation of a mobile authentication platform. In: 51st IEEE international Carnahan conference on security technology, Oct 2017, Madrid, Spain
15. Trewin S, Swart C, Koved L, Martino J, Singh K, Ben-David S (2012) Biometric authentication on a mobile device: a study of user effort, error and task disruption. In: 28th annual computer security applications conference
16. Nielsen J (1993) Usability engineering. Morgan Kaufmann Publishers Inc., San Francisco
17. Nielsen J (2010) Website response times
18. Anjos A, Günther M, de Freitas Pereira T, Korshunov P, Mohammadi A, Marcel S (2017) Continuously reproducing toolchains in pattern recognition and machine learning experiments. In: International conference on machine learning (ICML)
19. Anjos A, Shafey LE, Wallace R, Günther M, McCool C, Marcel S (2012) Bob: a free signal processing and machine learning toolbox for researchers. In: 20th ACM conference on multimedia systems (ACMMM), Nara, Japan
20. Galbally J, Marcel S, Fierrez J (2014) Image quality assessment for fake biometric detection: application to iris, fingerprint, and face recognition. IEEE Trans Image Process 23(2):710–724. https://doi.org/10.1109/TIP.2013.2292332
21. Jain AK, Klare B, Ross A (2015) Guidelines for best practices in biometrics research. In: International conference on biometrics, ICB 2015, Phuket, Thailand, 19–22 May 2015, pp 541–545

Chapter 13
Remote Blood Pulse Analysis for Face Presentation Attack Detection

Guillaume Heusch and Sébastien Marcel

Abstract In this chapter, the usage of Remote Photoplethysmography (rPPG) as a mean for face presentation attack detection is investigated. Remote photoplethysmography consists in retrieving the heart-rate of a subject from a video sequence containing some skin, and recorded at a distance. To get a pulse signal, such methods take advantage of subtle color variation on skin pixels due to the blood flowing through vessels. Since the inferred pulse signal gives information on the liveness of the recorded subject, it can be used for biometric presentation attack detection (PAD). Inspired by work made for speaker presentation attack detection, we propose to use long-term spectral statistical features of the pulse signal to discriminate real accesses from attack attempts. A thorough experimental evaluation, with different rPPG and classification algorithms is carried on four publicly available datasets containing a wide range of face presentation attacks. Obtained results suggest that the proposed features are effective for this task, and we empirically show that our approach performs better than state-of-the-art rPPG-based presentation attack detection algorithms.

13.1 Introduction

As face recognition systems are used for authentication purposes more and more, it is important to provide a mechanism to ensure that the biometric sample is genuine. Indeed, several studies showed that existing face recognition algorithms are not robust to simple spoofing attacks. Even simple display of a printed face photograph can fool biometric authentication systems. Nowadays, more sophisticated attacks could be performed by using high-quality silicone masks for instance [1]. Therefore, a remote authentication mechanism based on the face modality should take such

G. Heusch (✉) · S. Marcel
Idiap Research Institute, Rue Marconi 19, 1920 Martigny, Switzerland
e-mail: guillaume.heusch@idiap.ch

S. Marcel
e-mail: sebastien.marcel@idiap.ch

© Springer Nature Switzerland AG 2019
S. Marcel et al. (eds.), *Handbook of Biometric Anti-Spoofing*,
Advances in Computer Vision and Pattern Recognition,
https://doi.org/10.1007/978-3-319-92627-8_13

threats into account and provide a way to detect presentation attacks. In the last years, several methods to detect such attacks have been proposed, and are surveyed in both [2, 3]. Existing approaches can be roughly divided into two categories. The first category focuses on assessment of the liveliness of the presented biometric sample, by detecting blinking eyes [4] or exploiting motion information [5] for instance. The second category is concerned with finding the differences between images captured from real accesses and images coming from an attack. Representatives examples in this category include texture analysis [6], the usage of image quality measures [7] and frequency analysis [8]. However, current face presentation attacks methods suffers from their inability to generalize to different, or unknown attacks. Usually, existing approaches performs well on the same dataset they were trained on, but have difficulties when attack conditions are different [9]. However, a recent trend consists in deriving robust features that show better generalization abilities: examples can be found in [10, 11]. In the same spirit, presentation attack detection (PAD) based on remote blood pulse measurement is worth investigating: it should theoretically handle different attacks conditions well, since features does not depend on the type of attacks, but rather on properties of *bonafide* attempts.

13.1.1 Remote Photoplethysmography

Photoplethysmography (PPG) consists in measuring the variation in volume inside a tissue, using a light source. Since the heart pumps blood throughout the body, the volume of the arteria is changing with time. When a tissue is illuminated, the proportion of transmitted and reflected light varies accordingly, and the heart rate could thus be inferred from these variations. The aim of remote Photoplethysmography (rPPG) is to measure the same variations, but using ambient light instead of structured light and widely available sensors such as a simple webcam.

It has been empirically shown by Verkruysse et al. [12] that recorded skin colors (and especially the green channel) from a camera sensor contain subtle changes correlated to the variation in blood volumes. In their work, they considered the sequence of average color values in a manually defined region-of-interest (ROI) on the subject's forehead. After having filtered the obtained signals, they graphically showed that the green color signal main frequency corresponds to the heart rate of the subject.

Since then, there have been many attempts to infer the heart rate from video sequences containing skin pixels. Notable examples include the work by Poh et al. [13], where the authors proposed a technique where the color signals are processed by means of blind source separation (ICA), in order to isolate the component corresponding to the heart rate. In a similar trend, Lewandowska et al. [14] applied Principal Component Analysis (PCA) to the color signals and then manually selected the principal component that contains the variation due to blood flow. These two early studies empirically showed that the heart rate could be retrieved from video sequences of faces, but also highlight important limitations: the subject should be motionless, and proper lighting conditions must be ensured during the capture.

According to a recent survey [15], research in remote heart rate measurement has considerably increased in the last few years, most of which focuses on robustness to subject motion and illumination conditions. Since a large number of approaches have been proposed recently, they will not be discussed here. We refer the interested reader to [15, 16] for a comprehensive survey of existing algorithms. Current challenges in rPPG consists mainly of finding methods robust to a wide range of variability. For instance, de Haan et al. specifically devised a method to cope with subject motion in a fitness setting [17]. Also, it has been noted in [16] that different skin color tone affect the retrieved pulse signal. Lin et al. study the effect of different illumination conditions in [18]. Besides, video compression has also been identified as a limitations to retrieve reliable pulse signals [19].

13.1.2 rPPG and Face Presentation Attack Detection

Remote photoplethysmography is still an active research area, and that may explain that it has not been widely used in the context of face presentation attack detection yet. Moreover, and as noted in the previous section, main challenges to be addressed in this field (i.e. subject motion, illumination conditions and video quality) are usually present in a face recognition framework.

Despite its aforementioned limitations, rPPG has some potential for face presentation attack detection, as evidenced by previous work [20–22]. In this work, we thus propose to study pulse-based frequency features, as retrieved by rPPG algorithms, as a mean to discriminate real biometric accesses from presentation attacks. Indeed, in a legitimate, *bonafide* attempt, a consistent pulse signal should be detected, whereas such a signal should mostly consists of noise in case of a presentation attack. Furthermore, such approaches may have the desirable property to detect a wide range of attacks, since they do not rely on attack-specific information. They have the potential to overcome current limitations of classical PAD systems—relying on image quality or texture—through their better generalization abilities. Moreover, they are convenient, since they do not require user cooperation in assessing its liveness (challenge-response) nor do they necessitate additional hardware, such as devices studied in [23].

The typical workflow of a rPPG-based face presentation attack detection system is depicted in Fig. 13.1. Although several aspects of the whole system are considered in this work, our main contribution lies in the usage of long-term statistical spectral features, inspired by a recent work on speaker presentation attack detection [24]. Since these features are not specifically tailored to speech signals and are quite generic, we propose to use them on a pulse signal in the context of face presentation attack detection. Additionally, different rPPG algorithms as well as different classification scheme are studied. Extensive experiments are performed on four publicly available PAD databases following strict evaluation protocols. Besides, all the code needed to

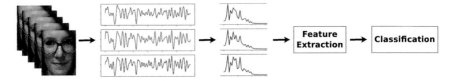

Fig. 13.1 Overview of a typical rPPG-based PAD system

reproduce presented results is made open-source and freely available to the research community.[1]

The rest of the paper is organized as follows: the next section presents prior work on remote physiological measurements for presentation attack detection. Then, proposed features are described, and considered rPPG algorithms as well as classification schemes are outlined. Databases and performances measures are presented in Sect. 13.4, before describing experiments and discussing obtained results. Finally, a conclusion is drawn and suggestions for future research are made in the last section.

13.2 Pulse-Based Approaches to Face Presentation Attack Detection

Remote Photoplethysmography has already been used in applications loosely related to face anti-spoofing. Gibert et al. [25] proposed a face detection algorithm, which builds a map of positive pulsatile response over an image sequence to detect the face. They even state that "Counterfeiting attempts using latex masks or images would be deceived if this map was taken into account". More recent work [26, 27] showed that detecting living skin using rPPG is feasible, at least in lab settings. However, using rPPG in the context of face PAD is still an emerging research area, as evidenced by the few number of previous works. At the time of writing, and to the best of our knowledge, only three studies using rPPG as a mean to detect presentation attack have been published. These previous relevant works are detailed below.

13.2.1 Liu et al.

Liu et al. [20] developed an algorithm based on local rPPG signals and their correlation. First, local pulse signals are extracted from different areas of the face. Usage of local signals is motivated for several reasons: first, it helps with robustness to acquisition conditions (illumination and subject's motion). Second, it can handle the case of a partially masked face, and finally, the strength of local rPPG signals are different

---

[1] Source code and results https://gitlab.idiap.ch/bob/bob.hobpad2.chapter13.

depending on the face area, but the strength pattern is the same across individuals. Local rPPG signals are extracted using the CHROM algorithm [28]. After having modeling the correlation of local pulse signals, a confidence map is learned and used for subsequent classification. Classification is done by feeding a Support Vector Machine (SVM) with local correlation models as features, and with an adapted RBF kernel using the confidence map as the metric. Their approach is evaluated on databases containing masks attacks only, namely 3DMAD [29] augmented with a supplementary dataset comprising six similar masks, plus two additional high-quality silicone masks. Obtained results on these different datasets, including cross dataset tests, show a good performance and hence validate the usage of pulse-based features to reliably detect masks presentation attacks. Unfortunately, the proposed algorithm is not assessed on traditionally used PAD databases, containing photo and video replay attacks.

13.2.2 Li et al.

Li et al. [21] suggest a relatively simple method to detect attacks using pulse-based features. First the pulse signal is retrieved using a simplified version of the algorithm presented in [30]. Three pulse signals—one for each color channel—are extracted by first considering the mean color value of pixels in a specific face area tracked along the sequence. Then, these colors signals are processed with three different temporal filters to finally get pulse signals, one in each color channel. Simple features are then extracted from each frequency spectra, and are concatenated before being fed to a linear SVM classifier. Experiments are again performed on 3DMAD, and also using the supplementary masks. Reported results show a better performance than [20], but do not seem to be directly comparable, since different experimental protocols were applied (training subjects were randomly chosen). An interesting point of this paper is that authors also report results on the MSU-MFSD database [7], and show that their method has difficulty to properly discriminate *bonafide* examples from video replay attacks.

13.2.3 PPGSecure

Nowara et al. [22] follow the same line of work as in [21], but considers the whole frequency spectrum derived from the intensity changes in the green color channel only. As in [20], this approach takes advantage of signals derived from different face areas, but also incorporates information from background areas (to achieve robustness to illumination fluctuations along the sequence). The final feature vector representing a video sequence is formed by concatenating the frequency spectra of pulse signals coming from five areas, three on the face (both cheeks and forehead) plus two on the background. Classification is then done either with a SVM or a random forest classifier. Experiments are performed on the widely used Replay-Attack database [6], but unfortunately, associated protocols have not been followed. Instead, the authors used a leave-one-subject-out cross validation scheme, which greatly increases the ratio of training to test data. Within this experimental framework, a perfect performance (i.e., 100%) accuracy is reported for both photographs and video attacks.

13.2.4 *Discussion and Motivation*

Although relevant, previous studies discussed here make it hard to objectively assess the effectiveness of rPPG-based approaches for face presentation attack detection. Indeed, performance is either reported on non-publicly available data or with different experimental protocols. As a consequence, it is difficult to compare published results with current state-of-the-art that relies on other means to detect attacks. A notable exception is [21], where authors reported results on the MSU-MFSD dataset. It also showed the limitation of such approaches, as compared to traditional face PAD approaches such as texture analysis.

In this work, we hope to help foster research in this area by adopting a reproducible research approach. All the data and the software to reproduce presented results are available to the research community, easing further development in this field. Moreover, our proposed approach is assessed on four publicly available datasets, containing a wide variety of attacks (print and video replays of different quality, and mask attacks). The software package also comprise our own implementation of two other similar approaches, [21, 22], to which our proposed approach is compared.

13.3 Proposed Approach

In this contribution, we suggest to use Long-term spectral statistics (LTSS) [24] as features for face presentation attack detection. This idea was first developed in the context of speaker PAD, and managed to successfully discriminate real speakers from recordings in a speaker authentication task. The main advantage of such features is their ability to deal with any kind of signal and not necessarily speech.

Also, and since there exists a wide variety of rPPG algorithms, it seems important to consider more than one approach since they differ in the way the pulse signal is computed. This results in features that may be more suited to the task of presentation attack detection. To illustrate the difference, the retrieved pulse signals for a *bonafide* video sequence using the three investigated algorithms are shown in Fig. 13.2. One can clearly see that the pulse signals are not the same, depending on the used algorithm.

Furthermore, different classification algorithms are also investigated. In addition to classical two-class discriminative approaches, the usage of one-class classifiers

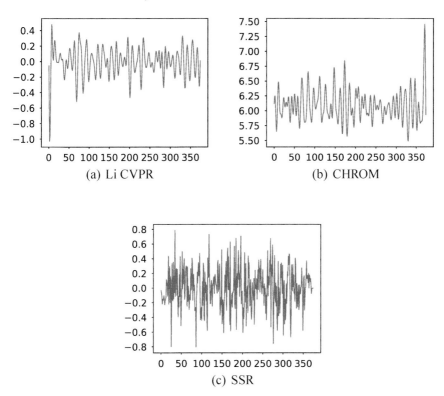

Fig. 13.2 Pulse signals obtained with different rPPG algorithms

considering face presentation attack detection as an *outlier* detection problem are considered. Indeed, recent studies [31, 32] using this paradigm for face presentation attack detection showed promising results. Besides, one-class classifiers have been successfully applied for PAD on other modalities, such as speech [33] or fingerprint [34] and showed better generalization abilities. Furthermore, modeling real samples may be well-suited to pulse-based features, where properties of *bonafide* attempts only are considered.

13.3.1   Long-Term Spectral Statistics

In the context of pulse-based face PAD, and on the contrary to other approaches, prior knowledge on the characteristics of attacks is generally unknown. For instance, LBP-based systems intrinsically assume that texture of faces coming from presentation attacks are different that the one present in *bonafide* face images. These differences are manifold: this could be a lack of texture details on a mask for instance, or undesirable effects such as Moiré patterns or print artifacts in the case of replay and print attacks.

In our framework, the nature of the "pulse" signal extracted from an attack is unknown a priori. Therefore, no prior assumption on the negative class can be made: it is only assumed that signals differ in their statistical characteristics, irrespective of their content (i.e. we do not look specifically for periodicity for instance). As suggested in [24], the means and variances of the energy in different frequency bins provides such a generic characterization.

Long-term spectral statistics are derived by processing the original signal using overlapping temporal windows. In each window w, a N-point Discrete Fourier Transform is computed, and yields a vector $\mathbf{X}_w$ of dimension $k = 0, \ldots, N/2 - 1$ containing DFT coefficients. The statistics of frequency bins of the spectrum are considered using its log-magnitude. As in [24], whenever a DFT coefficient $|X_w(k)|$ is lower than 1, it is clipped to 1 such that the log-magnitude remains positive.

Using the set of DFT coefficient vectors $\mathbf{X}_1, \mathbf{X}_2, \ldots, \mathbf{X}_W$, the first and second order statistics of frequency components are computed as

$$\mu(k) = \frac{1}{W} \sum_{i=1}^{W} \log |X_w(k)| \tag{13.1}$$

$$\sigma^2(k) = \frac{1}{W} \sum_{i=1}^{W} (\log |X_w(k)| - \mu(k)) \tag{13.2}$$

for $k = 0, \ldots, N/2 - 1$. The mean and variance vectors are then concatenated to represent the spectral statistics of a given signal. As a result, the rPPG-based feature for classifying a video sequence consists of a single feature vector. The presentation attack detection is thus performed on the whole video sequence. In other approaches (i.e. texture or image quality-based), detection is generally peformed at the frame level. Long-term spectral statistics feature vectors are then used in conjunction with a classifier to reach a final decision on whether the given video sequence is a *bonafide* example, or an attack.

13.3.2 Investigated rPPG Algorithms

In this section, selected algorithms to retrieve a pulse signal are presented. Two of them, one proposed by Li et al. [30] and CHROM [28] already served as basis for face presentation attack detection in [21] and [20] respectively. The third one, Spatial Subspace Rotation (SSR) [35], has been chosen for both its original analysis (it does not rely on mean skin color processing but rather considers the whole set of skin color pixels) and its potential effectiveness, as demonstrated in [16].

Li CVPR

In this work, a simplified version of the rPPG algorithm originally developed in [30] has been implemented. This simplification has already been used for presentation attack detection in [21]. In particular, the correction for illumination and for motion are ignored. Basically, the pulse signal is obtained by first accumulating the mean skin color value across the lower region of a face in each frame and then to filter the color signal to get the pulse signal. In this work, instead of tracking the lower face region from frame to frame, it is computed at each frame by using a pre-trained facial landmark detector [36].

CHROM

The CHROM approach [28] is relatively simple but has been shown to perform well. The algorithm first finds skin-colored pixels in a given frame and computes the mean skin color. Then, the mean skin color value is projected onto a specific color subspace, which aims to reveal subtle color variations due to blood flow. The final pulse signal is obtained by first bandpass filtering temporal signals in the proposed chrominance colorspace, and then by combining these two filtered signals into one. Note that in our implementation, the skin color filter described in [37] has been used.

SSR

The Spatial Subspace Rotation (SSR) algorithm has been proposed in [35]. It considers the subspace of skin pixels in the RGB space and derives the pulse signal by analyzing the rotation angle of the skin color subspace in consecutive frames. To do so, the eigenvectors of the skin pixels correlation matrix are considered. More precisely, the angle between the principal eigenvector and the hyperplane defined by the two others is analyzed across a temporal window. As claimed by the authors, this algorithm is able to directly retrieve a reliable pulse signal, and hence no post-processing step (i.e., bandpass filtering) is required. Again, skin color pixels are detected using the filter proposed in [37].

13.3.3 Classification

Previous work in rPPG-based face presentation attack detection all rely on SVM—a classical discriminative algorithm—to perform classification of pulse-derived features. Although successful, we believe that choosing a suitable classifier should not be overlooked given the unpredictable nature of attacks. Therefore, a comparison of classification scheme is also performed. Since PAD is inherently a two-class problem, any binary classifier could potentially be used. The literature contains many examples and we refer the interested reader to [2, 3] for a comprehensive overview of existing approaches. In this work, three binary classification algorithms are applied to the proposed features: Support Vector Machine (SVM), Multi-Layer Perceptron (MLP) and Linear Discriminant Analysis (LDA). This choice of algorithms has been motivated by the fact that SVM seems to be the *defacto* standard in face PAD, and because it is used in all the previous work using pulse-based features. MLP and LDA have been chosen since they are used in conjunction with the proposed features in [24].

Although presentation detection attack is usually viewed as a two-class classification problem, it can also be seen as an *outlier* detection problem. According to [3], modeling the genuine examples distribution is a promising research direction, since one cannot anticipate every possible attack type. One-class classification has already been applied in the context of face presentation detection in [31, 32], where one-class SVM and Gaussian Mixture Models (GMM) have been used. These two algorithms are hence also applied to the proposed features here.

13.4 Experiments and Results

13.4.1 Databases

Replay-Attack

The Replay-Attack database was first presented in [6] and contains both *bonafide* attempts and presentation attacks for 50 different subjects. For each subject, two real accesses were recorded under different conditions, referred to as controlled and adverse. Presentation attacks were generated according to three different scenarios:

1. print: high-resolution photographs printed on A4 paper
2. mobile: photos and videos are displayed on an iPhone
3. highdef: photos and videos are displayed on an iPad

Also, two different conditions have been used to display attacks: either held by hand by an operator or attached to a fixed support in order to avoid motion. In total, there are 1200 video sequences, divided into training (360 seq.), development (360 seq.) and evaluation sets (480 seq.). The average sequence length is around 10 s (real accesses are longer and last about 15 s, whereas attacks last around 9 s). Although several protocols have been defined to assess the performance of face PAD algorithms, only the *grandtest* is used here, since it contains all the different attacks and hence allows to test various approaches for a wider range of attacks.

Replay-Mobile

The Replay-Mobile database [38] has been built in the same spirit as of the Replay-Attack database, but with higher quality devices to forge the different attacks. Indeed, attacks are here performed using either high-resolution videos presented on a matte screen or high quality photographs displayed on matte paper. This is done in order to minimize specular reflections, and hence to be closer to real access attempts. This dataset contains 1030 video sequences of 40 subjects, again divided into training (312 seq.), development (416 seq.) and evaluation (302 seq.) sets. The average length of the sequences is 11.8 s, and real accesses and attacks are usually of the same length. Experimental protocols have also been devised in a similar way than in Replay-Mobile, and again, we will restrict ourselves to the *grandtest* protocol, for the same reasons.

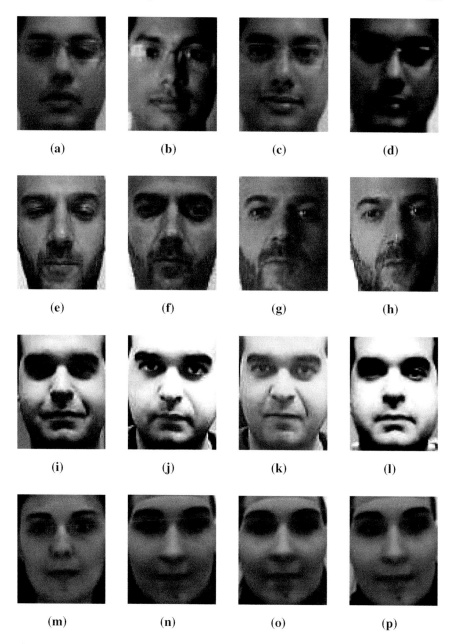

Fig. 13.3 Examples of frames extracted from both *bonafide* accesses (first column) and presentation attacks (column 2–4). The first row shows examples from the Replay-attack database, the second one from replay-mobile, the third one from MSU-MFSD, and the fourth one from 3DMAD

MSU-MFSD

The MSU Mobile Face Spoofing Database has been introduced in [7]. It contains a total of 440 video sequences of 55 subjects, but only a subset comprising 35 subjects has been provided to the research community. Video sequences last around 9 s in average. This database also contains two types of attacks, namely high-quality photograph and video sequences. The publicly available subset specifies 15 subjects used for training and 20 subjects to perform evaluation: these specifications have not been followed here, since no development set is provided. Instead, we build a training set and a development set with 80 video sequences from 10 subjects each, and an evaluation set containing 120 sequences coming from the 15 remaining subjects.

3DMAD

The 3D Mask Attack Database (3DMAD) [29] is the first publicly available database for 3D face presentation detection. It consists in 15 videos sequences of 17 subjects, recorded thanks to a Microsoft Kinect sensor. Note that here, only color sequences are used. The sequences, which all last exactly 10 s, were collected in three different sessions: the first two are *bonafide* accesses and the third one contains the mask attack for each subject. The recordings have been made in controlled conditions and with uniform background. As in [29], we divided the database into training (105 seq. from 7 subjects), development and evaluation sets (75 seq. from 5 subjects in each). However, the random splitting has not been applied here: the training set simply contains the first seven clients, the development set is made with subjects 8–12, and the evaluation set with subjects 13–17. Examples of frames extracted from both real attempts and attacks for all databases can be found in Fig. 13.3).

13.4.2 Performance Measures

Any face presentation attack detection algorithm encounters two type of errors: either an attack is misclassified as a real access, or the other way around, i.e., *bonafide* attempts are wrongly classified as attacks. As a consequence, performance is usually assessed using two metrics. The Attack Presentation Classification Error Rate (APCER) is defined as the expected probability of a successful attack and is defined as follows:

$$APCER = \frac{\# \text{ of accepted attacks}}{\# \text{ of attacks}} \tag{13.3}$$

Conversely, the Bona Fide Presentation Classification Error Rate (BPCER) is defined as the expected probability that a *bonafide* access will be falsely declared as a presentation attack. The BPCER is computed as:

$$BPCER = \frac{\# \text{ of rejected real accesses}}{\# \text{ of real accesses}} \tag{13.4}$$

Note that according to the ISO/IEC 30107-3 standard, each attack type should be taken into account separately. We did not follow this standard here, since our goal is to assess the robustness for a wide range of attacks. Note also that these PAD specific measures relate to the more traditionally used False Acceptance Rate (equivalent to APCER) and False Rejection Rate (equivalent to BPCER).

To provide a single number for the performance, results are typically presented using the Half Total Error Rate (HTER), which is basically the mean of the APCER and the BPCER:

$$HTER(\tau) = \frac{(APCER(\tau) + BPCER(\tau))}{2} \quad [\%] \qquad (13.5)$$

Note that the Half Total Error Rate depends on a threshold τ. Indeed, reducing the Attack Presentation Classification Error Rate will increase the Bonafide Presentation Classification Error Rate and vice-versa. The threshold τ is usually selected to minimize the Equal-Error Rate (EER, the operating point where APCER = BPCER) on the development set.

13.4.3   Experimental Results

In this section, the experimental framework and obtained results are presented. Implementation details are first discussed, before providing experimental results. In particular, a comparison of the proposed LTSS features is made with the spectral features proposed by both Li et al. [21] and Nowara et al. [22]. We then investigate the usage of different rPPG algorithms and classification schemes. Finally, an analysis of obtained results is made: it presents identified shortcomings and suggests directions for future research.

13.4.3.1   Implementation Details

For pulse retrieval, we used open-source implementation of selected rPPG algorithms[2] that have been compared for heart-rate retrieval in [39]. All algorithms have been used with their default parameters. Experiments have been performed on the four databases presented in Sect. 13.4.1, with their associated protocols. In particular, classifiers are trained using specified training sets, and various hyperparameters are optimized to minimize the EER on the development set. Finally, performance is assessed on the evaluation set. Experimental pipelines have been defined and performed using the bob toolbox [40, 41] and, as mentioned in Sect. 13.1, are reproducible by downloading the Python package associated with this contribution.

---

[2]https://pypi.python.org/pypi/bob.rppg.base.

Table 13.1 Performance of different features based on the frequency spectrum of the pulse signals. The HTER [%] is reported on the evaluation set of each databases

	Replay-Attack	Replay-Mobile	MSU-MFSD	3DMAD
Nowara et al. [22]	25.5	35.9	31.7	43.0
Li et al. [21]	27.3	30.7	23.3	29.0
Li CVPR + LTSS	**13.0**	**25.7**	**20.6**	**19.0**

13.4.3.2 Comparison with Existing Approaches

Here we present results for the proposed approach based on LTSS features and compare them with our own implementation of both previously published algorithms also using pulse frequency features [21, 22]. As features used in [21] come from pulse signals retrieved in three color channels, the only choice for the rPPG algorithm is Li CVPR [30]. The same approach has been made using the proposed LTSS features: they are computed from the frequency spectrum in each color channel and concatenated. Note that in the work of Nowara et al. [22], only the green channel has been considered.

For classification, a two-class SVM has been used to be consistent with previous studies. Therefore, the different systems mostly differs in the feature extraction step, making them easily comparable with each other. Table 13.1 shows the HTER performance of the different feature extraction approaches on the evaluation set of the different databases.

As can be seen, the proposed LTSS features achieve the best performance on all considered datasets, and provide a significant improvement over the similar investigated approaches. As compared to [21], where very simple statistics are used, long-term spectral statistics likely contain more information and are hence more suitable to reveal differences between pulse signals retrieved from real attempts and attacks. It also suggests that the temporal window-based analysis of frequency content is suitable: this is not surprising since pulse signals from real attempts should contain some periodicity, whereas pulse signals from attacks should not. Note finally that our implementation of Li's approach has a better performance on the MSU-MFSD dataset than the one reported in the original article [21]. Indeed, an EER of 20.0% is obtained, whereas authors reported an EER of 36.7% in [21].

When compared to features containing magnitude of the whole frequency spectrum in local areas [22], our proposed LTSS features performs consistently better, and by a large margin. This result is interesting for several reasons. First, features extracted from a single face region seem sufficient to retrieve valuable pulse information, as compared to features extracted from different local areas of the face. Second, embedding additional information (i.e features from the background) does not seem to help in this case. Finally, computing relevant statistics on the Fourier spectrum looks more suitable than using the whole spectrum as a feature.

Table 13.2 Performance when different algorithms are used to retrieve the pulse signal. The HTER [%] is reported on the evaluation set of each databases

	Replay-Attack	Replay-Mobile	MSU-MFSD	3DMAD
Li CVPR (green) + LTSS	16.1	**32.5**	**35.0**	17.0
CHROM + LTSS	20.9	38.1	50.6	29.0
SSR + LTSS	**5.9**	37.7	43.3	**13.0**

13.4.3.3 Comparison of Pulse Extraction Algorithms

In this section, we compare the different rPPG algorithms. Indeed, since they yield different pulse signals (see Fig. 13.2), it is interesting to see which one helps the most in discriminating *bonafide* attempts from presentation attacks. Since CHROM and SSR only retrieve a single pulse signal (and not three, as in [30]) LTSS features are derived from this single pulse signal only. For a fair comparison, and when using Li CVPR algorithm [30] for pulse extraction, only the pulse computed in the green channel is considered, since it has been shown that this color channel contains the most variation due to blood flow [16]. Table 13.2 reports the performance for different pulse extraction algorithms.

When comparing rPPG algorithms to retrieve the pulse signal, the SSR algorithm obtains the best performance on two out of four datasets. Actually, it has the overall best performance on both the Replay-Attack database with an HTER of 5.9% and on 3DMAD with an HTER of 13.0%. However, results on other databases do not show performance improvement as compared to the previous experiment, where LTSS features have been extracted and concatenated in three color channels. This suggests that in the context of PAD, all color channels carry valuable information.

13.4.3.4 Comparison of Classification Approaches

In this section, the different classifiers are compared. As mentioned in Sect. 13.3.3, several binary classification algorithms have been considered. The SVM is used with a classical RBF kernel in both two-class and one class settings. The MLP contains a single hidden layer and two outputs representing the two classes. Regarding Linear Discriminant Analysis, PCA is first applied to the features in order to ensure non-singularity of the within-class covariance matrix. Note also that in the LDA case features are projected to a single dimension, which is then directly used as a "score". Table 13.3 shows the obtained performance for different classification schemes.

It is clear from Table 13.3 that different classifiers obtain very different results on the same features. Within discriminant approaches, it is hard to define the most appropriate classifier for this task. They are quite close in terms of averaged performance over the four datasets: SVM has an average HTER of 19.8%, whereas MLP and LDA reach an average HTER of 21.8% and 22.4% respectively. Also, the optimal

Table 13.3 Performance of both tow-class and one-class classifiers. The HTER [%] is reported on the evaluation set of each databases

	Replay-Attack	Replay-Mobile	MSU-MFSD	3DMAD
SVM	**13.4**	26.0	20.6	19.0
MLP	27.8	27.7	**16.1**	**15.0**
LDA	24.6	**24.1**	23.3	17.0
GMM	21.6	54.1	35.6	44.0
OC-SVM	19.6	44.8	31.7	38.0

Table 13.4 HTER [%] performance on evaluation sets, with a breakdown on photo and video replay attacks

	Photo	Video
Replay-Attack	11.3	6.6
Replay-Mobile	19.0	26.5
MSU-MFSD	20.0	15.8

choice for the classifier is dependent on the database: this suggest that fusing different classifiers may be an interesting direction to investigate.

Table 13.3 also shows the poor performance obtained using the *outlier* detection approach. This may be explained by the lack of training data. Actually, modeling the distribution (GMM), or the support region (one-class SVM) of *bonafide* examples may be hard with few examples.

13.4.4 Discussion

In this section, a breakdown is made on the different attack types. This allows to better understand the behavior of our pulse-based face PAD approach, as well as to identify shortcomings, where future efforts should be made.

Table 13.4 shows the HTER of the proposed Li CVPR + LTSS system for two widely-used types of attack: photo and video replays. On the MSU-MFSD database, our approach performs better when dealing with video attacks, and this contradicts the result presented in [21]. Indeed, in the case of a photo attack, the image of the face is the same along the replayed sequence, therefore no pulse signal should be detected. Note that the same remark applies to the Replay-Attack database. This could maybe be explained by the motion introduced when the operator is holding the photograph in front of the camera, which may pollute the retrieved pulse signal. Also, some of the results reported on the Replay-Attack database in [22] exhibit the same trend: a better accuracy is sometimes observed on video attacks than on photo attacks.

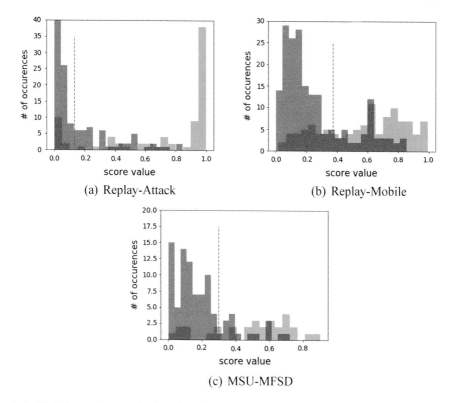

Fig. 13.4 Score values distribution of both *bonafide* accesses (green) and presentation attacks (red) on the evaluation set of the different databases. The dashed-line represents the decision threshold τ selected a priori on the development set. Note that for visualization purposes, the graph for Replay-Attack has been truncated. Actually the leftmost bin goes up to 300, meaning that most of the attacks have a very low score

Finally, the distribution of the scores obtained on the evaluation sets of the three databases containing both photo and video attacks are shown in Fig. 13.4 and provides two interesting insights:

1. Extracting reliable features from pulse signals is still a challenging problem for *bonafide* attempts. This is evidenced by the more uniform distribution of scores for genuine access (depicted in green in Fig. 13.4). This is especially true for both Replay-Mobile and MSU-MFSD databases. As a consequence, the BPCER is usually higher than the APCER.
2. On the other hand, proposed features are able to handle attacks pretty well: the distribution of attack scores (depicted in red in Fig. 13.4) spreads around a relatively low value on the left hand side of the histogram.

To further illustrate these observations, Fig. 13.5 shows example images, corresponding pulses and their respective frequency spectra for both *bonafide* examples

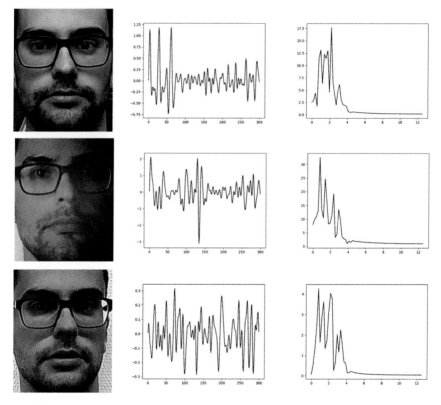

Fig. 13.5 Examples of images, retrieved pulses and their frequency spectrum for both real accesses and attacks from the Replay-Mobile database. The first row shows a legitimate access, the last two rows corresponds to a photo and a video attack respectively

(first row) and different presentation attacks (last two rows) of the Replay-Mobile database. One cannot clearly see differences in the frequency content between attacks and the real example. One would expect that for a real access, the corresponding rPPG signal would have a clear peak in the frequency spectrum that corresponds to the heart rate. In the example depicted in Fig. 13.5, it is actually the opposite: the pulse signal retrieved from the real access has more energy in high frequency components than the one in the photo attack. Note that high-frequency components are not present since the pulse signal is bandpassed; this may discard useful information to identify attacks, but recall that our goal is more oriented toward characterizing real accesses.

The same analysis has been made with mask attacks in the 3DMAD dataset and the score distribution is shown in Fig. 13.6. In this case, different observations can be made. Scores corresponding to *bonafide* examples are not that uniformly distributed and mainly lie on the right handside of the histogram, which is desirable. It means that for this dataset, extracted pulse-based features are more reliable than in previous case. This is not surprising, since sequences have been recorded under clean conditions

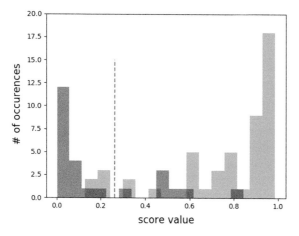

Fig. 13.6 Score values distribution of both *bonafide* accesses (green) and presentation attacks (red) on the evaluation set of the 3DMAD database. The dashed-line represents the decision threshold τ selected a priori on the development set

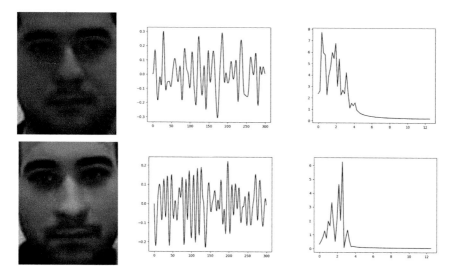

Fig. 13.7 Examples of images, retrieved pulses and their frequency spectrum for both real accesses and attacks from the 3DMAD database. The first row shows a legitimate access, and the second one is an attack

and do not contain as much variations as in other databases. Again, this suggest that illumination is an important factor for reliable pulse extraction.

Also, Fig. 13.7 shows example images, with their retrieved pulses and corresponding spectra for the 3DMAD database. Note here that the difference is easier to spot than in examples from the Replay-Mobile database (Fig. 13.5) and corresponds to expectations. In this case, one can clearly see the frequency component corresponding to the probable heart-rate of the subject (the leftmost peak of the spectrum) for the *bonafide* example. On the contrary, the signal retrieved from the attack is composed of higher frequencies, meaning that in this case, color variations should mainly be due to noise.

Although the proposed approach performs well as compared to other rPPG-based presentation attack detection, it does not reach state-of-the-art performance on these benchmarking datasets yet. Nevertheless, we believe that rPPG-based presentation attack detection systems have the potential to become successful, since there exists room for improvement.

First, and as evidenced in the previous analysis, a reliable pulse signal should be obtained. Current limitations of rPPG algorithms, and in particular illumination condition and compression have been identified and much effort is put on coping with this in current rPPG research. Second, existing approaches—including this one—consider relatively simple, hand-crafted features and progress can also be made here. For instance, Wang et al. successfully used more advanced spectral features in [27] to detect living skin. Moreover, recent advances in speaker presentation attack detection using convolutional neural networks (CNN) [42] show the superiority of such models over hand-crafted features. Finally, other classification approaches are to be studied yet. In particular, taking advantage of the temporal nature of the data using algorithms dedicated to time series, such as Hidden Markov Models or Recurrent Neural Networks, should be worth considering.

13.5   Conclusion

In this work, we studied the usage of remote photoplethysmography for face presentation attack detection. New features containing long term spectral statistics of pulse signals were proposed and successfully applied to this task. Experiments performed on four datasets containing a wide variety of attacks show that the proposed approach outperforms state-of-the-art pulse-based face PAD approaches by a large margin. Analysis of the results revealed that the greatest challenge for such systems is their ability to retrieve reliable pulse signals for *bonafide* attempts. This suggest that future work should first be directed towards improving rPPG algorithms in conditions suitable for PAD, where video quality is not necessarily sufficient for current approaches, and where both illumination variations and subject motion are present. Besides, there is also room for improvement in several other steps of the system. Automatically deriving pulse-based features, using convolutional neural networks for instance, and applying classification schemes tailored for time-series are, in our opinion, research directions worth investigating. Finally, such approaches have the potential to circumvent current limitations of face PAD systems. Actually, they may be well-suited to handle unknown attacks, since they only rely on properties exhibited in *bonafide* accesses, as opposed to approaches based on image quality or texture analysis.

Acknowledgements Part of this research is based upon work supported by the Office of the Director of National Intelligence (ODNI), Intelligence Advanced Research Projects Activity (IARPA), via IARPA R&D Contract No. 2017-17020200005. The views and conclusions contained herein are those of the authors and should not be interpreted as necessarily representing the official policies or endorsements, either expressed or implied, of the ODNI, IARPA, or the U.S. Government. The U.S. Government is authorized to reproduce and distribute reprints for Governmental purposes notwithstanding any copyright annotation thereon.

References

1. Liu S, Yang B, Yuen PC, Zhao G (2016) A 3D mask face anti-spoofing database with real world variations. In: IEEE conference computer vision and pattern recognition workshops (CVPRW), pp 1551–1557
2. Galbally J, Marcel S, Fierrez J (2014) Biometric antispoofing methods: a survey in face recognition. IEEE Access 2:1530–1552
3. Li L, Correia PL, Hadid A (2018) Face recognition under spoofing attacks: countermeasures and research directions. IET Biom 7(1):3–14
4. Pan G, Sun L, Wu Z, Lao S (2007) Eyeblink-based anti-spoofing in face recognition from a generic webcamera. In: International conference on computer vision (ICCV), pp 1–8
5. Anjos A, Marcel S (2011) Counter-measures to photo attacks in face recognition: a public database and a baseline. In: International joint conference on biometrics, pp 1–7
6. Chingovska I, Anjos A, Marcel S (2012) On the effectiveness of local binary patterns in face anti-spoofing. In: International conference of the biometrics special interest group, pp 1–7. IEEE
7. Wen D, Han H, Jain AK (2015) Face spoof detection with image distortion analysis. IEEE Trans Inf Forensics Secur 10(4):746–761
8. Caetano Garcia D, de Queiroz R (2015) Face-spoofing 2D-detection based on Moire-pattern analysis. IEEE Trans Inf Forensics Secur 10(4):778–786
9. de Freitas Pereira T, Anjos A, Martino JMD, Marcel S (2013) Can face anti-spoofing counter-measures work in a real world scenario? In: International conference on biometrics (ICB), pp 1–8
10. Patel K, Han H, Jain AK (2016) Cross-database face antispoofing with robust feature representation. In: Chinese conference on biometric recognition. Lecture Notes in Computer Science (LNCS), vol 9967. LNCS, pp 611–619
11. Patel K, Han H, Jain AK (2016) Secure face unlock: spoof detection on smartphones. IEEE Trans Inf Forensics Secur 11(10):2268–2283
12. Verkruysse W, Svaasand L, Nelson J (2008) Remote plethysmographic imaging using ambient light. Opt Express 16(26):21434–21445
13. Poh M, McDuff D, Picard R (2010) Non-contact, automated cardiac pulse measurements using video imaging and blind source separation. Opt Express 18(10)
14. Lewandowska M, Ruminski J, Kocejko T, Nowak J (2011) Measuring pulse rate with a webcam—a non-contact method for evaluating cardiac activity. In: Proceedings federated conference on computer science and information systems, pp 405–410
15. McDuff D, Estepp J, Piasecki A, Blackford E (2015) A survey of remote optical photoplethysmographic imaging methods. In: IEEE international conference of the engineering in medicine and biology society (EMBC), pp 6398–6404
16. Wang W, den Brinker AC, Stuijk S, de Haan G (2017) Algorithmic principles of remote PPG. IEEE Trans Biomed Eng 64:1479–1491
17. de Haan G, van Leest A (2014) Improved motion robustness of remote-PPG by using the blood volume pulse signature. Physiol Meas 35(9):1913

18. Lin YC, Lin YH (2017) A study of color illumination effect on the SNR of rPPG signals. In: International conference on engineering in medicine and biology society (EMBC), pp 4301–4304

19. McDuff DJ, Blackford EB, Estepp JR (2017) The impact of video compression on remote cardiac pulse measurement using imaging photoplethysmography. In: IEEE international conference on automatic face and gesture recognition (AFGR), pp 63–70

20. Liu S, Yuen P, Zhang S, Zhao G (2016) 3D mask face anti-spoofing with remote photoplethysmography. In: European conference on computer vision (ECCV), pp 85–100

21. Li X, Komulainen J, Zhao G, Yuen PC, Pietikäinen M (2016) Generalized face anti-spoofing by detecting pulse from face videos. In: International conference on pattern recognition (ICPR), pp 4244–4249

22. Nowara EM, Sabharwal A, Veeraraghavan A (2017) PPGSecure: biometric presentation attack detection using photopletysmograms. In: IEEE international conference on automatic face and gesture recognition (AFGR), pp 56–62

23. Bhattacharjee S, Marcel S (2017) What you can't see can help you—extended-range imaging for 3D-mask presentation attack detection. In: International conference of the biometrics special interest group, pp 1–7

24. Muckenhirn H, Korshunov P, Magimai-Doss M, Marcel S (2017) Long-term spectral statistics for voice presentation attack detection. IEEE/ACM Trans Audio Speech Lang Process 25(11):2098–2111

25. Gibert G, D'Alessandro D, Lance F (2013) Face detection method based on photoplethysmography. In: IEEE international conference on advanced video and signal based surveillance, pp 449–453

26. Wang W, Stuijk S, de Haan G (2015) Unsupervised subject detection via remote PPG. IEEE Trans Biomed Eng 62(11):2629–2637

27. Wang W, Stuijk S, de Haan G (2017) Living-skin classification via remote-PPG. IEEE Trans Biomed Eng 64(12):2781–2792

28. de Haan G, Jeanne V (2013) Robust pulse rate from chrominance based rPPG. IEEE Trans Biomed Eng 60(10):2878–2886

29. Erdogmus N, Marcel S (2013) Spoofing in 2D face recognition with 3D masks and anti-spoofing with kinect. In: Proceedings of biometrics: theory, applications and systems (BTAS)

30. Li X, Chen J, Zhao G, Pietikainen M (2014) Remote heart rate measurement from face videos under realistic situations. In: IEEE Conference on computer vision and pattern recognition (CVPR)

31. Arashloo S, Kittler J (2017) An anomaly detection approach to face spoofing detection: a new formulation and evaluation protocol. IEEE Access, 80–89

32. Nikisins O, Mohammadi A, Anjos A, Marcel S (2018) On effectiveness of anomaly detection approaches against unseen presentation attacks in face anti-spoofing. In: International conference on biometrics (ICB)

33. Alegre F, Amehraye A, Evans N (2013) A one-class classification approach to generalised speaker verification spoofing countermeasures using local binary patterns. In: IEEE international conference on biometrics: theory, applications and systems (BTAS)

34. Ding Y, Ross A (2016) An ensemble of one-class SVMs for fingerprint spoof detection across different fabrication materials. IEEE international workshop on information forensics and security (WIFS)

35. Wang W, Stuijk S, de Haan G (2015) A novel algorithm for remote photoplethysmography: spatial subspace rotation. IEEE Trans Biomed Eng

36. King DE (2009) Dlib-ml: a machine learning toolkit. J Mach Learn Res 10:1755–1758

37. Taylor M, Morris T (2014) Adaptive skin segmentation via feature-based face detection. In: SPIE proceedings, real-time image and video processing, vol 9139

38. Costa-Pazo A, Bhattacharjee S, Vazquez-Fernandez E, Marcel S (2016) The replay-mobile face presentation-attack database. In: International conference of the biometrics special interest group

39. Heusch G, Anjos A, Marcel S (2017) A reproducible study on remote heart rate measurement. arXiv
40. Anjos A, Günther M, de Freitas Pereira T, Korshunov P, Mohammadi A, Marcel S (2017) Continuously reproducing toolchains in pattern recognition and machine learning experiments. In: International conference on machine learning (ICML)
41. Anjos A, El Shafey L, Wallace R, Günther M, McCool C, Marcel S (2012) Bob: a free signal processing and machine learning toolbox for researchers. In: ACM conference on multimedia systems (ACMMM)
42. Korshunov P, Gonçalves A, Violato R, Simões F, Marcel S (2018) On the use of convolutional neural networks for speech presentation attack detection. In: International conference on identity, security and behavior analysis (ISBA)

Chapter 14
Review of Face Presentation Attack Detection Competitions

Jukka Komulainen, Zinelabidine Boulkenafet and Zahid Akhtar

Abstract Face presentation attack detection has received increasing attention ever since the vulnerabilities to spoofing have been widely recognized. The state of the art in software-based face anti-spoofing has been assessed in three international competitions organized in conjunction with major biometrics conferences in 2011, 2013, and 2017, each introducing new challenges to the research community. In this chapter, we present the design and results of the three competitions. The particular focus is on the latest competition, where the aim was to evaluate the generalization abilities of the proposed algorithms under some real-world variations faced in mobile scenarios, including previously unseen acquisition conditions, presentation attack instruments, and sensors. We also discuss the lessons learnt from the competitions and future challenges in the field in general.

14.1 Introduction

Spoofing (or presentation attacks as defined in the recent ISO/IEC 30107-3 standard [1]) poses serious security issue to biometric systems in general but face recognition systems in particular are easy to be deceived using images of the targeted person published in the web or captured from distance. Many works (e.g., [2–4]) have concluded that face biometric systems, even those presenting a high recognition performance, are vulnerable to attacks launched with different Presentation Attack Instruments (PAI), such as prints, displays and wearable 3D masks. The vulnerability to presentation attacks (PA) is one of the main reasons to the lack of public confidence in

J. Komulainen (✉) · Z. Boulkenafet
Center for Machine Vision and Signal Analysis, University of Oulu, Oulu, Finland
e-mail: jukka.komulainen@iki.fi

Z. Boulkenafet
e-mail: zboulkenafet@gmail.com

Z. Akhtar
INRS-EMT, University of Quebec, Quebec City, Canada
e-mail: zahid.akhtar.momin@emt.inrs.ca

© Springer Nature Switzerland AG 2019
S. Marcel et al. (eds.), *Handbook of Biometric Anti-Spoofing*,
Advances in Computer Vision and Pattern Recognition,
https://doi.org/10.1007/978-3-319-92627-8_14

(face) biometrics. Also, face recognition based user verification is being increasingly deployed even in high-security level applications, such as mobile payment services. This has created a necessity for robust solutions to counter spoofing.

One possible solution is to include a specific Presentation Attack Detection (PAD) component into a biometric system. PAD (commonly referred to also as anti-spoofing, spoof detection or liveness detection) aims at automatically differentiating whether the presented biometric sample originates from a living legitimate subject or not. PAD schemes can be broadly categorized into two groups: hardware-based and software-based methods. Hardware-based methods introduce some custom sensor into the biometric system that is designed specifically for capturing specific intrinsic differences between a valid living biometric trait and others. Software-based techniques exploit either only the same data that is used for the actual biometric purposes or additional data captured with the standard acquisition device.

Ever since the vulnerabilities of face based biometric systems to PAs have been widely recognized, face PAD has received significant attention in the research community and remarkable progress has been made. Still, it is hard to tell what are the best or most promising practices for face PAD, because extensive objective evaluation and comparison of different approaches is challenging. While it is relatively cheap for an attacker to exploit a known vulnerability of a face authentication system (a "golden fake"), such as a realistic 3D mask, manufacturing a huge amount of face artifacts and then simulating various types of attack scenarios (e.g. use-cases) for many subjects is extremely time-consuming and expensive. This is true especially in the case of hardware-based approaches because capturing new sensor-specific data is always required. Consequently, hardware-based techniques have been usually evaluated just to demonstrate a proof of concept, which makes direct comparison between different systems impossible.

Software-based countermeasures, on the other hand, can be assessed on common protocol benchmark datasets or, even better, if any new data is collected, it can be distributed to the research community. The early works in the field of software-based face PAD were utilizing mainly small proprietary databases for evaluating the proposed approaches but nowadays there exist several common public benchmark datasets, such as [5–10]. The public databases have been indispensable tools for the researchers for developing and assessing the proposed approaches, which has had a huge impact on the amount of papers on data-driven countermeasures during the recent years. However, even if standard benchmarks are used, objective evaluation between different methods is not straightforward. First, the used benchmark datasets may vary across different works. Second, not all the datasets have unambiguously defined evaluation protocols, for example for training and tuning the methods, that provide the possibility for fair and unbiased comparison between different works.

Competitions play a key role in advancing the research on face PAD. It is important to organize collective evaluations regularly in order to assess, or ascertain, the current state of the art and gain insight on the robustness of different approaches using a common platform. Also, new more challenging public datasets are often collected and introduced within such collective efforts to the research community for future development and benchmarking use. The quality of PAIs keeps improving as technology (i.e., printers and displays) gets cheaper and better, which is another reason why benchmark datasets need to be updated regularly. Open contests are likely to inspire researchers and engineers beyond the field to participate, and their outside the box thinking may lead to new ideas on the problem of face PAD and novel countermeasures.

In the context of software-based face PAD, three international competitions [11–13] have been organized in conjunction with major biometric conferences in 2011, 2013 and 2017, each introducing new challenges to the research community. The first competition on countermeasures to 2D face spoofing attacks [12] provided an initial assessment of face PAD by introducing a precisely defined evaluation protocol and evaluating the performance of the proposed face PAD systems under print attacks. The second competition on countermeasures to 2D face spoofing attacks [13] utilized the same evaluation protocol but assessed the effectiveness of the submitted systems in detecting a variety of attacks, introducing display attacks (digital photos and video-replays) in addition to print attacks. While the first two contests considered only known operating conditions, the latest international competition on face PAD [11] aimed to compare the generalization capabilities of the proposed algorithms under some real-world variations faced in mobile scenarios, including unseen acquisition conditions, PAIs and input sensors.

This chapter introduces the state of the art in face PAD with particular focus on the three international competitions. The remainder of the chapter is organized as follows. First, we will give a brief overview on face PAD approaches proposed in the literature in Sect. 14.2. In Sect. 14.3, we will recapitulate the first two international competitions on face PAD, while Sect. 14.4 provides more comprehensive analysis on the latest competition focusing on generalized face PAD in mobile scenarios. In Sect. 14.5, we will discuss the lessons learnt from the competitions and future challenges in the field of face PAD in general. Finally, Sect. 14.6 summarizes the chapter, and presents conclusions drawn from the competitions discussed here.

14.2 Literature Review on Face PAD Methods

There exists no universally accepted taxonomy for the different face PAD approaches. In this chapter, we categorize the methods into two very broad groups: hardware-based and software-based methods.

Hardware-based methods are probably the most robust ones for PAD because the dedicated sensors are able to directly capture or emphasize specific intrinsic differences between genuine and artificial faces in 3D structure [14, 15] and (multi-spectral) reflectance [15–18] properties. For instance, planar PAI detection becomes rather trivial if depth information is available [14], whereas near-infrared (NIR) or thermal cameras are efficient in display attack detection as most of the displays in consumer electronics emit only visible light. On the other hand, these kinds of unconventional sensors are usually expensive and not compact, thus not (yet) available in personal devices, which prevents their wide deployment.

It is rather appealing to perform face PAD by further analyzing only the same data that is used for face recognition or additional data captured with the standard acquisition device (e.g., challenge-response approach). These kinds of software-based methods can be broadly divided into active (requiring user collaboration) and passive approaches. Additional user interaction can be very effectively used for face PAD because we humans tend to be interactive, whereas a photo or video-replay attack cannot respond to randomly specified action requirements. Furthermore, it is very difficult to perform liveness detection or facial 3D structure estimation by relying only on spontaneous facial motion. Challenge-response based methods aim at performing face PAD detection based on whether the required action (challenge), for example facial expression [19, 20], mouth movement [19, 21] or head rotation (3D structure) [22–24], was observed within a predefined time window (response). Also, active software-based methods are able to generalize well across different acquisition conditions and attack scenarios but at the cost of usability due to increased authentication time and system complexity.

Passive software-based methods are preferable for face PAD because they are faster and less intrusive than active countermeasures. Due to the increasing number of public benchmark databases, numerous passive software-based approaches have been proposed for face PAD. In general, passive methods are based on analyzing different facial properties, such as frequency content [8, 25], texture [6, 14, 26–29] and quality [30–32], or motion cues, such as eye blinking [33–36], facial expression changes [19, 33, 35, 36], mouth movements [19, 33, 35, 36], or even color variation due to blood circulation (pulse) [13, 37, 38], to discriminate face artifacts from genuine ones. Passive software-based methods have shown impressive results on the publicly available datasets but the preliminary cross-database tests, such as [24, 39], revealed that the performance is likely to degrade drastically when operating in unknown conditions.

Recently, the research focus on software-based face PAD has been gradually moving towards assessing and improving the generalization capabilities of the proposed and existing methods in a cross-database setup instead of operating solely on single databases. Among hand-crafted feature-based approaches, color texture analysis [40–43], image distortion analysis [9, 31, 32], combination of texture and image quality analysis with interpupillary distance (IPD) based reject option [44], dynamic spectral domain analysis [45] and pulse detection [37] have been applied in the context of generalized face PAD but with only moderate success.

The initial studies using deep CNNs have resulted in excellent intra-test performance but the cross-database results have still been unsatisfactory [44, 46]. This is probably due to the fact that the current publicly available datasets may not provide enough data for training well-known deep neural network architectures from scratch or even for fine-tuning pre-trained networks. As a result, the CNN models have been suffering from overfitting to specific data and learning database-specific information instead of generalized PAD related representations. In order to improve the generalization of CNNs with limited data, more compact feature representations or novel methods for cross-domain adaptation are needed. In [47], deep dictionary learning based formulation was proposed to mitigate the requirement of large amounts of training data with very promising intra-test results but the generalization capability was again unsatisfying. In any case, the potential of application-specific learning needs to be further explored when more comprehensive face PAD databases are available.

14.3 First and Second Competitions on Countermeasures to 2D Face Spoofing Attacks

In this section, we recapitulate the first [12] and second [13] competitions on countermeasures to 2D face spoofing attacks, which were held in conjunction with International Joint Conference on Biometrics (IJCB) in 2011 and International Conference on Biometrics (ICB) in 2013, respectively. Both competitions focused on assessing the stand-alone PAD performance of the proposed algorithms in restricted acquisition conditions, thus integration with actual face verification stage was not considered.

In 2011, the research on software-based face PAD was still in its infancy mainly due to lack of public datasets. Since there were no comparative studies on the effectiveness of different PAD methods under the same data and protocols, the goal of the first competition on countermeasures to 2D facial spoofing attacks [12] was to provide a common platform to compare software-based face PAD using a standardized testing protocol. The performance of different algorithms was evaluated under print attacks using a unique evaluation method. The used PRINT-ATTACK database [48] defines a precise protocol for fair and unbiased algorithm evaluation as it provides a fixed development set to calibrate the countermeasures, while the actual test data is used solely for reporting the final results.

While the first competition [12] provided an initial assessment of face PAD, the 2013 edition of the competition on countermeasures to 2D face spoofing attacks [13] aimed at consolidating the recent advances and trends in the state of the art by evaluating the effectiveness of the proposed algorithms in detecting a variety of attacks. The contest was carried out using the same protocol on the newly collected video REPLAY-ATTACK database [6], introducing display attacks (digital photos and video-replays) in addition to print attacks.

Both competitions were open to all academic and industrial institutions. A noticeable increase in the number of participants between the two competitions can be seen. Particularly, six different competitors from universities participated in the first contest, while eight different teams participated in the second competition. The affiliation and corresponding algorithm name of the participating teams for the two competitions are summarized in Tables 14.1 and 14.2.

In the following, we summarize the design and main results of the first and second competitions on countermeasures to 2D face spoofing attacks. The reader can refer to [12, 13] for more detailed information on the competitions.

Table 14.1 Names and affiliations of the participating systems in the first competition on counter-measures to 2D facial spoofing attacks

Algorithm name	Affiliations
AMILAB	Ambient Intelligence Laboratory, Italy
CASIA	Chinese Academy of Sciences, China
IDIAP	Idiap Research Institute, Switzerland
SIANI	Universidad de Las Palmas de Gran Canaria, Spain
UNICAMP	University of Campinas, Brazil
UOULU	University of Oulu, Finland

Table 14.2 Names and affiliations of the participating systems in the second competition on countermeasures to 2D face spoofing attacks

Algorithm name	Affiliations
CASIA	Chinese Academy of Sciences, China
IGD	Fraunhofer Institute for Computer Graphics, Germany
MaskDown	Idiap Research Institute, Switzerland
	University of Oulu, Finland
	University of Campinas, Brazil
LNMIIT	LNM Institute of Information Technology, India
MUVIS	Tampere University of Technology, Finland
PRA Lab	University of Cagliari, Italy
ATVS	Universidad Autonoma de Madrid, Spain
UNICAMP	University of Campinas, Brazil

14.3.1 Datasets

The first face PAD competition [12] utilized PRINT-ATTACK [48] database consisting of 50 different subjects. The real access and attack videos were captured with a 320×240 pixels (QVGA) resolution camera of a MacBook laptop. The database includes 200 videos of real accesses and 200 videos of print attack attempts. The PAs were launched by presenting hard copies of high-resolution photographs printed on A4 papers with a Triumph-Adler DCC 2520 color laser printer. The videos were recorded under controlled (uniform background) and adverse (non-uniform background with day-light illumination) conditions.

The second competition on face PAD [13] was conducted using an extension of the PRINT-ATTACK database, named as REPLAY-ATTACK database [6]. The database consists of video recordings of real accesses and attack attempts corresponding to 50 clients. The videos were acquired using the built-in camera of a MacBook Air 13 inch laptop under controlled and adverse conditions. Under the same conditions, high-resolution pictures and videos were taken for each person using a Canon PowerShot SX150 IS camera and an iPhone 3GS camera, later to be used for generating the

Fig. 14.1 Sample images from the PRINT-ATTACK [48] and REPLAY-ATTACK [6] databases. Top and bottom rows correspond to controlled and adverse conditions, respectively. From left to right columns: real accesses, print, mobile phone and tablet attacks

attacks. Three different attacks were considered: (i) *print attacks* (i.e., high resolution pictures were printed on A4 paper and displayed to the camera); (ii) *mobile attacks* (i.e., attacks were performed by displaying pictures and videos on the iPhone 3GS screen); (iii) *high definition attacks* (i.e., the pictures and the videos were displayed on an iPad screen with 1024 × 768 pixels resolution). Moreover, attacks were launched with hand-held and fixed support modes for each PAI. Figure 14.1 shows sample images of real and fake faces from both PRINT-ATTACK and REPLAY-ATTACK databases.

14.3.2 Performance Evaluation Protocol and Metrics

The databases used in both competition editions are divided into train, development and test sets with no overlap between them (in terms of subjects or samples). During the system development phase of the first competition, the participants were given access to the labeled videos of the training and the development sets that were used to train and calibrate the devised face PAD methods. In the evaluation phase, the performances of the developed systems were reported on anonymized and unlabeled test video files. In the course of the second competition, the participants had access to all subsets because the competition was conducted on the publicly available REPLAY-ATTACK database. The final test data consisted of anonymized videos of 100 successive frames cut from the original test set videos starting from a random time.

The first and second competitions considered a face PAD method to be prone to two types of errors: either a real access attempt is rejected (false rejection) or a PA is accepted (false acceptance). Both competitions employed Half Total Error Rate (HTER) as principal performance measure metric, which is the average of False Rejection Rate (FRR) and False Acceptance Rate (FAR) at a given threshold τ:

$$HTER(\tau) = \frac{FAR(\tau) + FRR(\tau)}{2} \qquad (14.1)$$

For evaluating the proposed approaches, the participants were asked to provide two files containing a score value for each video in the development and test sets, respectively. The HTER is measured on the test set using the threshold τ corresponding to the Equal Error Rate (EER) operating point on the development set.

14.3.3 Results and Discussion

The algorithms proposed in the first competition on face PAD and the corresponding performances are summarized in Table 14.3. The participated teams used either single or multiple types of visual cues among motion, texture and liveness. Almost every system managed to obtain nearly perfect performance on both development and test sets of the PRINT-ATTACK database. The methods using facial texture analysis dominated because the photo attacks in the competition dataset suffered from obvious print quality defects. Particularly, two teams, IDIAP and UOULU, achieved zero percent error rates on both development and test sets relying solely on local binary pattern (LBP) [49] based texture analysis, while CASIA achieved perfect classification rates on the test set using combination of texture and motion analysis. Assuming that the attack videos usually are noisier than those of real videos, the texture analysis component in CASIA's system is based on estimating the difference in noise variance between the real and attack videos using first order Haar wavelet decomposition. Since the print attacks are launched with fixed and hand-held printouts with incorporated background (see Fig. 14.1), the motion analysis component measures the amount of non-rigid facial motion and face-background motion correlation.

Table 14.3 Overview and performance (in %) of the algorithms proposed in the first face PAD competition (F stands for feature-level and S for score-level fusion)

Team	Features	Fusion	Development			Test		
			FAR	FRR	HTER	FAR	FRR	HTER
AMILAB	Texture, motion and liveness	S	0.00	0.00	0.00	0.00	1.25	0.63
CASIA	Texture, motion	S	1.67	1.67	1.67	0.00	0.00	0.00
IDIAP	Texture	–	0.00	0.00	0.00	0.00	0.00	0.00
SIANI	Motion	–	1.67	1.67	1.67	0.00	21.25	10.63
UNICAMP	Texture, motion and liveness	F	1.67	1.67	1.67	1.25	0.00	0.63
UOULU	Texture	–	0.00	0.00	0.00	0.00	0.00	0.00

Table 14.4 Overview and performance (in %) of the algorithms proposed in the second face PAD competition (F stands for feature-level and S for score-level fusion)

Team	Features	Fusion	Development			Test		
			FAR	FRR	HTER	FAR	FRR	HTER
CASIA	Texture and motion	F	0.00	0.00	0.00	0.00	0.00	0.00
IGD	Liveness	–	5.00	8.33	6.67	17.00	1.25	9.13
MaskDown	Texture and motion	S	1.00	0.00	0.50	0.00	5.00	2.50
LNMIIT	Texture and motion	F	0.00	0.00	0.00	0.00	0.00	0.00
MUVIS	Texture	F	0.00	0.00	0.00	0.00	2.50	1.25
PRA Lab	Texture	S	0.00	0.00	0.00	0.00	2.50	1.25
ATVS	Texture	–	1.67	0.00	0.83	2.75	21.25	12.00
Unicamp	Texture	–	13.00	6.67	9.83	12.50	18.75	15.62

Table 14.4 gives an overview of the algorithms proposed within the second competition on face PAD and the corresponding performance figures for both development and test sets. The participating teams developed face PAD methods based on texture, frequency, image quality, motion and liveness (pulse) features. Again, the use of texture was popular as seven out of eight teams adopted some sort of texture analysis in the proposed systems. More importantly, since the attack scenarios in the second competition were more diverse and challenging, a common approach was combining several complementary concepts together (i.e., information fusion at feature or score level). The category of the used features did not influence the choice of fusion strategy. The best-performing systems were based on feature-level fusion but it is more likely that the high level of robustness is largely based on the feature design rather than the used fusion approach.

From Table 14.4, it can be seen that the two PAD techniques proposed by CASIA and LNMIIT achieved perfect discrimination between the real accesses and the spoofing attacks (i.e., 0.00% error rates on the development and test sets). Both of these top-performing algorithms employ a hybrid scheme combining the features of both texture and motion-based methods. Specifically, the used facial texture descriptions are based on LBP, while motion analysis components again measure the amount of non-rigid facial motion and face-background motion consistency as the new display attacks are inherently similar to the "scenic" print attacks of the previous competition (see Fig. 14.1). The results on the competition dataset suggested that face PAD methods relying on a single cue are not able to detect all types of attacks, and the generalizing capability of the hybrid approaches is higher but with high computational cost. On the other hand, MUVIS and PRA Lab managed to achieve excellent performance on the development and test sets using solely texture analysis. However, it is worth pointing out that both systems compute the texture features over whole video frame (i.e., including background region), thus the methods are severely overfitting to the scene context information that matches across the train, development, and test data. All in all, the astonishing results also on the REPLAY-ATTACK dataset conclude that more challenging configurations are needed before the research on face PAD can reach the next level.

14.4 Competition on Generalized Face Presentation Attack Detection in Mobile Scenarios

The vulnerabilities of face based biometric systems to PAs have been widely recognized but still we lack generalized software-based PAD methods performing robustly in practical (mobile) authentication scenarios. In recent years, many face PAD methods have been proposed and remarkable results have been reported on the existing benchmark datasets. For instance, as seen in Sect. 14.3, several methods achieved perfect error rates in the first [12] and second [13] face PAD competitions. More recent studies, such as [9, 39, 40, 43, 46], have revealed that the existing methods are not able to generalize well in more realistic scenarios, thus software-based face PAD is still an unsolved problem in unconstrained operating conditions.

Focused large scale evaluations on the generalization of face PAD had not been conducted or organized after the issue was first pointed out by de Freitas Pereira et al. [39] in 2013. To address this issue, we organized a competition on mobile face PAD [11] in conjunction with IJCB 2017 to assess the generalization abilities of state-of-the-art algorithms under some real-world variations, including unseen input sensors, PAIs, and illumination conditions. In the following, we will introduce the design and results of this competition in detail.

14.4.1 Participants

The competition was open to all academic and industrial institutions. The participants were required register for the competition and sign the end user license agreement (EULA) of the used OULU-NPU database [5] before obtaining the data for developing the PAD algorithms. Over 50 organizations registered for the competition and 13 teams submitted their systems in the end for evaluation. The affiliation and corresponding algorithm name of the participating teams are summarized in Table 14.5. Compared with the previous competitions, the number of participants increased significantly from six and eight in the first and second competitions, respectively. Moreover, in the previous competitions, all the participated teams were from academic institutes and universities, whereas in this competition, we had registered the participation of three companies as well, which highlights the importance of the topic for both academia and industry.

Table 14.5 Names and affiliations of the participating systems

Algorithm name	Affiliations
Baseline	University of Oulu, Finland
MBLPQ	University of Ouargla, Algeria
PML	University of Biskra, Algeria
	University of the Basque Country, Spain
	University of Valenciennes, France
Massy_HNU	Changsha University of Science and Technology
	Hunan University, China
MFT-FAS	Indian Institute of Technology Indore, India
GRADIANT	Galician Research and Development Center in Advanced Telecommunications, Spain
Idiap	Ecole Polytechnique Federale de Lausanne Idiap Research Institute, Switzerland
VSS	Vologda State University, Russia
SZUCVI	Shenzhen University, China
MixedFasNet	FUJITSU laboratories LTD, Japan
NWPU	Northwestern Polytechnical University, China
HKBU	Hong Kong Baptist University, China
Recod	University of Campinas, Brazil
CPqD	CPqD, Brazil

14.4.2   Dataset

The competition was carried out on the recently published[1] OULU-NPU face presentation attack database [5]. The dataset and evaluation protocols were designed particularly for evaluating the generalization of face PAD methods in more realistic mobile authentication scenarios by considering three covariates: unknown environmental conditions (namely illumination and background scene), PAIs and acquisition devices, separately and at once.

The OULU-NPU database consists of 4950 short video sequences of real access and attack attempts corresponding to 55 subjects (15 female and 40 male). The real access attempts were recorded in three different sessions separated by a time interval of one week. During each session, a different illumination condition and background scene were considered (see Fig. 14.2):

- *Session 1*: The recordings were taken in an open-plan office where the electric light was switched on, the windows blinds were open, and the windows were located behind the subjects.
- *Session 2*: The recordings were taken in a meeting room where the electric light was the only source of illumination.

---

[1]The dataset was not yet released at the time of the competition.

Fig. 14.2 Sample images of a real subject highlighting the illumination conditions across the three different scenarios

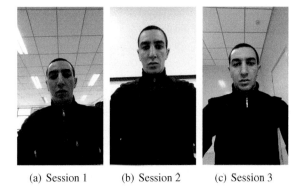

(a) Session 1   (b) Session 2   (c) Session 3

- *Session 3*: The recordings were taken in a small office where the electronic light was switched on, the windows blinds were open, and the windows were located in front of the subjects.

During each session, the subjects recorded the videos of themselves using the front-facing cameras of the mobile devices. In order to simulate realistic mobile authentication scenarios, the video length was limited to five seconds. Furthermore, the subjects were asked to use the device naturally while ensuring that the whole face is visible through the whole video sequence.

Six smartphones with high-quality front-facing cameras in the price range from €250 to €600 were used for the data collection:

- Samsung Galaxy S6 edge with 5 MP frontal camera (Phone 1).
- HTC Desire EYE with 13 MP frontal camera (Phone 2).
- MEIZU X5 with 5 MP frontal camera (Phone 3).
- ASUS Zenfone Selfie with 13 MP frontal camera (Phone 4).
- Sony XPERIA C5 Ultra Dual with 13 MP frontal camera (Phone 5).
- OPPO N3 with 16 MP rotating camera (Phone 6).

The videos were recorded at Full HD resolution (i.e., 1920×1080) using the same camera software[2] installed on each device. Even though the nominal camera resolution of some mobile devices is the same, such as Phone 2, Phone 4 and Phone 5 (13 MP), significant differences can be observed in the quality of the resulting videos as demonstrated in Fig. 14.3.

During each of the three sessions, a high-resolution photo and a video of each user was captured using the back camera of the Phone 1 capable of taking 16 MP still images and Full HD videos. These high resolution photos and videos were then used to create the PAs. The attack types considered in this database are print and video-replay attacks:

---

[2]http://opencamera.sourceforge.net/.

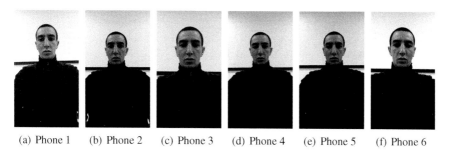

(a) Phone 1   (b) Phone 2   (c) Phone 3   (d) Phone 4   (e) Phone 5   (f) Phone 6

Fig. 14.3 Sample images showing the image quality of the different camera devices

(a) Print 1   (b) Print 2   (c) Display 1   (d) Display 2

Fig. 14.4 Samples of print and display attacks taken with the front camera of Sony XPERIA C5 Ultra Dual

- *Print attacks*: The high resolution photos were printed on A3 glossy paper using two different printers: a Canon imagePRESS C6011 (Printer 1) and a Canon PIXMA iX6550 (Printer 2).
- *Video-replay attacks*: The high-resolution videos were replayed on two different display devices: a 19″ Dell UltraSharp 1905FP display with 1280×1024 resolution (Display 1) and an early 2015 Macbook 13″ laptop with Retina display of 2560×1600 resolution (Display 2).

The print and video-replay attacks were then recorded using the front-facing cameras of the six mobile phones. While capturing the print attacks, the facial prints were held by the operator and captured with stationary capturing devices in order to maximize the image quality but still introduce some noticeable motion in the print attacks. In contrast, when recording the video-replay attacks both of the capturing devices and PAIs were stationary. Furthermore, we paid special attention that the background scene of the attacks matched that of the real accesses during each session and that the attack videos did not include the bezels of the screens or borders of the prints. Figure 14.4 shows samples of the attacks captured using the Phone 5.

14.4.3 Performance Evaluation Protocol and Metrics

During the system development phase of 2 months, the participants were given access to the labeled videos of the training and the development sets that were used to train and tune the devised face PAD methods. In addition to the provided training set, the participants were allowed to use external data to train their algorithms. In the evaluation phase of two weeks, the performances of the developed systems were reported on anonymized and unlabeled test video files. To assess the generalization of the developed face PAD methods, four protocols have been used:

Protocol I: This protocol is designed to evaluate the generalization of the face PAD methods under previously unseen environmental conditions, namely illumination and background scene. As the database is recorded in three sessions with different illumination condition and location, the train, development and evaluation sets are constructed using video recordings taken in different sessions.

Protocol II: This protocol is designed to evaluate the effect of attacks created with different printers or displays on the performance of the face PAD methods as they may suffer from new kinds of artifacts. The effect of attack variation is assessed by introducing previously unseen print and video-replay attacks in the test set.

Protocol III: One of the critical issues in face PAD and image classification in general is sensor interoperability. To study the effect of the input camera variation, a Leave One Camera Out (LOCO) protocol is used. In each iteration, the real and the attack videos recorded with five smartphones are used to train and tune the algorithms, and the generalization of the models is assessed using the videos recorded with the remaining smartphone.

Protocol IV: In the most challenging protocol, all above three factors are considered simultaneously and generalization of face PAD methods are evaluated across previously unseen environmental conditions, attacks, and sensors.

Table 14.6 gives detailed information about the video recordings used in the train, development and test sets of each test scenario. For every protocol, the participants were asked to provide separate score files for the development and test sets containing a single score for each video.

For the performance evaluation, we selected the recently standardized ISO/IEC 30107-3 metrics [1], Attack Presentation Classification Error Rate (APCER) and Bona Fide Presentation Classification Error Rate (BPCER):

$$APCER_{PAI} = \frac{1}{N_{PAI}} \sum_{i=1}^{N_{PAI}} (1 - Res_i) \qquad (14.2)$$

$$BPCER = \frac{\sum_{i=1}^{N_{BF}} Res_i}{N_{BF}} \qquad (14.3)$$

where, N_{PAI}, is the number of the attack presentations for the given PAI, N_{BF} is the total number of the bona fide presentations. Res_i takes the value 1 if the ith

Table 14.6 The detailed information about the video recordings in the train, development and test sets of each protocol (P stands for print and D for display attack)

Protocol	Subset	Session	Phones	Subjects	Attacks	Real/Attack videos
Protocol I	Train	1, 2	6	1–20	P 1, 2; D 1, 2	240/960
	Dev	1, 2	6	21–35	P 1, 2; D 1, 2	180/720
	Test	3	6	36–55	P 1, 2; D 1, 2	120/480
Protocol II	Train	1, 2, 3	6	1–20	P 1; D 1	360/720
	Dev	1, 2, 3	6	21–35	P 1; D 1	270/540
	Test	1, 2, 3	6	36–55	P 2; D 2	360/720
Protocol III	Train	1, 2, 3	5	1–20	P 1, 2; D 1, 2	300/1200
	Dev	1, 2, 3	5	21–35	P 1, 2; D 1, 2	225/900
	Test	1, 2, 3	1	36–55	P 1, 2; D 1, 2	60/240
Protocol IV	Train	1, 2	5	1–20	P 1; D 1	200/400
	Dev	1, 2	5	21–35	P 1; D 1	150/300
	Test	3	1	36–55	P 2; D 2	20/40

presentation is classified as an attack presentation and 0 if classified as bona fide presentation. These two metrics correspond to the False Acceptance Rate (FAR) and False Rejection Rate (FRR) commonly used in the PAD related literature. However, $APCER_{PAI}$ is computed separately for each PAI (e.g., print or display) and the overall PAD performance corresponds to the attack with the highest APCER (i.e., the "worst case scenario").

To summarize the overall system performance in a single value, the Average Classification Error Rate (ACER) is used, which is the average of the APCER and the BPCER at the decision threshold defined by the Equal Error Rate (EER) on the development set:

$$ACER = \frac{\max_{PAI=1...S} (APCER_{PAI}) + BPCER}{2} \tag{14.4}$$

where S is the number of the PAIs. In Protocols III and IV, these measures (i.e., APCER, BPCER and ACER) are computed separately for each mobile phone, and the average and standard deviation are taken over the folds to summarize the results. Since the attack potential of the PAIs may vary across the different folds, the overall APCER does not necessarily correspond to the highest mean $APCER_{PAI}$.

14.4.4   Baseline

In addition to the training and development data, the participants were given the source code[3] of the baseline face PAD method that could be freely improved or used

---

[3]The source code for baseline can be downloaded along with the OULU-NPU database.

as it is in the final systems. The color texture based method [40] was as the baseline because it has shown promising generalization abilities. In this method, the texture features are extracted from the color images instead of the grayscale representation that has been more commonly used in face PAD, for example in [14, 27–29]. The key idea behind color texture based face PAD is that an image of an artificial face is actually an image of a face which passes through two different camera systems and a printing system or a display device, thus it can be referred to in fact as a recaptured image. As a consequence, the observed artificial face image is likely to suffer from different kinds of quality issues, such as printing defects, video artifacts, PAI dependent (local) color variations and limited color reproduction (gamut), that can be captured by analyzing the texture content of both luminance and chrominance channels.

The steps of the baseline method are the following. First, the face is detected, cropped and normalized into 64 × 64 pixels. Then, the RGB face image is converted into HSV and YCbCr color spaces. The local binary pattern (LBP) texture features [49] are extracted from each channel of the color spaces. The resulting feature vectors are concatenated into an enhanced feature vector which is fed into a Softmax classifier. The final score for each video is computed by averaging the output scores of ten random frames.

14.4.5 Results and Discussion

In this competition, typical "liveness detection" was not adopted as none of the submitted systems is explicitly aiming at detecting physiological signs of life, such as eye blinking, facial expression changes and mouth movements. Instead, every proposed face PAD algorithm relies on one or more types of feature representations extracted from the face and/or the background regions. The used descriptors can be categorized into three groups (see Table 14.7): hand-crafted, learned and hybrid (fusion of hand-crafted and learned). The performances of the submitted systems under the four test protocols are reported in Tables 14.8, 14.9, 14.10 and 14.11.

It appears that the analysis of mere grayscale or even RGB images does not result in particularly good generalization. In the case of hand-crafted features, every algo-

Table 14.7 Categorization of the proposed systems based on hand-crafted, learned and hybrid features

Category	Teams
Hand-crafted features	Baseline, MBLPQ, PML, Massy_HNU, MFT-FAS, GRADIANT, Idiap
Learned features	VSS, SZCVI, MixedFASNet
Hybrid features	NWPU, HKBU, Recod, CPqD

Table 14.8   The performance (%) of the proposed methods under different illumination and location conditions (Protocol I)

Methods	Dev	Test				
	EER	Display	Print	Overall		
		APCER	APCER	APCER	BPCER	ACER
GRADIANT_extra	0.7	7.1	3.8	7.1	5.8	**6.5**
CPqD	0.6	1.3	2.9	2.9	10.8	6.9
GRADIANT	1.1	**0.0**	1.3	1.3	12.5	6.9
Recod	2.2	3.3	0.8	3.3	13.3	8.3
MixedFASNet	1.3	**0.0**	**0.0**	**0.0**	17.5	8.8
PML	0.6	7.5	11.3	11.3	9.2	10.2
Baseline	4.4	5.0	1.3	5.0	20.8	12.9
Massy_HNU	1.1	5.4	3.3	5.4	20.8	13.1
HKBU	4.3	9.6	7.1	9.6	18.3	14.0
NWPU	**0.0**	8.8	7.5	8.8	21.7	15.2
MFT-FAS	2.2	0.4	3.3	3.3	28.3	15.8
MBLPQ	2.2	31.7	44.2	44.2	**3.3**	23.8
Idiap	5.6	9.6	13.3	13.3	40.0	26.7
VSS	12.2	20.0	12.1	20.0	41.7	30.8
SZUCVI	16.7	11.3	**0.0**	11.3	65.0	38.1
VSS_extra	24.0	9.6	11.3	11.3	73.3	42.3

rithm is based on the recently proposed color texture analysis [40] in which RGB images are converted into HSV and/or YCbCr color spaces prior feature extraction. The only well-generalizing feature learning based method, MixedFASNet, uses HSV images as input, whereas the networks operating on grayscale or RGB images do not generalize well. On the other hand, it is worth mentioning that VSS and SZCVI architectures consist only of five convolutional layers, whereas the MixedFASNet, consisting of over 30 layers, is much deeper. The best performing hybrid methods, Recod and CPqD, fuse the scores of their deep learning based method and the provided baseline in order to increase the generalization capabilities. Since only the scores of hybrid systems were provided, the robustness of the proposed fine-tuned CNN models operating on RGB images remains unclear. Among the methods solely based on RGB image analysis, HKBU fusing IDA, LBP and deep features is the only one that generalizes fairly well across the four protocols.

In general, the submitted systems process each (selected) frame of a video sequence independently then the final score for a given video is obtained by averaging the resulting scores of individual frames. None of the deep learning or hybrid methods exploited temporal variations but in the case of hand-crafted features two different temporal aggregation approaches were proposed for encoding the dynamic information within a video sequence, for example motion. MBLPQ and PML averaged the feature vectors over the sampled frames, whereas GRADIANT and MFT-FAS

Table 14.9 The performance (%) of the proposed methods under novel attacks (Protocol II)

Methods	Dev	Test				
	EER	Display	Print	Overall		
		APCER	APCER	APCER	BPCER	ACER
GRADIANT	0.9	**1.7**	3.1	**3.1**	**1.9**	**2.5**
GRADIANT_extra	0.7	6.9	**1.1**	6.9	2.5	4.7
MixedFASNet	1.3	6.4	9.7	9.7	2.5	6.1
SZUCVI	4.4	3.9	3.3	3.9	9.4	6.7
MFT-FAS	2.2	10.0	11.1	11.1	2.8	6.9
PML	0.9	11.4	9.4	11.4	3.9	7.6
CPqD	2.2	9.2	14.7	14.7	3.6	9.2
HKBU	4.6	13.9	12.5	13.9	5.6	9.7
Recod	3.7	13.3	15.8	15.8	4.2	10.0
MBLPQ	1.9	5.6	19.7	19.7	6.1	12.9
Baseline	4.1	15.6	22.5	22.5	6.7	14.6
Massy_HNU	1.3	16.1	26.1	26.1	3.9	15.0
Idiap	8.7	21.7	7.5	21.7	11.1	16.4
NWPU	**0.0**	12.5	5.8	12.5	26.7	19.6
VSS	14.8	25.3	13.9	25.3	23.9	24.6
VSS_extra	23.3	36.1	33.9	36.1	33.1	34.6

map the temporal variations into a single image prior feature extraction [11]. The approach by GRADIANT turned out to be particularly successful as the achieved performance was simply the best and most consistent across all the four protocols.

In this competition, the simple color texture based face descriptions were very powerful compared to deep learning based methods, of which the impressive results achieved by GRADIANT are a good example. On the other hand, the current (public) datasets may not probably provide enough data for training CNNs from scratch or even fine-tuning the pre-trained models to their full potential. NWPU extracted LBP features from convolutional layers in order to reduce the number of trainable parameters, thus avoiding the need for enormous training sets. Unfortunately, the method did not generalize well on the evaluation set.

Few teams used additional public and/or proprietary datasets for training and tuning their algorithms. The VSS team augmented the subset of real subjects with CASIA-WebFace and collected their own attack samples. The usefulness of these external datasets remains unclear because their grayscale image analysis based face PAD method did not perform well. Recod used publicly available datasets for fine-tuning the pre-trained network but the resulting generalization was comparable to similar method, CPqD, that did not use any extra-data. GRADIANT submitted two systems with and without external training data. Improved BPCER was obtained in unseen acquisition conditions but APCER was much better in general when using only the provided OULU-NPU training data.

Table 14.10 The performance (%) of the proposed methods under input camera variations (Protocol III)

Methods	Dev	Test				
	EER	Display	Print	Overall		
		APCER	APCER	APCER	BPCER	ACER
GRADIANT	0.9 ± 0.4	1.0 ± 1.7	2.6 ± 3.9	2.6 ± 3.9	**5.0 ± 5.3**	**3.8 ± 2.4**
GRADIANT_extra	0.7 ± 0.2	1.4 ± 1.9	**1.4 ± 2.6**	**2.4 ± 2.8**	5.6 ± 4.3	4.0 ± 1.9
MixedFASNet	1.4 ± 0.5	1.7 ± 3.3	5.3 ± 6.7	5.3 ± 6.7	7.8 ± 5.5	6.5 ± 4.6
CPqD	0.9 ± 0.4	4.4 ± 3.4	5.0 ± 6.1	6.8 ± 5.6	8.1 ± 6.4	7.4 ± 3.3
Recod	2.9 ± 0.7	4.2 ± 3.8	8.6 ± 14.3	10.1 ± 13.9	8.9 ± 9.3	9.5 ± 6.7
MFT-FAS	0.8 ± 0.4	**0.8 ± 0.9**	10.8 ± 18.1	10.8 ± 18.1	9.4 ± 12.8	10.1 ± 9.9
Baseline	3.9 ± 0.7	9.3 ± 4.3	11.8 ± 10.8	14.2 ± 9.2	8.6 ± 5.9	11.4 ± 4.6
HKBU	3.8 ± 0.3	7.9 ± 5.8	9.9 ± 12.3	12.8 ± 11.0	11.4 ± 9.0	12.1 ± 6.5
SZUCVI	7.0 ± 1.6	10.0 ± 8.3	7.5 ± 9.5	12.1 ± 10.6	16.1 ± 8.0	14.1 ± 4.4
PML	1.1 ± 0.3	8.2 ± 12.5	15.3 ± 22.1	15.7 ± 21.8	15.8 ± 15.4	15.8 ± 15.1
Massy_HNU	1.9 ± 0.6	5.8 ± 5.4	19.0 ± 26.7	19.3 ± 26.5	14.2 ± 13.9	16.7 ± 10.9
MBLPQ	2.3 ± 0.6	5.8 ± 5.8	12.9 ± 4.1	12.9 ± 4.1	21.9 ± 22.4	17.4 ± 10.3
NWPU	**0.0 ± 0.0**	1.9 ± 0.7	1.9 ± 3.3	3.2 ± 2.6	33.9 ± 10.3	18.5 ± 4.4
Idiap	7.9 ± 1.9	8.3 ± 3.0	9.3 ± 10.0	12.9 ± 8.2	26.9 ± 24.4	19.9 ± 11.8
VSS	14.6 ± 0.8	21.4 ± 7.7	13.8 ± 7.0	21.4 ± 7.7	25.3 ± 9.6	23.3 ± 2.3
VSS_extra	25.9 ± 1.7	25.0 ± 11.4	32.2 ± 27.9	40.3 ± 22.2	35.3 ± 27.4	37.8 ± 6.8

Since unseen attack scenarios will be definitely experienced in operation, the problem of PAD could be easily ideally solved using one-class classifiers for modeling the variations of the only known class (i.e., bona-fide). Idiap method is based on the idea of anomaly detection but it lacked generalization mainly because the individual grayscale image analysis based methods were performing poorly.[4] Thus, one-class modeling would be worth investigating when combined with more robust feature representations.

Several general observations can be made based on the results of protocols I, II and III assessing the generalization of the PAD method across unseen conditions (i.e., acquisition conditions, attack types and sensors, separately):

Protocol I: In general, a significant increase in BPCER can be noticed compared to APCER when the PAD systems are operating in new acquisition conditions. The reason behind this may be in the data collection principles of the OULU-NPU dataset. Legitimate users have to be verified in various conditions, while attackers aim probably at high-quality attack presentation in order to increase the chance of successfully fooling a face biometric system. The bona-fide samples were collected in three sessions with different illumination. In contrast, the bona-fide data corresponding to each session was used to create face artifacts but the attacks themselves were always

---

[4]Idiap submitted also the scores of the individual sub-systems.

Table 14.11 The performance (%) of the proposed methods under environmental, attack and camera device variations (Protocol IV)

Methods	Dev	Test				
	EER	Display	Print	Overall		
		APCER	APCER	APCER	BPCER	ACER
GRADIANT	1.1 ± 0.3	**0.0 ± 0.0**	5.0 ± 4.5	5.0 ± 4.5	15.0 ± 7.1	**10.0 ± 5.0**
GRADIANT_extra	1.1 ± 0.3	27.5 ± 24.2	5.8 ± 4.9	27.5 ± 24.2	**3.3 ± 4.1**	15.4 ± 11.8
Massy_HNU	1.0 ± 0.4	20.0 ± 17.6	26.7 ± 37.5	35.8 ± 35.3	8.3 ± 4.1	22.1 ± 17.6
CPqD	2.2 ± 1.7	16.7 ± 16.0	24.2 ± 39.4	32.5 ± 37.5	11.7 ± 12.1	22.1 ± 20.8
Recod	3.7 ± 0.7	20.0 ± 19.5	23.3 ± 40.0	35.0 ± 37.5	10.0 ± 4.5	22.5 ± 18.2
MFT-FAS	1.6 ± 0.7	**0.0 ± 0.0**	12.5 ± 12.9	12.5 ± 12.9	33.3 ± 23.6	22.9 ± 8.3
MixedFASNet	2.8 ± 1.1	10.0 ± 7.7	4.2 ± 4.9	10.0 ± 7.7	35.8 ± 26.7	22.9 ± 15.2
Baseline	4.7 ± 0.6	19.2 ± 17.4	22.5 ± 38.3	29.2 ± 37.5	23.3 ± 13.3	26.3 ± 16.9
HKBU	5.0 ± 0.7	16.7 ± 24.8	21.7 ± 36.7	33.3 ± 37.9	27.5 ± 20.4	30.4 ± 20.8
VSS	11.8 ± 0.8	21.7 ± 8.2	9.2 ± 5.8	21.7 ± 8.2	44.2 ± 11.1	32.9 ± 5.8
MBLPQ	3.6 ± 0.7	35.0 ± 25.5	45.0 ± 25.9	49.2 ± 27.8	24.2 ± 27.8	36.7 ± 4.7
NWPU	**0.0 ± 0.0**	30.8 ± 7.4	6.7 ± 11.7	30.8 ± 7.4	44.2 ± 23.3	37.5 ± 9.4
PML	0.8 ± 0.3	59.2 ± 24.2	38.3 ± 41.7	61.7 ± 26.4	13.3 ± 13.7	37.5 ± 14.1
SZUCVI	9.1 ± 1.6	**0.0 ± 0.0**	**0.8 ± 2.0**	**0.8 ± 2.0**	80.8 ± 28.5	40.8 ± 13.5
Idiap	6.8 ± 0.8	26.7 ± 35.2	13.3 ± 8.2	33.3 ± 30.4	54.2 ± 12.0	43.8 ± 20.4
VSS_extra	21.1 ± 2.7	13.3 ± 17.2	15.8 ± 21.3	25.8 ± 20.8	70.0 ± 22.8	47.9 ± 12.1

launched with short standoff and captured in the same laboratory setup. Thus, the intrinsic properties of the attacks do not vary too much across the different sessions. **Protocol II**: In most cases, previously unseen attack leads into dramatic increase in APCER, which is expected as only one PAI of each print and video-replay attacks is provided for training and tuning purposes.

Protocol III: It is also interesting to notice that the standard deviation of APCER across different sensors is much larger in the case of print attacks compared to video-replay attacks, which suggests that the nature of print attacks seems to vary more although both attack types can be detected equally well on average.

Based on the results of the protocol IV, it is much harder to make general conclusions because all the factors are combined and different approaches seem to be more robust to different covariates. The last protocol reveals, however, that none of the methods is able to achieve a reasonable trade-off between usability and security. For instance, in the case of GRADIANT, either the APCER or BPCER of the two systems is too high for practical applications. Nevertheless, the overall performance of GRADIANT, MixedFASNET, CPqD and Recod is very impressive considering the challenging conditions of the competition and the OULU-NPU dataset.

14.5   Discussion

All the three competitions on face PAD were very successful in consolidating and benchmarking the current state of the art. In the following, we provide general observations and further discussion on the lessons learnt and potential future challenges.

14.5.1   General Observations

It can be noticed that the used datasets and evaluation protocols, and also the recent advances in the state of the art reflect the face PAD scheme trends seen in the different contests. The algorithms proposed in the first and second competitions on countermeasures to 2D face spoofing attacks exploited the evident visual cues that we humans can observe in the videos of the PRINT-ATTACK and REPLAY-ATTACK databases, including localized facial movements, global motion, face-background motion correlation, print quality defects and other degradations in facial texture quality. While simple texture analysis was sufficient for capturing the evident printing artifacts in the PRINT-ATTACK database, fusion of multiple visual cues was needed for achieving robust performance under variety of attacks of the REPLAY-ATTACK database. The perfect error rates of the best-performing PAD schemes in homogeneous development and test conditions indicated that more challenging configurations were needed for future benchmarks.

In the competition on generalized face PAD, typical liveness detection and motion analysis were hardly used. In general, the proposed solutions relied on one or more types of feature representations extracted from the face and/or background regions using hand-crafted and/or learned descriptors, which is not surprising considering the recent trends in (face) PAD. Color texture analysis had shown promising generalization capabilities in preliminary studies [40–42]. This explains why most teams proposed new facial color texture representations or used the provided baseline as a complementary PAD method. Although it was nice to see a diverse set of deep learning based systems and further improved versions of the provided baseline method, it was bit disappointing that entirely novel generalized face PAD solutions were not proposed. While the best-performing approaches were able to generalize remarkably well under the individual unknown conditions, no major breakthrough in generalized face PAD was achieved as the none of the methods was able to achieve satisfying performance under the most challenging test protocol, Protocol IV.

14.5.2   Lessons Learnt

The competitions have given valuable lessons on designing databases and test protocols, and competitions in general. In the second competition on countermeasures to

2D face spoofing attacks, two teams managed to achieve perfect discrimination on the REPLAY-ATTACK database, and consequently PRINT-ATTACK database, by computing texture features over the whole video frame. The two background conditions in the REPLAY-ATTACK dataset are the same across the training, development and test sets and the corresponding scene is incorporated in the attack presentations (see Fig. 14.1). Thus, also the differences in background scene texture between the real access and attack videos match between the development and test data, while only the facial texture is unknown due to previously unseen subjects. It is also worth mentioning that the original video encoder of the REPLAY-ATTACK dataset was not used for creating the randomly sampled test videos. The resulting video encoding artifacts and noise patterns did not match between the development and test phases, which might explain the increase in FRR for the methods relying largely on static and dynamic texture analysis.

In the third competition, focusing on generalization in face PAD, the time between the release of test data and submission of results was two weeks. The labeled test set of OULU-NPU database was not yet publicly available during the competition. However, we humans are apt in differentiating attack videos from real ones and the test subset of the OULU-NPU database contains still only 1800 videos. Therefore, it was feasible to label the anonymized and unlabeled test data by hand for "data peeking", that is calibrating, or even training, the systems on the test data. This kind of cheating could be prevented by hiding some "anchor" videos from the development set (with randomized file names) in the evaluation data and releasing the augmented test set once the development set scores have been submitted (fixed), as done in the BTAS 2016 Speaker Anti-spoofing Competition [50]. The scores of the anchor videos could be used for checking whether the scores for the development and test sets have been generated by the same system.

An even more serious concern with the third competition is that the data provided for system development contained all variations in attacks, input sensors and acquisition conditions that the generalization was tested for. While only a specific subset defined in the test protocols (see Table 14.6) was supposed to be used for training, no measures were taken to prevent cheating by training and calibrating a single system on all data (containing also the unknown scenarios) and using it for reporting the scores for the according development and test sets of the individual protocols. In this case, only the test subjects would be unknown to the system. Since only the integrity of the participants was trusted, the overall conclusions should be handled with care. However, it is worth pointing out that none of the submitted algorithms managed to achieve satisfying PAD performance on the OULU-NPU dataset even though cheating was possible. Although promising generalization was achieved across the different protocols, the best results are far from perfect, unlike in the previous competitions.

The best solution to prevent "data peeking" or cheating in general would be to keep the test data, including unknown scenarios, inaccessible during algorithm development phase and to conduct independent (third-party) evaluations, in which the organizers run the provided executables or source codes of the submitted systems on the competition data. The results of the iris liveness detection competitions (LivDet-Iris)

[51, 52] have shown already that the resulting performances can be far from satisfactory even for the winning methods, thus probably reflecting better the true generalization capabilities. It is worth highlighting that any later comparison to this kind of competition results should be treated with caution because it is impossible to reproduce the "blind" evaluation conditions any more and, consequently, to achieve a fair comparison.

Over 50 organizations registered for the competition on generalized face PAD but only 13 teams made a final submission. Among the remaining 37 registered, there were also many companies. In general, the industrial participants should be encouraged to make an "anonymous" submission, for example if the results might be unsatisfactory or details of the used algorithm cannot be revealed, as the results can still provide extremely useful additional information on the performance of the state of the art. For instance, in the LivDet-Iris 2017 competition [51], the best-performing algorithm was submitted anonymously.

14.5.3   Future Challenges

The test cases in the OULU-NPU database measuring the generalization across the different covariates are still very limited. The video sequences have been captured with six different mobile devices but the attacks consists of only two different print attacks and display attacks and the acquisition conditions are quite controlled and restricted to three indoor office locations. Also, the variability in user demographics could be increased. The results of the third competition suggest that among the three tested covariates previously unseen acquisition conditions cause the most significant degradation in performance due to increase in BPCER, whereas unknown attacks have huge impact in APCER, especially in the case of print attacks. This observation is consistent with cross-dataset experiments conducted in other studies (e.g., [7, 9, 43]). While there is still plenty of room for improvement in the results obtained on the OULU-NPU dataset, more comprehensive datasets for investigating face presentation attack detection "in the wild" will be eventually needed.

In general, the evaluation of biometric systems under presentations attacks can be conducted either at algorithm or system level [53]. In the first case, the robustness of the PAD modules is evaluated independently of the performance of the rest of the system, for instance, the face recognition stage. System level evaluation considers the performance of the biometric system as a whole. The advantage of system-based evaluations is that it provides better insight into the overall robustness of the whole system to spoofing attacks, and how a proposed PAD technique affects the overall system accuracy (in terms of FRR). All three competitions have considered only stand-alone face PAD. Therefore, a possible future study would be combining match scores with both PAD and quality measures to improve the resilience of face verification systems [54, 55]. So far, the competitions have assessed the proposed PAD algorithms based on single liveness score values that have been assigned to each video after processing all or some of its frames. It would be also useful to measure the complexity, speed and latency of the participating systems, for example by computing the error rates over time.

Due to the recent advances in technology and vulnerabilities to spoofing, manufacturers, such as Microsoft, Apple, and Samsung, have introduced new sensors (e.g., active NIR and depth cameras) for face verification purposes on personal devices. The dedicated imaging solutions are better capable of capturing the intrinsic differences between bona-fide samples and face artifacts than conventional cameras (hardware-based PAD). Since the new sensors are emerging in consumer devices, algorithm-based evaluations on sensor-specific data would be valuable addition in upcoming competitions. Alternatively, system based evaluations of complete biometric systems with novel sensors and PAD modules could be assessed on the spot, as conducted already in LivDet-Iris 2013 [52], for instance. Naturally, this kind of arrangement requires careful competition design and execution, let alone significant efforts compared to algorithm level evaluation.

14.6 Conclusions

Competitions play a vital role in consolidating the recent trends and assessing the state of the art in face PAD. This chapter introduced the design and results of the three international competitions on software-based face PAD. These contests have been important milestones in advancing the research on face PAD to the next level as each competition has offered new challenges to the research community and resulted in novel countermeasures and new insight. The number of participants has grown in each successive competition, which indicates the increasing interest and importance of the research problem. The first and second competitions had six and eight participants from academic institutes, while the latest contest had 13 entries including three companies.

The first two competitions provided initial assessments of the state of the art by introducing a precisely defined test protocol and evaluating the performance of the systems under print and display attacks in homogeneous conditions. The best-performing teams achieved perfect results in the first two competitions, because the test data did not introduce conditions (e.g., sensors, illumination, or attacks) not seen during the algorithm development phase. Despite significant progress in the field, existing face PAD methods have shown lack of generalization in real-world operating conditions. Therefore, the latest contest considered a more unconstrained setup than in previous competitions, and aimed at measuring the generalization capabilities of the proposed algorithms under some real-world variations faced in mobile scenarios, including unknown acquisition conditions, PAIs and sensors. While the best results were promising, no major breakthrough in generalized face PAD was achieved even though the use of external training data was allowed.

Although none of the systems proposed in the latest competition managed to achieve satisfying PAD performance on the recent OULU-NPU database, more comprehensive datasets on presentation attack detection are still needed, especially considering the needs of data-hungry deep learning algorithms. So far, the competitions have focused only on stand-alone PAD, thus joint-operation with face verification would be worth investigating in future. Since new imaging solutions, such as NIR and depth cameras, are already emerging in consumer devices, it would be important to include these kinds of sensors in the upcoming benchmark datasets and competitions.

Acknowledgements The financial support from the Finnish Foundation for Technology Promotion and Infotech Oulu Doctoral Program is acknowledged.

References

1. International Organization for Standardization (2016) ISO/IEC JTC 1/SC 37 biometrics: information technology—biometric presentation attack detection—part 1: framework. Technical report
2. Chingovska I, Erdogmus N, Anjos A, Marcel S (2016) Face recognition systems under spoofing attacks. In: Bourlai T (ed) Face recognition across the imaging spectrum. Springer International Publishing, pp 165–194
3. Li Y, Li Y, Xu K, Yan Q, Deng R (2016) Empirical study of face authentication systems under OSNFD attacks. IEEE Trans Dependable Secur Comput
4. Mohammadi A, Bhattacharjee S, Marcel S (2018) Deeply vulnerable: a study of the robustness of face recognition to presentation attacks. IET Biom 7(1):15–26
5. Boulkenafet Z, Komulainen J, Li L, Feng X, Hadid A (2017) OULU-NPU: a mobile face presentation attack database with real-world variations. In: IEEE International conference on automatic face and gesture recognition
6. Chingovska I, Anjos A, Marcel S (2012) On the effectiveness of local binary patterns in face anti-spoofing. In: International conference of the biometrics special interest group (BIOSIG), pp 1–7
7. Costa-Pazo A, Bhattacharjee S, Vazquez-Fernandez E, Marcel S (2016) The REPLAY-MOBILE face presentation-attack database. In: International conference on biometrics special interests group (BIOSIG)
8. Tan X, Li Y, Liu J, Jiang L (2010) Face liveness detection from a single image with sparse low rank bilinear discriminative model. Springer, Berlin, Heidelberg, pp 504–517. https://doi.org/10.1007/978-3-642-15567-3_37
9. Wen D, Han H, Jain A (2015) Face spoof detection with image distortion analysis. Trans Inf Forensics Secur 10(4):746–761
10. Zhang Z, Yan J, Liu S, Lei Z, Yi D, Li SZ (2012) A face antispoofing database with diverse attacks. In: International conference on biometrics (ICB), pp 26–31
11. Boulkenafet Z, Komulainen J, Akhtar Z, Benlamoudi A, Samai D, Bekhouche S, Ouafi A, Dornaika F, Taleb-Ahmed A, Qin L, Peng F, Zhang L, Long M, Bhilare S, Kanhangad V, Costa-Pazo A, Vazquez-Fernandez E, Perez-Cabo D, Moreira-Perez JJ, Gonzalez-Jimenez D, Mohammadi A, Bhattacharjee S, Marcel S, Volkova S, Tang Y, Abe N, Li L, Feng X, Xia Z, Jiang X, Liu S, Shao 0R, Yuen PC, Almeida WR, Andalo F, Padilha R, Bertocco G, Dias W, Wainer J, Torres R, Rocha A, Angeloni MA, Folego G, Godoy A, Hadid A (2017) A competition on generalized software-based face presentation attack detection in mobile scenarios. In: IEEE international joint conference on biometrics (IJCB)
12. Chakka M, Anjos A, Marcel S, Tronci R, Muntoni D, Fadda G, Pili M, Sirena N, Murgia G, Ristori M, Roli F, Yan J, Yi D, Lei Z, Zhang Z, Li S, Schwartz W, Rocha A, Pedrini H, Lorenzo-Navarro J, Castrillon-Santana M, Määttä J, Hadid A, Pietikäinen M (2011) Competition on counter measures to 2-D facial spoofing attacks. In: International joint conference on biometrics (IJCB)

13. Chingovska I, Yang J, Lei Z, Yi D, Li SZ, Kähm O, Glaser C, Damer N, Kuijper A, Nouak A, Komulainen J, Pereira T, Gupta S, Khandelwal S, Bansal S, Rai A, Krishna T, Goyal D, Waris MA, Zhang H, Ahmad I, Kiranyaz S, Gabbouj M, Tronci R, Pili M, Sirena N, Roli F, Galbally J, Fierrez J, Pinto A, Pedrini H, Schwartz WS, Rocha A, Anjos A, Marcel S (2013) The 2nd competition on counter measures to 2D face spoofing attacks. In: International conference on biometrics (ICB)
14. Erdogmus N, Marcel S (2013) Spoofing attacks to 2D face recognition systems with 3D masks. In: IEEE international conference of the biometrics special interest group
15. Raghavendra R, Raja KB, Busch C (2015) Presentation attack detection for face recognition using light field camera. IEEE Trans Image Process 24(3):1060–1075
16. Pavlidis I, Symosek P (2000) The imaging issue in an automatic face/disguise detection system. In: IEEE workshop on computer vision beyond the visible spectrum: methods and applications (CVBVS), pp 15–24
17. Rudd EM, Günther M, Boult TE (2016) PARAPH: presentation attack rejection by analyzing polarization hypotheses. In: IEEE conference on computer vision and pattern recognition workshops (CVPRW), pp 171 – 178. https://doi.org/10.1109/CVPRW.2016.28
18. Zhang Z, Yi D, Lei Z, Li SZ (2011) Face liveness detection by learning multispectral reflectance distributions. In: International conference on face and gesture, pp 436–441
19. Kollreider K, Fronthaler H, Faraj MI, Bigun J (2007) Real-time face detection and motion analysis with application in liveness assessment. IEEE Trans Inf Forensics Secur 2(3):548–558
20. Ng ES, Chia AYS (2012) Face verification using temporal affective cues. In: International conference on pattern recognition (ICPR), pp 1249–1252
21. Chetty G, Wagner M (2004) Liveness verification in audio-video speaker authentication. In: Australian international conference on speech science and technology, pp 358–363
22. Frischholz RW, Werner A (2003) Avoiding replay-attacks in a face recognition system using head-pose estimation. In: IEEE international workshop on analysis and modeling of faces and gestures
23. De Marsico M, Nappi M, Riccio D, Dugelay JL (2012) Moving face spoofing detection via 3D projective invariants. In: IAPR international conference on biometrics (ICB)
24. Wang T, Yang J, Lei Z, Liao S, Li SZ (2013) Face liveness detection using 3D structure recovered from a single camera. In: International conference on biometrics (ICB)
25. Li J, Wang Y, Tan T, Jain AK (2004) Live face detection based on the analysis of Fourier spectra. In: Proceedings of biometric technology for human identification, pp 296–303
26. Bai J, Ng TT, Gao X, Shi YQ (2010) Is physics-based liveness detection truly possible with a single image? In: IEEE international symposium on circuits and systems (ISCAS), pp 3425–3428
27. Kose N, Dugelay JL (2013) Countermeasure for the protection of face recognition systems against mask attacks. In: International conference on automatic face and gesture recognition (FG)
28. Määttä J, Hadid A, Pietikäinen M (2011) Face spoofing detection from single images using micro-texture analysis. In: Proceedings of international joint conference on biometrics (IJCB). https://doi.org/10.1109/IJCB.2011.6117510
29. Yang J, Lei Z, Liao S, Li SZ (2013) Face liveness detection with component dependent descriptor. In: International conference on biometrics (ICB)
30. Feng L, Po LM, Li Y, Xu X, Yuan F, Cheung TCH, Cheung KW (2016) Integration of image quality and motion cues for face anti-spoofing: a neural network approach. J Vis Commun Image Represent 38:451–460
31. Galbally J, Marcel S (2014) Face anti-spoofing based on general image quality assessment. In: IAPR/IEEE international conference on pattern recognition (ICPR), pp 1173–1178
32. Galbally J, Marcel S, Fierrez J (2014) Image quality assessment for fake biometric detection: application to iris, fingerprint, and face recognition. IEEE Trans Image Process 23(2):710–724
33. Bharadwaj S, Dhamecha TI, Vatsa M, Richa S (2013) Computationally efficient face spoofing detection with motion magnification. In: IEEE Conference on Computer Vision and Pattern Recognition Workshops (CVPRW)

34. Pan G, Wu Z, Sun L (2008) Liveness detection for face recognition. In: Proceedings of recent advances in face recognition, pp 109–124. In-Teh
35. Siddiqui T, Bharadwaj S, Dhamecha T, Agarwal A, Vatsa M, Singh R, Ratha N (2016) Face anti-spoofing with multifeature videolet aggregation. In: International conference on pattern recognition (ICPR)
36. Tirunagari S, Poh N, Windridge D, Iorliam A, Suki N, Ho ATS (2015) Detection of face spoofing using visual dynamics. IEEE Trans Inf Forensics Secur 10(4):762–777
37. Li X, Komulainen J, Zhao G, Yuen PC, Pietikäinen M (2016) Generalized face anti-spoofing by detecting pulse from face videos. In: International conference on pattern recognition (ICPR)
38. Liu S, Yuen PC, Zhang S, Zhao G (2016) 3D mask face anti-spoofing with remote photo-plethysmography. In: European conference on computer vision (ECCV). Springer International Publishing, pp 85–100
39. de Freitas Pereira T, Anjos A, De Martino J, Marcel S (2013) Can face anti-spoofing counter-measures work in a real world scenario? In: International conference on biometrics (ICB)
40. Boulkenafet Z, Komulainen J, Hadid A (2015) Face anti-spoofing based on color texture analysis. In: IEEE international conference on image processing (ICIP)
41. Boulkenafet Z, Komulainen J, Hadid A (2016) Face antispoofing using speeded-up robust features and fisher vector encoding. IEEE Sig Process Lett 24(2):141–145
42. Boulkenafet Z, Komulainen J, Hadid A (2016) Face spoofing detection using colour texture analysis. IEEE Trans Inf Forensics Secur 11(8):1818–1830
43. Boulkenafet Z, Komulainen J, Hadid A (2018) On the generalization of color texture-based face anti-spoofing. Image Vis Comput
44. Patel K, Han H, Jain AK (2016) Cross-database face antispoofing with robust feature representation. In: Chinese conference on biometric recognition (CCBR), pp 611–619
45. Pinto A, Pedrini H, Robson Schwartz W, Rocha A (2015) Face spoofing detection through visual codebooks of spectral temporal cubes. IEEE Trans Image Process 24(12):4726–4740
46. Yang J, Lei Z, Li SZ (2014) Learn convolutional neural network for face anti-spoofing. CoRR. http://arxiv.org/abs/1408.5601
47. Manjani I, Tariyal S, Vatsa M, Singh R, Majumdar A (2017) Detecting silicone mask based presentation attack via deep dictionary learning. IEEE Trans Inf Forensics Secur
48. Anjos A, Marcel S (2011) Counter-measures to photo attacks in face recognition: a public database and a baseline. In: Proceedings of IAPR IEEE international joint conference on biometrics (IJCB)
49. Ojala T, Pietikäinen M, Mäenpaa T (2002) Multiresolution gray-scale and rotation invariant texture classification with local binary patterns. IEEE Trans Pattern Anal Mach Intell 24(7):971–987
50. Korshunov P, Marcel S, Muckenhirn H, Gonçalves AR, Mello AGS, Violato RPV, Simoes FO, Neto MU, de Assis Angeloni M, Stuchi JA, Dinkel H, Chen N, Qian Y, Paul D, Saha G, Sahidullah M (2016) Overview of BTAS 2016 speaker anti-spoofing competition. In: IEEE international conference on biometrics theory, applications and systems (BTAS)
51. Yambay D, Becker B, Kohli N, Yadav D, Czajka A, Bowyer KW, Schuckers S, Singh R, Vatsa M, Noore A, Gragnaniello D, Sansone C, Verdoliva L, He L, Ru Y, Li H, Liu N, Sun Z, Tan T (2017) LivDet-Iris 2017—iris liveness detection competition 2017. In: IEEE international joint conference on biometrics (IJCB)
52. Yambay D, Doyle JS, Bowyer KW, Czajka A, Schuckers S (2014) LivDet-iris 2013—iris liveness detection competition 2013. In: IEEE international joint conference on biometrics (IJCB), pp 1–8
53. Galbally J, Marcel S, Fiérrez J (2014) Biometric antispoofing methods: a survey in face recognition. IEEE Access 2:1530–1552
54. Chingovska I, Anjos A, Marcel S (2013) Anti-spoofing in action: joint operation with a verification system. In: IEEE conference on computer vision and pattern recognition workshops (CVPRW), pp 98–104
55. Rattani A, Poh N, Ross A (2013) A bayesian approach for modeling sensor influence on quality, liveness and match score values in fingerprint verification. In: IEEE international workshop on information forensics and security (WIFS), pp 37–42

Part IV
Voice Biometrics

Chapter 15
Introduction to Voice Presentation Attack Detection and Recent Advances

Md Sahidullah, Héctor Delgado, Massimiliano Todisco, Tomi Kinnunen, Nicholas Evans, Junichi Yamagishi and Kong-Aik Lee

Abstract Over the past few years, significant progress has been made in the field of presentation attack detection (PAD) for automatic speaker recognition (ASV). This includes the development of new speech corpora, standard evaluation protocols and advancements in front-end feature extraction and back-end classifiers. The use of standard databases and evaluation protocols has enabled for the first time the meaningful benchmarking of different PAD solutions. This chapter summarises the progress, with a focus on studies completed in the last 3 years. The article presents a summary of findings and lessons learned from two ASVspoof challenges, the first community-led benchmarking efforts. These show that ASV PAD remains an unsolved problem and that further attention is required to develop generalised PAD solutions which have potential to detect diverse and previously unseen spoofing attacks.

M. Sahidullah (✉) · T. Kinnunen
School of Computing, University of Eastern Finland, Kuopio, Finland
e-mail: sahid@cs.uef.fi

T. Kinnunen
e-mail: tkinnu@cs.uef.fi

H. Delgado · M. Todisco · N. Evans
Department of Digital Security, EURECOM, Biot Sophia Antipolis, France
e-mail: hector.delgado@eurecom.fr

M. Todisco
e-mail: massimiliano.todisco@eurecom.fr

N. Evans
e-mail: evans@eurecom.fr

J. Yamagishi
National Institute of Informatics, Tokyo, Japan
e-mail: jyamagis@nii.ac.jp

J. Yamagishi
University of Edinburgh, Edinburgh, Scotland

K.-A. Lee
Data Science Research Laboratories, NEC Corporation (Japan), Tokyo, Japan
e-mail: k-lee@ax.jp.nec.com

© Springer Nature Switzerland AG 2019
S. Marcel et al. (eds.), *Handbook of Biometric Anti-Spoofing*,
Advances in Computer Vision and Pattern Recognition,
https://doi.org/10.1007/978-3-319-92627-8_15

321

15.1   Introduction

Automatic speaker verification (ASV) technology aims to recognise individuals using samples of the human voice signal [1, 2]. Most ASV systems operate on estimates of the spectral characteristics of voice in order to recognise individual speakers. ASV technology has matured in recent years and now finds application in a growing variety of real-world authentication scenarios involving both *logical* and *physical* access. In logical access scenarios, ASV technology can be used for remote person authentication via the Internet or traditional telephony. In many cases, ASV serves as a convenient and efficient alternative to more conventional password-based solutions, one prevalent example being person authentication for Internet and mobile banking. Physical access scenarios include the use of ASV to protect personal or secure/sensitive facilities, such as domestic and office environments. With the growing, widespread adoption of smartphones and voice-enabled smart devices, such as intelligent personal assistants all equipped with at least one microphone, ASV technology stands to become even more ubiquitous in the future.

Despite its appeal, the now-well-recognised vulnerability to manipulation through presentation attacks (PAs), also known as spoofing, has dented confidence in ASV technology. As identified in ISO/IEC 30107-1 standard [3], the possible locations of presentation attack points in a typical ASV system are illustrated in Fig. 15.1. Two of the most vulnerable places in an ASV system are marked by 1 and 2, corresponding to physical access and logical access. This work is related to these two types of attacks.

Unfortunately, ASV is arguably more prone to PAs than other biometric systems based on traits or characteristics that are less easily acquired; samples of a given person's voice can be collected readily by fraudsters through face-to-face or telephone conversations and then replayed in order to manipulate an ASV system. Replay attacks are furthermore only one example of ASV PAs. More advanced voice

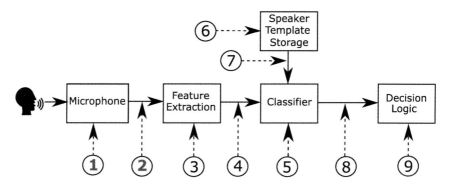

Fig. 15.1 Possible attack locations in a typical ASV system. 1: microphone point, 2: transmission point, 3: override feature extractor, 4: modify probe to features, 5: override classifier, 6: modify speaker database, 7: modify biometric reference, 8: modify score and 9: override decision

conversion or speech synthesis algorithms can be used to generate particularly effective PAs using only modest amounts of voice data collected from a target person.

There are a number of ways to prevent PA problems. The first one is based on a text-prompted system which uses an utterance verification process [4]. The user needs to utter a specific text prompted for authentication by the system which requires a text-verification system. Second, as human can never reproduce an identical speech signal, some countermeasures use template matching or audio fingerprinting to verify whether the speech utterance was presented to the system earlier [5]. Third, some work looks into statistical acoustic characterisation of authentic speech and speech created with presentation attack methods or spoofing techniques [6]. Our focus is on the last category, which is more convenient in a practical scenario for both text-dependent and text-independent ASV. In this case, given a speech signal, S, PA detection here, the determination of whether S is a natural or PA speech can be formulated as a hypothesis test:

- H_0: S is natural speech.
- H_1: S is created with PA methods.

A likelihood ratio test can be applied to decide between H_0 and H_1. Suppose that $\mathbf{X} = \{\mathbf{x}_1, \mathbf{x}_2, ..., \mathbf{x}_N\}$ are the acoustic feature vectors of N speech frames extracted from S, then the logarithmic likelihood ratio score is given by

$$\Lambda(\mathbf{X}) = \log p(\mathbf{X}|\lambda_{H_0}) - \log p(\mathbf{X}|\lambda_{H_1}) \tag{15.1}$$

In (15.1), λ_{H_0} and λ_{H_1} are the acoustic models to characterise the hypotheses correspondingly for natural speech and PA speech. The parameters of these models are estimated using training data for natural and PA speech. A typical PAD system is shown in Fig. 15.2. A test speech can be accepted as natural or rejected as PA speech with help of a threshold, θ computed on some development data. If the score is greater than or equal to the threshold, it is accepted; otherwise, rejected. The performance of the PA system is assessed by computing the *Equal Error Rate* (EER) metric. This is the error rate for a specific value of a threshold where two error rates, i.e. the probability of a PA speech detected as being natural speech (known as false acceptance rate or FAR) and the probability of a natural speech being misclassified as a PA speech (known as false rejection rate or FRR), are equal. Sometimes *Half Total Error Rate* (HTER) is also computed [7]. This is the average of FAR and FRR which are computed using a decision threshold obtained with the help of the development data.

Awareness and acceptance of the vulnerability to PAs have generated a growing interest in developing solutions to presentation attack detection (PAD), also referred to as spoofing countermeasures. These are typically dedicated auxiliary systems which function in tandem to ASV in order to detect and deflect PAs. The research in this direction has progressed rapidly in the last three years, due partly to the release of several public speech corpora and the organisation of PAD challenges for ASV. This article, a continuation of the chapter [8] in the first edition of the

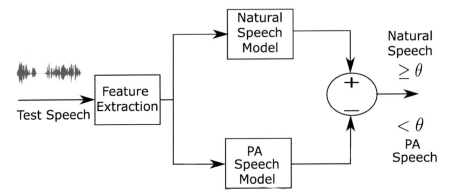

Fig. 15.2 Block diagram of a typical presentation attack detection system

Handbook for Biometrics [9] presents an up-to-date review of the different forms of voice presentation attacks, broadly classified in terms of impersonation, replay, speech synthesis and voice conversion. The primary focus is nonetheless on the progress in PAD. The chapter reviews the most recent work involving a variety of different features and classifiers. Most of the work covered in the chapter relates to that conducted using the two most popular and publicly available databases, which were used for the two ASVspoof challenges co-organised by the authors. The chapter concludes with a discussion of research challenges and future directions in PAD for ASV.

15.2 Basics of ASV Spoofing and Countermeasures

Spoofing or presentation attacks are performed on a biometric system at the sensor or acquisition level to bias score distributions towards those of genuine clients, thus provoking increases in the false acceptance rate (FAR). This section reviews four well-known ASV spoofing techniques and their respective countermeasures: impersonation, replay, speech synthesis and voice conversion. Here, we mostly review the work in the pre-ASVspoof period, as well as some very recent studies on presentation attacks.

15.2.1 Impersonation

In speech impersonation or mimicry attacks, an intruder speaker intentionally modifies his or her speech to sound like the target speaker. Impersonators are likely to copy lexical, prosodic and idiosyncratic behaviour of their target speakers presenting a potential point of vulnerability concerning speaker recognition systems.

15.2.1.1   Spoofing

There are several studies about the consequences of mimicry on ASV. Some studies concern attention to the voice modifications performed by professional impersonators. It has been reported that impersonators are often particularly able to adapt the fundamental frequency (F0) and occasionally also the formant frequencies towards those of the target speakers [10–12]. In studies, the focus has been on analysing the vulnerability of speaker verification systems in the presence of voice mimicry. The studies by Lau et al. [13, 14] suggest that if the target of impersonation is known in advance and his or her voice is "similar" to the impersonator's voice (in the sense of automatic speaker recognition score), then the chance of spoofing an automatic recognizer is increased. In [15], the experiments indicated that professional impersonators are potentially better impostors than amateur or naive ones. Nevertheless, the voice impersonation was not able to spoof the ASV system. In [10], the authors attempted to quantify how much a speaker is able to approximate other speakers' voices by selecting a set of prosodic and voice source features. Their prosodic and acoustic-based ASV results showed that two professional impersonators imitating known politicians increased the identification error rates.

More recently, a fundamentally different study was carried out by Panjwani et al. [16] using crowdsourcing to recruit both amateur and more professional impersonators. The results showed that impersonators succeed in increasing their average score, but not in exceeding the target speaker score. All of the above studies analysed the effects of speech impersonation either at the acoustic or speaker recognition score level, but none proposed any countermeasures against impersonation. In a recent study [17], the experiments aimed to evaluate the vulnerability of three modern speaker verification systems against impersonation attacks and to further compare these results to the performance of non-expert human listeners. It is observed that, on average, the mimicry attacks lead to increased error rates. The increase in error rates depends on the impersonator and the ASV system.

The main challenge, however, is that no large speech corpora of impersonated speech exists for the quantitative study of impersonation effects on the same scale as for other attacks, such as text-to-speech synthesis and voice conversion, where generation of simulated spoofing attacks as well as developing appropriate countermeasures are more convenient.

15.2.1.2   Countermeasures

While the threat of impersonation is not fully understood due to limited studies involving small datasets, it is perhaps not surprising that there is no prior work investigating countermeasures against impersonation. If the threat is proven to be genuine, then the design of appropriate countermeasures might be challenging. Unlike the spoofing attacks discussed below, all of which can be assumed to leave traces of the physical properties of the recording and playback devices, or signal processing artefacts from synthesis or conversion systems, impersonators are live human beings who produce entirely natural speech.

15.2.2 Replay

Replay attacks refer to the use of pre-recorded speech from a target speaker, which is then replayed through some playback device to feed the system microphone. These attacks require no specific expertise nor sophisticated equipment, thus they are easy to implement. Replay is a relatively low-technology attack within the grasp of any potential attacker even without specialised knowledge in speech processing. Several works in the earlier literature report significant increases in error rates when using replayed speech. Even if replay attacks may present a genuine risk to ASV systems, the use of prompted-phrase has the potential to mitigate the impact.

15.2.2.1 Spoofing

The study on the impact of replay attack on ASV performance was very limited until recently before the release of AVspoof [18] and ASVspoof 2017 corpus. The earlier studies were conducted either on simulated or on real replay recording from far-field.

The vulnerability of ASV systems to replay attacks was first investigated in a text-dependent scenario [19], where the concatenation of recorded digits was tested against a hidden Markov model (HMM)-based ASV system. Results showed an increase in the FAR from 1 to 89% for male speakers and from 5 to 100% for female speakers.

The work in [20] investigated text-independent ASV vulnerabilities through the replaying of far-field recorded speech in a mobile telephony scenario where signals were transmitted by analogue and digital telephone channels. Using a baseline ASV system based on *joint factor analysis* (JFA), the work showed an increase in the EER of 1% to almost 70% when impostor accesses were replaced by replayed spoof attacks.

A physical access scenario was considered in [21]. While the baseline performance of the Gaussian mixture model-universal background model (GMM-UBM) ASV system was not reported, experiments showed that replay attacks produced a FAR of 93%.

The work in [18] introduced audio-visual spoofing (AVspoof) database for replay attack detection where the replayed signals are collected and played back using different low-quality (phones and laptop) and high-quality (laptop with loudspeakers) devices. The study reported that FARs for replayed speech was 77.4 and 69.4% for male and female, respectively, using a total variability system speaker recognition system. In this study, the EER for bona fide trials was 6.9 and 17.5% for those conditions. This study also includes presentation attack where speech signals created with voice conversion and speech synthesis were used in playback attack. In that case, higher FAR was observed, particularly when high-quality device is used for playback.

15.2.2.2 Countermeasures

A countermeasure for replay attack detection in the case of text-dependent ASV was reported in [5]. The approach is based upon the comparison of new access samples with stored instances of past accesses. New accesses which are deemed too similar to previous access attempts are identified as replay attacks. A large number of different experiments, all relating to a telephony scenario, showed that the countermeasures succeeded in lowering the EER in most of the experiments performed. While some form of text-dependent or challenge response countermeasure is usually used to prevent replay attacks, text-independent solutions have also been investigated. The same authors in [20] showed that it is possible to detect replay attacks by measuring the channel differences caused by far-field recording [22]. While they show spoof detection error rates of less than 10% it is feasible that today's state-of-the-art approaches to channel compensation will render some ASV systems still vulnerable.

Two different replay attack countermeasures are compared in [21]. Both are based on the detection of differences in channel characteristics expected between licit and spoofed access attempts. Replay attacks incur channel noise from both the recording device and the loudspeaker used for replay and thus the detection of channel effects beyond those introduced by the recording device of the ASV system thus serves as an indicator of replay. The performance of a baseline GMM-UBM system with an EER of 40% under spoofing attack falls to 29% with the first countermeasure and a more respectable EER of 10% with the second countermeasure.

In another study [23], a speech database of 175 subjects have been collected for different kinds of replay attack. Other than the use of genuine voice samples for the legitimate speakers in playback, the voice samples recorded over the telephone channel was also used for unauthorised access. Further, a far-field microphone is used to collect the voice samples as eavesdropped (covert) recording. The authors proposed an algorithm motivated from music recognition system used for comparing recordings on the basis of the similarity of the local configuration of maxima pairs extracted from spectrograms of verified and reference recordings. The experimental results show the EER of playback attack detection to be as low as 1.0% on the collected data.

15.2.3 Speech Synthesis

Speech synthesis, commonly referred to as text-to-speech (TTS), is a technique for generating intelligible, natural sounding artificial speech for any arbitrary text. Speech synthesis is used widely in various applications including in-car navigation systems, e-book readers, voice-over for the visually impaired and communication aids for the speech impaired. More recent applications include spoken dialogue systems, communicative robots, singing speech synthesisers and speech-to-speech translation systems.

Typical speech synthesis systems have two main components [24]: text analysis followed by speech waveform generation, which is sometimes referred to as the front-end and back-end, respectively. In the text analysis component, input text is converted into a linguistic specification consisting of elements such as phonemes. In the speech waveform generation component, speech waveforms are generated from the produced linguistic specification. There are emerging end-to-end frameworks that generate speech waveforms directly from text inputs without using any additional modules.

Many approaches have been investigated, but there have been major paradigm shifts every ten years. In the early 1970s, the speech waveform generation component used very low-dimensional acoustic parameters for each phoneme, such as formants, corresponding to vocal tract resonances with hand-crafted acoustic rules [25]. In the 1980s, the speech waveform generation component used a small database of phoneme units called *diphones* (the second half of one phoneme plus the first half of the following) and concatenated them according to the given phoneme sequence by applying signal processing such as linear predictive (LP) analysis, to the units [26]. In the 1990s, larger speech databases were collected and used to select more appropriate speech units that matched both phonemes and other linguistic contexts such as lexical stress and pitch accent in order to generate high-quality natural sounding synthetic speech with the appropriate prosody. This approach is generally referred to as *unit selection*, and is nowadays used in many speech synthesis systems [27–31].

In the late 2000s, several machine learning based data-driven approaches emerged. 'Statistical parametric speech synthesis' was one of the more popular machine learning approaches [32–35]. In this approach, several acoustic parameters are modelled using a time-series stochastic generative model, typically an HMM. HMMs represent not only the phoneme sequences but also various contexts of the linguistic specification. Acoustic parameters generated from HMMs and selected according to the linguistic specification are then used to drive a vocoder, a simplified speech production model in which speech is represented by vocal tract parameters and excitation parameters in order to generate a speech waveform. HMM-based speech synthesisers [36, 37] can also learn speech models from relatively small amounts of speaker-specific data by adapting background models derived from other speakers based on the standard model adaptation techniques drawn from speech recognition, i.e. maximum likelihood linear regression (MLLR) [38, 39].

In the 2010s, deep learning has significantly improved the performance of speech synthesis and led to a significant breakthrough. First, various types of deep neural networks are used to improve the prediction accuracy of the acoustic parameters [40, 41]. Investigated architectures include recurrent neural network [42–44], residual/highway network [45, 46], autoregressive network [47, 48] and generative adversarial networks (GAN) [49–51]. Furthermore, in the late 2010s, conventional waveform generation modules that typically used signal processing and text analysis modules that used natural language processing were substituted by neural networks. This allows for neural networks capable of directly outputting the desired speech waveform samples from the desired text inputs. Successful architectures for direct waveform modelling include dilated convolutional autoregressive neural network, known as 'Wavenet' [52] and hierarchical recurrent neural network, called 'SampleRNN' [53]. Finally, we have also seen successful architectures that totally remove the hand-crafted linguistic features obtained through text analysis by relying in sequence-to-sequence systems. This system is called Tacotron [54]. As expected, the combination of these advanced models results in a very high-quality end-to-end TTS synthesis system [55, 56] and recent results reveal that the generated synthetic speech sounds as natural as human speech [56].

For more details and technical comparisons, please see the results of Blizzard Challenge, which annually compares the performance of speech synthesis systems built on the common database over decades [57, 58].

15.2.3.1 Spoofing

There is a considerable volume of research in the literature which has demonstrated the vulnerability of ASV to synthetic voices generated with a variety of approaches to speech synthesis. Experiments using formant, diphone, and unit selection based synthetic speech in addition to the simple cut-and-paste of speech waveforms have been reported [19, 20, 59].

ASV vulnerabilities to HMM-based synthetic speech were first demonstrated over a decade ago [60] using an HMM-based, text-prompted ASV system [61] and an HMM-based synthesiser where acoustic models were adapted to specific human speakers [62, 63]. The ASV system scored feature vectors against speaker and background models composed of concatenated phoneme models. When tested with human speech, the ASV system achieved a FAR of 0% and a false rejection rate (FRR) of 7%. When subjected to spoofing attacks with synthetic speech, the FAR increased to over 70%, however, this work involved only 20 speakers.

Larger scale experiments using the Wall Street Journal corpus containing in the order of 300 speakers and 2 different ASV systems (GMM-UBM and SVM using Gaussian supervectors) was reported in [64]. Using an HMM-based speech synthesiser, the FAR was shown to rise to 86 and 81% for the GMM-UBM and SVM systems, respectively, representing a genuine threat to ASV. Spoofing experiments using HMM-based synthetic speech against a forensics speaker verification tool *BATVOX* was also reported in [65] with similar findings. Therefore, the above speech synthesisers were chosen as one of spoofing methods in the ASVspoof 2015 database.

Spoofing experiments using the above advanced DNNs or using spoofing-specific strategies such as GAN have not yet been properly investigated. Only a relatively small-scale spoofing experiment against a speaker recognition system using Wavenet, SampleRNN and GAN is reported in [66].

15.2.3.2 Countermeasures

Only a small number of attempts to discriminate synthetic speech from natural speech had been investigated before the ASVspoof challenge started. Previous work has demonstrated the successful detection of synthetic speech based on prior knowledge of the acoustic differences of specific speech synthesisers such as the dynamic ranges of spectral parameters at the utterance level [67] and variance of higher order parts of mel-cepstral coefficients [68].

There are some attempts which focus on acoustic differences between vocoders and natural speech. Since the human auditory system is known to be relatively insensitive to phase [69], vocoders are typically based on a minimum-phase vocal tract model. This simplification leads to differences in the phase spectra between human and synthetic speech, differences which can be utilised for discrimination [64, 70].

Based on the difficulty in reliable prosody modelling in both unit selection and statistical parametric speech synthesis, other approaches to synthetic speech detection use F0 statistics [71, 72]. F0 patterns generated for the statistical parametric speech synthesis approach tend to be oversmoothed and the unit selection approach frequently exhibits 'F0 jumps' at concatenation points of speech units.

After the ASVspoof challenges took place, various types of countermeasures that work for both speech synthesis and voice conversion have been proposed. Please read the next section for the details of the recently developed countermeasures.

15.2.4 Voice Conversion

Voice conversion, in short, VC, is a spoofing attack against automatic speaker verification using an attacker's natural voice which is converted towards that of the target. It aims to convert one speaker's voice towards that of another and is a sub-domain of voice transformation [73]. Unlike TTS, which requires text input, voice conversion operates directly on speech inputs. However, speech waveform generation modules such as vocoders may be the same as or similar to those for TTS.

A major application of VC is to personalise and create new voices for TTS synthesis systems and spoken dialogue systems. Other applications include speaking aid devices that generate more intelligible voice sounds to help people with speech disorders, movie dubbing, language learning, and singing voice conversion. The field has also attracted increasing interest in the context of ASV vulnerabilities for almost two decades [74].

Most voice conversion approaches require a parallel corpus where source and target speakers read out identical utterances and adopt a training phase which typically requires frame- or phone-aligned audio pairs of the source and target utterances and estimates transformation functions that convert acoustic parameters of the source speaker to those of the target speaker. This is called 'parallel voice conversion'. Frame alignment is traditionally achieved using dynamic time warping (DTW) on the source target training audio files. Phone alignment is traditionally achieved using *automatic speech recognition* (ASR) and phone-level forth alignment. The estimated conversion function is then applied to any new audio files uttered by the source speaker [75].

A large number of estimation methods for the transformation functions have been reported starting in the late 1980s. In the late 1980s and 90s, simple techniques employing vector quantisation (VQ) with codebooks [76] or segmental codebooks [77] of paired source-target frame vectors were proposed to represent the transformation functions. However, these VQ methods introduced frame-to-frame discontinuity problems.

In the late 1990 and 2000s, *joint density Gaussian mixture model* (JDGMM) based transformation methods [78, 79] were proposed and have since then been actively improved by many researchers [80, 81]. This method still remains popular even now. Although this method achieves smooth feature transformations using a locally linear transformation, this method also has several critical problems such as oversmoothing [82–84] and overfitting [85, 86] which leads to muffled quality of speech and degraded speaker similarity.

Therefore, in the early 2010, several alternative linear transformation methods were developed. Examples are partial least square (PLS) regression [85], tensor representation [87], a trajectory HMM [88], mixture of factor analysers [89], local linear transformation [82] or noisy channel models [90].

In parallel to the linear-based approaches, there have been studies on nonlinear transformation functions such as support vector regression [91], kernel partial least square [92] and conditional restricted Boltzmann machines [93], neural networks [94, 95], highway network [96] and RNN [97, 98]. Data-driven frequency warping techniques [99–101] have also been studied.

Recently, deep learning has changed the above standard procedures for voice conversion and we can see many different solutions now. For instance, variational auto-encoder or sequence-to-sequence neural networks enable us to build VC systems without using frame level alignment [102, 103]. It has also been shown that a cycle-consistent adversarial network called 'CycleGAN' [104] is one possible solution for building VC systems without using a parallel corpus. Wavenet can also be used as a replacement for the purpose of generating speech waveforms from converted acoustic features [105].

The approaches to voice conversion considered above are usually applied to the transformation of spectral envelope features, though the conversion of prosodic features such as fundamental frequency [106–109] and duration [107, 110] has also been studied.

For more details and technical comparisons, please see results of Voice Conversion Challenges that compare the performance of VC systems built on a common database [111, 112].

15.2.4.1 Spoofing

When applied to spoofing, the aim with voice conversion is to synthesise a new speech signal such that the extracted ASV features are close in some sense to the target speaker. Some of the first works relevant to text-independent ASV spoofing were reported in [113, 114]. The work in [113] showed that baseline EER increased from 16 to 26% thanks to a voice conversion system which also converted prosodic aspects not modelled in typical ASV systems. This work targeted the conversion of spectral-slope parameters and showed that the baseline EER of 10% increased to over 60% when all impostor test samples were replaced with converted voices. Moreover, signals subjected to voice conversion did not exhibit any perceivable artefacts indicative of manipulation.

The work in [115] investigated ASV vulnerabilities to voice conversion based on JDGMMs [78] which requires a parallel training corpus for both source and target speakers. Even if the converted speech could be easily detectable by human listeners, experiments involving five different ASV systems showed their universal susceptibility to spoofing. The FAR of the most robust, JFA system increased from 3% to over 17%. Instead of vocoder-based waveform generation, unit selection approaches can be applied directly to feature vectors coming from the target speaker to synthesise converted speech [116]. Since they use target speaker data directly, unit selection approaches arguably pose a greater risk to ASV than statistical approaches [117]. In the ASVspoof 2015 challenge, we therefore had chosen these popular VC methods as spoofing methods.

Other work relevant to voice conversion includes attacks referred to as artificial signals. It was noted in [118] that certain short intervals of converted speech yield extremely high scores or likelihoods. Such intervals are not representative of intelligible speech but they are nonetheless effective in overcoming typical ASV systems which lack any form of speech quality assessment. The work in [118] showed that artificial signals optimised with a genetic algorithm provoke increases in the EER from 10% to almost 80% for a GMM-UBM system and from 5% to almost 65% for a factor analysis (FA) system.

15.2.4.2 Countermeasures

Here, we provide an overview of countermeasure methods developed for the VC attacks before the ASVspoof challenge began.

Some of the first works to detect converted voice draws on related work in synthetic speech detection [119]. In [70, 120], cosine phase and modified group delay function (MGDF) based countermeasures were proposed. These are effective in detecting converted speech using vocoders based on minimum phase. In VC, it is, however, possible to use natural phase information extracted from a source speaker [114]. In this case, they are unlikely to detect converted voice.

Two approaches to artificial signal detection are reported in [121]. Experimental work shows that supervector-based SVM classifiers are naturally robust to such attacks, and that all the spoofing attacks they used could be detected by using an utterance-level variability feature, which detected the absence of the natural and dynamic variabilities characteristic of genuine speech. A related approach to detect converted voice is proposed in [122]. Probabilistic mappings between source and target speaker models are shown to typically yield converted speech with less short-term variability than genuine speech. Therefore, the thresholded, average pair-wise distance between consecutive feature vectors was used to detect converted voice with an EER of under 3%.

Due to the fact that majority of VC techniques operate at the short-term frame level, more sophisticated long-term features such as temporal magnitude and phase modulation feature can also detect converted speech [123]. Another experiment reported in [124] showed that local binary pattern analysis of sequences of acoustic vectors can also be used for successfully detecting frame-wise JDGMM-based converted voice. However, it is unclear whether these features are effective in detecting recent VC systems that consider long-term dependency such as recurrent or autoregressive neural network models.

After the ASVspoof challenges took place, new countermeasures that works for both speech synthesis and voice conversion were proposed and evaluated. See the next section for a detailed review of the recently developed countermeasures.

15.3 Summary of the Spoofing Challenges

A number of independent studies confirm the vulnerability of ASV technology to spoofed voice created using voice conversion, speech synthesis and playback [6]. Early studies on speaker anti-spoofing were mostly conducted on in-house speech corpora created using a limited number of spoofing attacks. The development of countermeasures using only a small number of spoofing attacks may not offer the generalisation ability in the presence of different or unseen attacks. There was a lack of publicly available corpora and evaluation protocol to help with comparing the results obtained by different researchers.

The ASVspoof[1] initiative aims to overcome this bottleneck by making available standard speech corpora consisting of a large number of spoofing attacks, evaluation protocols and metrics to support a common evaluation and the benchmarking of different systems. The speech corpora were initially distributed by organising an evaluation challenge. In order to make the challenge simple and to maximise participation, the ASVspoof challenges so far involved only the detection of spoofed speech; in effect, to determine whether a speech sample is genuine or spoofed. A training set and development set consisting of several spoofing attacks were first shared with the challenge participants to help them develop and tune their anti-spoofing algorithm. Next, the evaluation set without any label indicating genuine or spoofed speech was distributed, and the organisers asked the participants to submit scores within a specific deadline. Participants were allowed to submit scores of multiple systems. One of these systems was designated as the primary submission. Spoofing detectors for all primary submissions were trained using only the training data in the challenge corpus. Finally, the organisers evaluated the scores for benchmarks and ranking.

---

[1] http://www.asvspoof.org/.

Table 15.1 Summary of the datasets used in ASVspoof challenges

	ASVspoof 2015 [125]	ASVspoof 2017 [126]
Theme	Detection of artificially generated speech	Detection of replay speech
Speech format	$F_s = 16$ kHz, 16 bit PCM	$F_s = 16$ kHz, 16 bit PCM
Natural speech	Recorded using high-quality microphone	Recorded using different smart phones
Spoofed speech	Created with seven VC and three SS methods	Collected 'in the wild' by crowdsourcing using different microphone and playback devices from diverse environments
Spoofing types in train/dev/eval	5/5/10	3/10/57
No of speakers in train/dev/eval	25/35/46	10/8/24
No of genuine speech files in train/dev/eval	3750/3497/9404	1508/760/1298
No of spoofed speech files in train/dev/eval	12625/49875/184000	1508/950/12008

The evaluation keys were subsequently released to the challenge participants. The challenge results were discussed with the participants in a special session in INTERSPEECH conferences, which also involved sharing knowledge and receiving useful feedback. To promote further research and technological advancements, the datasets used in the challenge are made publicly available.

The ASVspoof challenges have been organised twice so far. The first was held in 2015 and the second in 2017. A summary of the speech corpora used in the two challenges are shown in Table 15.1. In both the challenges, EER metric was used to evaluate the performance of spoofing detector. The EER is computed by considering the scores of genuine files as positive scores and those of spoofed files as negative scores. A lower EER means more accurate spoofing countermeasures. In practice, the EER is estimated using a specific *receiver operating characteristics convex hull* (ROCCH) technique with an open-source implementation[2] originating from outside the ASVspoof consortium. In the following subsections, we briefly discuss the two challenges. For more interested readers, [125] contains details of the 2015 edition while [126] discusses the results of the 2017 edition.

---

[2]https://sites.google.com/site/bosaristoolkit/.

15.3.1   ASVspoof 2015

The first ASVspoof challenge involved detection of artificial speech created using a mixture of voice conversion and speech synthesis techniques [125]. The dataset was generated with ten different artificial speech generation algorithms. The ASVspoof 2015 was based upon a larger collection spoofing and anti-spoofing (SAS) corpus (v1.0) [127] that consists of both natural and artificial speech. Natural speech was recorded from 106 human speakers using a high-quality microphone and without significant channel or background noise effects. In a speaker disjoint manner, the full database was divided into three subsets called the training, development, and evaluation set. Five of the attacks (S1–S5), named as *known attacks*, were used in the training and development set. The other five attacks, S6-S10, called *unknown attacks*, were used only in the evaluation set, along with the known attacks. Thus, this provides the possibility of assessing the generalisability of the spoofing detectors. The detailed evaluation plan is available in [128], describing the speech corpora and challenge rules.

Ten different spoofing attacks used in the ASVspoof 2015 are listed below:

- **S1**: a simplified frame selection (FS) based voice conversion algorithm, in which the converted speech is generated by selecting target speech frames.
- **S2**: the simplest voice conversion algorithm which adjusts only the first mel-cepstral coefficient (C1) in order to shift the slope of the source spectrum to the target.
- **S3**: a speech synthesis algorithm implemented with the HMM-based speech synthesis system (HTS3) using speaker adaptation techniques and only 20 adaptation utterances.
- **S4**: the same algorithm as S3, but using 40 adaptation utterances.
- **S5**: a voice conversion algorithm implemented with the voice conversion toolkit and with the Festvox system.[3]
- **S6**: a VC algorithm based on joint density Gaussian mixture models (GMMs) and maximum likelihood parameter generation considering global variance.
- **S7**: a VC algorithm similar to S6, but using line spectrum pair (LSP) rather than mel-cepstral coefficients for spectrum representation.
- **S8**: a tensor-based approach to VC, for which a Japanese dataset was used to construct the speaker space.
- **S9**: a VC algorithm which uses kernel-based partial least square (KPLS) to implement a nonlinear transformation function.
- **S10**: an SS algorithm implemented with the open-source MARY text-to-tpeech system (MaryTTS).[4]

More details of how the SAS corpus was generated can be found in [127].

The organisers also confirmed the vulnerability to spoofing by conducting speaker verification experiments with this data and demonstrating considerable performance

---

[3]http://www.festvox.org/.

[4]http://mary.dfki.de/.

degradation in the presence of spoofing. With a state-of-the-art probabilistic linear discriminant analysis (PLDA) based ASV system, it is shown that in presence of spoofing, the average EER for ASV increases from 2.30 to 36.00% for male and 2.08 to 39.53% for female [125]. This motivates the development of the anti-spoofing algorithm.

For ASVspoof 2015, the challenge evaluation metric was the average EER. It is computed by calculating EERs for each attack and then taking average. The dataset was requested by 28 teams from 16 countries, 16 teams returned primary submissions by the deadline. A total of 27 additional submissions were also received. Anonymous results were subsequently returned to each team, who were then invited to submit their work to the ASVspoof special session for INTERSPEECH 2015.

Table 15.2 shows the performance of the top five systems in the ASVspoof 2015 challenge. The best performing system [129] uses a combination of *mel cepstral* and *cochlear filter cepstral coefficients plus instantaneous frequency* features with GMM back-end. In most cases, the participants have used fusion of multiple feature based systems to get better recognition accuracy. Variants of cepstral features computed from the magnitude and phase of short-term speech are widely used for the detection of spoofing attacks. As a back-end, GMM was found to outperform

Table 15.2 Performance of top five systems in ASVspoof 2015 challenge (ranked according to the average % EER for all attacks) with respective features and classifiers

System	Avg. EER for			System
Identifier	Known	Unknown	All	Description
A [129]	0.408	2.013	1.211	*Features*: mel-frequency cepstral coefficients (MFCC), Cochlear filter cepstral coefficients plus instantaneous frequency (CFCCIF). *Classifier*: GMM
B [130]	0.008	3.922	1.965	*Features*: MFCC, MFPC, cosine phase principal coefficients (CosPhasePCs). *Classifier*: Support vector machine (SVM) with i-vectors
C [131]	0.058	4.998	2.528	*Feature*: DNN-based with filterbank output and their deltas as input. *Classifier*: Mahalanobis distance on s-vectors
D [132]	0.003	5.231	2.617	*Features*: log magnitude spectrum (LMS), residual log magnitude spectrum (RLMS), group delay (GD), modified group delay (MGD), instantaneous frequency derivative (IF), baseband phase difference (BPD), and pitch synchronous phase (PSP), *Classifier*: Multilayer perceptron (MLP)
E [133]	0.041	5.347	2.694	*Features*: MFCC, product spectrum MFCC (PS-MFCC), MGD with and without energy, weighted linear prediction group delay, cepstral coefficients (WLP-GDCCs), and MFCC cosine-normalised phase-based cepstral coefficients (MFCC-CNPCCs) *Classifier*: GMM

more advanced classifiers like i-vectors, possibly due to the use of short segments of high-quality speech not requiring treatment for channel compensation and background noise reduction. All the systems submitted in the challenge are reviewed in more detail [134].

15.3.2 ASVspoof 2017

The ASVspoof 2017 is the second automatic speaker verification anti-spoofing and countermeasures challenge. Unlike the 2015 edition that used very high-quality speech material, the 2017 edition aims to assess spoofing attack detection with 'out in the wild' conditions. It focuses exclusively on replay attacks. The corpus originates from the recent *text-dependent RedDots* corpus,[5] whose purpose was to collect speech data over mobile devices, in the form of smartphones and tablet computers, by volunteers from across the globe.

The replayed version of the original *RedDots* corpus was collected through a crowdsourcing exercise using various replay configurations consisting of varied devices, loudspeakers, and recording devices, under a variety of different environments across four European countries within the EU Horizon 2020-funded OCTAVE project,[6] (see [126]). Instead of covert recording, we made a "short-cut" and took the digital copy of the target speakers' voice to create the playback versions. The collected corpus is divided into three subsets: for training, development and evaluation. Details of each are presented in Table 15.1. All three subsets are disjoint in terms of speakers and data collection sites. The training and development subsets were collected at three different sites. The evaluation subset was collected at the same three sites and also included data from two new sites. Data from the same site include different recordings and replaying devices and from different acoustic environments. The evaluation subset contains data collected from 161 replay sessions in 62 unique replay configurations.[7] More details regarding replay configurations can be found in [126, 135].

The primary evaluation metric is 'pooled' EER. In contrast to the ASVspoof 2015 challenge, the EER is computed from scores pooled across all the trial segments rather than condition averaging. A baseline[8] system based on common GMM backend classifier with constant-Q cepstral coefficient (CQCC) [136, 137] features were provided to the participants. This configuration is chosen as baseline as it has shown best recognition performance on ASVspoof 2015. The baseline is trained using either combined training and development data (B01) or training data (B02) alone. The baseline system does not involve any kind of optimisation or tuning with respect

---

[5]https://sites.google.com/site/thereddotsproject/.

[6]https://www.octave-project.eu/.

[7]A **replay configuration** refers to a unique combination of room, replay device and recording device while a **session** refers to a set of source files, which share the same replay configuration.

[8]See *Appendix A.2. Software packages.*

Table 15.3 Summary of top 10 primary submissions to ASVspoof 2017. Systems' IDs are the same received by participants in the evaluation. The column 'Training' refers to the part of data used for training: train (T) and/or development (D)

ID	Features	Post-proc.	Classifiers	Fusion	#Subs.	Training	Performances on eval subset (EER%)
S01 [138]	Log-power Spectrum, LPCC	MVN	CNN, GMM, TV, RNN	Score	3	T	6.73
S02 [139]	CQCC, MFCC, PLP	WMVN	GMM-UBM, TV-PLDA, GSV-SVM, GSV-GBDT, GSV RF	Score	–	T	12.34
S03	MFCC, IMFCC, RFCC, LFCC, PLP, CQCC, SCMC, SSFC	–	GMM, FF-ANN	Score	18	T+D	14.03
S04	RFCC, MFCC, IMFCC, LFCC, SSFC, SCMC	–	GMM	Score	12	T+D	14.66
S05 [140]	Linear filterbank feature	MN	GMM, CT-DNN	Score	2	T	15.97
S06	CQCC, IMFCC, SCMC, Phrase one-hot encoding	MN	GMM	Score	4	T+D	17.62
S07	HPCC, CQCC	MVN	GMM, CNN, SVM	Score	2	T+D	18.14
S08 [141]	IFCC, CFCCIF, Prosody	–	GMM	Score	3	T	18.32
S09	SFFCC	No	GMM	None	1	T	20.57
S10 [142]	CQCC	–	ResNet	None	1	T	20.32

to [136]. The dataset was requested by 113 teams, of which 49 returned primary submissions by the deadline. The results of the challenge were disseminated at a special session consisting of two slots at INTERSPEECH 2017.

Most of the systems are based on standard spectral features, such as CQCCs, MFCCs and *perceptual linear prediction* (PLP). As a back-end, in addition to the classical GMM to model the replay and non-replay classes, it has also exploited the power of deep classifiers, such as *convolutional neural network* (CNN) or *recurrent neural network* (RNN). A fusion of multiple features and classifiers is also widely adopted by the participants. A summary of the top-10 primary systems is provided in Table 15.3. Results in terms of EER of the 49 primary systems and the baseline B01 and B02 are shown in Fig. 15.3.

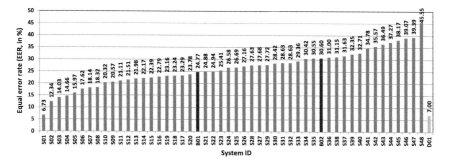

Fig. 15.3 Performance of the two baseline systems (B01 and B02) and the 49 primary systems (S01–S48 in addition to late submission D01) for the ASVspoof 2017 challenge. Results are in terms of the replay/non-replay EER (%)

15.4 Advances in Front-End Features

The selection of appropriate features for a given classification problem is an important task. Even if the classic boundary to think between a feature extractor (front-end) and a classifier (back-end) as separate components is getting increasingly blurred with the use of end-to-end deep learning and other similar techniques, research on the 'early' components in a pipeline remains important. In the context of anti-spoofing for ASV, this allows the utilisation of one's domain knowledge to guide the design of new discriminative features. For instance, earlier experience suggests that lack of spectral [70] and temporal [123] detail is characteristic of synthetic or voice-coded (vocoded) speech, and that low-quality replayed signals tend to experience loss of spectral details [143]. These initial findings sparked further research into developing advanced front-end features with improved robustness, generalisation across datasets, and other desideratum. As a matter of fact, in contrast to classic ASV (without spoofing attacks) where the most significant advancements have been in the back-end modelling [2], in ASV anti-spoofing, the features seem to make the difference. In this section, we take a brief look at a few such methods emerging from the ASVspoof evaluations. The list is by no means exhaustive and the interested reader is referred to [134] for further discussion.

15.4.1 Front-Ends for Detection of Voice Conversion and Speech Synthesis Spoofing

The front-ends described below have been shown to provide good performance on the ASVspoof 2015 database of spoofing attacks based on voice conversion and speech synthesis. The first front-end was used in the ASVspoof 2015 challenge, while the rest were proposed later after the evaluation.

Cochlear Filter Cepstral Coefficients with Instantaneous Frequency (CFC-CIF). These features were introduced in [129] and successfully used as part of the top-ranked system in the ASVspoof 2015 evaluation. They combine cochlear filter cepstral coefficients (CFCC), proposed in [144], with instantaneous frequency [69]. CFCC is based on wavelet transform-like auditory transform and on some mechanisms of the cochlea of the human ear such as hair cells and nerve spike density. To compute CFCC with instantaneous frequency (CFCCIF), the output of the nerve spike density envelope is multiplied by the instantaneous frequency, followed by the derivative operation and logarithm nonlinearity. Finally, the Discrete Cosine Transform (DCT) is applied to decorrelate the features and obtain a set of cepstral coefficients.

Linear Frequency Cepstral Coefficients (LFCC). LFCCs are very similar to the widely used mel-frequency cepstral coefficients (MFCCs) [145], though the filters are placed in equal sizes for linear scale. This front-end is widely used in speaker recognition and has been shown to perform well in spoofing detection [146]. This technique performs a windowing on the signal, computes the magnitude spectrum using the short-time Fourier transform (STFT), followed by logarithm nonlinearity and the application of a filterbank of linearly spaced N triangular filters to obtain a set of N log-density values. Finally, the DCT is applied to obtain a set of cepstral coefficients.

Constant-Q Cepstral Coefficients (CQCC). This feature was proposed in [136, 137] for spoofing detection and it is based on the Constant-Q Transform (CQT) [147]. The CQT is an alternative time–frequency analysis tool to the STFT that provides variable time and frequency resolution. It provides greater frequency resolution at lower frequencies but greater time resolution at higher frequencies. Figure 15.4 illustrates the CQCC extraction process. The CQT spectrum is obtained, followed by logarithm nonlinearity and by a linearisation of the CQT geometric scale. Finally, cepstral coefficients are obtained through the DCT.

As an alternative to CQCC, infinite impulse response constant-Q transform cepstrum (ICQC) features [148] use the infinite impulse response—constant-Q transform [149], an efficient constant-Q transform based on the IIR filtering of the fast Fourier transform (FFT) spectrum. It delivers multiresolution time–frequency analysis in a linear scale spectrum which is ready to be coupled with traditional cepstral analysis. The IIR-CQT spectrum is followed by the logarithm and decorrelation, either through the DCT or principal component analysis.

Deep Features for Spoofing Detection. All of the above three features sets are handcrafted and consist of a fixed sequence of standard digital signal processing

Fig. 15.4 Block diagram of CQCC feature extraction process

operations. An alternative approach, seeing increased popularity across different machine learning problems, is to learn the feature extractor from a given data by using deep learning techniques [150, 151]. In speech-related applications, these features are widely employed for improving recognition accuracy [152–154]. The work in [155] uses deep neural network to generate bottleneck features for spoofing detection; that is, the activations of a hidden layer with a relatively small number of nodes compared to the size of other layers. The study in [156] investigates various features based on deep learning techniques. Different feed-forward DNNs are used to obtain frame level deep features. Input acoustic features consisting of filterbank outputs with their first derivatives are used to train the network to discriminate between the natural and spoofed speech classes, and output of hidden layers are taken as deep features which are then averaged to obtain an utterance-level descriptor. RNNs are also proposed to estimate utterance-level features from input sequences of acoustic features. In another recent work [157], the authors have investigated deep features based on filterbank trained with the natural and artificial speech data. A feed-forward neural network architecture called here as filter bank neural network (FBNN) is used here that includes a linear hidden layer, a sigmoid hidden layer and a softmax output layer. The number of nodes in the output is six; and of them, five are for the number of spoofed classes in the training set, and the remaining one is for natural speech. The filter banks are learned using the stochastic gradient descent algorithm. The cepstral features extracted using these DNN-based features are shown to be better than the hand-crafted cepstral coefficients.

Scattering Cepstral Coefficients. This feature for spoofing detection was proposed in [158]. It relies upon *scattering spectral decomposition* [159, 160]. This transform is a hierarchical spectral decomposition of a signal based on wavelet filter banks (constant-Q filters), modulus operator, and averaging. Each level of decomposition processes the input signal (either the input signal for the first level of decomposition, or the output of a previous level of decomposition) through the wavelet filterbank and takes the absolute value of filter outputs, producing a scalogram. The scattering coefficients at a certain level are estimated by windowing the scalogram signals and computing the average value within these windows. A two-level scattering decomposition has been shown to be effective for spoofing detection [158]. The final feature vector is computed by taking the DCT of the vector obtained by concatenating the logarithms of the scattering coefficients from all levels and retaining the first a few coefficients. The 'interesting' thing about scattering transform is its stability to small signal deformation and more details of the temporal envelopes than MFCCs [158, 159].

Fundamental Frequency Variation Features. The prosodic features are not as successful as cepstral features in detecting artificial speech on ASVspoof 2015, though some earlier results on PAs indicate that pitch contours are useful for such tasks [6]. In a recent work [161], the author uses fundamental frequency variation (FFV) for this. The FFV captures pitch variation at the frame level and provides complementary information on cepstral features [162]. The combined system gives a very promising performance for both known and unknown conditions on ASVspoof evaluation data.

Phase-based Features. The phase-based features are also successfully used in PAD systems for ASVspoof 2015. For example, relative phase shift (RPS) and modified group delay (MGD) based features are explored in [163]. The authors in [164] have investigated relative phase information (RPI) features. Though the performances on seen attacks are promising with these phase-based features, the performances noticeably degrade for unseen attacks, particularly for S10.

General Observations Regarding Front-Ends for Artificial Speech Setection. Beyond the feature extraction method used, there are two general findings common to any front- end [129, 137, 146, 148]. The first refers to the use of dynamic coefficients. The first and second derivatives of the static coefficients, also known as velocity and acceleration coefficients, respectively, are found important to achieve good spoofing detection performance. In some cases, the use of only dynamic features is superior to the use of static plus dynamic coefficients [146]. This is not entirely surprising, since voice conversion and speech synthesis techniques may fail to model the dynamic properties of the speech signals, introducing artefacts that help the discrimination of spoofed signals. The second finding refers to the use of speech activity detection. In experiments with ASVspoof 2015 corpus, it appears that the silence regions also contain useful information for discriminating between natural and synthetic speech. Thus, retaining non-speech frames turns out to be a better choice for this corpus [146]. This is likely due to the fact that non-speech regions are usually replaced with noise during the voice conversion or speech synthesis operation. However, this could be a database-dependent observation, thus detailed investigations are required.

15.4.2 Front-Ends for Replay Attack Detection

The following front-ends have been proposed for the task of replay spoofing detection, and evaluated in replayed speech databases such as the BTAS 2016 and ASVspoof 2017. Many standard front-ends, such as MFCC, LFCC and PLP, have been combined to improve the performance of replay attack detection. Other front-ends proposed for synthetic and converted speech detection (CFCCIF, CQCC) have been successfully used for the replay detection task. In general, and in opposition to the trend for synthetic and converted speech detection, the use of static coefficients has been shown to be crucial for achieving good performance. This may be explained by the nature of the replayed speech detection task, where detecting changes in the channel captured by static coefficients help with the discrimination of natural and replayed speech. Two additional front-ends are described next.

Inverted Mel-Frequency Cepstral Coefficients (IMFCC). This front-end is relatively simple and similar to the standard MFCC. The only difference is that the filterbank follows an inverted mel scale; that is, it provides an increasing frequency resolution (narrower filters) when frequency increases, and a decreased frequency resolution (wider filters) for decreasing frequency, unlike the mel scale [165]. This front-end was used as part of the top-ranked system of the Biometrics: Theory, Applications, and Systems (BTAS) 2016 speaker anti-spoofing competition [7].

Features Based on Convolutional Neural Networks. In the recent ASVspoof 2017 challenge, the use of deep learning frameworks for feature learning was proven to be key in achieving good replay detection performance. In particular, convolutional neural networks have been successfully used to learn high-level utterance-level features which can later be classified with simple classifiers. As part of the top-ranked system [138] in the ASVspoof 2017 challenge, a light convolutional neural network architecture [166] is fed with truncated normalised FFT spectrograms (to force fixed data dimensions). The network consists of a set of convolutional layers, followed by a fully connected layer. The last layer contains two outputs with softmax activation corresponding to the two classes. All layers use the max-feature-map activation function [166], which acts as a feature selector and reduces the number of feature maps by half on each layer. The network is then trained to discriminate between the natural and spoofed speech classes. Once the network is trained, it is used to extract a high-level feature vector which is the output of the fully connected layer. All the test utterances are processed to obtain high-level representations, which are later classified with an external classifier.

Other Hand-Crafted Features. Many other features have also been used for replayed speech detection in the context of the ASVspoof 2017 database. Even if the performances of single systems using such features are not always high, they are shown to be complementary when fused at the score level [167], similar to conventional ASV research outside of the spoofing detection. These features include MFCC, IMFCC, rectangular filter cepstral coefficients (RFCCs), PLP, CQCC, spectral centroid magnitude coefficients (SCMC), subband spectral flux coefficient (SSFC) and variable length Teager energy operator energy separation algorithm-instantaneous frequency cosine coefficients (VESA-IFCC). Though, of course, one usually then has to further train the fusion system, which makes the system more involved concerning practical applications.

15.5 Advances in Back-End Classifiers

In the natural versus spoof classification problem, two main families of approaches have been adopted, namely generative and discriminative. Generative approaches include those of GMM-based classifiers and i-vector representations combined with support vector machines (SVMs). As for discriminative approaches, deep learning-based techniques have become more popular. Finally, new deep learning end-to-end solutions are emerging. Such techniques perform the typical pipeline entirely through deep learning, from feature representation learning and extraction to the final classification. While including such approaches into the traditional classifiers category may not be the most precise, they are included in this classifiers section for simplicity.

15.5.1 Generative Approaches

Gaussian Mixture Model (GMM) Classifiers. Considering two classes, namely natural and spoofed speech, one GMM can be learned for each class using appropriate training data. In the classification stage, an input utterance is processed to obtain its likelihoods with respect to the natural and spoofed models. The resulting classification score is the log-likelihood ratio between the two competing hypotheses; in effect, those of the input utterance belonging to the natural and to the spoofed classes. A high score supports the former hypothesis, while a low score supports the latter. Finally, given a test utterance, classification can be performed by thresholding the obtained score. If the score is above the threshold, the test utterance is classified as natural, and otherwise, it is classified as spoof. Many proposed anti-spoofing systems use GMM classifiers [129, 136, 146, 148, 155, 158, 168].

I-vector. The state-of-the-art i-vector paradigm for speaker verification [169] has been explored for spoofing detection [170, 171]. Typically, an i-vector is extracted from an entire speech utterance and used as a low-dimensional, high-level feature which is later classified by means of a binary classifier, commonly cosine distance measure or support vector machine (SVM). Different amplitude- and phase-based front-ends [130, 138] can be employed for the estimation of i-vectors. A recent work shows that data selection for i-vector extractor training (also known as $\mathbf{T}$ matrix) is an important factor for achieving completive recognition accuracy [172].

15.5.2 Discriminative Approaches

DNN Classifiers. Deep learning-based classifiers have been explored for use in the task of natural and spoofed speech discrimination. In [155, 173], several front-ends are evaluated with neural network classifier consisting of several hidden layers with sigmoid nodes and softmax output, which is used to calculate utterance posteriors. However, the implementation detail of the DNNs—such the number of nodes, the cost function, the optimization algorithm and the activation functions—is not precisely mentioned in those work and the lack of this very relevant information makes it difficult to reproduce the results.

In a recent work [174], a five-layer DNN spoofing detection system is investigated for ASVspoof 2015 which uses a novel scoring method, termed in the paper as *human log-likelihoods* (HLLs). Each of the hidden layers has 2048 nodes with a sigmoid activation function. The network has six softmax output layers. The DNN is implemented using a computational network toolkit[9] and trained with stochastic gradient descent methods with dynamics information of acoustic features, such as spectrum-based cepstral coefficients (SBCC) and CQCC as input. The cross entropy function is selected as the cost function and the maximum training epoch is chosen as 120. The mini-batch size is set to 128. The proposed method shows considerable PAD detection performance. The author obtain an EER for S10 of 0.255% and average EER for all attacks of 0.045- when used with CQCC acoustic features. These are the best reported performance in ASVspoof 2015 so far.

DNN-Based End-to-End Approaches. End-to-end systems aim to perform all the stages of a typical spoofing detection pipeline, from feature extraction to classification, by learning the network parameters involved in the process as a whole. The advantage of such approaches is that they do not explicitly require prior knowledge of the spoofing attacks as required for the development of acoustic features. Instead, the parameters are learned and optimised from the training data. In [175], a convolutional long short-term memory (LSTM) deep neural network (CLDNN) [176] is used as an end-to-end solution for spoofing detection. This model receives input in the form of a sequence of raw speech frames and outputs a likelihood for the whole sequence. The CLDNN performs time–frequency convolution through CNN to reduce spectral variance, long-term temporal modelling by using a LSTM, and classification using a DNN. Therefore, it is entirely an end-to-end solution which does not rely on any external feature representation. The works in [138, 177] propose other end-to-end solutions by combining convolutional and recurrent layers, where the first act as a feature extractor and the second models the long-term dependencies and acts as a classifier. Unlike the work in [175], the input data is the FFT spectrogram of the speech utterance and not the raw speech signal. In [178], the authors have investigated CNN-based end-to-end system for PAD where the raw speech is used to jointly learn the feature extractor and classifier. Score level combination of this CNN system with standard long-term spectral statistics based system shows considerable overall improvement.

15.6   Other PAD Approaches

While most of the studies in voice PAD detection research focus on algorithmic improvements for discriminating natural and artificial speech signals, some recent studies have explored utilising additional information collected using special additional hardware to protect ASV system from presentation attacks [179–182]. Since an intruder can easily collect voice samples for the target speakers using covert recording; the idea here is to detect and recognise supplementary information related to the speech production process. Moreover, by its nature, that supplementary information is difficult, if not impossible, to mimic using spoofing methods in the practical

---

[9]https://github.com/Microsoft/CNTK.

scenario. These PAD techniques have shown excellent recognition accuracy in the spoofed condition, at the cost of additional setup in the data acquisition step.

The work presented in [180, 181] utilises the phenomenon of *pop noise*, which is a distortion in human breath when it reaches a microphone [183]. During natural speech production, the interactions between the airflow and the vocal cavities may result in a sort of plosive burst, commonly know as pop noise, which can be captured via a microphone. In the context of professional audio and music production, pop noise is unwanted and is eliminated during the recording or mastering process. In the context of ASV, however, it can help in the process of PAD. The basic principle is that a replay sound from a loudspeaker does not involve the turbulent airflow generating the pop noise as in the natural speech. The authors in [180, 181] have developed a pop noise detector which eventually distinguishes natural speech from playback recording as well as synthetic speech generated using VC and SS methods. In experiments with 17 female speakers, a tandem detection system that combines both single- and double-channel pop noise detection gives the lowest ASV error rates in the PA condition.

The authors in [179] have introduced the use of a smartphone-based *magnetometer* to detect voice presentation attack. The conventional loudspeakers, which are used for playback during access of the ASV systems, generate sound using acoustic transducer and generate a magnetic field. The idea, therefore, is to capture the use of loudspeaker by sensing the magnetic field which would be absent from human vocals. Experiments were conducted using playback from 25 different conventional loudspeakers, ranging from low-end to high-end and placed in different distances from the smartphone that contains the ASV system. A speech corpus of five speakers was collected for the ASV experiments executed using an open-source ASV toolkit, SPEAR.[10] Experiments were conducted with other datasets, using a similarly limited number of speakers. The authors demonstrated that the magnetic field based detection can be reliable for the detection of playback within 6–8 cm from the smartphone. They further developed a mechanism to detect the size of the sound source to prevent the use of small speakers such as earphones.

The authors in [184, 185] utilise certain acoustics concepts to prevent ASV systems from PAs. They first introduced a method [184] that estimates dynamic sound source position (articulation position within mouth) of some speech sounds using a small array using *microelectromechanical systems* (MEMS) microphones embedded in mobile devices and compare it with loudspeakers, which have a flat sound source. In particular, the idea is to capture the dynamics of *time-difference-of-arrival* (TDOA) in a sequence of speech sounds to the microphones of the smartphone. Such unique TDOA changes, which do not exist under replay conditions, are used for detecting replay attacks. The similarities between the TDOAs of test speech and user templates are measured using probability function under Gaussian assumption and correlation measure as well as their combinations. Experiments involving 12 speakers and 3 different types of smartphone demonstrate a low EER and high PAD accuracy. The

---

[10]https://www.idiap.ch/software/bob/docs/bob/bob.bio.spear/stable/index.html.

proposed method is seen to remain robust despite the change of smartphones during the test and the displacements.

In [185], the same research group has used the idea of the *Doppler effect* to detect the replay attack. The idea here is to capture the *articulatory gestures* of the speakers when they speak a passphrase. The smartphone acts as a Doppler radar and transmits a high-frequency tone at 20 kHz from the built-in speaker and senses the reflections using the microphone during authentication process. The movement of the speaker's articulators during vocalisation creates a speaker-dependent Doppler frequency shift at around 20 kHz, which is stored along with the speech signal during the speaker-enrolment process. During a playback attack, the Doppler frequency shift will be different due to the lack of articulatory movements. Energy-based frequency features and frequency-based energy features are computed from a band of 19.8 and 20.2 kHz. These features are used to discriminate between the natural and replayed voice, and the similarity scores are measured in terms of Pearson correlation coefficient. Experiments are conducted with a dataset of 21 speakers and using three different smartphones. The data also includes test speech for replay attack with different loudspeakers and for impersonation attack with four different impersonators. The proposed system was demonstrated to be effective in achieving low EER for both types of attacks. Similar to [184], the proposed method indicated robustness to the phone placement.

The work in [182] introduces the use of a specific non-acoustic sensor, *throat microphone* (TM), or laryngophone, to enhance the performance of the voice PAD system. An example of such microphones is shown in Fig. 15.5. The TM is used with a conventional acoustic microphone (AM) in a dual-channel framework for robust speaker recognition and PAD. Since this type of microphone is attached to the speaker's neck, it would be difficult for the attacker to obtain a covert recording

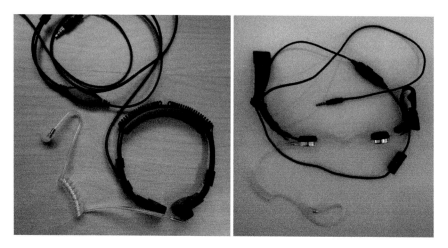

Fig. 15.5 Throat-microphones used in [182]. (Reprinted with permission from IEEEACM Transactions on (T-ASL) Audio, Speech, and Language Processing)

of the target speaker's voice. Therefore, one possibility for the intruder is to use the stolen recording from an AM and to try to record it back using a TM for accessing the ASV system. A speech corpus of 38 speakers were collected for the ASV experiments. The dual-channel setup yielded considerable ASV for both licit and spoofed conditions. The performance is further improved when this ASV system is integrated with the dual-channel based PAD. The authors show zero FAR for replay imposters by decision fusion of ASV and PAD.

All of the above new PAD methods deviating from the 'mainstream' of PAD research in ASV are reported to be reliable and useful in specific application scenarios for identifying presentation attacks. The methods are also fundamentally different and difficult to compare in the same settings. Since the authors focus on the methodological aspects, experiments are mostly conducted on a dataset of limited number of speakers. Extensive experiments with more subjects from diverse environmental conditions should be performed to assess their suitability for real-world deployment.

15.7 Future Directions of Anti-spoofing Research

The research in ASV anti-spoofing is becoming popular and well recognised in the speech processing and voice biometric community. The state-of-the-art spoofing detector gives promising accuracy in the benchmarking of spoofing countermeasures. Further work is needed to address a number of specific issues regarding its practical use. A number of potential topics for consideration in further work are now discussed.

- **Noise, reverberation and channel effect**. Recent studies indicate that spoofing countermeasures offer little resistance to additive noise [186, 187], reverberation [188] and channel effect [189] even though their performances on 'clean' speech corpus is highly promising. The relative degradation of performance is actually much worse than the degradation of a typical ASV system under the similar mismatch condition. One reason could be that, at least until the ASVspoof 2017 evaluation, the methodology developed has been driven in clean, high-quality speech. In other words, the community might have developed its methods implicitly for laboratory testing. The commonly used speech enhancement algorithms also fail to reduce the mismatch due to environmental differences, though multi-condition training [187] and more advanced training methods [190] have been found useful. The study presented in [189] shows considerable degradation of PAD performance even in *matched* acoustic conditions. The feature settings used for the original corpus gives lower accuracy when both training and test data are digitally processed with the telephone channel effect. These are probably because the spoofing artefacts themselves act as extrinsic variabilities which degrade the speech quality in some way. Since the task of spoofing detection is related to detecting those artefacts, the problem becomes more difficult in the presence of small external effects due to variation in environment and channel. These suggest further investigations need to be carried out for the development of robust spoofing countermeasures.

- **Generalisation of spoofing Countermeasures**. The generalisation property of spoofing countermeasures for detecting new kinds of speech presentation attack is an important requirement for their application in the wild. Study explores that countermeasure methods trained with a class of spoofing attacks fail to generalise this for other classes of spoofing attack [167, 191]. For example, PAD systems trained with VC- and SS-based spoofed speech give a very poor performance for playback detection [192]. The results of the first two ASVspoof challenges also reveal that detecting the converted speech created with an "unknown" method or the playback voice recording in a new replay session are difficult to detect. These clearly indicate the overfitting of PAD systems with available training data. Therefore, further investigation should be conducted to develop attack-independent universal spoofing detector. Other than the unknown attack issue, generalisation is also an important concern for cross-corpora evaluation of the PAD system [193]. This specific topic is discussed in chapter 19 of this book.

- **Investigations with new spoofing methods**. The studies of converted spoof speech mostly focused on methods based on classical signal processing and machine learning techniques. Recent advancements in VC and SS research with deep learning technology show significant improvements in creating high-quality synthetic speech [52]. The GAN [194] can be used to create (generator) spoofed voices with relevant feedback from the spoofing countermeasures (discriminator). Some preliminary studies demonstrate that the GAN-based approach can make speaker verification systems more vulnerable to presentation attacks [66, 195]. More detailed investigations should be conducted on this direction for the development of countermeasure technology to guard against this type of advanced attack.

- **Joint operations of PAD and ASV**. The ultimate goal of developing PAD system is to protect the recogniser, the ASV system from imposters with spoofed speech. So far, the majority of the studies focused on the evaluation of standalone countermeasures. The integration of these two systems is not trivial number of reasons. First, standard linear output score fusion techniques, being extensively used to combine homogenous ASV system, are not appropriate since the ASV and its countermeasures are trained to solve two different tasks. Second, an imperfect PAD can increase the false alarm rate by rejecting genuine access trials [196]. Third, and more fundamentally, it is not obvious whether improvements in standalone spoofing countermeasures should improve the overall system as a whole: a nearly perfect PAD system with close to zero EER may fail to protect ASV system in practice if not properly calibrated [197]. In a recent work [198], the authors propose a modification in a GMM-UBM based ASV system to make it suitable for both licit and spoofed conditions. The joint evaluation of PAD and ASV, as well as their combination techniques, certainly deserves further attention. Among other feedback received from the attendees of the ASVspoof 2017 special session organised during INTERSPEECH 2017, it was proposed that the authors of this chapter consider shifting the focus from standalone spoofing to more ASV-centric solutions in future. We tend to agree. In our recent work [199], we propose a new cost function for joint assessment of PAD and ASV system. In another work [200], we propose a new fusion method for combining scores of countermeasures and recognisers. This work also explores speech features which can be used both for PAD and ASV.

15.8 Conclusion

This contribution provides an introduction to the different voice presentation attacks and their detection methods. It then reviews previous works with a focus on recent progress in assessing the performance of PAD systems. We have also briefly reviewed two recent ASVspoof challenges organised for the detection of voice PAs. This study includes discussion of recently developed features and the classifiers which are predominantly used in ASVspoof evaluations. We further include an extensive survey on alternative PAD methods. Apart from the conventional voice-based systems that use statistical properties of natural and spoofed speech for their discrimination, these recently developed methods utilise a separate hardware for the acquisition of other signals such as pop noise, throat signal and extrasensory signals with smartphones for PAD. The current status of these non-mainstream approaches to PAD detection are somewhat similar to the status of the now more-or-less standard methods for artificial speech and replay PAD detection some 3–4 years ago: they are innovative and show promising results, but the pilot experiments have been carried out on relatively small and/or proprietary datasets, leaving an open question as to how scalable or generalisable these solutions are in practice. Nonetheless, in the long run and noting especially the rapid development of speech synthesis technology, it is likely that the quality of artificial/synthetic speech will eventually be indistinguishable from that of natural human speech. Such future spoofing attacks therefore could not be detected using the current mainstream techniques that focus on spectral or temporal details of the speech signal, but will require novel ideas that benefit from auxiliary information, rather than just the acoustic waveform.

In the past three years, the progress in voice PAD research has been accelerated by the development and free availability of speech corpus such as the ASVspoof series, SAS, BTAS 2016, AVSpoof. The work discussed several open challenges which show that this problem requires further attention to improving robustness due to mismatch condition, generalisation to a new type of presentation attacks, and so on. Results from joint evaluations with integrated ASV system are also an important requirement for practical applications of PAD research. We think, however, that this extensive review will be of interest not only to those involved in voice PAD research but also to voice biometrics researchers in general.

Appendix A. Action Towards Reproducible Research

A.1. Speech Corpora

1. Spoofing and Anti-Spoofing (SAS) database v1.0: This database presents the first version of a speaker verification spoofing and anti-spoofing database, named SAS corpus [201]. The corpus includes nine spoofing techniques, two of which are speech synthesis, and seven are voice conversion.
   Download link: http://dx.doi.org/10.7488/ds/252
2. ASVspoof 2015 database: This database has been used in the first Automatic Speaker Verification Spoofing and Countermeasures Challenge (ASVspoof 2015). Genuine speech is collected from 106 speakers (45 male, 61 female) and with no significant channel or background noise effects. Spoofed speech is generated from the genuine data using a number of different spoofing algorithms. The full dataset is partitioned into three subsets, the first for training, the second for development and the third for evaluation.
   Download link: http://dx.doi.org/10.7488/ds/298
3. ASVspoof 2017 database: This database has been used in the Second Automatic Speaker Verification Spoofing and Countermeasuers Challenge: ASVspoof 2017. This database makes an extensive use of the recent text-dependent RedDots corpus, as well as a replayed version of the same data. It contains a large amount of speech data from 42 speakers collected from 179 replay sessions in 62 unique replay configurations.
   Download link: http://dx.doi.org/10.7488/ds/2313

A.2. Software Packages

1. Feature extraction techniques for anti-spoofing: This package contains the MAT-LAB implementation of different acoustic feature extraction schemes as evaluated in [146].
   Download link: http://cs.joensuu.fi/~sahid/codes/AntiSpoofing_Features.zip
2. Baseline spoofing detection package for ASVspoof 2017 corpus: This package contains the MATLAB implementations of two spoofing detectors employed as baseline in the official ASVspoof 2017 evaluation. They are based on constant-Q cepstral coefficients (CQCC) [137] and Gaussian mixture model classifiers.
   Download link: http://audio.eurecom.fr/software/ASVspoof2017_baseline_countermeasures.zip

References

1. Kinnunen T, Li H (2010) An overview of text-independent speaker recognition: From features to supervectors. Speech Commun 52(1):12–40. https://doi.org/10.1016/j.specom.2009. 08.009. http://www.sciencedirect.com/science/article/pii/S0167639309001289
2. Hansen J, Hasan T (2015) Speaker recognition by machines and humans: a tutorial review. IEEE Signal Process Mag 32(6):74–99
3. ISO/IEC 30107: Information technology—biometric presentation attack detection. International Organization for Standardization (2016)
4. Kinnunen T, Sahidullah M, Kukanov I, Delgado H, Todisco M, Sarkar A, Thomsen N, Hautamäki V, Evans N, Tan ZH (2016) Utterance verification for text-dependent speaker recognition: a comparative assessment using the reddots corpus. In: Proceedings of Interspeech, pp 430–434
5. Shang, W, Stevenson, M. (2010). Score normalization in playback attack detection. In: Proceedings of ICASSP. IEEE, pp 1678–1681
6. Wu Z, Evans N, Kinnunen T, Yamagishi J, Alegre F, Li H (2015) Spoofing and countermeasures for speaker verification: a survey. Speech Commun 66:130–153
7. Korshunov P, Marcel S, Muckenhirn H, Gonçalves A, Mello A, Violato R, Simoes F, Neto M, de Angeloni AM, Stuchi J, Dinkel H, Chen N, Qian Y, Paul D, Saha G, Sahidullah M. (2016). Overview of BTAS 2016 speaker anti-spoofing competition. In: 2016 IEEE 8th international conference on biometrics theory, applications and systems (BTAS), pp 1–6 (2016)
8. Evans N, Kinnunen T, Yamagishi J, Wu Z, Alegre F, DeLeon P (2014) Speaker recognition anti-spoofing. In: Marcel S, Li, SZ, Nixon M (eds) Handbook of biometric anti-spoofing. Springer
9. Marcel S, Li SZ, Nixon M (eds) Handbook of biometric anti-spoofing: trusted biometrics under spoofing attacks. Springer (2014)
10. Farrús Cabeceran M, Wagner M, Erro D, Pericás H (2010) Automatic speaker recognition as a measurement of voice imitation and conversion. The Int J Speech Lang Law 1(17):119–142
11. Perrot P, Aversano G, Chollet G (2007) Voice disguise and automatic detection: review and perspectives. Progress in nonlinear speech processing, pp. 101–117
12. Zetterholm E (2007) Detection of speaker characteristics using voice imitation. In: Speaker Classification II. Springer, pp 192–205
13. Lau Y, Wagner M, Tran D (2004) Vulnerability of speaker verification to voice mimicking. In: Proceedings of 2004 international symposium on intelligent multimedia, video and speech processing, 2004. IEEE, pp 145–148
14. Lau Y, Tran D, Wagner M (2005) Testing voice mimicry with the YOHO speaker verification corpus. In: International conference on knowledge-based and intelligent information and engineering systems. Springer, pp 15–21
15. Mariéthoz J, Bengio S (2005) Can a professional imitator fool a GMM-based speaker verification system? Technical report, Idiap Research Institute
16. Panjwani S, Prakash A (2014) Crowdsourcing attacks on biometric systems. In: Symposium on usable privacy and security (SOUPS 2014), pp 257–269
17. Hautamäki R, Kinnunen T, Hautamäki V, Laukkanen AM (2015) Automatic versus human speaker verification: the case of voice mimicry. Speech Commun 72:13–31
18. Ergunay S, Khoury E, Lazaridis A, Marcel S (2015) On the vulnerability of speaker verification to realistic voice spoofing. In: IEEE international conference on biometrics: theory, applications and systems, pp 1–8
19. Lindberg J, Blomberg M (1999) Vulnerability in speaker verification-a study of technical impostor techniques. Proceedings of the European conference on speech communication and technology 3:1211–1214
20. Villalba J, Lleida E (2010) Speaker verification performance degradation against spoofing and tampering attacks. In: FALA 10 workshop, pp 131–134

21. Wang ZF, Wei G, He QH (2011) Channel pattern noise based playback attack detection algorithm for speaker recognition. In: 2011 International conference on machine learning and cybernetics, vol 4, pp 1708–1713
22. Villalba J, Lleida E (2011) Preventing replay attacks on speaker verification systems. In: 2011 IEEE International Carnahan Conference on Security Technology (ICCST). IEEE, pp 1–8
23. Gałka J, Grzywacz M, Samborski R (2015) Playback attack detection for text-dependent speaker verification over telephone channels. Speech Commun 67:143–153
24. Taylor P (2009) Text-to-speech synthesis. Cambridge University Press
25. Klatt DH (1980) Software for a cascade/parallel formant synthesizer. J Acoust Soc Am 67:971–995
26. Moulines E, Charpentier F (1990) Pitch-synchronous waveform processing techniques for text-to-speech synthesis using diphones. Speech Commun 9:453–467
27. Hunt A, Black AW (1996) Unit selection in a concatenative speech synthesis system using a large speech database. In: Proceedings ICASSP, pp 373–376
28. Breen A, Jackson P (1998) A phonologically motivated method of selecting nonuniform units. In: Proceedings of ICSLP, pp 2735–2738
29. Donovan RE, Eide EM (1998) The IBM trainable speech synthesis system. In: Proceedings of ICSLP, pp 1703–1706
30. Beutnagel B, Conkie A, Schroeter J, Stylianou Y, Syrdal A (1999) The AT&T Next-Gen TTS system. In: Proceedigns of joint ASA, EAA and DAEA meeting, pp 15–19
31. Coorman G, Fackrell J, Rutten P, Coile B (2000) Segment selection in the L & H realspeak laboratory TTS system. In: Proceedings of ICSLP, pp 395–398
32. Yoshimura T, Tokuda K, Masuko T, Kobayashi T, Kitamura T (1999) Simultaneous modeling of spectrum, pitch and duration in HMM-based speech synthesis. In: Proceedings of Eurospeech, pp 2347–2350
33. Ling ZH, Wu YJ, Wang YP, Qin L, Wang RH (2006) USTC system for Blizzard Challenge 2006 an improved HMM-based speech synthesis method. In: Proceedings of the Blizzard challenge workshop
34. Black A (2006) CLUSTERGEN: a statistical parametric synthesizer using trajectory modeling. In: Proceedings of Interspeech, pp 1762–1765
35. Zen H, Toda T, Nakamura M, Tokuda K (2007) Details of the Nitech HMM-based speech synthesis system for the Blizzard challenge 2005. IEICE Trans Inf Syst E90-D(1):325–333
36. Zen H, Tokuda K, Black AW (2009) Statistical parametric speech synthesis. Speech Commun 51(11):1039–1064
37. Yamagishi J, Kobayashi T, Nakano Y, Ogata K, Isogai J (2009) Analysis of speaker adaptation algorithms for HMM-based speech synthesis and a constrained SMAPLR adaptation algorithm. IEEE Trans Speech Audio Lang Process 17(1), 66–83 (2009)
38. Leggetter CJ, Woodland PC (1995) Maximum likelihood linear regression for speaker adaptation of continuous density hidden Markov models. Comput Speech Lang 9:171–185
39. Woodland PC (2001) Speaker adaptation for continuous density HMMs: a review. In: Proceedings of ISCA workshop on adaptation methods for speech recognition, p 119
40. Ze H, Senior A, Schuster M (2013) Statistical parametric speech synthesis using deep neural networks. In: Proceedings of ICASSP, pp 7962–7966
41. Ling ZH, Deng L, Yu D (2013) Modeling spectral envelopes using restricted boltzmann machines and deep belief networks for statistical parametric speech synthesis. IEEE Trans Audio Speech Lang Process 21(10):2129–2139
42. Fan Y, Qian Y, Xie FL, Soong F (2014) TTS synthesis with bidirectional LSTM based recurrent neural networks. In: Proceedings of Interspeech, pp 1964–1968
43. Zen H, Sak H (2015) Unidirectional long short-term memory recurrent neural network with recurrent output layer for low-latency speech synthesis. In: Proceedings of ICASSP, pp 4470–4474
44. Wu Z, King S (2016) Investigating gated recurrent networks for speech synthesis. In: Proceedings of ICASSP, pp 5140–5144 (2016)

45. Wang X, Takaki S, Yamagishi J (2016) Investigating very deep highway networks for parametric speech synthesis. In: 9th ISCA speech synthesis workshop, pp 166–171
46. Wang X, Takaki S, Yamagishi J (2018) Investigating very deep highway networks for parametric speech synthesis. Speech Commun 96:1–9
47. Wang X, Takaki S, Yamagishi J (2017) An autoregressive recurrent mixture density network for parametric speech synthesis. In: Proceedings of ICASSP, pp 4895–4899
48. Wang X, Takaki S, Yamagishi J (2017) An RNN-based quantized F0 model with multi-tier feedback links for text-to-speech synthesis. In: Proceedings of Interspeech, pp 1059–1063 (2017)
49. Saito, Y., Takamichi, S., Saruwatari, H.: Training algorithm to deceive anti-spoofing verification for DNN-based speech synthesis. In: Proc. ICASSP, pp 4900–4904 (2017)
50. Saito Y, Takamichi S, Saruwatari H (2018) Statistical parametric speech synthesis incorporating generative adversarial networks. IEEE/ACM Trans Audio Speech Lang Process 26(1):84–96
51. Kaneko T, Kameoka H, Hojo N, Ijima Y, Hiramatsu K, Kashino K (2017) Generative adversarial network-based postfilter for statistical parametric speech synthesis. In: Proceedings of ICASSP, pp 4910–4914
52. Van Oord D, Dieleman A, Zen S, Simonyan H, Vinyals K, Graves O, Kalchbrenner A, Senior N, Kavukcuoglu AK (2016) Wavenet: a generative model for raw audio. arXiv:1609.03499
53. Mehri S, Kumar K, Gulrajani I, Kumar R, Jain S, Sotelo J, Courville A, Bengio Y (2016) Samplernn: an unconditional end-to-end neural audio generation model. arXiv:1612.07837
54. Wang Y, Skerry-Ryan R, Stanton D, Wu Y, Weiss R, Jaitly N, Yang Z, Xiao Y, Chen Z, Bengio S, Le Q, Agiomyrgiannakis Y, Clark R, Saurous R (2017) Tacotron: towards end-to-end speech synthesis. In: Proceedings of Interspeech, pp 4006–4010
55. Gibiansky A, Arik S, Diamos G, Miller J, Peng K, Ping W, Raiman J, Zhou Y (2017) Deep voice 2: multi-speaker neural text-to-speech. In: Advances in neural information processing systems, pp 2966–2974
56. Shen J, Schuster M, Jaitly N, Skerry-Ryan R, Saurous R, Weiss R, Pang R, Agiomyrgiannakis Y, Wu Y, Zhang Y, Wang Y, Chen Z, Yang Z (2018) Natural tts synthesis by conditioning wavenet on mel spectrogram predictions. In: Proceedigns of ICASSP
57. King S (2014) Measuring a decade of progress in text-to-speech. Loquens 1(1):006
58. King S, Wihlborg L, Guo W (2017) The blizzard challenge 2017. In: Proceedings of Blizzard Challenge Workshop, Stockholm, Sweden
59. Foomany F, Hirschfield A, Ingleby M (2009) Toward a dynamic framework for security evaluation of voice verification systems. In: 2009 IEEE toronto international conference science and technology for humanity (TIC-STH), pp 22–27
60. Masuko T, Hitotsumatsu T, Tokuda K, Kobayashi T (1999) On the security of HMM-based speaker verification systems against imposture using synthetic speech. In: Proceedings of EUROSPEECH
61. Matsui T, Furui S (1995) Likelihood normalization for speaker verification using a phoneme- and speaker-independent model. Speech Commun 17(1–2):109–116
62. Masuko T, Tokuda K, Kobayashi T, Imai S (1996) Speech synthesis using HMMs with dynamic features. In: Proceedings of ICASSP
63. Masuko T, Tokuda K, Kobayashi T, Imai S (1997) Voice characteristics conversion for HMM-based speech synthesis system. In: Proceedings of ICASSP
64. De Leon PL, Pucher M, Yamagishi J, Hernaez I, Saratxaga I (2012) Evaluation of speaker verification security and detection of HMM-based synthetic speech. IEEE Trans Audio Speech Lang Process 20(8):2280–2290
65. Galou G (2011) Synthetic voice forgery in the forensic context: a short tutorial. In: Forensic speech and audio analysis working group (ENFSI-FSAAWG), pp 1–3
66. Cai W, Doshi A, Valle R (2018) Attacking speaker recognition with deep generative models. arXiv:1801.02384
67. Satoh T, Masuko T, Kobayashi T, Tokuda K (2001) A robust speaker verification system against imposture using an HMM-based speech synthesis system. In: Proceedings of Eurospeech (2001)

68. Chen LW, Guo W, Dai LR (2010) Speaker verification against synthetic speech. In: 2010 7th International symposium on Chinese spoken language processing (ISCSLP), pp 309–312
69. Quatieri TF (2002) Discrete-time speech signal processing: principles and practice. Prentice-Hall, Inc
70. Wu Z, Chng E, Li H (2012) Detecting converted speech and natural speech for anti-spoofing attack in speaker recognition. In: Proceedings of Interspeech
71. Ogihara A, Unno H, Shiozakai A (2005) Discrimination method of synthetic speech using pitch frequency against synthetic speech falsification. IEICE Trans Fund Electron Commun Comput Sci 88(1):280–286
72. De Leon P, Stewart B, Yamagishi J (2012) Synthetic speech discrimination using pitch pattern statistics derived from image analysis. In: Proceedings of Interspeech 2012. Portland, Oregon, USA
73. Stylianou Y (2009) Voice transformation: a survey. In: Proceedings of ICASSP, pp 3585–3588
74. Pellom B, Hansen J (1999) An experimental study of speaker verification sensitivity to computer voice-altered imposters. In: Proceedings of ICASSP, vol 2, pp 837–840
75. Mohammadi S, Kain A (2017) An overview of voice conversion systems. Speech Commun 88:65–82
76. Abe M, Nakamura S, Shikano K, Kuwabara H (1988) Voice conversion through vector quantization. In: Proceedigns of ICASSP, pp 655–658
77. Arslan L (1999) Speaker transformation algorithm using segmental codebooks (STASC). Speech Commun 28(3):211–226
78. Kain A, Macon M (1998) Spectral voice conversion for text-to-speech synthesis. In: Proceedings of ICASSP vol 1, pp 285–288
79. Stylianou Y, Cappé O, Moulines E (1998) Continuous probabilistic transform for voice conversion. IEEE Trans Speech Audio Process 6(2):131–142
80. Toda T, Black A, Tokuda K (2007) Voice conversion based on maximum-likelihood estimation of spectral parameter trajectory. IEEE Trans Audio Speech Lang Process 15(8):2222–2235
81. Kobayashi K, Toda T, Neubig G, Sakti S, Nakamura S (2014) Statistical singing voice conversion with direct waveform modification based on the spectrum differential. In: Proceedings of Interspeech
82. Popa V, Silen H, Nurminen J, Gabbouj M (2012) Local linear transformation for voice conversion. In: Proceedigns of ICASSP. IEEE, pp 4517–4520
83. Chen Y, Chu M, Chang E, Liu J, Liu R (2003) Voice conversion with smoothed GMM and MAP adaptation. In: Proceedings of EUROSPEECH, pp 2413–2416
84. Hwang HT, Tsao Y, Wang HM, Wang YR, Chen SH (2012) A study of mutual information for GMM-based spectral conversion. In: Proceedigns of Interspeech
85. Helander E, Virtanen T, Nurminen J, Gabbouj M (2010) Voice conversion using partial least squares regression. IEEE Trans Audio Speech Lang Process 18(5):912–921
86. Pilkington N, Zen H, Gales M (2011) Gaussian process experts for voice conversion. In: Proceedings of Interspeech
87. Saito D, Yamamoto K, Minematsu N, Hirose K (2011) One-to-many voice conversion based on tensor representation of speaker space. In: Proceedings of Interspeech, pp 653–656
88. Zen H, Nankaku Y, Tokuda K (2011) Continuous stochastic feature mapping based on trajectory HMMs. IEEE Trans Audio Speech Lang Process 19(2):417–430
89. Wu Z, Kinnunen T, Chng E, Li H (2012) Mixture of factor analyzers using priors from non-parallel speech for voice conversion. IEEE Signal Process Lett 19(12)
90. Saito D, Watanabe S, Nakamura A, Minematsu N (2012) Statistical voice conversion based on noisy channel model. IEEE Trans Audio Speech Lang Process 20(6):1784–1794
91. Song P, Bao Y, Zhao L, Zou C (2011) Voice conversion using support vector regression. Electron Lett 47(18):1045–1046
92. Helander E, Silén H, Virtanen T, Gabbouj M (2012) Voice conversion using dynamic kernel partial least squares regression. IEEE Trans Audio Speech Lang Process 20(3):806–817
93. Wu Z, Chng E, Li H (2013) Conditional restricted boltzmann machine for voice conversion. In: The first IEEE China summit and international conference on signal and information processing (ChinaSIP). IEEE

94. Narendranath M, Murthy H, Rajendran S, Yegnanarayana B (1995) Transformation of formants for voice conversion using artificial neural networks. Speech Commun 16(2):207–216
95. Desai S, Raghavendra E, Yegnanarayana B, Black A, Prahallad K (2009) Voice conversion using artificial neural networks. In: Proceedings of ICASSP. IEEE, pp 3893–3896
96. Saito Y, Takamichi S, Saruwatari H (2017) Voice conversion using input-to-output highway networks. IEICE Transactions on Inf Syst E100.D(8):1925–1928
97. Nakashika T, Takiguchi T, Ariki Y (2015) Voice conversion using RNN pre-trained by recurrent temporal restricted boltzmann machines. IEEE/ACM Trans Audio Speech Lang Process (TASLP) 23(3):580–587
98. Sun L, Kang S, Li K, Meng H (2015) Voice conversion using deep bidirectional long shortterm memory based recurrent neural networks. In: Proceedings of ICASSP, pp 4869–4873
99. Sundermann D, Ney H (2003) VTLN-based voice conversion. In: Proceedings of the 3rd IEEE international symposium on signal processing and information technology, 2003. ISSPIT 2003. IEEE
100. Erro D, Moreno A, Bonafonte A (2010) Voice conversion based on weighted frequency warping. IEEE Trans Audio Speech Lang Process 18(5):922–931
101. Erro D, Navas E, Hernaez I (2013) Parametric voice conversion based on bilinear frequency warping plus amplitude scaling. IEEE Trans Audio Speech Lang Process 21(3):556–566
102. Hsu CC, Hwang HT, Wu YC, Tsao Y, Wang HM (2017) Voice conversion from unaligned corpora using variational autoencoding wasserstein generative adversarial networks. In: Proceedings of Interspeech, vol 2017, pp 3364–3368
103. Miyoshi H, Saito Y, Takamichi S, Saruwatari H (2017) Voice conversion using sequence-to-sequence learning of context posterior probabilities. Proceedings of Interspeech, vol 2017, pp 1268–1272
104. Fang F, Yamagishi J, Echizen I, Lorenzo-Trueba J (2018) High-quality nonparallel voice conversion based on cycle-consistent adversarial network. In: Proceedings of ICASSP 2018
105. Kobayashi K, Hayashi T, Tamamori A, Toda T (2017) Statistical voice conversion with wavenet-based waveform generation. In: Proceedings of Interspeech, pp 1138–1142
106. Gillet B, King S (2003) Transforming F0 contours. In: Proceedings of EUROSPEECH, pp 101–104 (2003)
107. Wu CH, Hsia CC, Liu TH, Wang JF (2006) Voice conversion using duration-embedded bi-HMMs for expressive speech synthesis. IEEE Trans Audio Speech Lang Process 14(4):1109–1116
108. Helander E, Nurminen J (2007) A novel method for prosody prediction in voice conversion. In: Proceedings of ICASSP, vol 4. IEEE, pp IV–509
109. Wu Z, Kinnunen T, Chng E, Li H (2010) Text-independent F0 transformation with non-parallel data for voice conversion. In: Proceedings of Interspeech
110. Lolive D, Barbot N, Boeffard O (2008) Pitch and duration transformation with non-parallel data. Speech Prosody 2008:111–114
111. Toda T, Chen LH, Saito D, Villavicencio F, Wester M, Wu Z, Yamagishi J (2016) The voice conversion challenge 2016. In: Proceedings of Interspeech, pp 1632–1636
112. Wester M, Wu Z, Yamagishi J (2016) Analysis of the voice conversion challenge 2016 evaluation results. In: Proceedings of Interspeech, pp 1637–1641
113. Perrot P, Aversano G, Blouet R, Charbit M, Chollet G (2005) Voice forgery using ALISP: indexation in a client memory. In: Proceedings of ICASSP, vol 1. IEEE, pp 17–20
114. Matrouf D, Bonastre JF, Fredouille C (2006) Effect of speech transformation on impostor acceptance. In: Proceedings of ICASSP, vol 1. IEEE, pp I–I
115. Kinnunen T, Wu Z, Lee K, Sedlak F, Chng E, Li H (2012) Vulnerability of speaker verification systems against voice conversion spoofing attacks: the case of telephone speech. In: Proceedings of ICASSP. IEEE, pp 4401–4404
116. Sundermann D, Hoge H, Bonafonte A, Ney H, Black A, Narayanan S (2006) Text-independent voice conversion based on unit selection. In: Proceedings of ICASSP, vol 1, pp I–I
117. Wu Z, Larcher A, Lee K, Chng E, Kinnunen T, Li H (2013) Vulnerability evaluation of speaker verification under voice conversion spoofing: the effect of text constraints. In: Proceedings of Interspeech, Lyon, France (2013)

118. Alegre F, Vipperla R, Evans N, Fauve B (2012) On the vulnerability of automatic speaker recognition to spoofing attacks with artificial signals. In: 2012 EURASIP conference on european conference on signal processing (EUSIPCO)

119. De Leon PL, Hernaez I, Saratxaga I, Pucher M, Yamagishi J (2011) Detection of synthetic speech for the problem of imposture. In: Proceedings of ICASSP, Dallas, USA, pp 4844–4847

120. Wu Z, Kinnunen T, Chng E, Li H, Ambikairajah E (2012) A study on spoofing attack in state-of-the-art speaker verification: the telephone speech case. In: Proceedings of Asia-Pacific signal information processing association annual summit and conference (APSIPA ASC). IEEE, pp 1–5

121. Alegre F, Vipperla R, Evans,N (2012) Spoofing countermeasures for the protection of automatic speaker recognition systems against attacks with artificial signals. In: Proceedings of Interspeech

122. Alegre F, Amehraye A, Evans N (2013) Spoofing countermeasures to protect automatic speaker verification from voice conversion. In: Proceedings of ICASSP

123. Wu Z, Xiao X, Chng E, Li H (2013) Synthetic speech detection using temporal modulation feature. In: Proceedings of ICASSP

124. Alegre F, Vipperla R, Amehraye A, Evans N (2013) A new speaker verification spoofing countermeasure based on local binary patterns. In: Proceedings of Interspeech, Lyon, France

125. Wu Z, Kinnunen T, Evans N, Yamagishi J, Hanilçi C, Sahidullah M, Sizov A (2015) ASVspoof 2015: the first automatic speaker verification spoofing and countermeasures challenge. In: Proceedings of Interspeech

126. Kinnunen T, Sahidullah M, Delgado H, Todisco M, Evans N, Yamagishi J, Lee K (2017) The ASVspoof 2017 challenge: assessing the limits of replay spoofing attack detection. In: INTERSPEECH

127. Wu Z, Khodabakhsh A, Demiroglu C, Yamagishi J, Saito D, Toda T, King S (2015) SAS: a speaker verification spoofing database containing diverse attacks. In: Proceedings of IEEE International Conference on Acoustics, Speech, and Signal Processing (ICASSP)

128. Wu Z, Kinnunen T, Evans N, Yamagishi J (2014) ASVspoof 2015: automatic speaker verification spoofing and countermeasures challenge evaluation plan. http://www.spoofingchallenge. org/asvSpoof.pdf

129. Patel T, Patil H (2015) Combining evidences from mel cepstral, cochlear filter cepstral and instantaneous frequency features for detection of natural vs. spoofed speech. In: Proceedings of Interspeech

130. Novoselov S, Kozlov A, Lavrentyeva G, Simonchik K, Shchemelinin V (2016) STC anti-spoofing systems for the ASVspoof 2015 challenge. In: Proceedings of IEEE international conference on acoustics, speech, and signal processing (ICASSP), pp 5475–5479

131. Chen N, Qian Y, Dinkel H, Chen B, Yu K (2015) Robust deep feature for spoofing detection-the SJTU system for ASVspoof 2015 challenge. In: Proceedings of Interspeech

132. Xiao X, Tian X, Du S, Xu H, Chng E, Li H (2015) Spoofing speech detection using high dimensional magnitude and phase features: the NTU approach for ASVspoof 2015 challenge. In: Proceedings of Interspeech

133. Alam M, Kenny P, Bhattacharya G, Stafylakis T (2015) Development of CRIM system for the automatic speaker verification spoofing and countermeasures challenge 2015. In: Proceedings of Interspeech

134. Wu Z, Yamagishi J, Kinnunen T, Hanilçi C, Sahidullah M, Sizov A, Evans N, Todisco M, Delgado H (2017) Asvspoof: the automatic speaker verification spoofing and countermeasures challenge. IEEE J Sel Top Signal Process 11(4):588–604

135. Delgado H, Todisco M, Sahidullah M, Evans N, Kinnunen T, Lee K, Yamagishi J (2018) ASVspoof 2017 version 2.0: meta-data analysis and baseline enhancements. In: Proceedings of Odyssey 2018 the speaker and language recognition workshop, pp 296–303

136. Todisco M, Delgado H, Evans N (2016) A new feature for automatic speaker verification anti-spoofing: constant Q cepstral coefficients. In: Proceedings of Odyssey: the speaker and language recognition workshop, Bilbao, Spain, pp 283–290

137. Todisco M, Delgado H, Evans N (2017) Constant Q cepstral coefficients: a spoofing counter-measure for automatic speaker verification. Comput Speech Lang 45:516–535
138. Lavrentyeva G, Novoselov S, Malykh E, Kozlov A, Kudashev O, Shchemelinin V (2017) Audio replay attack detection with deep learning frameworks. In: Proceedings of Interspeech, pp 82–86
139. Ji Z, Li Z, Li P, An M, Gao S, Wu D, Zhao F (2017) Ensemble learning for countermeasure of audio replay spoofing attack in ASVspoof2017. In: Proceedings of Interspeech, pp 87–91
140. Li L, Chen Y, Wang D, Zheng T (2017) A study on replay attack and anti-spoofing for automatic speaker verification. In: Proceedings of Interspeech, pp 92–96
141. Patil H, Kamble M, Patel T, Soni M (2017) Novel variable length teager energy separation based instantaneous frequency features for replay detection. In: Proceedings of Interspeech, pp 12–16
142. Chen Z, Xie Z, Zhang W, Xu X (2017) ResNet and model fusion for automatic spoofing detection. In: Proceedings of Interspeech, pp 102–106
143. Wu Z, Gao S, Cling E, Li H (2014) A study on replay attack and anti-spoofing for text-dependent speaker verification. In: Proceedings of Asia-Pacific signal information processing association annual summit and conference (APSIPA ASC). IEEE, pp 1–5
144. Li Q (2009) An auditory-based transform for audio signal processing. In: 2009 IEEE workshop on applications of signal processing to audio and acoustics. IEEE, pp 181–184
145. Davis S, Mermelstein P (1980) Comparison of parametric representations for monosyllabic word recognition in continuously spoken sentences. IEEE Trans Acoust Speech Signal Process 28(4):357–366
146. Sahidullah M, Kinnunen T, Hanilçi C (2015) A comparison of features for synthetic speech detection. In: Proceedings of Interspeech. ISCA, pp 2087–2091
147. Brown J (1991) Calculation of a constant Q spectral transform. J Acoust Soc Am 89(1):425–434
148. Alam M, Kenny P (2017) Spoofing detection employing infinite impulse response—constant Q transform-based feature representations. In: Proceedings of European signal processing conference (EUSIPCO)
149. Cancela P, Rocamora M, López E (2009) An efficient multi-resolution spectral transform for music analysis. In: Proceedings of international society for music information retrieval conference, pp 309–314
150. Bengio Y (2009) Learning deep architectures for AI. Found Trends Mach Learn 2(1):1–127
151. Goodfellow I, Bengio Y, Courville A, Bengio Y (2016) Deep learning. MIT Press, Cambridge
152. Tian Y, Cai M, He L, Liu J (2015) Investigation of bottleneck features and multilingual deep neural networks for speaker verification. In: Proceedings of Interspeech, pp 1151–1155
153. Richardson F, Reynolds D, Dehak N (2015) Deep neural network approaches to speaker and language recognition. IEEE Signal Process Lett 22(10):1671–1675
154. Hinton G, Deng L, Yu D, Dahl GE, Mohamed RA, Jaitly N, Senior A, Vanhoucke V, Nguyen P, Sainath TN, Kingsbury B (2012) Deep neural networks for acoustic modeling in speech recognition: the shared views of four research groups. IEEE Signal Process Mag 29(6):82–97
155. Alam M, Kenny P, Gupta V, Stafylakis T (2016) Spoofing detection on the ASVspoof2015 challenge corpus employing deep neural networks. In: Proceedings of Odyssey: the Speaker and Language Recognition Workshop, Bilbao, Spain, pp 270–276
156. Qian Y, Chen N, Yu K (2016) Deep features for automatic spoofing detection. Speech Commun 85:43–52
157. Yu H, Tan ZH, Zhang Y, Ma Z, Guo J (2017) DNN filter bank cepstral coefficients for spoofing detection. IEEE Access 5:4779–4787
158. Sriskandaraja K, Sethu V, Ambikairajah E, Li H (2017) Front-end for antispoofing counter-measures in speaker verification: scattering spectral decomposition. IEEE J Sel Top Signal Process 11(4):632–643. https://doi.org/10.1109/JSTSP.2016.2647202
159. Andén J, Mallat S (2014) Deep scattering spectrum. IEEE Trans Signal Process 62(16):4114–4128
160. Mallat S (2012) Group invariant scattering. Commun Pure Appl Math 65:1331–1398

161. Pal M, Paul D, Saha G (2018) Synthetic speech detection using fundamental frequency variation and spectral features. Comput Speech Lang 48:31–50
162. Laskowski K, Heldner M, Edlund J (2008) The fundamental frequency variation spectrum. Proc FONETIK 2008:29–32
163. Saratxaga I, Sanchez J, Wu Z, Hernaez I, Navas E (2016) Synthetic speech detection using phase information. Speech Commun 81:30–41
164. Wang L, Nakagawa S, Zhang Z, Yoshida Y, Kawakami Y (2017) Spoofing speech detection using modified relative phase information. IEEE J Sel Top Signal Process 11(4):660–670
165. Chakroborty S, Saha G (2009) Improved text-independent speaker identification using fused MFCC & IMFCC feature sets based on Gaussian filter. Int J Signal Process 5(1):11–19
166. Wu X, He R, Sun Z, Tan T (2018) A light CNN for deep face representation with noisy labels. IEEE Trans Inf Forensics Secur 13(11):2884–2896
167. Goncalves AR, Violato RPV, Korshunov P, Marcel S, Simoes FO (2017) On the generalization of fused systems in voice presentation attack detection. In: 2017 International conference of the biometrics special interest group (BIOSIG), pp 1–5. https://doi.org/10.23919/BIOSIG. 2017.8053516
168. Paul D, Pal M, Saha G (2016) Novel speech features for improved detection of spoofing attacks. In: Proceedings of annual IEEE India conference (INDICON)
169. Dehak N, Kenny P, Dehak R, Dumouchel P, Ouellet P (2011) Front-end factor analysis for speaker verification. IEEE Trans Audio Speech Lang Process 19(4):788–798
170. Khoury E, Kinnunen T, Sizov A, Wu Z, Marcel S (2014) Introducing i-vectors for joint anti-spoofing and speaker verification. In: Proceedings of Interspeech
171. Sizov A, Khoury E, Kinnunen T, Wu Z, Marcel S (2015) Joint speaker verification and antispoofing in the i-vector space. IEEE Trans Inf Forensics Secur 10(4):821–832
172. Hanilçi C (2018) Data selection for i-vector based automatic speaker verification antispoofing. Digit Signal Process 72:171–180
173. Tian X, Wu Z, Xiao X, Chng E, Li H (2016) Spoofing detection from a feature representation perspective. In: Proceedings of IEEE international conference on acoustics, speech, and signal processing (ICASSP), pp 2119–2123
174. Yu H, Tan ZH, Ma Z, Martin R, Guo J (2018) Spoofing detection in automatic speaker verification systems using dnn classifiers and dynamic acoustic features. IEEE Trans Neural Netw Learn Syst PP(99):1–12
175. Dinkel H, Chen N, Qian Y, Yu K (2017) End-to-end spoofing detection with raw waveform cldnns. In: Proceedings of IEEE International Conference on Acoustics, Speech, and Signal Processing (ICASSP), pp 4860–4864
176. Sainath T, Weiss R, Senior A, Wilson K, Vinyals O (2015) Learning the speech front-end with raw waveform CLDNNs. In: Proceedigns of Interspeech
177. Zhang C, Yu C, Hansen JHL (2017) An investigation of deep-learning frameworks for speaker verification antispoofing. IEEE J Sel Top Signal Process 11(4):684–694
178. Muckenhirn H, Magimai-Doss M, Marcel S (2017) End-to-end convolutional neural network-based voice presentation attack detection. In: 2017 IEEE international joint conference on biometrics (IJCB), pp 335–341
179. Chen S, Ren K, Piao S, Wang C, Wang Q, Weng J, Su L, Mohaisen A (2017) You can hear but you cannot steal: Defending against voice impersonation attacks on smartphones. In: 2017 IEEE 37th international conference on distributed computing systems (ICDCS). IEEE, pp 183–195
180. Shiota S, Villavicencio F, Yamagishi J, Ono N, Echizen I, Matsui T (2015) Voice liveness detection algorithms based on pop noise caused by human breath for automatic speaker verification. In: Proceedings of Interspeech
181. Shiota S, Villavicencio F, Yamagishi J, Ono N, Echizen I, Matsui T (2016) Voice liveness detection for speaker verification based on a tandem single/double-channel pop noise detector. In: ODYSSEY
182. Sahidullah M, Thomsen D, Hautamäki R, Kinnunen T, Tan ZH, Parts R, Pitkänen M (2018) Robust voice liveness detection and speaker verification using throat microphones. IEEE/ACM Trans Audio Speech Lang Process 26(1):44–56

183. Elko G, Meyer J, Backer S, Peissig J (2007) Electronic pop protection for microphones. In: 2007 IEEE workshop on applications of signal processing to audio and acoustics. IEEE, pp 46–49

184. Zhang L, Tan S, Yang J, Chen Y (2016) Voicelive: a phoneme localization based liveness detection for voice authentication on smartphones. In: Proceedings of the 2016 ACM SIGSAC conference on computer and communications security. ACM, pp 1080–1091

185. Zhang L, Tan S, Yang J (2017) Hearing your voice is not enough: An articulatory gesture based liveness detection for voice authentication. In: Proceedings of the 2017 ACM SIGSAC conference on computer and communications security. ACM, pp 57–71

186. Hanilçi C, Kinnunen T, Sahidullah M, Sizov A (2016) Spoofing detection goes noisy: an analysis of synthetic speech detection in the presence of additive noise. Speech Commun 85:83–97

187. Yu H, Sarkar A, Thomsen D, Tan ZH, Ma Z, Guo J (2016) Effect of multi-condition training and speech enhancement methods on spoofing detection. In: Proceedings of international workshop on sensing, processing and learning for intelligent machines (SPLINE)

188. Tian X, Wu Z, Xiao X, Chng E, Li H (2016) An investigation of spoofing speech detection under additive noise and reverberant conditions. In: Proceedings of Interspeech (2016)

189. Delgado H, Todisco M, Evans N, Sahidullah M, Liu W, Alegre F, Kinnunen T, Fauve B (2017) Impact of bandwidth and channel variation on presentation attack detection for speaker verification. In: 2017 International conference of the biometrics special interest group (BIOSIG), pp 1–6

190. Qian Y, Chen N, Dinkel H, Wu Z (2017) Deep feature engineering for noise robust spoofing detection. IEEE/ACM Trans Audio Speech Lang Process 25(10):1942–1955

191. Korshunov P, Marcel S (2016) Cross-database evaluation of audio-based spoofing detection systems. In: Proceedings of Interspeech

192. Paul D, Sahidullah M, Saha G (2017) Generalization of spoofing countermeasures: a case study with ASVspoof 2015 and BTAS 2016 corpora. In: Proceedigns of IEEE international conference on acoustics, speech, and signal processing (ICASSP). IEEE pp 2047–2051

193. Lorenzo-Trueba J, Fang F, Wang X, Echizen I, Yamagishi J, Kinnunen T (2018) Can we steal your vocal identity from the Internet?: Initial investigation of cloning Obama's voice using GAN, WaveNet and low-quality found data. In: Proceedings of Odyssey: the speaker and language recognition workshop

194. Goodfellow I, Pouget-Abadie J, Mirza M, Xu B, Warde-Farley D, Ozair S, Courville A, Bengio Y (2014) Generative adversarial nets. In: Advances in neural information processing systems, pp 2672–2680

195. Kreuk F, Adi Y, Cisse M, Keshet J (2018) Fooling end-to-end speaker verification by adversarial examples. arXiv:1801.03339

196. Sahidullah M, Delgado H, Todisco M, Yu H, Kinnunen T, Evans N, Tan ZH (2016) Integrated spoofing countermeasures and automatic speaker verification: an evaluation on ASVspoof 2015. In: Proceedings of Interspeech

197. Muckenhirn H, Korshunov P, Magimai-Doss M, Marcel S (2017) Long-term spectral statistics for voice presentation attack detection. IEEE/ACM Trans Audio Speech Lang Process 25(11):2098–2111

198. Sarkar A, Sahidullah M, Tan ZH, Kinnunen T (2017) Improving speaker verification performance in presence of spoofing attacks using out-of-domain spoofed data. In: Proceedings of Interspeech

199. Kinnunen T, Lee K, Delgado H, Evans N, Todisco M, Sahidullah M, Yamagishi J, Reynolds D (2018) t-DCF: a detection cost function for the tandem assessment of spoofing countermeasures and automatic speaker verification. In: Proceedings of Odyssey: the speaker and language recognition workshop

200. Todisco M, Delgado H, Lee K, Sahidullah M, Evans N, Kinnunen T, Yamagishi J (2018) Integrated presentation attack detection and automatic speaker verification: common features and Gaussian back-end fusion. In: Proceedings of Interspeech

201. Wu Z, De Leon P, Demiroglu C, Khodabakhsh A, King S, Ling ZH, Saito D, Stewart B, Toda T, Wester M, Yamagishi Y (2016) Anti-spoofing for text-independent speaker verification: an initial database, comparison of countermeasures, and human performance. IEEE/ACM Trans Audio Speech Lang Process 24(4):768–783

Chapter 16
A Cross-Database Study of Voice Presentation Attack Detection

Pavel Korshunov and Sébastien Marcel

Abstract Despite an increasing interest in speaker recognition technologies, a significant obstacle still hinders their wide deployment—their high vulnerability to spoofing or presentation attacks. These attacks can be easy to perform. For instance, if an attacker has access to a speech sample from a target user, he/she can replay it using a loudspeaker or a smartphone to the recognition system during the authentication process. The ease of executing presentation attacks and the fact that no technical knowledge of the biometric system is required to make these attacks especially threatening in practical application. Therefore, late research focuses on collecting data databases with such attacks and on development of presentation attack detection (PAD) systems. In this chapter, we present an overview of the latest databases and the techniques to detect presentation attacks. We consider several prominent databases that contain bona fide and attack data, including ASVspoof 2015, ASVspoof 2017, AVspoof, voicePA, and BioCPqD-PA (the only proprietary database). Using these databases, we focus on the performance of PAD systems in the cross-database scenario or in the presence of "unknown" (not available during training) attacks, as these scenarios are closer to practice, when pretrained systems need to detect attacks in unforeseen conditions. We first present and discuss the performance of PAD systems based on handcrafted features and traditional Gaussian mixture model (GMM) classifiers. We then demonstrate whether the score fusion techniques can improve the performance of PADs. We also present some of the latest results of using neural networks for presentation attack detection. The experiments show that PAD systems struggle to generalize across databases and mostly unable to detect unknown attacks, with systems based on neural networks demonstrating better performance compared to the systems based on handcrafted features.

P. Korshunov (✉) · S. Marcel
Idiap Research Institute, Martigny, Switzerland
e-mail: pavel.korshunov@idiap.ch

S. Marcel
e-mail: sebastien.marcel@idiap.ch

© Springer Nature Switzerland AG 2019
S. Marcel et al. (eds.), *Handbook of Biometric Anti-Spoofing*,
Advances in Computer Vision and Pattern Recognition,
https://doi.org/10.1007/978-3-319-92627-8_16

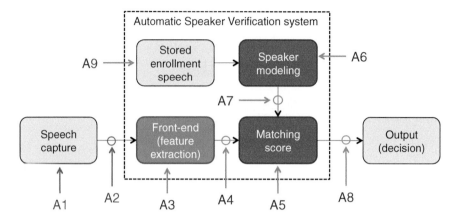

Fig. 16.1 Possible attack places in a typical ASV system

16.1 Introduction

Given the complexity of a practical Automatic Speaker Verification system (ASV), several different modules of the system are prone to attacks, as it is identified in ISO/IEC 30107-1 standard [1] and illustrated in Fig. 16.1. Depending on the usage scenario, two of the most vulnerable places for spoofing attacks in an ASV system are marked by A1 (aka "physical access" as defined in [2] or presentation attacks) and A2 (aka "logical access" attacks as defined in [2]) in the figure. In this chapter, we are considering A1 and A2 attacks, where the system can be attacked by presenting a spoofed signal as input. For the other points of attacks from A3 to A9, the attacker needs to have privileged access rights and know the operational details of the biometric system. Prevention or countering such attacks is more related to system security and is thus out of the scope of this chapter.

There are three prominent methods through which A1 and A2 attacks can be carried out: (a) by recording and replaying the target speakers' speech, (b) by synthesizing speech that carries target speaker characteristics, and (c) by applying voice conversion methods to convert impostor speech into target speaker speech. Among these three, replay attack is the most viable attack, as the attacker mainly needs a recording and playback device. In the literature, there is evidence that ASV systems might be immune to "zero-effort" impostor claims and mimicry attacks [3], however, they are still vulnerable to such presentation attacks (PAs) [4]. One of the reasons for such vulnerability is a built-in ability of biometric systems in general, and ASV systems, in particular, to handle undesirable variabilities. Since spoofed speech can exhibit the undesirable variabilities that ASV systems are robust to, the attacks can pass undetected.

Therefore, developing mechanisms for the detection of presentation attacks is gaining interest in the speech community [5]. At first, researchers were mostly focusing on logical access attacks, largely thanks to the "Automatic Speaker Verification

Spoofing and Countermeasures Challenge" [2], which provided a large benchmark corpus ASVspoof 2015,[1] containing voice conversion-based and speech synthesis-based attacks. In the literature, development of PAD systems has largely focused on investigating handcrafted features, such as short-term speech processing-based features that can aid in discriminating genuine speech from spoofed signal. Typical detection methods use features based on audio spectrogram, such as spectral [6, 7] and cepstral-based features with temporal derivatives [8, 9], phase spectrum-based features [10], the combination of amplitude and phase features [11], recently proposed constant Q cepstral coefficients (CQCCs) [12], extraction of local binary patterns in the cepstral domain [13], and audio quality-based features [14]. A survey by Wu et al. [5] provides a comprehensive overview of the attacks based on synthetic speech and the detection methods tailored to those types of attacks.

Besides determining "good features for detecting presentation attacks", it is also important to correctly classify the computed feature vectors as belonging to bona fide or spoofed data. Choosing a reliable classifier is especially important given a possibly unpredictable nature of attacks in a practical system, since it is unknown what kind of attack the perpetrator may use when spoofing the verification system. Different methods use different classifiers but the most common choices include logistic regression, support vector machine (SVM), and Gaussian mixture model (GMM) classifiers. The benchmarking study on logical access attacks [15] finds GMMs to be more successful compared to two-class SVM (combined with an LBP-based feature extraction from [13]) in detecting synthetic spoofing attacks. Therefore, in this book chapter, we focus on GMM-based classifiers as the best representatives of the "traditional" approaches. Deep learning networks are also showing promising performance in simultaneous feature selection and classification [16] and therefore are also addressed in this chapter.

The most common approach to detect presentation attacks is to pretrain the classifier on the examples of both bona fide and spoofed data. To simulate realistic environments, the classifier can be trained on a subset of the attacks, termed *known attacks*, and tested on a larger set of attacks that include both known and *unknown attacks*.

Generalization ability of the PAD systems based on handcrafted features has been assessed recently with [12] showing the degradation in performance when specific features optimized using one database are used unchanged on another database. In [17], cross-database experiments demonstrated the inability of current techniques to deal with unforeseen conditions. However, it did not include strict presentation attacks, which can be considered one of the hardest attacks to be detected. The authors of [18, 19] focused on presentation attacks in cross-database and cross-attack scenarios, and demonstrated that current state-of-the-art PAD systems do not generalize well, with especially poor performance on presentation attacks. In this chapter, we will discuss the performance of several such systems on unknown attacks and in cross-database evaluations.

---

[1] http://datashare.is.ed.ac.uk/handle/10283/853.

To solve the problems of the PAD systems and improve their performance, especially, in unseen conditions, many turned to score fusion techniques, as a straightforward and convenient way to combine the outputs of several PAD systems into one joint PAD ensemble. However, the studies have shown [19, 20] that although fusion can improve the performance, even large fusion ensembles, e.g., fusion of many different systems (we coin them as "mega-fusion" systems, in this chapter), are not very helpful outside of controlled academic challenges.

Neural nets are also promising for detection of replay or presentation attacks [16, 21]. The latest study by Muckenhirn et al. [22] demonstrated the high accuracy of convolutional neural networks (CNNs) compared to systems based on handcrafted features for attack detection. However, little is known how CNNs perform on unknown presentation attacks, and whether they can generalize across different databases with presentation attacks. The impact of the neural net's depth on the performance is also not well understood.

In this chapter, we consider most of the recently proposed types of PAD systems, including those based on handcrafted features, fusion-based systems, and CNN-based PAD systems, which learn features from raw speech data (similar to systems in [22]). The main focus of the chapter is on the performance on unknown attacks or in cross-database settings. For systems based on handcrafted features, we consider eight well-performing methods based on GMM classifier that use cepstral-based features with rectangular (RFCC), mel-scale triangular (MFCC) [23], inverted mel-scale triangular (IMFCC), linear triangular (LFCC) filters [24], spectral flux-based feature (SSFC) [25], subband centroid frequency (SCFC) [26], and subband centroid magnitude (SCMC) [26] features. We also included recently proposed constant Q cepstral coefficients (CQCCs) [27], which were shown good performance on ASVspoof 2015 database.[2] We also discuss joint PAD systems obtained by fusing several systems via score fusion approach, using mean, logistic regression, and polynomial logistic regression fusion methods. The correct performance of the fusion is ensured by using scores pre-calibrated with logistic regression. For CNN-based PAD systems, we evaluate two network architectures (with one and three CNN layers) of the system originally proposed in [22], which learn features from raw speech data.

To compare with previous work, we use ASVspoof 2015 database when evaluating systems based on handcrafted features and their fusion-based derivatives. We also evaluate these systems on AVspoof.[3] since it is the first database with presentation attacks [4]. So-called mega-fusion-based systems are evaluated using ASVspoof 2017 database, because they were first introduced and performed well (third and fourth places) in the latest ASVspoof 2017 grand challenge. CNN-based systems need a lot of data for training; hence, for these systems, we use two databases with large number of different presentation attacks. One is voicePA[4] database, which is an extension of the AVspoof but with more attacks, and another is BioCPqD-PA [28] database (i.e., a biometric database with presentation attacks by CPqD) of

---

[2]Precomputed CQCC features were provided by the authors.

[3]https://www.idiap.ch/dataset/avspoof.

[4]https://www.idiap.ch/dataset/voicepa.

Portuguese speakers and many high-quality *unknown* presentation attacks recorded in an acoustically isolated room. Note that, although proprietary, BioCPqD-PA database will be publicly available for machine learning experiments on a web-based BEAT platform.[5] which allows to perform experiments on private databases in secure and reproducible way.

In summary, this chapter has the following main contributions:

- Overview of the latest comprehensive speech databases with spoofing (presentation) attacks.
- Overview of the state-of-the-art PAD systems, including systems based on hand-crafted features, fusion-based systems, and based on neural networks.
- Open-source implementations of the databases and the reviewed PAD systems.[6]

16.2   Databases

Appropriate databases are necessary for testing different presentation attack detection approaches. These databases need to contain a set of practically feasible presentation attacks and also data for speaker verification task, so that a verification system can be tested for both issues: the accuracy of speaker verification and the resistance to the attacks.

In this chapter, we present experiments on several prominent publicly available databases and one proprietary database that are used for evaluation of PAD methods: ASVspoof 2015, AVspoof 2015, voicePA 2016, BioCPqD-PA 2016, and ASVspoof 2017.[7] ASVspoof 2015 was created as part of the 2015 Interspeech anti-spoofing challenge and contains only synthetically generated and converted speech attacks. These attacks are assumed to be fed into a verification system directly bypassing its microphone, and are also coined as logical access attacks [2]. AVspoof contains both logical access attacks (LAs) and presentation attacks (PAs). For the ease of comparison with ASVspoof 2015, the set of attacks in AVspoof is split into LA and PA subsets (see Table 16.1). voicePA is an extension of the AVspoof database with more added attacks and is, therefore, suitable for training neural nets. A proprietary BioCPqD-PA [28] database of Portuguese speakers and many high-quality *unknown* presentation attacks recorded in an acoustically isolated room. Note that, although proprietary, BioCPqD-PA database will be publicly available for machine learning experiments on a web-based BEAT platform. ASVspoof 2017, similarly to ASVspoof 2015, was developed for the Automatic Speaker Verification Spoofing and Countermeasures Challenge (ASVspoof 2017) [29] and was used for evaluation of the mega-fusion systems in this chapter.

---

[5]https://www.beat-eu.org/platform/.

[6]Source code: https://gitlab.idiap.ch/bob/bob.hobpad2.chapter16.

[7]https://datashare.is.ed.ac.uk/handle/10283/3017.

16.2.1 ASVspoof 2015 Database

The ASVspoof 2015[1] database contains genuine and spoofed samples from 45 male and 61 female speakers. This database contains only speech synthesis and voice conversion attacks produced via logical access, i.e., they are directly injected in the system. The attacks in this database were generated with 10 different speech synthesis and voice conversion algorithms. Only 5 types of attacks are in the training and development set ($S1$ to $S5$), while 10 types are in the evaluation set ($S1$ to $S10$). Since last five attacks appear in the evaluation set only and PAD systems are not trained on them, they are considered "unknown" attacks (see Table 16.1). This split of attacks allows to evaluate the systems on known and unknown attacks. The full description of the database and the evaluation protocol are given in [2]. This database was used for the ASVspoof 2015 Challenge and is a good basis for system comparison as several systems have already been tested on it.

16.2.2 AVspoof Database

AVspoof[2] database contains bona fide (genuine) speech samples from 44 participants (31 males and 13 females) recorded over the period of 2 months in four sessions, each

Table 16.1 Details of AVspoof, ASVspoof 2015, ASVspoof 2017, voicePA, and BioCPqD-PA databases. For each separate set, the number of utterances is given

Database	Type of data	Train	Dev	Eval
AVspoof	Bona fide	4,973	4,995	5,576
	LA attacks	17,890	17,890	20,060
	PA attacks	38,580	38,580	43,320
	Total	61,443	61,465	68,956
ASVspoof 2015	Bona fide	3,750	3,497	9,404
	Known attacks	12,625	49,875	92,000
	Unknown attacks	–	–	92,000
	Total	16,375	53,372	193,404
ASVspoof 2017	Bona fide	1,508	760	1,298
(original release)	Attacks	1,508	950	12,008
	Total	3,016	1,710	13,306
VoicePA	Bona fide	4,973	4,995	5,576
	Attacks	115,740	115,740	129,988
	Total	120,713	120,735	135,564
BioCPqD-PA	Bona fide	6,857	12,455	7,941
	Attacks	98,562	179,005	114,111
	Total	105,419	191,460	122,052

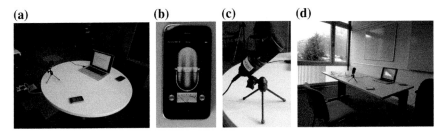

Fig. 16.2 AVspoof database recording setup. The images show different devices and locations used for bona fide data collection and for creating the attacks

scheduled several days apart in different setups and environmental conditions such as background noises. The recording devices, including microphone AT2020USB+, Samsung Galaxy S4 phone, and iPhone 3GS, and the environments are shown in Fig. 16.2. The first session was recorded in the most controlled conditions.

From the recorded genuine data, two major types of attacks were created for AVspoof database: logical access attacks, similar to those in ASVspoof 2015 database [2], and presentation attacks. Logical access attacks are generated using (i) a statistical parametric-based speech synthesis algorithm [30] and (ii) a voice conversion algorithm from Festvox.[8]

When generating presentation attacks, the assumption is that a verification system is installed on a laptop (with an internal built-in microphone) and an attacker is trying to gain access to this system by playing back to it a pre-recorded genuine data or an automatically generated synthetic data using some playback device. In AVspoof database, presentation attacks consist of (i) replay attacks when a genuine data is played back using a laptop with internal speakers, a laptop with external high-quality speakers, Samsung Galaxy S4 phone, and iPhone 3G; (ii) synthesized speech replayed with a laptop; and (iii) converted voice attacks replayed with a laptop.

The data in AVspoof database is split into three nonoverlapping subsets, ensuring that the same speaker does not appear in different sets (see Table 16.1 for details): training or *train* (bona fide and spoofed samples from 4 female and 10 male participants), development or *dev* (bona fide and spoofed samples from 4 female and 10 male participants), and evaluation or *eval* (bona fide and spoofed samples from 5 female and 11 male participants). For more details on AVspoof database, please refer to [4].

16.2.3 ASVspoof 2017

The ASVspoof 2017 was created for the grand challenge with the main focus on presentation attacks. To this end, the challenge makes use of the RedDots corpus

---

[8]http://festvox.org/.

Table 16.2 Attack types in voicePA database

Laptop replay	Phone replay	Synthetic replay
Laptop speakers,	Samsung Galaxy S3,	Speech synthesis
High-quality speakers	iPhone 3GS & 6S	Voice conversion

[31] and a replayed version of the same data [32]. While the former serves as genuine samples, the latter is used as spoof samples, collected by replaying a subset of the original RedDots corpus utterances using different loudspeakers and recording devices, in different environments, through a crowdsourcing approach.

The database was split into three subsets: *train* for training, *dev* for development, and *eval* for evaluation. In the challenge, it was also allowed to use both *train* and *dev* subsets to train the final system for score submission. A more detailed description of the challenge and the database can be found in [29].

16.2.4 VoicePA Database

The voicePA[3] database inherits bona fide (genuine) speech samples from AVspoof database, which is described in Sect. 16.2.2.

The presentation attacks for voicePA were generated with assumption that a verification system, which is considered to be attacked, is installed either on a laptop (with an internal built-in microphone), on Samsung Galaxy S3, or iPhone 3GS. The attacker is trying to gain access to this system by playing back to it a pre-recorded bona fide data or an automatically generated synthetic data using some playback device.

The following devices were used to playback the attacks (see Table 16.2): (i) replay attacks using a laptop with internal speakers and a laptop with external high-quality speaker; (ii) replay attacks using Samsung Galaxy S3, iPhone 3G, and iPhone 6S phones; and (iii) replay of synthetic speech generated with text to speech and voice conversion algorithms. Attacks targeting verification system on the laptop are the same as the attacks in AVspoof database (see Sect. 16.2.2), while the attacks on Samsung Galaxy S3 and iPhone 3G phones are newer and are contained only in voicePA database.

The attacks were also recorded into three different noise environments: a large conference room, an empty office with window open, and a typical lab with closed windows. In total, voicePA contains 24 different types of presentation attacks, including 16 attacks replayed by iPhone 3GS and Samsung Galaxy S3 in two different environments (4 by one phone in one environment), and 8 by the laptop in another environment.

Similarly to AVspoof, all utterances (see Table 16.1) in voicePA database are split into three nonoverlapping subsets: training or *train* (bona fide and spoofed samples

Fig. 16.3 Example of BioCPqD-PA database recording setup. All attacks were recorded in an acoustically isolated room

from 4 male and 10 female participants), development or *dev* (bona fide and spoofed samples from 4 male and 10 female participants), and evaluation or *eval* (bona fide and spoofed samples from 5 male and 11 female participants).

16.2.5 BioCPqD-PA Database

BioCPqD-PA [28] is a proprietary database, and it contains video (audio and image) of 222 participants (124 males and 98 females) speaking different types of content, including free speech, read text, and read numbers (credit card, telephone, personal ID, digits sequences, and other numbers set). Recordings used different devices (laptops and smartphones) and were performed in different environments in Portuguese language.

The subset used in this paper as bona fide samples consists of only the laptop part and includes all participants. The recordings used four different laptops, took place at three different environments, including a quiet garden, an office, and a noisy restaurant, and were performed during five recording sessions.[9] In each session, 27 utterances with variable content were recorded.

The presentation attacks were recorded in an acoustically isolated room (see Fig. 16.3) using 3 different microphones and 8 different loudspeakers, resulting in 24 configurations (see Table 16.3 for details). The total number of bona fide recordings is 27, 253, and presentation attacks is 391, 678. This database was split in three nonoverlapping subsets (see Table 16.1), isolating pairs of microphones and loudspeakers in each subset (each microphone and loudspeaker pair belongs to only one subset), thus providing a protocol to evaluate the ability of a PAD system to generalize to unseen configurations. As shown in Table 16.3, *train* set contains 44 pairs of microphone and loudspeaker, *dev* set contains 12 pairs, and *eval* set 8 pairs. Additionally, the protocol guarantees that train and eval sets do not contain any repeated microphone–loudspeaker pairs. There is no split among speakers, meaning that samples from all speakers are present in all subsets. Such split was done on

---

[9]Not all subjects recorded five sessions, due to scheduling difficulties.

Table 16.3 Microphone/speaker pairs forming attack types in BioCPqD-PA database. **(T)**, **(D)**, and **(E)** indicate train, dev, and eval sets

Mic/Speak	Genius	Megaw.	Dell A225	Edifier	Log S-150	SBS20	Dell XPS	Mackie
1. Genius	A1-1 (T)	A1-2 (T)	A1-3 (T)	A1-4 (T)	A1-5 (D)	A1-6 (D)	A1-7 (D)	A1-8 (D)
2. Dell XPS	A2-1 (D)	A2-2 (D)	A2-3 (D)	A2-4 (D)	A2-5 (E)	A2-6 (E)	A2-7 (E)	A2-8 (E)
3. Log. USB	A3-1 (D)	A3-2 (D)	A3-3 (D)	A3-4 (D)	A3-5 (E)	A3-6 (E)	A3-7 (E)	A3-8 (E)

purpose to study the effect of different recording–playback device pairs on PAD systems.

16.2.6 Evaluation Protocol

In a single-database evaluation, the *train* set of a given database is used for training PAD system, the *dev* set is used for selecting hyperparameters, and *eval* set is used for testing. In a cross-database evaluation, typically, the *train* and *dev* sets are taken from one database, while the *eval* set is taken from another database. In some scenarios, however, it is also possible that PAD is trained on *train* set from one database by both *dev* and *eval* sets taken from another database.

For evaluation of PAD systems, the following metrics are recommended [33]: attack presentation classification error rate (APCER) and bona fide presentation classification error rate (BPCER). APCER is the number of attacks misclassified as bona fide samples divided by the total number of attacks, and is defined as follows:

$$
\text{APCER} = \frac{1}{N_{AT}} \sum_{i=1}^{N_{AT}} (1 - Res_i), \tag{16.1}
$$

where N_{AT} is the number of attack presentations. Res_i takes value 1 if the ith presentation is classified as an attack, and value 0 if classified as bona fide. Thus, APCER can be considered as an equivalent to FAR for PAD systems, as it represents the ratio of falsely accepted attack samples in relation to the total number of attacks.

BPCER is the number of incorrectly classified bona fide samples divided by the total number of bona fide samples:

$$
\text{BPCER} = \frac{1}{N_{BF}} \sum_{i=1}^{N_{BF}} Res_i, \tag{16.2}
$$

where N_{BF} is the number of bona fide presentations, and Res_i is defined similar to APCER. Hence, BPCER can be considered as an equivalent to FRR for PAD systems.

In this chapter's evaluations, when testing PADs on each database and in cross-database scenarios, we report EER rates on *dev* set (when BPCER and APCER are equal) and separate BPCER and APCER values on *eval* set using the EER threshold computed on the *dev* set.

16.3   Presentation Attack Detection Approaches

Usually, PAD system consists of a feature extractor and a binary classifier (see Fig. 16.4 for an overview), which is trained to distinguish bona fide data from attacks. In this section, we present the most commonly used recent approaches for PAD, discuss feature extraction and classification components, explore performance-enhancement score fusion techniques, and evaluate CNN-based systems.

16.3.1   Handcrafted Features

Based on the overview of the methods for synthetic speech detection by Sahidullah et al. [15], we selected eight-system-based handcrafted-based features to present in this chapter.

These systems rely on GMM-based classifier (two models for bona fide and attacks, 512 Gaussians components with diagonal covariances, using 10 expectation–maximization iterations for each model), since it has demonstrated improved performance compared to support vector machine (SVM) on the data from ASVspoof 2015 database [15]. Four cepstral-based features with mel-scale, i.e., mel-frequency cepstral coefficients (MFCC) [23], rectangular (RFCC), inverted mel-scale (IMFCC), and linear (LFCC) filters [24], were selected. These features are computed from a power spectrum (power of magnitude of 512-sized fast Fourier transform) by applying one of the above filters of a given size (we use size 20 as per [15]). Spectral flux-based features, i.e., subband spectral flux coefficients (SSFCs) [25], which are Euclidean distances between power spectrums (normalized by the maximum value) of two consecutive frames, subband centroid frequency (SCFC) [26], and subband centroid magnitude (SCMC) [26] coefficients are considered as well. A discrete cosine transform (DCT-II) is applied to these above features, except for SCFC, and first 20 coefficients are taken. Before computing selected features, a given audio sample is first split into overlapping 20-ms-long speech frames with 10ms overlap. The

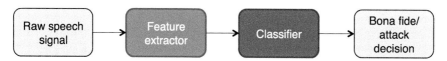

Fig. 16.4   Presentation attack detection system

frames are pre-emphasized with 0.97 coefficient and preprocessed by applying Hamming window. Then, for all features, deltas and double-deltas [34] are computed and only these derivatives (40 in total) are used by the classifier. Only deltas and delta–deltas are kept, because [15] reported that static features degraded performance of PAD systems.

In addition to the above features, we also consider recently proposed CQCC [27], which are computed using constant Q transform instead of FFT. To be consistent with the other features and fair in the system comparison, we used also only delta and delta–deltas (40 features in total) derived from 19 plus C_0 coefficients.

A survey by Wu et al. [5] provides a comprehensive overview of both the existing spoofing attacks and the available attack detection approaches. An overview of the methods for synthetic speech detection by Sahidullah et al. [15] benchmarks several existing feature extraction methods and classifiers on ASVspoof 2015 database.

16.3.2   *Fusion and Large Fusion Ensembles*

When stand-alone PAD systems do not work well, researchers turn to fusion techniques as a way to increase the overall accuracy. In this chapter, we focus on a score-level fusion due to its relative simplicity and evidence that it leads to a better performance. The score fusion is performed by combining scores from each of the N systems into a new feature vector of length N that needs to be classified. For classification, we consider three different algorithms: (i) a logistic regression (LR), i.e., a linear classifier that is trained using a logistic loss; (ii) a multilayer perceptron (MLP); and (iii) a simple mean function (Mean), which is taken on scores of the fused systems. For LR and MLP fusion, the classifier is pretrained on the score feature vectors from the training set.

When analyzing, comparing, and especially fusing PAD systems, it is important to have calibrated scores. Raw scores can be mapped to log-likelihood ratio scores with logistic regression, and an associated cost of calibration C_{llr} together with a discrimination loss C_{llr}^{min} is then used as application-independent performance measures of calibrated PAD or ASV systems. Calibration cost C_{llr} can be interpreted as a scalar measure that summarizes the quality of the calibrated scores. A well-calibrated system has $0 \leq C_{llr} < 1$ and produces well-calibrated likelihood ratio. Discrimination loss C_{llr}^{min} can be viewed as the theoretically best C_{llr} value of an optimally calibrated systems. We refer to [35] for a discussion on the score calibration and C_{llr} and C_{llr}^{min} metrics.

An extreme example of score fusion is the recent tendency to fuse many different systems, with hope that the resulted "mega-fusion" system will generalize better, especially on "unseen" or "unknown" attacks. In this chapter, we consider two examples of such systems, which won third and fourth places in the latest ASVspoof 2017 challenge, and we demonstrate how well they generalize on unknown attacks.

Two PAD systems, simply referred to as *System-1* and *System-2*, are essentially the ensembles of different combinations of features and classifiers. Table 16.4 shows

Table 16.4 Description of the submitted systems: *System-1* and *System-2*

	System-1	System-2
Subsystems	**GMM** with: $RFCC_{all}$, $RFCC_{\Delta s}$, $LFCC_{all}$, $LFCC_{\Delta s}$	**GMM** with: $RFCC_{all}$, $RFCC_{\Delta s}$
	$MFCC_{all}$, $MFCC_{\Delta s}$, $IMFCC_{all}$, $MFCC_{\Delta s}$, $SSFC_{all}$	$LFCC_{all}$, $LFCC_{\Delta s}$, $MFCC_{all}$
	$SSFC_{\Delta s}$, $SCMC_{all}$, $SCMC_{\Delta s}$	$MFCC_{\Delta s}$, $IMFCC_{all}$, $IMFCC_{\Delta s}$
	MLP with: $IMFCC_{all}$, $LFCC_{all}$, $MFCC_{all}$	$SSFC_{all}$, $SSFC_{\Delta s}$, $SCMC_{all}$
	PLP-Cepstral$_{all}$, $RFCC_{all}$, $SCMC_{all}$	$SCMC_{\Delta s}$
Fusion	Logistic regression	Logistic regression

the set of subsystems and the fusion method used for each PAD system. Features are presented with a subscript "*all*" or "Δs", where "*all*" means that all static and dynamic (delta and delta–delta) features were used, while "Δs" indicates that only the dynamic features were considered. The choice of the set of subsystems was based on their performances measured on the *dev* set provided within ASVspoof 2017 challenge.

16.3.3 Convolution Networks

The popularity of neural networks has reached PAD community, and therefore, in this chapter, we present and evaluate two examples of convolutional neural networks (CNNs) designed and trained for speech presentation attack detection. First system is a smaller network (denoted as "CNN-Shallow"), and the second system is a deeper model (denoted as "CNN-Deep") with more layers stacked up. The CNNs are implemented using TensorFlow framework.[10] The architecture of both CNNs is presented in Fig. 16.5. The number of neurons is shown at the top of each layer.

These networks are by no means the best possible architectures for PAD, as it is not our goal to present such. We simply aim to understand whether CNNs, even such simple ones, would be better alternatives to the systems based on handcrafted features. Hence, all the parameters of the considered CNNs were chosen empirically from a small number of experiments, i.e., in a semi-arbitrary fashion.

Unlike the traditional MFCC–GMM model, in a CNN model, the discriminative features are learned jointly with the classification model. Hence, a raw waveform is used as an input to the model and the convolutional layers are responsible to build relevant features.

---

[10]https://www.tensorflow.org/.

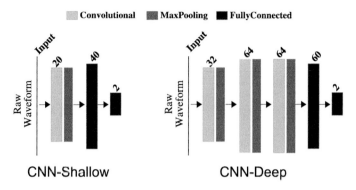

Fig. 16.5 Architecture of the two CNNs designed for speech presentation attack detection. Two more convolutional layers and more neurons are added in CNN-Deep model

In our CNN networks, the raw input audio is split into 20-ms-long speech frames. The feature vector consists of each frame plus its 20 left and right neighbors, resulting in 41 input frames.

In the CNN-Shallow network, the only convolutional layer contains 20 neurons, each with kernel = 300 and stride = 200, followed by a fully connected layer composed of 40 neurons. Both layers use hard tangent as an activation function. The output of convolutional layer is flattened for the fully connected layer input. The last layer has two neurons corresponding to the two output classes (bona fide and attacks). LogSoftMax function is applied to the output of the network before a negative log-likelihood loss function is computed. Gradient descent with constant learning rate 0.0001 is used to optimize the loss.

A deeper CNN (CNN-Deep) is a slightly larger network with three convolutional layers and we added it to analyze how increasing depth of CNN architecture impacts the PAD performance. The same raw data input and activation function are used as in the shallow network. The first convolutional layer has 32 neurons, each with kernel = 160 and stride = 20, followed by a max pooling layer (kernel = 2 and stride = 1). A second convolutional layer has 64 neurons (kernel = 32 and stride = 2) and the same max pooling layer. The third convolutional layer contains 64 neurons (kernel = 1 and stride = 1) followed by the same max pooling layer. The output of the last max pooling is flattened and connected to a fully connected layer of 60 neurons. The last layer is an output layer with 2 neuron classes. Similarly to the shallow network, LogSoftMax function is applied to the outputs. For all convolutional layers, hard tangent activation function is used. Gradient descent with exponentially decay learning rate with base rate of 0.001 and decaying step 10000 is used for optimizing the negative log-likelihood loss function.

16.4 Evaluation

Following previous Sect. 16.3, in this section, we first focus on the performance of PAD systems based on handcrafted features in single and cross-database scenarios, followed by the score fusion-based techniques. Then, we present the evaluation details of the mega-fusion systems from ASVspoof 2017 challenge, and we conclude with comparing the performances of two examples of CNN-based systems.

16.4.1 PADs Based on Handcrafted Features

As discussed in Sect. 16.3.1, we have selected several methods based on handcrafted features for presentation attacks detection, which were recently evaluated by Sahidullah et al. [15] on ASVspoof 2015 database with an addition of CQCC feature-based method [27].

The selected PAD systems are evaluated on each ASVspoof 2015 and AVspoof database and in cross-database scenario. To keep results comparable with current state-of-the-art work [15, 36], we computed average EER (*eval* set) for single-database evaluations and APCER with BPCER for cross-database evaluations. APCER with BPCER is computed for *eval* set of a given dataset using the EER threshold obtained from the *dev* set from another dataset (see Table 16.6).

To avoid bias, prior to the evaluations, the raw scores from each individual PAD system are pre-calibrated with logistic regression based on Platts sigmoid method [37] by modeling scores of the training set and applying the model on the scores from development and evaluation sets. The calibration cost C_{llr} and the discrimination loss C_{llr}^{min} of the resulted calibrated scores are provided.

In Table 16.5, the results for known and unknown attacks (see Table 16.1) of *eval* set of ASVspoof 2015 are presented separately to demonstrate the differences between these two types of attacks provided in ASVspoof 2015 database. Also, since the main contribution to the higher EER for unknown attacks is given by a more challenging attack "S10", we separately present the EER for this attack in Table 16.5.

Since AVspoof contains both logical access (LA for short) and presentation attacks (PAs), the results for these two types of attacks are also presented separately. Hence, it allows to compare the performance on ASVspoof 2015 database (it has logical access attacks only) with AVspoof-LA attacks.

From the results in Table 16.5, we can note that (i) LA set of AVspoof is less challenging compared to ASVspoof 2015 for almost all methods; (ii) unknown attacks and, especially, "S10" attack, for which PADs are not trained and are more challenging; and (iii) presentation attacks are also more challenging compared to LA attacks.

It can be also noted that PAD systems fused using a simple mean fusion are *on par* or sometimes performing even better than systems fused with LR (though, LR

Table 16.5 Performance of PAD systems based on handcrafted features (average EER %, C_{llr}, and C_{llr}^{min} of calibrated scores) on ASVspoof 2015 and AVspoof databases

| PADs | ASVspoof (Eval) | | | | | | | AVspoof (Eval) | | | | | |
| | Known | | | S10 | Unknown | | | LA | | | PA | | |
	EER	C_{llr}	C_{llr}^{min}	EER	EER	C_{llr}	C_{llr}^{min}	EER	C_{llr}	C_{llr}^{min}	EER	C_{llr}	C_{llr}^{min}
SCFC	0.11	0.732	0.006	23.92	5.17	0.951	0.625	0.00	0.730	0.000	5.34	0.761	0.160
RFCC	0.14	0.731	0.009	6.34	1.32	0.825	0.230	0.04	0.729	0.001	3.27	0.785	0.117
LFCC	0.13	0.730	0.005	5.56	1.20	0.818	0.211	0.00	0.728	0.000	4.73	0.811	0.153
MFCC	**0.47**	**0.737**	**0.023**	**14.03**	**2.93**	**0.877**	**0.435**	**0.00**	**0.727**	**0.000**	**5.43**	**0.812**	**0.165**
IMFCC	0.20	0.730	0.007	5.11	1.57	0.804	0.192	0.00	0.728	0.000	4.09	0.797	0.137
SSFC	0.27	0.733	0.016	7.15	1.60	0.819	0.251	0.70	0.734	0.027	4.70	0.800	0.160
SCMC	0.19	0.731	0.009	6.32	1.37	0.812	0.229	0.01	0.728	0.000	3.95	0.805	0.141
CQCC	0.10	0.732	0.008	1.59	0.58	0.756	0.061	0.66	0.733	0.028	3.84	0.796	0.128
8-fused-PADs	**0.04**	**0.732**	**0.003**	**1.74**	**0.37**	**0.828**	**0.077**	**0.00**	**0.729**	**0.000**	**3.10**	**0.793**	**0.111**
CQCC–MFCC	0.08	0.734	0.006	2.18	0.47	0.811	0.085	0.02	0.730	0.001	4.14	0.802	0.132
LFCC–MFCC	0.13	0.733	0.005	7.08	1.46	0.845	0.249	0.00	0.728	0.000	5.08	0.811	0.153
IMFCC–MFCC	0.15	0.734	0.006	6.26	1.29	0.838	0.219	0.00	0.728	0.000	4.09	0.803	0.133
SCFC–SCMC	0.08	0.732	0.004	7.00	1.47	0.876	0.249	0.00	0.729	0.000	3.84	0.780	0.144
SCFC–CQCC	0.03	0.732	0.002	1.82	0.50	0.844	0.071	0.05	0.732	0.002	3.72	0.775	0.129

generally leads to lower C_{llr} compared to mean). A probable reason for this is the performed pre-calibration of the scores using logistic regression. Calibration insures that the scores are well distributed within [0, 1] range, leading to similar EER-based thresholds among individual PAD systems. Hence, mean, which can be considered as a special case of LR, leads to "good enough" fusion results.

Table 16.6 presents the cross-database results when a given PAD system is trained and tuned using training and development sets from one database but is tested using evaluation set from another database. For instance, results in the second column of the table are obtained by using training and development sets from ASVspoof 2015 database but evaluation set from AVspoof-LA. Also, we evaluated the effect of using one type of attacks (e.g., logical access from AVspoof-LA) for training and another type (e.g., presentation attacks of AVspoof-PA) for testing (the results are in the last column of the table).

From Table 16.6, we can note that all methods generalize poorly across different datasets with BPCER reaching 100%, for example, especially, CQCC-based PAD showing poor performance for all cross-database evaluations. It is also interesting to note that even similar methods, for instance, RFCC- and LFCC-based, have very different accuracies in cross-database testing, even though they showed less drastic difference in single-database evaluations (see Table 16.5).

Based on the results in individual and cross-database evaluations, we have selected 2 PAD systems that performed the most well consistently across all databases and attacks: *8-fused-PADs* fused via mean score fusion and a simple MFCC-GMM PAD, which is also based on a very commonly used MFCC features. These systems are highlighted in bold in Tables 16.5 and 16.6.

16.4.2   Mega-Fusion Systems

The evaluation results of two mega-fusion systems (see Table 16.4 for overview) from ASVspoof 2017 challenge (third and fourth place in the challenge), described in Sect. 16.3.2, are presented in Table 16.7. The table shows the performance of "System-1" and "System-2" in terms of EER, both for the *dev* and the *eval* sets. The results obtained for the *dev* set are based on the systems trained exclusively on the *train* set of ASVspoof 2017 database, while to obtain the results for *eval* set, the systems were trained on the aggregated *train + dev* set.

Additionally, the table shows the results of a baseline system provided by the challenge organizers, which is based on CQCC front-end and two-class GMMs back end. *Best individual* system corresponds to a single IMFCC-based subsystem trained using GMM, which demonstrated the best performance during pre-submission evaluations.

The only difference between baseline and best individual system is the features used, as the classifier is the same. An interesting result is the one obtained with best individual system. While on the *dev* set it provides comparable performance to the fusion-based systems, on the *eval* set it performs dramatically worse.

Table 16.6 Performance of PAD systems based on handcrafted features (average APCER %, BPCER %, and C_{llr} of calibrated scores) in cross-database testing on ASVspoof 2015 and AVspoof databases

| PADs | ASVspoof (Train/Dev) | | | | | | AVspoof-LA (Train/Dev) | | | | | |
| | AVspoof-LA (Eval) | | | AVspoof-PA (Eval) | | | ASVspoof (Eval) | | | AVspoof-PA (Eval) | | |
	APCER	BPCER	C_{llr}	APCER	BPCER	C_{llr}	APCER	BPCER	C_{llr}	APCER	BPCER	C_{llr}
SCFC	0.10	2.76	0.751	10.20	2.76	0.809	15.12	0.00	0.887	39.62	0.35	0.970
RFCC	0.29	69.57	0.887	7.51	69.57	0.927	26.39	0.00	0.902	48.32	2.86	0.988
LFCC	1.30	0.13	0.740	21.03	0.13	0.868	17.70	0.00	0.930	37.49	0.02	0.958
MFCC	**1.20**	**2.55**	**0.764**	**17.09**	**2.55**	**0.838**	**10.60**	**0.00**	**0.819**	**19.72**	**1.22**	**0.870**
IMFCC	4.57	0.00	0.761	92.98	0.00	1.122	99.14	0.00	1.164	43.00	0.60	0.966
SSFC	4.81	64.47	0.899	18.89	64.47	0.973	71.84	0.68	1.047	63.45	23.54	1.070
SCMC	0.75	1.70	0.750	22.61	1.70	0.866	15.94	0.00	0.861	45.97	0.01	0.978
CQCC	13.99	57.05	0.968	66.29	57.05	1.191	44.65	0.61	1.009	0.86	100.00	1.009
8-fused-PADs	**0.41**	**12.73**	**0.804**	**12.46**	**12.73**	**0.930**	**19.71**	**0.00**	**0.944**	**26.97**	**5.25**	**0.959**
CQCC–MFCC	0.93	49.71	0.855	21.81	49.71	0.997	20.77	0.00	0.908	1.22	99.74	0.914
LFCC–MFCC	0.88	0.52	0.751	16.78	0.52	0.851	10.92	0.00	0.872	21.33	0.55	0.911
IMFCC–MFCC	1.36	0.34	0.761	25.91	0.34	0.967	13.29	0.00	0.978	21.82	0.81	0.914
SCFC–SCMC	0.13	0.82	0.750	11.51	0.82	0.835	17.59	0.00	0.873	41.39	0.03	0.971
SCFC–CQCC	0.17	49.86	0.848	12.94	49.86	0.980	28.70	0.02	0.945	1.45	99.91	0.960

Table 16.7 EER results of two mega-fusion systems of ASVspoof 2017 challenge, the baseline system, and the best individual model (GMM with IMFCC) trained and evaluated on different sets

Trained on	Tested on	System-1	System-2	Best individual	Baseline
Train	dev	**4.09**	4.32	4.86	11.17
Train + dev	Eval	**14.31**	14.93	29.41	24.65

Table 16.8 EER results for the cross-database experiments. In the first case, the systems were trained on *train + dev* set of ASVspoof 2017 and tested on BioCPqD-PA. In the second case, systems were trained on BioCPqD-PA and tested on *eval* set of ASVspoof 2017

	Fusion type	ASVspoof (train + dev) ↓ BioCPqD-PA	BioCPqD-PA ↓ ASVspoof (eval)
System 1	Mean	23.35	31.86
	LR	**21.35**	**26.58**
	MLP	22.34	30.77
System 2	Mean	22.23	27.74
	LR	**21.28**	27.96
	MLP	22.41	28.37
Best Indiv.	–	37.24	27.77

To asses the ability of the mega-fusion systems to generalize on unknown attacks and unseen data, we also evaluated them to the completely unrelated BioCPqD-PA database.

Table 16.8 shows that systems, "System-1" and "System-2", trained on the ASVspoof 2017 challenge database (*train + dev*) and tested on BioCPqD-PA database led to twice larger EER compared to when the same systems were tested on the *eval* set of ASVspoof 2017, as reported in Table 16.7. This finding confirms the limited generalization power of the systems. The performance degradation in cross-database experiments is not unprecedented: it has been observed in previous anti-spoofing evaluations [12, 17, 18].

Three different versions of "System-1" and "System-2" were tested by using mean, LR, and MLP algorithms for score fusion (see different rows in Table 16.8). LR led to a slightly better performance, especially for *System-1* trained on BioCPqD-PA database and evaluated on ASVspoof 2017 . Comparing the best individual sub-systems against fused systems, although fusion did not improve results for systems trained on BioCPqD-PA database, there is a significant improvement when it is trained on ASVspoof 2017 database. Thus, we can reason that, in practice, when the scenario is unknown, fusion adds robustness to the system performance.

Observing the non-negligible difference between the two crossing possibilities in Table 16.8, one can arguably say that training data diversity matters. While ASVspoof 2017 database has few speakers (only male) and a limited number of utterances, it contains presumably more diverse conditions (devices and recording environments) than BioCPqD-PA, due to the crowdsourcing data collection. On the other hand, BioCPqD-PA is larger, both in terms of speakers and number of utterances, but recording conditions are more restricted.

16.4.3 Convolutional Networks

To evaluate the performance of CNN-based PAD systems, described in Sect. 16.3.3, we first trained both CNN-Shallow and CNN-Deep networks, presented in the previous section, on training sets of voicePA and BioCPqD-PA databases. The two trained models (one for each database) were then used in two different capacities: (i) use pretrained models directly as classifiers on development and evaluation sets; and (ii) use models as feature extractors, by taking the output of the fully connected layer. The performance of CNN-based systems was compared with an MFCC–GMM systems, which represents systems based on handcrafted features.

When CNN systems are used as feature extractors, the layers before the last are used as feature vectors (see Fig. 16.5), resulting in 40 values for CNN-Shallow model and 60 values for CNN-Deep model, and two GMM classifiers are trained (one for bona fide and one for attacks) in the same fashion as for MFCC-based PAD. Using the same GMM classifier allows us to understand the effectiveness of self-learned CNN-based features compared to the handcrafted MFCC features (with CNN-Shallow model, the number of features is also the same 40 as in MFCC-based PAD).

Table 16.9 demonstrates the evaluation results of four versions of CNN-based PAD systems and baseline MFCC–GMM-based PAD using two databases voicePA and BioCPqD-PA. The first column of the table describes the combinations of the datasets used in each evaluation scenario and other columns contain the evaluation results (EER for *dev* set with APCER and BPCER for *eval* set) for each of the considered PAD system.

For instance, in the first row of Table 16.9, "voicePA (Train/Dev/Eval)" means that the training set of voicePA was used to train the model of each evaluated PAD, the development set of voicePA was used to compute the EER value and the corresponding threshold, and this threshold was used to compute APCER and BPCER values on evaluation set from the same voicePA database. In the second row of Table 16.9, "voicePA (Train/Dev) → BioCPqD-PA (Eval)" means that training and computation for development set were performed in the same way as for the system in the first row (hence, EER rate for *dev* set is the same as in the first row), but the evaluation was done on the *eval* set of BioCPqD-PA database instead. This cross-database evaluation simulates a practical scenario when a PAD system is built and tuned on one type of data but is deployed, as a black box, in a different setting and environment with different data. The last cross-database scenario is when only a pretrained model is built using some pre-existing data (a common situation in recognition), for instance, from voicePA as in row "voicePA (Train) → BioCPqD-PA (Dev/Eval)" of Table 16.9, but the system is tuned and evaluated on another data, e.g., from BioCPqD-PA.

Table 16.9 Performance of PAD systems in terms of EER (%) on *dev* set, APCER (%) on *eval* set, and BPCER (%) on *eval* set of the scores for each voicePA and BioCPqD-PA databases and for different cross-database scenarios. **T**, **D**, and **E** stand for *train*, *dev*, and *eval* sets

Origin of train, dev, and eval sets	GMM-MFCC			GMM-CNN-Shallow			CNN-Shallow		
	EER	APCER	BPCER	EER	APCER	BPCER	EER	APCER	BPCER
voicePA (T/D/E)	4.28	4.07	4.45	1.26	1.40	0.47	1.25	1.41	0.52
voicePA (T/D) → BioCPqD-PA (E)	4.28	96.18	8.89	1.26	48.65	54.40	1.25	79.59	3.49
voicePA (T) → BioCPqD-PA (D/E)	41.00	70.55	41.71	34.89	56.11	34.43	37.05	57.07	36.90
BioCPqD-PA (T/D/E)	41.00	70.55	41.71	11.39	22.45	11.09	11.69	23.73	11.48
BioCPqD-PA (T/D) → voicePA (E)	41.00	81.57	29.16	11.39	0.00	100.00	11.69	11.84	85.28
BioCPqD-PA (T) → voicePA (D/E)	50.19	37.73	47.31	22.86	24.83	18.49	33.29	34.02	26.54

Origin of train, dev, and eval sets	GMM-CNN-Shallow			GMM-CNN-Deep			CNN-Deep		
	EER	APCER	BPCER	EER	APCER	BPCER	EER	APCER	BPCER
voicePA (T/D/E)	1.26	1.40	0.47	**0.26**	**0.18**	**0.39**	**0.30**	**0.22**	**0.38**
voicePA (T/D) → BioCPqD-PA (E)	1.26	48.65	54.40	0.26	50.45	13.76	0.30	75.69	1.73
voicePA (T) → BioCPqD-PA (D/E)	34.89	56.11	34.43	**19.98**	**46.84**	**19.93**	25.20	43.73	25.22
BioCPqD-PA (T/D/E)	11.39	22.45	11.09	7.34	24.09	7.14	**7.01**	**23.81**	**6.89**
BioCPqD-PA (T/D) → voicePA (E)	11.39	0.00	100.00	7.34	0.00	100.00	7.01	11.97	86.48
BioCPqD-PA (T) → voicePA (D/E)	**22.86**	**24.83**	**18.49**	37.04	39.59	32.08	32.23	32.52	26.15

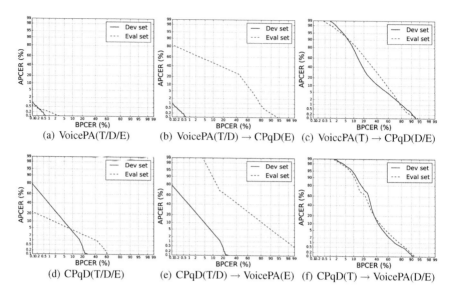

Fig. 16.6 DET curves of calibrated scores of CNN-Deep system in different evaluation scenarios (see the corresponding rows in Table 16.9)

The results in Table 16.9 demonstrate several important findings. First, it is clear that CNN-based PADs perform significantly better compared to MFCC-based PAD. This is especially evident in individual database evaluations, with "CNN-Deep" variants showing more than 10 times lower error rates compared to MFCC-based PAD for voicePA database and a few times lower for BioCPqD-PA database. Then, deeper CNN models perform generally better compared to shallow variants. Also, using CNN models as feature extractors coupled with a GMM classifier can be beneficial and can lead to an increase in accuracy, though the increase is not as significant compared to the larger computational resources GMM–CNN-based systems require.

To illustrate the performance of CNN-based systems in more detail, we also plot detection error tradeoff (DET) curves for a "CNN-Deep" system in Fig. 16.6. You can notice the large gap between the curves for *dev* and *eval* sets in Fig. 16.6b, e, when both training and threshold tuning are performed on one database but evaluation is done on another.

Although none of the considered CNN-based PAD systems generalizes well across different databases, it is also important to understand how they perform on different types of attacks, including *unknown* attacks, for which the systems were not trained. This analysis can help us understand which types of presentation attacks are more challenging. In this scenario, PAD systems are trained, tuned, and evaluated on the same database; only error rates are computed for specific attacks. Thus, we computed APCER value separately for each type of attacks in *eval* sets of voicePA and BioCPqD-PA database. Note that EER and BPCER values do not change, since

Table 16.10 Per attack APCER results for eval sets of voicePA and BioCPqD-PA databases

Types of attacks	GMM-MFCC	CNN-Shallow	CNN-Deep
VoicePA, laptop replay	74.19	20.12	8.94
VoicePA, phone replay	51.00	2.73	0.91
VoicePA, synthetic replay	0.01	1.08	0.06
BioCPqD-PA, A2-5	71.42	3.93	28.91
BioCPqD-PA, A2-6	42.65	1.04	23.22
BioCPqD-PA, A2-7	77.01	0.00	0.31
BioCPqD-PA, A2-8	76.93	60.60	13.68
BioCPqD-PA, A3-5	73.96	1.59	4.41
BioCPqD-PA, A3-6	36.67	0.02	0.10
BioCPqD-PA, A3-7	68.72	43.87	73.67
BioCPqD-PA, A3-8	70.17	0.86	0.63

EER is computed on the whole development set and BPCER only measures the detection of bona fide utterances.

The results for different types of attacks of the database detailed in Tables 16.2 and 16.3 are shown in Table 16.10. It is important to note that in case of voicePA, the same attacks are present in all training, development, and evaluation sets (data is split by speakers), so voicePA does not contain *unknown* attacks. However, in BioCPqD-PA, different types of attacks are distributed into *train*, *dev*, and *eval* sets differently (see Table 16.3), so that all attacks in *eval* set are basically *unknown* to the PAD systems.

The results in Table 16.10 for voicePA database demonstrate that using high-quality speakers as a replay device (see "voicePA, laptop replay" row of the table) lead to significantly more challenging attacks compared to attacks replayed with mobile phone (see row "voicePA, phone replay"). Also, synthetic speech poses considerably lesser challenge to PAD systems compared to the replay of natural speech. Also, note that we did not consider different environments and ASV systems (different microphones) for each of these types of attacks in voicePA; we only separate different speakers and natural speech from synthetic.

The attacks in BioCPqD-PA, however, are formed by combining different pairs of speakers (attack devices) and microphones of ASV systems, while influence of environment and types of speech were excluded, since acoustically isolate room was used and attacks were recorded by replaying natural speech only. Results in Table 16.10 for BioCPqD-PA show the significance of the choice for both speakers, with which attacks are made, and the microphone of the attacked ASV system. For instance, the same microphone is used in attacks "A3-6" and "A3-7" (see attack details in Table 16.3) but the difference in speakers can lead to drastically different detection results, as "A3-6' is easily detected by all CNN-based PAD systems, while all were spoofed by "A3-7". Similarly, the results of the CNN-Shallow and the CNN-Deep substantially vary across different pairs of speakers and microphones, e.g., for pairs "A2-5", "A2-6", "A2-8", and "A3-7". These differences may be due to different features learned by each neural network, as the model learns the features directly from the audio signal. Therefore, changing the neural network architecture will possibly affect the features learned and consequently the results.

16.5 Discussion

Based on all presented experiments, it is clear that the question of the generalization of PAD systems to completely unseen conditions (including different languages) remains open. Such situation is more likely to happen in practical PAD systems, where the system is trained on a given database and the attacks come from completely unknown conditions.

Training a system with good generalization capability might require a larger and more diverse database. Modern algorithms based on deep learning [38] approaches, for instance, which have proven to beat standard approaches in different kinds of tasks, such as speech recognition and computer vision, need massive amounts of data to provide state-of-the-art performance. In cases when the acquisition of such an amount of data is unfeasible, data augmentation strategies, such as [39], should be considered.

Another point that leads to a controversy is the use of so-called *mega-fusion* strategies. Although the fusion of many systems, sometimes more than a dozen (e.g., the submitted *System-1* is a fusion of 18 systems), usually leads to a better performance, its practical use is questionable. Mega-fusion has also been frequently used for the speaker recognition task, holding the current state-of-the-art results. However, its computational burden makes it unacceptable in practical cases, specially when system's response time is crucial.

It is surprising to note that even CNN-based PAD systems do not generalize well across databases, although, in the scenario when only a model is pretrained on another database, CNNs are more stable and significantly more accurate compared to PAD based on handcrafted features. However, if the system is both trained and tuned (threshold is chosen) on the same database but is evaluated on another database, CNN-based systems completely fail just as MFCC-based systems.

It is worth pointing out that the cross-database experiments for mega-fusion and CNN-based systems were designed for an extremely mismatched situation, when even the language is different between databases. It is expected that a PAD system should not be sensitive to language mismatch, however, that might not be the case in practice, as most speech features represent acoustic properties of speech that are indeed affected by the language spoken. This has been a concern for the speaker recognition community as well: the effect of language mismatch has been evaluated in speaker recognition tasks within NIST SRE over the past few years.

16.6 Conclusions

In this chapter, we provide an overview of the existing presentation attack detection (PAD) systems for voice biometrics and present evaluation results for selected eight systems that are based on handcrafted features, fusion-based systems, and CNN-based systems. We used several comprehensive publicly available databases: ASVspoof from 2015 and 2017, AVspoof and its extension voicePA, as well as proprietary BioCPqD-PA database. The cross-database evaluation results of the PAD systems demonstrate that none of the state-of-the-art systems generalizes well across different databases and data.

Presentation attack detection in voice biometrics is far from being solved, as currently proposed methods do not generalize well across different data. It means that no effective method is yet proposed that would make speaker verification system resistant even to trivial replay attacks, which prevents the wide adoption of ASV systems in practical applications, especially in security-sensitive areas. Deep learning methods for PAD are showing some promise and may eventually[11] evolve into systems that can detect even unseen attacks.

Acknowledgements This work has been supported by the European H2020-ICT project TeSLA (grant agreement no. 688520), the project on Secure Access Control over Wide Area Networks (SWAN) funded by the Research Council of Norway (grant no. IKTPLUSS 248030/O70), and by the Swiss Center for Biometrics Research and Testing.

References

1. ISO/IEC JTC 1/SC 37 Biometrics. (2016) DIS 30107-1, information technology biometrics presentation attack detection. American National Standards Institute
2. Wu Z, Kinnunen T, Evans N, Yamagishi J, Hanilçi C, Sahidullah M, Sizov A (2015) ASVspoof 2015: The first automatic speaker verification spoofing and countermeasures challenge. In: INTERSPEECH, pp 2037–2041
3. Mariéthoz J, Bengio S (2005) Can a professional imitator fool a GMM-based speaker verification system?. Tech Rep Idiap-RR-61-2005, Idiap Research Institute
4. Kucur Ergunay S, Khoury, E., Lazaridis, A., Marcel, S.: On the vulnerability of speaker verification to realistic voice spoofing. In: IEEE international conference on biometrics: Theory, applications and systems, pp 1–6
5. Wu Z, Evans N, Kinnunen T, Yamagishi J, Alegre F, Li H (2015) Spoofing and countermeasures for speaker verification: A survey. Speech Commun 66:130–153
6. Wu Z, Xiao X, Chng ES, Li H (2013) Synthetic speech detection using temporal modulation feature. In: IEEE international conference on acoustics, speech and signal processing (ICASSP), pp 7234–7238. https://doi.org/10.1109/ICASSP.2013.6639067
7. Gałka J, Grzywacz M, Samborski R (2015) Playback attack detection for text-dependent speaker verification over telephone channels. Speech Commun 67:143–153
8. Shiota S, Villavicencio F, Yamagishi J, Ono N, Echizen I, Matsui T (2015) Voice liveness detection algorithms based on pop noise caused by human breath for automatic speaker verification.

---

[11]http://www.tesla-project.eu.

In: Sixteenth annual conference of the international speech communication association, pp 239–243

9. Wu Z, Siong CE, Li H (2012) Detecting converted speech and natural speech for anti-spoofing attack in speaker recognition. In: INTERSPEECH

10. De Leon P, Pucher M, Yamagishi J, Hernaez I, Saratxaga I (2012) Evaluation of speaker verification security and detection of hmm-based synthetic speech. IEEE Trans Audio Speech Lang Process 20(8):2280–2290. https://doi.org/10.1109/TASL.2012.2201472

11. Patel TB, Patil HA (2015) Combining evidences from mel cepstral, cochlear filter cepstral and instantaneous frequency features for detection of natural vs. spoofed speech. In: INTER-SPEECH, pp 2062–2066

12. Todisco M, Delgado H, Evans N (2017) Constant Q cepstral coefficients: A spoofing counter-measure for automatic speaker verification. Comput Speech Lang

13. Alegre F, Amehraye A, Evans N (2013) A one-class classification approach to generalised speaker verification spoofing countermeasures using local binary patterns. In: IEEE interna-tional conference on biometrics: Theory, applications and systems, pp 1–8

14. Janicki A (2015) Spoofing countermeasure based on analysis of linear prediction error. In: Sixteenth annual conference of the international speech communication association, pp 2077–2081

15. Sahidullah M, Kinnunen T, Hanilçi C (2015) A comparison of features for synthetic speech detection. In: INTERSPEECH, pp 2087–2091

16. Luo D, Wu H, Huang J (2015) Audio recapture detection using deep learning. In: 2015 IEEE China summit and international conference on signal and information processing (ChinaSIP), pp 478–482. https://doi.org/10.1109/ChinaSIP.2015.7230448

17. Paul D, Sahidullah M, Saha G (2017) Generalization of spoofing countermeasures: A case study with ASVspoof 2015 and BTAS 2016 corpora. In: ICASSP, pp 2047–2051

18. Korshunov P, Marcel S (2016) Cross-database evaluation of audio-based spoofing detection systems. In: Interspeech, pp 1705–1709

19. Korshunov P, Marcel S (2017) Impact of score fusion on voice biometrics and presentation attack detection in cross-database evaluations. IEEE J Sel Top Sig Process 11(4):695–705. https://doi.org/10.1109/JSTSP.2017.2692389

20. Goncalves AR, Korshunov P, Violato RPV, Simões FO, Marcel S (2017) On the generalization of fused systems in voice presentation attack detection. In: Brömme A, Busch C, Dantcheva A, Rathgeb C, Uhl A (eds) 16th international conference of the biometrics special interest group. Darmstadt, Germany

21. Korshunov P, Marcel S, Muckenhirn H, Gonçalves AR, Mello AGS, Violato RPV, Simoes FO, Neto MU, de, (2016) In: Assis Angeloni M, Stuchi JA, Dinkel H, Chen N, Qian Y, Paul D, Saha G, Sahidullah, M (eds) Overview of BTAS 2016 speaker anti-spoofing competition. IEEE international conference on biometrics: Theory applications and systems. Niagara Falls, NY, USA, pp 1–6

22. Muckenhirn H, Magimai-Doss M, Marcel S (2017) End-to-end convolutional neural network-based voice presentation attack detection. In: International joint conference on biometrics

23. Davis S, Mermelstein P (1980) Comparison of parametric representations for monosyllabic word recognition in continuously spoken sentences. IEEE Trans Acoust Speech Signal Process 28(4):357–366. https://doi.org/10.1109/TASSP.1980.1163420

24. Furui S (1981) Cepstral analysis technique for automatic speaker verification. IEEE Trans Acoust Speech Signal Process 29(2):254–272. https://doi.org/10.1109/TASSP.1981.1163530

25. Scheirer E, Slaney M (1997) Construction and evaluation of a robust multifeature speech/music discriminator. In: IEEE international conference on acoustics, speech, and signal processing (ICASSP-97), vol 2, pp 1331–1334. https://doi.org/10.1109/ICASSP.1997.596192

26. Le PN, Ambikairajah E, Epps J, Sethu V, Choi EHC (2011) Investigation of spectral centroid features for cognitive load classification. Speech Commun 53(4):540–551

27. Todisco M, Delgado H, Evans N (2016) A new feature for automatic speaker verification anti-spoofing: Constant Q cepstral coefficients. In: Odyssey, pp 283–290

28. Violato R, Neto MU, Simões F, Pereira T, Angeloni M (2013) BioCPqD: uma base de dados biométricos com amostras de face e voz de indivíduos brasileiros. Cad CPqD Tecnolo 9(2):7–18
29. Kinnunen T, Sahidullah M, Delgado H, Todisco M, Evans N, Yamagishi J, Lee KA (2017) The ASVspoof 2017 challenge: Assessing the limits of replay spoofing attack detection. In: INTERSPEECH 2017, annual conference of the international speech communication association, 20–24 August 2017, Stockholm, Sweden
30. Zen H, Tokuda K, Black AW (2009) Statistical parametric speech synthesis. Speech Commun 51(11):1039–1064
31. Lee K, Larcher A, Wang G, Kenny P, Brümmer N, van Leeuwen DA, Aronowitz H, Kockmann M, Vaquero C, Ma B, Li H, Stafylakis T, Alam MJ, Swart A, Perez J (2015) The reddots data collection for speaker recognition. In: Interspeech, pp 2996–2091
32. Kinnunen T, Sahidullah M, Falcone M, Costantini L, Hautamäki RG, Thomsen D, Sarkar A, Tan Z, Delgado H, Todisco M, Evans N, Hautamäki V, Lee K (2017) RedDots replayed: a new replay spoofing attack corpus for text-dependent speaker verification research. In: ICASSP, pp 5395–5399
33. ISO/IEC JTC 1/SC 37 Biometrics (2016) DIS 30107-3:2016, information technology—biometrics presentation attack detection—part 3: Testing and reporting. American National Standards Institute
34. Soong FK, Rosenberg AE (1988) On the use of instantaneous and transitional spectral information in speaker recognition. IEEE Trans Acoust Speech Signal Process 36(6):871–879. https://doi.org/10.1109/29.1598
35. Mandasari MI, Gnther M, Wallace R, Saeidi R, Marcel S, van Leeuwen DA (2014) Score calibration in face recognition. IET Biom 3(4):246–256. https://doi.org/10.1049/iet-bmt.2013.0066
36. Scherhag U, Nautsch A, Rathgeb C, Busch C (2016) Unit-selection attack detection based on unfiltered frequency-domain features. In: INTERSPEECH, San Francisco, USA pp 2209–2212
37. Platt JC (1999) Probabilistic outputs for support vector machines and comparisons to regularized likelihood methods. In: Advances in large margin classifiers. MIT Press, pp 61–74
38. Goodfellow I, Bengio Y, Courville A (2016) Deep learning. MIT Press. http://www.deeplearningbook.org
39. Gonçalves AR, Uliani Neto M, Yehia HC (2015) Accelerating replay attack detector synthesis with loudspeaker characterization. In: 7th symposium of instrumentation and medical images /6th symposium of signal processing of UNICAMP

Chapter 17
Voice Presentation Attack Detection Using Convolutional Neural Networks

Ivan Himawan, Srikanth Madikeri, Petr Motlicek, Milos Cernak,
Sridha Sridharan and Clinton Fookes

Abstract Current state-of-the-art automatic speaker verification (ASV) systems are prone to spoofing. The security and reliability of ASV systems can be threatened by different types of spoofing attacks using voice conversion, synthetic speech, or recorded passphrase. It is therefore essential to develop countermeasure techniques which can detect such spoofed speech. Inspired by the success of deep learning approaches in various classification tasks, this work presents an in-depth study of convolutional neural networks (CNNs) for spoofing detection in automatic speaker verification (ASV) systems. Specifically, we have compared the use of three different CNNs architectures: AlexNet, CNNs with max-feature-map activation, and an ensemble of standard CNNs for developing spoofing countermeasures, and discussed their potential to avoid overfitting due to small amounts of training data that is usually available in this task. We used popular deep learning toolkits for the system implementation and have released the implementation code of our methods publicly. We have evaluated the proposed countermeasure systems for detecting replay attacks on recently released spoofing corpora ASVspoof 2017, and also provided in-depth visual analyses of CNNs to aid for future research in this area.

I. Himawan (✉) · S. Sridharan · C. Fookes
Queensland University of Technology, Brisbane, Australia
e-mail: i.himawan@qut.edu.au

S. Sridharan
e-mail: s.sridharan@qut.edu.au

C. Fookes
e-mail: c.fookes@qut.edu.au

S. Madikeri · P. Motlicek
Idiap Research Institute, Martigny, Switzerland
e-mail: srikanth.madikeri@idiap.ch

P. Motlicek
e-mail: petr.motlicek@idiap.ch

M. Cernak
Logitech, Lausanne, Switzerland
e-mail: mcernak@logitech.com

© Springer Nature Switzerland AG 2019
S. Marcel et al. (eds.), *Handbook of Biometric Anti-Spoofing*,
Advances in Computer Vision and Pattern Recognition,
https://doi.org/10.1007/978-3-319-92627-8_17

17.1 Introduction

The ability to authenticate a person's identity through their speech is now a reality after recent years of advances in automatic speaker verification technology. With a growing range of applications in security, commerce, and mobile, ASV systems are required to be very secure under malicious attacks and expected to thwart unauthorized access. In order to reach the level of mass-market adoption, ASV systems should target low false positive rate (not easily fooled) and, therefore, the technology needs to be capable of thwarting every possible spoofing attempt.

Attacks to gain illegitimate acceptance from an ASV system can be performed by tampering with the enrolled person's voice. The four major threats that present a real threat to ASV systems are impersonation [1], voice conversion [2, 3], speaker-adapted speech synthesis [4, 5], and replay attacks. Among these attacks, impersonation is believed to be easiest to detect since it aims to mimic the prosodic or stylistic cues rather than those aspects related to human speech production [6]. A voice conversion system (which converts one speaker's voice to the target user's speech) and a text-to-speech (TTS) system (which involves generating synthetic speech) are the two most sophisticated attacks to date, requiring specific expertise and technology for implementation. The vulnerability of an ASV system against voice conversion and TTS has been the subject of many studies in the past including the first Automatic Speaker Verification Spoofing and Countermeasures Challenge (ASVspoof 2015). These attacks assume that the spoofing signal is injected directly into the verification system bypassing its microphone, which is referred to as *logical access attack* [7]. Replay attacks referred to as *presentation attack* or *physical access attack* [8], on the other hand, can be performed with ease since it is uncommon for an attacker to have access to the system's internal software. In a typical scenario, a fraudster plays back a recorded utterance, which can be obtained directly from the real person or artificially produced speech using TTS or voice conversion systems to a microphone of the ASV system using high-quality audio equipment [9–11].

In this work, we aim to develop generalized countermeasures for biometric anti-spoofing research by focusing on developing back-end classifiers using convolutional neural networks (CNNs) to detect spoofing attacks. Specifically, our focus is on replay attacks detection using ASVspoof 2017[1] corpus. An investigation on back-end classifiers will be beneficial in addition to handcrafted feature studies when developing countermeasures.

A new trend emerging from the ASVspoof 2017 challenge is the use of an end-to-end-representation learning framework based on deep learning architectures. Specifically, the success of CNNs in image classification and recognition tasks has inspired many studies for detecting spoofed speech from genuine speech. For example, light CNN and the combined CNN with recurrent neural network (RNN) architectures were shown to yield equal error rate (EER) between 6 and 11% in the evaluation set of ASVspoof 2017 [12]. Also, other variants of CNNs such as residual network (ResNet) were also shown to achieve promising results [13, 14]. Interestingly, some

---

[1]http://www.asvspoof.org/.

of the best performing systems used a spectrogram as an input to CNNs rather than handcrafted features or other features' selection approaches.

Although advanced classifiers such as CNNs can deliver good performance on the classification tasks, it has been shown from the challenge results that model overfitting has a strong impact on performance of submitted systems when evaluated on unseen data. This is often caused by a lack of training data rather than ineffective feature representation when training a deeper network [15–17]. Without the ability to reproduce and verify the state-of-the-art results, unreliable comparisons can be made which make it difficult to focus on a fruitful research direction. In this case, it is not, for instance, clear whether the use of similar or better feature representation or acoustic modeling would offer the best results for the detection. To this end, we assume that the well-trained networks—irrespective of whether they are trained on images or acoustic spectrograms' input—can be used as a feature extractor, and the classification is accomplished by means of the support vector machine (SVM).

In this work, we used pretrained AlexNet [18], a convolutional neural network (CNN) trained on more than 1 million pictures, which can classify images into one thousand categories. From this perspective, a new benchmarking criterion is proposed, which can facilitate a systematic assessment of automatic feature learning and the effectiveness of conventional handcrafted feature methods for detecting spoofing attacks. Having developed a general indication of baseline performance, we can then focus on the investigation of back-end classifiers with a reasonable assumption that the learned feature descriptor and classifier can capture patterns in the data, presumably with no model overfitting. For this consideration, we implemented the state-of-the-art light CNN with max-feature-map (MFM) activation function that allows for the reduction of CNN parameters and comparison of such architectures with an ensemble of CNNs to improve the accuracy of prediction. These two methods are assumed to avoid overfitting problem.

In summary, the three main contributions of our study are as follows: (1) satisfying the reproducibility aspects that allow other researchers to apply state-of-the-art methods and efficiently build upon the previous work. Hence, we implemented the proposed algorithms in this paper using two popular open-source deep learning frameworks: Tensorflow [19] and PyTorch [20]. We emphasize "reproducibility" aspects of the proposed technologies evaluated in this paper; (2) investigating CNNs architectures for spoofing detection and methods to address overfitting the model; and (3) understanding and visualizing the CNNs to gain an insight of common error types when distinguishing genuine speech from spoofed speech and how they compare. Evaluation of the proposed techniques is performed on ASVspoof 2017 corpus and compared to so far best known results.

The rest of paper is organized as follows: Sect. 17.2 discusses related work. Section 17.3 describes the proposed architecture of the deep learning framework for replay attack detection. Section 17.4 reports the experimental setup and results. Section 17.5 discusses the key findings. Finally, Sect. 17.6 concludes the paper.

17.2   Related Work

Our objective in this paper is to investigate efficiency of currently very popular deep learning architectures (especially CNNs) for the voice spoofing detection problem. CNNs, exploited either as a feature extraction technique or as end-to-end models, have shown large potential in many other tasks, such as image classification or face recognition. ASV anti-spoofing tasks have been therefore largely motivated by those architectures, since the input representation to the model can be given in a form of acoustic spectrogram (2-D representation), similar to image processing tasks [16].

In terms of recent contributions to the ASV anti-spoofing task, the proposed works can be divided into two problems: (a) finding appropriate feature extraction methods (front-end), and (b) final classifiers (back-end).

17.2.1   Front-End

The selection of proper features for parameterizing the speech signal has been widely investigated and is motivated by works from automatic speech and speaker recognition tasks. The standard method for analyzing sound signals is to perform short-time Fourier transform (STFT), assuming that the signal is locally stationary (i.e., the statistics of the process are unchanging across the small length of a window). However, this raw time–frequency representation of a signal is often assumed to contain redundant information (due to their high dimensionality) and needs to be further processed to obtain the best signal representation depending on the applications. Often, this involves characterizing the spectra by means of a filter bank method, where each coefficient represents the signal's overall intensity over the frequency range of interest. For example, Mel-frequency cepstral coefficients (MFCCs) use a nonuniform set of frequency ranges (or bands) in a mel scale to obtain coefficients that produce a better discrimination of speech signal. Various representations of audio signals derived from STFT include measuring the frame-by-frame change in the spectrum in the case of spectral-flux-based coefficients [21], or capturing centroid frequencies of subbands in the case of subband spectral centroid-based features [22]. Also, there are other time–frequency analyses in the literature that do not exploit STFT such as constant Q transform (CQT) [23] and wavelet transform (WT) [24].

The features for anti-spoofing can be broadly classified into two groups, which are either magnitude based or phase based. Previous studies have demonstrated that both the magnitude and phase spectrums, extracted from the Fourier transform, contain detailed speech information that is useful for spoofing speech detection [25]. Recent studies by Sahidullah et al. [26], which benchmarked several feature sets for characterizing real and synthetic speech, found that a filter bank used in the computation of cepstral features could have significant effects on the performance of anti-spoofing systems. Specifically, linear-frequency cepstral coefficients (LFCCs) where the filters are spaced in linear scale [27], and inverted Mel-frequency cepstral coefficients

(IMFCCs) where the filters are linearly spaced on an "inverted-Mel" scale [28], have been found to perform reasonably well compared to other filter bank features. Other approaches, which employ magnitude information including log magnitude spectrum and residual log magnitude spectrum features [26, 29], have also been shown to be able to capture important artifacts of the magnitude. The features derived from the phase spectrum are also popular for synthetic speech detection because spoofed speech does not retain natural phase information [7, 30]. Phase-based countermeasures used for anti-spoofing include group delay (GD) [31] and modified group delay (MGD) [31], cosine-phase function [26], and relative phase shift (RPS) [30]. However, the phase-based methods fail to detect spoofed speech when a little modification is made to the phase of natural speech, such as the unit selection method, where natural speech signal is preserved outside the concatenation points when synthesizing speech [30].

17.2.2   Back-End

The ASV anti-spoofing baseline system has proposed to apply a conventional GMM classifier. In this case, all the genuine speech segments were used to train a genuine speaker model and the spoofed recordings were used to train the spoofed model. Traditional log-likelihood ratio (LLR) per frame is then calculated and summed, and normalized for the whole test utterance as criterion of assessment.

The current trend in acoustic modeling is to exploit deep learning architectures as they have shown tremendous capabilities in a variety of machine learning tasks. The first approaches in ASV anti-spoofing task have focused on fully connected feed-forward architectures combined with the proposed front-ends [32–35]. For spoofing detection, a deep neural network (DNN) is typically employed as a classifier to estimate the posterior probability of a particular class given the input utterance. Different types of handcrafted features have been used with DNN such as RPS and log-filter-bank features [27], the Teager Energy Operator Critical Band Autocorrelation Envelope (TEO-CB-Auto-Env) and Perceptual Minimum Variance Distortionless Response (PMVDR) features [36], and dynamic acoustic features [37, 38]. However, the performance gain of DNN systems is marginal compared to GMM classifiers, and the systems have difficulties when detecting unknown attacks [39].

As an alternative, DNN can be incorporated into spoofing detection as a feature extractor to extract deep features where the activations of the hidden layers are used as features to train other classifiers (i.e., GMM and SVM) [32, 34, 39–41]. For example, [40] investigated deep features extracted from DNN and RNN, and found that deep features' combination offers EER of almost 0.0% on attacks S1 to S9 but fail on the S10 attack (a very efficient spoofing algorithm, based on MARY TTS system [7]) using the ASVspoof 2015 corpus. Features extracted from deep architectures including the DNN, CNN, and bidirectional long short-term memory RNN (BLSTM-RNN) are also shown to perform better than the baseline features for spoofing detection under noisy and reverberant conditions [34]. To enhance the robustness of deep features and avoid overfitting, several training strategies can be employed such as multi-condition training, noise-aware training, and annealed dropout training [34]. In [39], deep neural network filter bank cepstral coefficient (DNN-FBCC) features are introduced which outperform both handcrafted and DNN-based features. To produce these features, a non-negative constraint and a band-limiting mask matrix are applied to the weight matrix between the input layer and the first hidden layer where the learned weight matrix can be considered as a special type of filter bank. The DNN-FBCC features are then extracted from the activation of the first hidden layer of a trained DNN.

Addressing the sensitivity of initial DNN parameters is a crucial point to reach better performance over the GMM baseline. One of the proposed solutions applies batch normalization and dropout regularization on each hidden layer implemented in a standard backpropagation algorithm. DNNs are usually trained with a softmax output layer, allowing to classify input speech into two classes [27].

CNN architectures are similar to the conventional feed-forward neural networks, but they consist of learnable shared filter weights which exploit local correlation of the input in two-dimensional space. This allows CNNs to consider its input as an image and learn the same patterns in different positions of the image, which is lacking in DNN architectures due to its fixed connections from input to the next layer. Other approaches have considered ResNet architectures, which allow to train deeper neural networks, while avoiding gradient vanishing problems [13, 14, 42]. Recent works were motivated by findings on fairly similar tasks focused on image recognition. In general, ResNets allow deeper networks to be trained, resulting in models' outperforming conventional DNN architectures [43].

Some recent studies in anti-spoofing suggest that raw features with CNNs [35, 44] can perform competitively with the feature engineering approaches that are often trained with GMM as a classifier. Hence, both features and classifiers are learnt directly from the raw signal in an end-to-end fashion. The comparison of different CNN architectures reveals that a deeper CNN network attains better performance than a shallow network. However, CNN-based systems do not generalize well across different databases [41]. Incorporating longer temporal context information is also shown to be beneficial for playback spoofing attack where two special implementations of RNNs, the long short-term memory (LSTM) and the gated recurrent unit (GRU), outperform DNN model [45].

The fusion of scores from individual systems is one of the mechanisms allowing the combination of different modeling architectures, usually resulting in significantly improved performance over individual systems. In the anti-spoofing context, score fusion allows different architectures or features excelling only in detecting specific attacks to work collectively to thwart multiple forms of spoofing attacks. In a typical scenario, first, the scores from multiple systems are normalized to bring them into a common domain since the scores from individual systems may fall in different ranges. Finally, the scores can be fused using a simple mean function or via a classifier-based solution where score fusion is considered as a classification problem (i.e., discriminate genuine and spoof scores) [46]. For the latter, a logistic-regression calibration (for normalizing the scores) and fusion can be employed to give specific weights $\{w_k\}$ to each class in the form $s_{fused} = w_0 + w_1 s_1 + \cdots + w_K s_K$, where s_{fused} denotes the fused score, $\{s_k\}$ are the base classifier scores, and w_0 indicates a bias term [31]. For DNN-based classifiers that use the softmax function at the output layer, and the number of outputs is the same across the classifiers, the score ranges will be the same and comparable. Thus, score fusion can be implemented through a weighted combination of scores. The weights can be optimized using a validation set.

17.2.3   Summary of Previous Work

One recent successful countermeasure for anti-spoofing is based on the constant Q cepstral coefficients (CQCCs) [47, 48]. CQCC features are extracted with the CQT analysis of speech. The CQCC has been shown to perform competitively better than other features in utterance verification [49] and speaker verification tasks [26]. However, the performance of CQCC for detecting replay attacks has not been thoroughly investigated. A good countermeasure approach must generalize to detect spoofed speech for any given attack, regardless of text content.

Considering different distortions are introduced by attacks that are different in nature, one effective strategy would be to build specialized countermeasures for each kind of attack, rather than to find a single generalized countermeasure that behaves well for different kinds of attacks. In fact, one finding from the ASVspoof 2017 challenge is that the CQCC features with a standard Gaussian mixture model (GMM) classifier do not perform very well when detecting the replay version of speech [50–52]. This means that one may need to focus on finding salient features for replay attacks. However, one disadvantage is that these features may not be optimal for the subsequent classification tasks and the use of a sophisticated classifier will not help to improve the anti-spoofing system performance.

In order to offer some context on the difficulty of comparing results between feature engineering and back-end modeling approaches, we provided examples from the experimental analysis from multiple papers which were published after the challenge. The official CQCC GMM baseline performance obtained an EER of 30.6% using the training data alone [50]. Moreover, one paper reported that the use of residual networks (i.e., very deep networks used in the field of computer vision) achieved an EER of 18.79% with CQCC and even better results are obtained with MFCC input (with an EER of 16.26%) [13]. Another work, which presents comparison of different features, revealed that CQCCs outperform MFCCs (17.43 vs. 26.13%) using a simple GMM classifier, albeit using more training data from external source [52]. To add to the confusion, another paper showed that the use of CQCC with GMM outperforms a residual network classifier [14]. While it is not possible to directly compare all published results as the different training regimes (e.g., number of filters in MFCCs, number of hidden layers, choice of optimization algorithms) affect the results significantly, this highlights that effective methods should be investigated when incorporating different handcrafted features into deep learning architectures and avoid model overfitting [50]. In this case, the use of CNN as a stand-alone feature extractor will be beneficial since it does not necessarily depend on a specific feature representation.

Currently, a large body of studies in the anti-spoofing literature focuses on countermeasure features with the aim of finding the best signal representations for characterizing genuine speech and spoofed speech. Even though some of these features are successful for discriminating between genuine and spoofed speech, prior information of spoofing attacks is often taken into account when designing the countermeasures, which is not realistic in practice. Meanwhile, machine learning techniques do not necessarily require an understandable representation as their input, since they have different requirements from what is being perceived by the human eye. A spectrogram contains high-dimensional information, such as message, speaker, channel, and environment. Thus, further processing may discard some important characteristics of the sound signal which otherwise would have been preserved in the spectrogram. In many recent classification tasks, the use of spectrogram or time-domain waveforms with sophisticated classifiers (i.e., CNNs) can produce a competitive performance compared to systems that utilize an efficient feature representation for their input [44]. Therefore, the same strategy (using a spectrogram) will be adopted for developing the spoofing detection systems in this paper.

17.3 Convolutional Neural Networks for Voice Spoofing Detection

This section presents the main contributions of this paper related to analysis and employment of convolutional neural networks (CNNs) for voice spoofing detection. Overview of results on ASVspoof 2017 challenge has indicated good performance of various CNN architectures [12–14]. In fact, the best performing system published on 2017 ASVspoof data is built around CNNs, exploiting 2D spectrogram derived from input audio. This work has been partially motivated by success of CNNs on various image classification tasks.

This section first provides a theoretical background for CNN, and then in detail describes selected architectures. More specifically, we will first focus on the CNN architecture based on the use of MFM, applying Max-Out activation. As this paper strongly targets the reproducibility aspects, we have re-implemented the proposed MFM architecture using two open-source deep learning platforms (TensorFlow and PyTorch). MFM CNN will represent a benchmark to the following experiments. Further, this section focuses on application of transfer learning, enabling a deep learning model trained for image classification to be exploited for detection of spoofing attacks from voice. Finally, we propose novel CNN-based models requiring much less computing resources than the benchmark systems, offering competitive performance.

17.3.1 Feature Matrix Construction

A unified time–frequency shape of features is created for the input data since the duration of each utterance is not the same in ASVspoof 2017 dataset. The first 400 frames are selected (5 s for window size $= 16$ ms, step size $= 8$ ms). Hence, we concatenate data for the utterance which is less than 400 frames or select the first 400 frames of consecutive data and truncate the rest for utterance which is more than 400 frames. As a final step of feature matrix construction, mean and variance normalization are applied on the log power spectrum. We use normalized time–frequency features as an image representation as input to the CNNs. Two different input features are investigated by setting different number of filters in the STFT analysis: (1) the high-resolution features (768×400) and the low-resolution features (128×400).

17.3.2 Convolutional Neural Networks Classifier

CNNs have been extensively applied in computer vision tasks where they are capable of processing large amounts of training data [16, 53, 54]. These networks consist of convolution layers, which compute convolutional transforms, followed by non-linearities and pooling operators that can be stacked into multiple layers, making them deep. The CNNs can be trained in end-to-end fashion and used for classification directly, or they can also be used to extract features which are then fed into a classifier such as an SVM. The main building block of convolutional layers is a set of learnable filters or kernels. Given an input image, each filter that covers a small subregion of the visual field (called a receptive field) is convolved (by sliding the filter) across the entire image, which results in the feature map. This filter is typically constrained to use the same weights and bias that allow the same features to be detected but at different positions (which also reduce the number of parameters). The kth feature map output can be computed for the input patch x_{ij} centered at location (i, j) (for linear rectifier nonlinearities) as [55]

$$f_{i,j}^k = \max(w_k^T x_{i,j}, 0), \qquad (17.1)$$

where (i, j) is the pixel index. In a typical configuration, a number of filters are employed where each of them will produce a separate feature map that collectively forms a rich representation of the input data.

One important extension of CNNs is the use of Network in Network (NiN) structure to enhance the ability of CNNs to learn feature representation. In the convolution layer, the convolved feature is obtained by sliding a filter over the input image and evaluating the dot product between a filter and the receptive field (equivalent to the filter size) location. This operation is equivalent of inputting feature vectors (i.e., small patch of the image) to a linear classifier. NiN replaces the linear filter for convolution with a multilayer perceptron (MLP) consisting of multiple fully connected layers with nonlinear functions. Therefore, a linear classifier has been replaced with tiny neural networks. The micro-network (i.e., MLP) used in NiN is defined by 1×1 convolutions in spatial domain in the CNN architecture. This is often followed by a pooling process. The resulting CNN is called NiN since the filter itself is a network. As a result of incorporating the micro-networks, the depth and subsequently the learning capacity of the network are increasing. This structure delivers better recognition performance compared to the classic CNN and at the same time reduces the total number of parameters that would have been required to design CNN models at similar depth.

In the next sections, we will describe the three CNN architectures which are investigated in this study: AlexNet, light CNN with max-feature-map activation, and a standard CNN architecture, because of their notable performance for solving image classification and recognition tasks.

17.3.3 AlexNet

AlexNet was one of the first deep networks to achieve a breakthrough in computer vision by winning the world-wide ImageNet Large-Scale Visual Recognition Challenge (ILSVRC) competition in 2012. The success of AlexNet rocked the computer vision community since it was the first time that a deep network was used to solve the problem of object detection and image classification at a large scale. After the competition, an enormous interest using deep learning to solve machine learning problems has catapulted the popularity of the CNN as the algorithmic approach to analyzing visual imagery. The network has eight layers where the first five are convolutional which are followed by three fully connected layers. For image recognition, AlexNet used rectified linear unit (ReLU) as the activation function, instead of Sigmoid which was the earlier standard for traditional neural networks. In addition, the cross-channel normalization is applied by creating competitions to the neuron outputs computed using different kernels [16]. The AlexNet architecture,[2] which is employed in this paper, is shown in Fig. 17.1 Since the pretrained AlexNet model requires the image size to be the same as the input size of the network, the spectrogram is produced such that it can be fed into the model (a three-channel input is created by replicating the spectrogram image three times and then stacked together).

17.3.4 Implementation for Reproducibility

The immense success and popularity of DNN-based modeling architectures have given rise to many toolkits to simplify training and extension of commonly used

---

[2]https://au.mathworks.com/help/nnet/ref/alexnet.html.

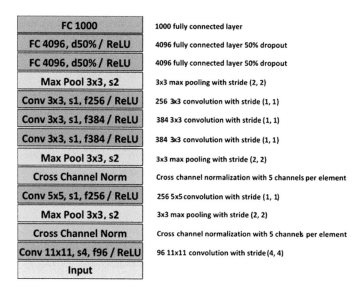

FC 1000	1000 fully connected layer
FC 4096, d50% / ReLU	4096 fully connected layer 50% dropout
FC 4096, d50% / ReLU	4096 fully connected layer 50% dropout
Max Pool 3x3, s2	3x3 max pooling with stride (2, 2)
Conv 3x3, s1, f256 / ReLU	256 3x3 convolution with stride (1, 1)
Conv 3x3, s1, f384 / ReLU	384 3x3 convolution with stride (1, 1)
Conv 3x3, s1, f384 / ReLU	384 3x3 convolution with stride (1, 1)
Max Pool 3x3, s2	3x3 max pooling with stride (2, 2)
Cross Channel Norm	Cross channel normalization with 5 channels per element
Conv 5x5, s1, f256 / ReLU	256 5x5 convolution with stride (1, 1)
Max Pool 3x3, s2	3x3 max pooling with stride (2, 2)
Cross Channel Norm	Cross channel normalization with 5 channels per element
Conv 11x11, s4, f96 / ReLU	96 11x11 convolution with stride (4, 4)
Input	

Fig. 17.1 AlexNet architecture [16][2]

Table 17.1 Given below is the list of software used to implement various components in our system for training and testing

Module	Software	Version
Feature extraction	Matlab	R2017b
DNN training	PyTorch (via Miniconda)	0.2.0
DNN training	TensorFlow	r.1.4.1

neural network architectures. TensorFlow (from Google), PyTorch (from Facebook), Microsoft Cognitive Toolkit (CNTK), Keras, etc. are among the commonly used libraries for DNN modeling. They differ in training speed, interface, features, flexibility, extensibility, and most critically, implementation of the underlying algorithms. These differences can lead to different performances with the same system configuration on the same data. We demonstrate this result in this paper with two toolkits: Tensorflow[3] and PyTorch.[4] First, the system presented in Table 17.1 and Sect. 17.4.2 of [12] is implemented in both libraries and their performances are compared.

This system contains five convolutional layers followed by two fully connected (FC) layers. Each convolutional layer (except for the first layer) is implemented through a NiN architecture. Max pooling is applied at the end of each convolutional layer. ReLU nonlinearity is used for the FC layers. The use of max-feature-map (MFM) to suppress neurons based on competing activations brings the novelty to this network's architecture. Functionally, it is equivalent to a max pooling operation

---

[3]https://www.tensorflow.org.

[4]https://www.pytorch.org.

across filter outputs. The max pooling is done between two adjacent filters. An element-by-element maximum is applied to obtain one filter output from two input filter activations. Specifically, given an input tensor $x \in \mathbb{R}^{h \times w \times n}$, the MFM operations perform $y_{ij}^k = \max(x_{ij}^k, x_{ij}^{k+\frac{n}{2}})$, where the channel of the input is n, $1 \leq i \leq h$, $1 \leq j \leq w$ [56]. Thus, we use 3D max pooling feature when such an implementation is accessible as is the case with PyTorch. In the TensorFlow implementation, the MFM function was explicitly implemented. The MFM operation allows CNN to perform feature selection of features and at the same time reduce its architecture [56]. Adam algorithm is used for optimization with a learning rate of 0.0001. Dropouts are not applied on the convolutional layer (it is not immediately clear from [12] if dropout was applied on convolutional layers). A dropout rate of 0.7 is used on the first hidden FC layer. Weights are initialized using Xavier initialization. In [12], the CNN is used as a feature extractor. The features obtained from this CNN are passed through a Gaussian classifier. In our experiments, we did not observe any improvement when applying such a classifier. Thus, our implementation does not include such a module. We consider the output of the CNN directly for classification of genuine and spoof recordings.

Next, in order to ensure the reproducibility of our systems, we ensure that our exact training environment is reproducible on an x86_64 Linux machine using Anaconda.[5] In Table 17.1, we list the software used at each stage in our systems. In PyTorch, all optimizations related to multithreading are switched off as they may lead to unexpected inconsistencies during training. This procedure can lead to slower training but will ensure reproducibility of system performance.

17.4 Experiments

17.4.1 Datasets

The accuracy of our countermeasure systems for replay attack detection is evaluated on RedDots Replayed database distributed with the ASVspoof 2017 challenge [50]. The dataset contains ten fixed-phrase sentences and recorded from six recording environments. A detailed information about the corpus derived from the re-recording of the RedDots database is available in [50]. The proposed systems are trained with the training subset only, and the results are reported on the development and evaluation utterances. The statistics of the ASVspoof 2017 is described in Table 17.2.

---

[5]https://anaconda.org/anaconda/python.

Table 17.2 Number of speakers, genuine, and spoof utterances in the training, development, and evaluation sets [50]

Subset	#Speakers	#Utterances	
		Genuine	Spoofed
Training	10	1508	1508
Development	8	760	950
Evaluation	24	1298	12008

17.4.2 Experimental Setup

The input to CNN is a spectrogram of the 16 kHz audio signal. The spectrogram features are extracted from speech frames with frame size 16 ms using a Hamming window function and of 50% overlap. Features with different frequency resolutions are produced by carrying out the STFT analysis with the N-point DFT. That is N set to twice of the specified resolution (i.e., N = 256 for 128 × 400 features). The speech frame is padded with trailing zeros to length N if the number of samples is less than N. The resolution of the spectrogram varies for different setups. This was tuned to obtain the best performance for each setup. For the CNN using MFM, the resolution was 768 × 400. That is, there are 768 frequency bins for each frame of speech and the number of frames for every recording is 400 irrespective of the length of the audio. Audios shorter than 400 frames were data-padded to obtain the necessary length. Audios longer than 400 frames were truncated to 400 frames. The resolution for the proposed ensemble CNN system is 128 × 400. Since the first layer of AlexNet requires input images of size 227 × 227 × 3, we created a custom spectrogram image with 227 frequency bins (STFT analysis is carried out with 454 DFT bins). The number of audio frames is fixed to 227 frames irrespective of the length of the audio. A three-channel input is created by replicating the spectrogram image three times and then stacked together.

The standard CNN consists of three convolutional layers followed by two fully connected (dense) layers. We use a receptive field of 3 × 3 followed by a max pooling operation for every convolutional layer, where ReLU is employed as an activation function. Dropout is applied to every convolutional layer and the subsequent fully connected layer to address overfitting. The network is trained using RMSProp optimizer [57] and cross-entropy as the cost function. Such simple CNN architecture is essentially a reduced version of AlexNet, where we employ three convolutional layers instead of five, and two fully connected layers with a reduced number of neurons (256 instead of 4096).

In order to do classification using SVM from AlexNet, the feature vectors are extracted from the activations on the fully connected layer (the second fully connected layer after convolutional layer). This yields 1 × 4096 feature vectors and then fed to SVM classifier. For the transfer learning process, we replaced the last fully connected layer in the pretrained AlexNet (originally configured for 1000 classes) with a new

Table 17.3 Differences in the variants of CNNs implemented in this work. Note that AlexNet is a pretrained network

Parameter/System	Light CNN	Standard CNN	AlexNet
Number of CNN layers	5	3	5
Training data	ASVspoof 2017	ASVspoof 2017	Pictures
Input feature dimension	128×400 or 768×400	128×400 or 768×400	$227 \times 227 \times 3$
Number of output nodes	2	2	1000
Optimization algorithm	Adam	RMSProp	Gradient descent
Learning rate	0.0001	0.001	0.01
Back-end classifier	None	None	SVM
Ensemble averaging	No	Yes	No
Implementation toolkit	PyTorch, TensorFlow	PyTorch, TensorFlow	C++/CUDA

layer with two output nodes (genuine vs. spoof). The network is then retrained with the maximum number of epochs set to 10 and mini-batch size of 32. We used Matlab implementation of SVM, *fitcsvm*, to train the SVM classifier using a linear kernel function.

In Table 17.3, we summarize the difference in the three CNN architectures presented. Source code is available at https://github.com/idiap/cnn-for-voice-antispoofing.

The performance evaluation metric for replay attack detection is reported in terms of equal error rate (EER) as suggested in the ASVspoof 2017 evaluation plan. First, the output score of the proposed architecture is computed from the estimated posterior probabilities which are transformed into a log-likelihood ratio [27, 58],

$$LLR = \log p(genuine|\mathbf{X}) - \log(1 - p(genuine|\mathbf{X})), \qquad (17.2)$$

where $\mathbf{X}$ is the feature vector and $p(genuine|\mathbf{X})$ is the output posterior with respect to the genuine model. EER corresponds to the threshold θ_{EER} at which the false alarm $P_{fa}(\theta_{EER})$ and the $P_{miss}(\theta_{EER})$ are equal. The $P_{fa}(\theta)$ and $P_{miss}(\theta)$ are defined as

$$P_{fa}(\theta) = \frac{\#\{\text{spoof trials with score} > \theta\}}{\#\{\text{Total spoof trials}\}},$$

$$P_{miss}(\theta) = \frac{\#\{\text{human trials with score} \leq \theta\}}{\#\{\text{Total human trials}\}}.$$

17.5 Results and Discussion

In this section, we present the results of our experiments performed on the different classifiers considered. As mentioned earlier, experiments are conducted on the ASVspoof 2017 dataset. All systems except AlexNet are trained, validated, and tested in the benchmark condition with training data, development data, and evaluation data, respectively. First, the results on development and evaluation data are presented. Next, a visual analysis of the CNN activations is presented that help better analyze the discriminative properties of the spectrogram that aid the classification of genuine and spoof speech.

17.5.1 Results on Evaluation Sets

Table 17.4 presents the results of the study on the evaluation set. First, we will discuss the results when training data alone are used to develop the systems. The use of a pretrained AlexNet network as a stand-alone feature extractor without retraining it, where the extracted features are fed into an SVM, yielded an EER of 24.27% on the evaluation set. This result is better than the baseline system, provided by the ASVspoof 2017 organizer, and shows that developing handcrafted features for anti-spoofing is not trivial. In fact, the network which never sees a spectrogram during training is capable of discovering features that are important for classification. If we consider a CNN as a pattern detector (consisting of a set of learned filters), unknown inputs which have different characteristics will result in different activations, and at the same time, the same detector will output similar activations for inputs which are considered to have similar representations. Hence, a linear classifier such as an SVM can be employed to perform classification using the CNN's extracted feature vectors. Using the transfer learning approach where we replace the output layer with a fully connected layer with two outputs (i.e., spoof and genuine), we achieve an EER of 20.82% (a relative improvement of 14.2%). For a comparison, using deep features extracted from the output of the first fully connected layer of a 3CNN+2FC model (256 neurons) with an SVM classifier, we achieve an EER of 15.11 and 11.21% on the evaluation and development sets, respectively.

Our implementation of a CNN with max-feature-map layers (5CNN+2FC+MFM) performs better than AlexNet with an EER of 17.60%. However, the performance gain obtained is lower compared to the baseline $LCNN_{FFT}$ reported in [12]. One significant reason in the performance gap between the expected and the obtained result is that the training never reached the reported performance on the development set. This may happen when the optimizer is stuck in local minima. This problem is exaggerated when training with TensorFlow where the best error rate on the development data was only 18.7%. Interestingly, the number of epochs required in the case of PyTorch and TensorFlow vary significantly. The former required more than 30 epochs for convergence, while the latter required only 4. Training with more epochs

Table 17.4 EER [%] on the ASVspoof 2017 database

Individual system	Toolkit	Input size	Dev dataset (with **only** train)	Eval dataset (with **only** train)	Eval dataset (with **train+dev**)
Baseline CQCC GMM [50]	Matlab	90 × 1	10.35	30.60	24.77
Baseline $LCNN_{FFT}$ [12]	Not mentioned	864 × 400	4.53	7.37	Unknown
AlexNet (as a feature extractor), SVM + transfer learning	Matlab	227 × 227 × 3	14.23	24.27	23.93
	Matlab	227 × 227 × 3	9.61	20.70	16.43
5CNN + 2FC + MFM	PyTorch	768 × 400	9.0	17.60	16.80
	TensorFlow	768 × 400	18.70	20.36	18.08
3CNN + 2FC, ensemble of 5 models	PyTorch	128 × 400	**8.42**	15.79	15.18
	TensorFlow	128 × 400	**9.93**	**12.60**	**11.10**
	PyTorch	768 × 400	16.71	20.57	18.90
	TensorFlow	768 × 400	16.07	18.16	15.80
3CNN + 2FC (as a feature extractor), SVM	Tensorflow	128 × 400	11.21	15.11	12.04

in TensorFlow did not improve results on the evaluation set. This reiterates our initial argument over the necessity to evaluate a given architecture on multiple toolkits.

The best performance is achieved by performing an ensemble averaging on the output probabilities provided by several models. In this case, five different models in standard CNN configurations described in Sect. 17.4.2 are trained with different weight initializations. The fused system obtained an EER of 12.60% with TensorFlow and 15.79% with PyTorch, thereby benefiting from the larger generalization power compared to the best single-model system on the evaluation data.

The use of pooled data (combined training and development data) improves the performance of the investigated classifiers. However, the performance gain varied across different systems, where substantial improvements are obtained using a 768 × 400 input resolution. The ensemble of five models (TensorFlow 3CNN+2FC) achieved the best result with an EER of 11.10%.

The best performance obtained by a single system, for example, yields an EER of 14.28% on the eval set (a TensorFlow 3CNN+2FC model (128 × 400)). To illustrate the impact of averaging multiple predictions from different models trained with different weight initializations, we calculated the EER for each individual system (a TensorFlow 3CNN+2FC system), and computed the mean and the standard error of

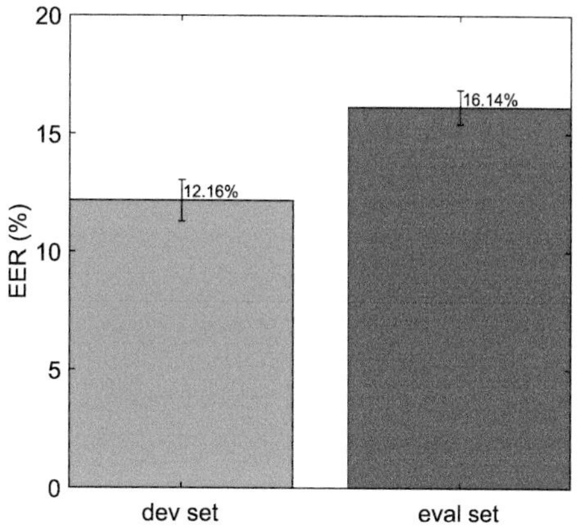

Fig. 17.2 Mean EER (%) with standard error bars on dev and eval sets using TensorFlow 3CNN+2FC models (128×400)

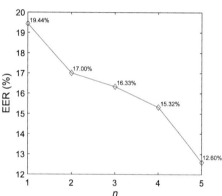

Fig. 17.3 Score fusion of n independently train CNNs on the eval set using TensorFlow 3CNN+2FC models (128×400)

the mean (standard deviation divided by the square root of the number of systems). The average EER including the error bars for both eval and dev sets is illustrated in Fig. 17.2. As mentioned in Sect. 17.2.2, the score-level fusion technique can be employed to improve the system performance over an individual system instead of averaging the EER. Figure 17.3 shows the mean-based fusion of n models on the evaluation set. The figure reveals that training several models and fusing the scores result in improved system performance. However, using a higher resolution input hurts the performance. Using a higher number of frequency bins for computing spectrogram is not effective for discriminating genuine and spoof speech without substantially modifying the time–frequency feature representations. In order to provide further analysis of the results and the interpretation of the model, the next section presents visual explanations to identify common error types and proposes possible research directions for developing robust CNN-based countermeasures.

17.5.2   *Visualization*

Despite the potential of convolutional neural networks in classifying spoofed and genuine speech, it still offers little insight into its operation and the behavior of this complex model. When the CNN is trained for classification, it gradually transforms the image input (using sets of convolution and pooling operations) into a high-level representation of its input. In a typical CNN architecture, the fully connected layer (a multilayer perceptron with the softmax as the activation function) is placed after the convolutional layers for classifying the input image into multiple possible outputs (classes) based on the training labels. Hence, a CNN can be regarded as a feature extractor, which extracts useful features as it learns the parts that are recognizable between classes.

To gain insight into what the network has learnt, we visualized the features from the trained model (using a TensorFlow implementation of 3Conv+2FC architecture) once the training was complete. Specifically, we extracted 1×256 feature vectors at the output of the first convolutional layer (after ReLU nonlinearity) and embed this data into a two-dimensional space using t-SNE algorithm [59]. The t-SNE plots are presented in Fig. 17.4. The high-dimensional features (before t-SNE are applied) are shown along with their corresponding spectrograms for genuine and spoof classes in Figs. 17.5 and 17.6, respectively. Similar patterns can be observed from feature vectors in the same class.

In our case, each point in the scatter plot of Fig. 17.4 (left) represents an evaluation utterance where nearby points represent similar objects in the high-dimensional space and vice versa. While the nice separation between spoof (light-green) and genuine (red) can be seen, they are not totally distinguishable. The lack of explicit "cluster centers" indicates that the problem of spoofing detection is challenging and there is still room for algorithm improvement to learn more complex nonlinear representations. To gain an insight of the common error types and for the sake of feature visualization, we set a threshold, θ_{EER}, to determine the estimated labels. A comparison

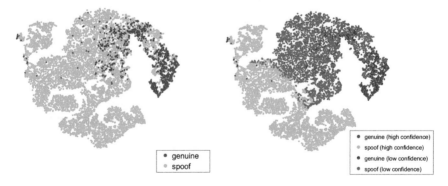

Fig. 17.4 Features extracted from a TensorFlow 3CNN+2FC model (128×400), visualized in a two-dimensional space

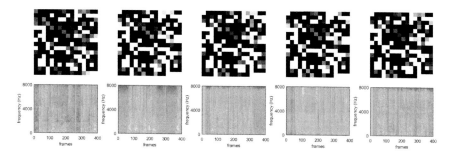

Fig. 17.5 Examples of feature representation in latent space for genuine class and their corresponding spectrograms. Features extracted from a TensorFlow 3CNN+2FC model (128 × 400)

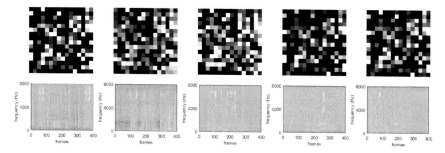

Fig. 17.6 Examples of feature representation in latent space for spoof class and their corresponding spectrograms. Features extracted from a TensorFlow 3CNN+2FC model (128 × 400)

between Fig. 17.4 (left) and (right) reveals that the spoof speech is likely to be predicted as genuine. From these low-dimensional visual representations, we can then select a few samples in order to understand why the CNN is struggling to classify some spectrogram images.

Several methods have been proposed to identify patterns or visualize the impact of the particular regions that are linked to a particular CNN's response [54, 60, 61]. Typically, a heatmap is constructed that allows for visualizing the "salient" regions (e.g., the impact of contribution of a single pixel to the CNN inference result for a given input image). In this paper, we adopted a gradient-weighted class activation mapping (Grad-CAM) to visualize the activation in our trained model [60]. Although this technique was originally developed for the image recognition task, the same principle can be applied to the spectrogram image. The Grad-CAM[6] computed the gradient of the predicted score for a particular class with respect to feature maps output of a final convolutional layer in order to highlight the importance of feature maps for a target class. Figures 17.7 and 17.8 show the Grad-CAM visualizations, which highlight the important regions in the genuine and spoof class, respectively. In the genuine class, low-frequency components seem to contribute more to classification. While in the spoof class, the CNN seems to highlight low-energy speech regions

---

[6]https://github.com/insikk/Grad-CAM-tensorflow.

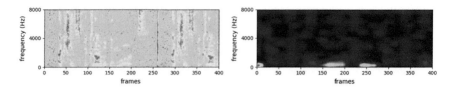

Fig. 17.7 Original spectrogram for genuine class (E_1000010.wav) (left) and the Grad-CAM visualization (right), using a TensorFlow 3CNN+2FC model (128 × 400). Note that the red regions correspond to high score for class

during speech activity in addition to a few low frequency components. In order to see what mistakes a CNN is making, especially when detecting the spoof class as genuine, we visualize the misclassified examples for both the spoof and genuine target using Grad-CAM. As shown in Fig. 17.9, it is apparent that the errors were made when the low-energy speech regions are considered to be important (which occurred in the low-frequency range); rather, the CNN should put more emphasis on high-frequency components to make the correct predictions. However, it is not possible to exactly pinpoint the reasons for misclassification by looking at the spectrogram, for example, removing the silence region has been shown to hurt the system performance [52]. In this case, we may be able to generalize that information containing high-frequency components are important to reduce the classification errors. To test this hypothesis, we applied a high-pass filter to retain only high-frequency components (≥ 4 kHz) on all utterances and trained CNN models. The reason behind this is that the channel artifacts are shown to be evident outside the voice band (300–3400 Hz) [62]. This operation obtained an EER of 22.8% on the evaluation set. The observation in Fig. 17.7 shows that some low-frequency components are important for predicting genuine class, hence, applying the proposed modification to all evaluation files will hurt the performance. Therefore, in the second experiment, we applied a low-pass filter to retain only low-frequency components (≤ 4 kHz), and trained CNN models. This low-pass filtering operation obtained an EER of 26.2% which is worse than performing high-pass filtering. This finding indicates that high-frequency components are more important than low-frequency regions for replay attack detection. The combined scores via mean fusion (ensemble of five models, two low-pass, and three high-pass models) yielded a substantial performance gain with an EER of 13.3% on the evaluation set. Although slightly worse performance is achieved compared to models trained using full bandwidth spectrogram (an EER of 12.6% in Table 17.4), this finding suggests that one should consider detailed feature representations not only in high- but also low-frequency regions when developing the countermeasures.

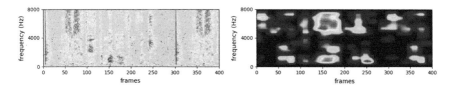

Fig. 17.8 Original spectrogram for spoof class (E_1000004.wav) (left) and the Grad-CAM visualization (right), using a TensorFlow 3CNN+2FC model (128 × 400). Note that the red regions correspond to high score for class

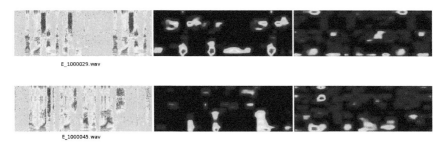

Fig. 17.9 Examples of misclassified spoof utterances with the spectrogram images shown on the left. The Grad-CAM visualizations using a TensorFlow 3CNN+2FC model (128 × 400) are shown for spoof (middle) and genuine (right) class

17.6 Conclusion

In this study, we investigated the use of CNNs architectures for replay attack detection. A common evaluation platform such as ASVspoof challenges means that the results can be compared reliably and allow possible directions of fruitful research. With a possible interlink between the feature engineering and the back-end classifiers, the use of pretrained AlexNet as a feature extractor can serve as a reliable baseline where the choice of the features and the classifier is reasonable. This motivates us to solely focus on developing robust classifiers rather than following the conventional feature-based methods, where preferably the features should be learned from the training data.

In this paper, we proposed a simple architecture of CNNs, which is a reduced version of AlexNet. Comparison of results from three different CNN architectures showed that an ensemble of standard CNNs could achieve better results compared to a sophisticated single-model classifier such as light CNN with MFM activations. Although our results appear to be worse than the best reported results for ASVspoof 2017, this could not be verified since we are not able to reproduce their results. This highlights several difficulties when training end-to-end CNN models such as model generalization, parameter optimization, and the reproducibility of reported research results. Specifically, we observed that same models (i.e., trained using the same input resolution but only different on the toolkit implementation) which perform well on the development set failed to generalize on the evaluation set. This indicates that CNNs overfit on the training data and urge us to conclude that as more diverse data becomes available, for example, by combining training and development data, the performance will improve. Further, without the knowledge of the exact training regimes and environments, it is difficult to reproduce previous works. Therefore, in order to ensure reproducibility of our work, we implemented our proposed methods using open-source deep learning frameworks: TensorFlow and PyTorch, and have made the source codes publicly available.

Our in-depth investigation through Grad-CAM visualization of the CNN reveals that emphasis on high-frequency components for spoofing utterances helps reduce an EER, and could serve as a valuable source of information for developing robust anti-spoofing systems. For real applications, spoofing detection systems should also work on noisy and reverberant environments, and to date only a few studies have been conducted on the development of robust countermeasures when speech is degraded due to noise, the speech acquisition system, and the channel artifacts. Adaptation techniques may also be devised to detect playback speech in a new replay session. In the future, our research will focus on employing advanced classifiers such as RNN for modeling the temporal dependency of speech and generative adversarial network (GAN) for generating synthetic spoofed-like data as a data augmentation method to address the overfitting problems.

Acknowledgements Computational (and/or data visualization) resources and services used in this work were provided by the HPC and Research Support Group, Queensland University of Technology, Brisbane, Australia. This project was supported in part by an Australian Research Council Linkage grant LP 130100110.

References

1. Hautamäki RS et al (2015) Automatic versus human speaker verification: the case of voice mimicry. Speech Commun 72:13–31
2. Toda T, Black AW, Tokuda K (2007) Voice conversion based on maximum-likelihood estimation of spectral parameter trajectory. IEEE Trans Audio, Speech, Lang Process 15(8):2222–2235
3. Erro D, Polyakova T, Moreno A (2008) On combining statistical methods and frequency warping for high-quality voice conversion. In: Proceedings of IEEE international conference on acoustic, speech, and signal processing, pp 4665–4668
4. Masuko T, Tokuda K, Kobayashi T (2008) Imposture using synthetic speech against speaker verification based on spectrum and pitch. In: Proceedings of international conference on spoken language processing, pp 302–305
5. Satoh T et al (2001) A robust speaker verification system against imposture using an HMM-based speech synthesis system. In: Proceedings of interspeech, pp 759–762
6. Zheng TF, Li L (2017) Robustness-related issues in speaker recognition. Springer, Singapore
7. Wu Z et al (2015) ASVspoof 2015: the first automatic speaker verification spoofing and countermeasures challenge. In: Proceedings of interspeech, pp 2037–2041
8. ISO/IEC JTC 1/SC 37 Biometrics: ISO/IEC 30107-1:2016, Information technology - Biometrics presentation attack detection - part 1: Framework. ISO/IEC Information Technology Task Force (ITTF) (2016)

9. Wu Z et al (2014) A study on replay attack and anti-spoofing for text-dependent speaker verification. In: Proceedings of asia-pacific signal and information processing association, annual summit and conference (APSIPA), pp 1–5
10. Gałka J, Grzywacz M, Samborski R (2015) Playback attack detection for text-dependent speaker verification over telephone channels. Speech Commun 67:143–153
11. Janicki A, Alegre F, Evans N (2016) An assessment of automatic speaker verification vulnerabilities to replay spoofing attacks. Sec Commun Netw 9:3030–3044
12. Lavrentyeva G et al (2017) Audio replay attack detection with deep learning frameworks. In: Proceedings of interspeech, pp 82–86
13. Chen Z et al (2017) ResNet and model fusion for automatic spoofing detection. In: Proceedings of interspeech, pp 102–106
14. Cai W et al (2017) Countermeasures for automatic speaker verification replay spoofing attack: on data augmentation, feature representation, classification and fusion. In: Proceedings of intespeech, pp 17–21
15. Hinton GE et al (2012) Improving neural networks by preventing co-adaption of feature detectors. arXiv:1207.0580
16. Krizhevsky A, Sutskever I, Hinton GE (2012) ImageNet classification with deep convolutional neural networks. Adv Neural Inf Process Syst 2:1097–1105
17. Simonyan K, Zisserman A (2014) Very deep convolutional networks for large-scale image recognition. arXiv:1409.1556
18. Russakovsky O et al (2015) ImageNet large scale visual recognition challenge. Int J Comput Vis 115(3):211–252
19. Abadi M et al (2016) Tensorflow: large-scale machine learning on heterogeneous distributed systems. arXiv:1603.04467
20. Paszke A et al (2017) Automatic differentiation in PyTorch. In: 31st conference on neural information processing systems
21. Scheirer E, Slaney M (1997) Construction and evaluation of a robust multifeature speech/music discriminator. In: IEEE international conference on acoustics, speech, and signal processing (ICASSP-97)vol 2, pp 1331–1334. https://doi.org/10.1109/ICASSP.1997.596192
22. Paliwal KK (1998) Spectral subband centroid features for speech recognition. Proc IEEE Int Conf Acoustic, Speech Signal Process 2:617–620
23. Youngberg J, Boll S (1978) Constant-Q signal analysis and synthesis. In: Proceedings of IEEE international conference on acoustic, speech, and signal processing, pp 375–378
24. Mallat S (2008) A wavelet tour of signal processing, 3rd edn. The sparse way. Academic press, New York
25. Liu Y, Tian Y, He L, Liu J, Johnson MT (2015) Simultaneous utilization of spectral magnitude and phase information to extract supervectors for speaker verification anti-spoofing. sign (gp-gc) 2:1
26. Sahidullah Md et al (2016) Integrated spoofing countermeasures and automatic speaker verification: an evaluation on ASVspoof 2015. In: Proceedings of interspeech, pp 1700–1704
27. Villalba J et al (2015) Spoofing detection with DNN and one-class SVM for the ASVspoof 2015 challenge. In: Proceedings of interspeech, pp 2067–2071
28. Chakroborty S, Roy A, Saha G (2007) Improved close set text-independent speaker identification by combining MFCC with evidence from flipped filter banks. Int J Signal Process 4(2):114–121
29. Xiao X, Tian X, Du S, Xu H, Chng ES, Li H (2015) Spoofing speech detection using high dimensional magnitude and phase features: The NTU approach for ASVspoof 2015 challenge. In: Proceedings of interspeech
30. Saratxaga I (2016) Synthetic speech detection using phase information. Speech Commun 81:30–41
31. Wu Z (2016) Anti-spoofing for text-independent speaker verification: An initial database, comparison of countermeasures, and human performance. IEEE/ACM Trans Audio, Speech Lang Process 24:768–783

32. Chen N, Qian Y, Dinkel H, Chen B, Yu K (2015) Robust deep feature for spoofing detection-the SJTU system for ASVspoof 2015 challenge. In: Proceedings of interspeech
33. Korshunov P, Marcel S, Muckenhirn H, Gonçalves AR, Mello AGS, Violato RPV, Simoes FO, Neto MU, de Assis Angeloni M, Stuchi JA, Dinkel H, Chen N, Qian Y, Paul D, Saha G, Sahidullah M (2016) Overview of BTAS 2016 speaker anti-spoofing competition. In: IEEE international conference on biometrics theory, applications and systems (BTAS)
34. Qian Y (2017) Deep feature engineering for noise robust spoofing detection. IEEE/ACM Trans Audio, Speech Lang Process 25(10):1942–1955
35. Dinkel H et al (2017) End-to-end spoofing detection with raw waveform CLDNNS. In: Proceedings of IEEE international conference on acoustic, speech, and signal processing, pp 4860–4864
36. Zhang C, Yu C, Hansen JHL (2017) An investigation of deep-learning frameworks for speaker verification antispoofing. IEEE J Sel Top Signal Process 11(4):684–694. https://doi.org/10.1109/JSTSP.2016.2647199
37. Alam MJ et al (2016) Spoofing detection on the ASVspoof 2015 challenge corpus employing deep neural networks. In: Proceedings of odessey, pp 270–276
38. Yu H et al (2017) Spoofing detection in automatic speaker verification systems using DNN classifiers and dynamic acoustic features. In: IEEE transactions on neural networks and learning systems, pp 1–12
39. Yu H et al (2017) DNN filter bank cepstral coefficients for spoofing detection. IEEE Access 5:4779–4787
40. Qian Y, Chen N, Yu K (2016) Deep features for automatic spoofing detection. Speech Commun 85(C):43–52. https://doi.org/10.1016/j.specom.2016.10.007
41. Korshunov P et al (2018) On the use of convolutional neural network for speech presentation attack detection. In: Proceedings of IEEE international conference on identity, security, and behavior analysis
42. He K, Zhang X, Ren S, Sun J (2016) Deep residual learning for image recognition. In: CVPR, pp 770–778
43. Veit A et al (2016) Residual networks behave like ensembles of relatively shallow networks. Adv Neural Inf Process Syst 550–558
44. Muckenhirn H, Magimai-Doss M, Marcel S (2017) End-to-end convolutional neural network-based voice presentation attack detection. In: Proceedings of international joint conference on biometrics
45. Chen Z et al (2018) Recurrent neural networks for automatic replay spoofing attack detection. In: Proceedings of IEEE international conference on acoustic, speech, and signal processing
46. Nandakumar K (2008) Likelihood ratio-based biometric score fusion. IEEE Trans Pattern Anal Mach Intell 30(2):342–347
47. Todisco M, Delgado H, Evans N (2016) Articulation rate filtering of CQCC features for automatic speaker verification. In: Proceeding of interspeech, pp 3628–3632
48. Todisco M, Delgado H, Evans N (2016) A new feature for automatic speaker verification anti-spoofing: Constant Q cepstral coefficients. In: Proceedings of odessey, pp 283–290
49. Kinnunen T et al (2016) Utterance verification for text-dependent speaker recognition: a comparative assessment using the RedDots corpus. In: Proceedings of interspeech, pp 430–434
50. Kinnunen T et al (2017) The ASVspoof 2017 challenge: Assesing the limits of replay spoofing attack detection. In: Proceedings of interspeech, pp 2–6
51. Wang X, Takaki S, Yamagishi J (2017) An RNN-based quantized F0 model with multi-tier feedback links for text-to-speech synthesis. In: Proceedings of interspeech, pp 1059–1063
52. Font R, Espín JM, Cano MJ (2017) Experimental analysis of features for replay attack detection-results on the ASVspoof 2017 challenge. In: Proceedings of interspeech, pp 7–11
53. Sermanet P et al (2014) Overfeat: integrated recognition, localization and detection using convolutional networks. In: Proceedings of international conference on learning representations
54. Zeiler MD, Fergus R (2014) Visualizing and understanding convolutional networks. In: European conference on computer vision, pp 818–833
55. Lin M, Chen Q, Yan S (2014) Network in network. In: Proceedings of international conference on learning representations

56. Wu X, He R, Sun Z (2015) A lightened CNN for deep face representation. arXiv:1511.02683v1
57. Tieleman T, Hinton G (2012) Lecture 6.5—RMSProp: Divide the gradient by a running average of its recent magnitude. COURSERA: neural networks for machine learning
58. Brümmer N, du Preez J (2006) Application-independent evaluation of speaker detection. Comput Speech Lang 20:230–275
59. van der Maaten LJP, Hinton GE (2008) Visualizing high-dimensional data using t-SNE. J Mach Learn Res 9:2579–2605
60. Selvaraju RR et al (2017) Grad-CAM: Visual explanations from deep networks via gradient-based localization. In: Proceedings of IEEE international conference on computer vision, pp 618–626
61. Samek W (2017) Evaluating the visualization of what a deep neural network has learned. IEEE Trans Neural Netw Learn Syst 28(11):2660–2673
62. Nagarsheth P et al (2017) Replay attack detection using DNN for channel discrimination. In: Proceedings of interspeech, pp 97–101

Part V
Other Biometrics

Chapter 18
An Introduction to Vein Presentation Attacks and Detection

André Anjos, Pedro Tome and Sébastien Marcel

Abstract   The domain of presentation attacks (PAs), including vulnerability studies and detection (PAD), remains very much unexplored by available scientific literature in biometric vein recognition. Contrary to other modalities that use visual spectral sensors for capturing biometric samples, vein biometrics is typically implemented with near-infrared imaging. The use of invisible light spectra challenges the creation of instruments, but does not render it impossible. In this chapter, we provide an overview of current landscape for PA manufacturing in possible attack vectors for vein recognition, describe existing public databases and baseline techniques to counter such attacks. The reader will also find material to reproduce experiments and findings for finger vein recognition systems. We provide this material with the hope that it will be extended to other vein recognition systems and improved in time.

18.1   Introduction

Even if other modalities gain in market acceptance and find more and more applications year after year, fingerprints probably remain one of the most commonly accepted biometric modes being present in a variety of applications around the world including smartphones, physical access and immigration control solutions. In fingerprint recognition, an image is captured from the user finger and matched against

---

A. Anjos (✉) · S. Marcel
Idiap Research Institute, Martigny, Switzerland
e-mail: andre.anjos@idiap.ch

S. Marcel
e-mail: marcel@idiap.ch

P. Tome
Universidad Autonoma de Madrid, Madrid, Spain
e-mail: pedro.tome@inv.uam.es

© Springer Nature Switzerland AG 2019   419
S. Marcel et al. (eds.), *Handbook of Biometric Anti-Spoofing*,
Advances in Computer Vision and Pattern Recognition,
https://doi.org/10.1007/978-3-319-92627-8_18

the stored biometric reference for authentication or identification. Because finger-prints are external traits belonging to individuals, they make very easy attack vectors. Latent prints can be lifted from objects we interact day to day. Attack instruments may be then manufactured without consent following one of the innumerous recipes currently available on the Internet. This is unfortunately also true for most biometric modalities relying on traits which can be easily captured from distance such as face, voice or gait.

Behind fingerprints, buried under the skin, human vein networks can also be used to verify the identity of a person with high performance and security [1]. Vein recog-nition is not limited to fingers   uses of the technology also include recognition using wrist and palm vein networks. The idea of employing the vascular patterns embodied in human fingers for authentication purposes originally comes from Hitachi Research Laboratory at the end of the nineties [2], which makes it a fairly new topic in bio-metrics. While studying the human brain's blood system using imaging techniques, researchers discovered that near-infrared (NIR) light is absorbed significantly more by haemoglobin transporting carbon dioxide (in veins) than by the surrounding tis-sues. Furthermore, transmitting NIR light around 850 nm through the finger was

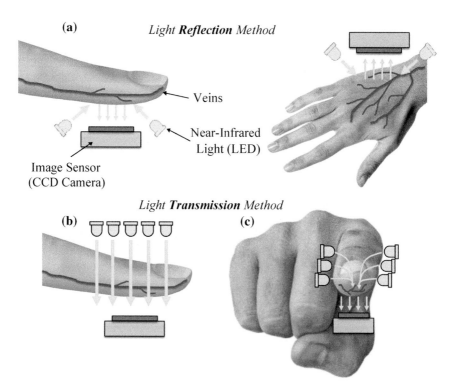

Fig. 18.1 Illustration of the most used capturing techniques in vein network imaging: and (i) reflection-based and (ii) transmission-based, divided on top (**b**) and side (**c**) illumination

found to optimize imaging of vein patterns [3]. The first patent application to finger vein recognition was soon filed in 2001 and granted by 2004 [4]. Commercial physical access control devices based on vein recognition were made available in 2002, followed by logical access solutions for automatic teller machines in 2005 [2]. Presently, the technology is widely used in the financial sector in Japan, China and Poland, where it proved to be accurate. More recently, the technology has been introduced in hospitals in Turkey for patient authentication.

Figure 18.1 shows setups for acquiring vein network imagery—here demonstrated on the finger region. Systems work in one of two models: reflection or transmission. In light reflection-based systems, the region from which the vein network is to be sampled is illuminated by NIR light which is reflected by the object and captured by a band-selective image sensor. In reflection-based systems, both the illumination and the capturing lie in the same side of the setup. These settings allow the creation

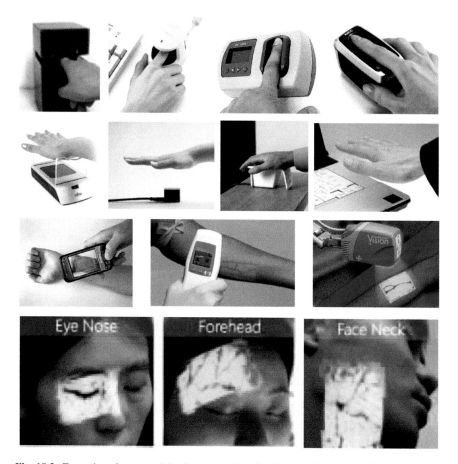

Fig. 18.2 Examples of commercial vein sensors, first line finger vein sensors, palm vein sensors, and the last two lines show medial vein sensors

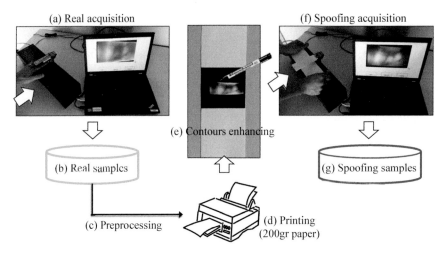

Fig. 18.3 Illustration of the recipe and attack presentation from a real finger vein sample (**a**) available from a *bona fide* dataset (**b**). The image is first preprocessed (**c**), and printed (**d**) on high-quality white paper with a grammage of 200 g/m^2 through a commercially available laser printer. The contours of the finger are enhanced using a black ink whiteboard marker (**e**) and the attack instrument is presented to the sensor (**f**) and a sample acquired (**g**)

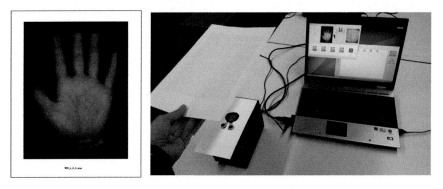

Fig. 18.4 The generated palm vein PAI (right) and its acquisition (left) using a reflection-based sensor

of very compact sensors. In light transmission-based systems, the sensor and illumination are on opposed sides of the setup. Light diffuses through the object with the vein networks to be captured. Transmission systems can be used for capturing finger veins as the finger is typically thin enough to support this mode of operation. Variants of transmission-based systems using side illumination exist, but are not mainstream. Capturing wrist or palm veins in transmission mode is often impractical and seldomly used. For illustration, Fig. 18.5 (top-left) shows examples of finger vein images captured in transmission mode, and Fig. 18.6a shows examples of palm vein imagery capture in reflection mode.

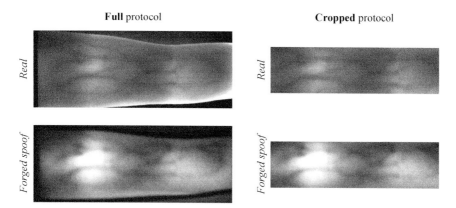

Fig. 18.5 Samples of the two *full* and *cropped* protocols available in the database to study the presentation attacks detection methods

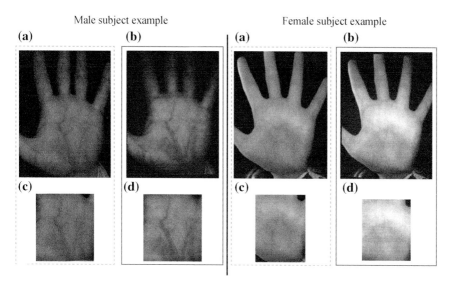

Fig. 18.6 Images of *bona fide* samples (**a** and **c**, inside of a dashed green line) from the original database and their correspond recaptured samples (**b** and **d**, inside of a solid red line). First row **a** and **b** shows the RAW images saved by the sensor and second row **c** and **d** shows the ROI-1 regions automatically extracted by the sensor during the acquisition process

The security advantages of vein networks to other biometrics lie in a couple of factors: (i) vein networks do not leave latent marks nor can be captured without dedicated equipment, at close distance, and (ii) technologies for extracting vein network images from a person's hand or finger require the use of non-visible light (typically near-infrared) which challenges the creation of efficient attack instruments. Therefore, vein networks are generally perceived as difficult to forge (Fig. 18.2).

However, it remains possible to obtain images from a person's vein patterns, and, therefore, interest in manufacturing efficient attack instruments for this biometric modality are bound only by the amount of resources being protected. In this chapter, we present two major contributions: (i) we demonstrate the vulnerability if current vein recognition systems to presentation attacks through a simple set of recipes and methodology, and (ii) based on a recently published dataset with presentation attacks, how to possibly counter such vectors to create more robust systems. We study finger vein recognition systems, though most of what we present can be easily generalized to other related modes for palm and wrist-based vein recognition.

This chapter is organized as follows: Sect. 18.2 introduces the only two publicly available datasets for vulnerability analysis and creation of countermeasures to presentation attacks in finger and palm vein recognition. We explain how to create attacks for NIR-based vein recognition systems with examples. Section 18.3 discusses the current state-of-the-art in presentation attack detection for both finger vein and palm vein recognition with results from baselines [5, 6] and the first competition on countermeasures to finger vein spoofing attacks [7]. For reproducibility purposes, we provide scores and analysis routines for this competition and baselines. Section 18.4 shows more in-depth reproducible results on the vulnerability analysis and baselines for presentation attack detection (PAD) in finger vein recognition systems. In this section, we introduce an extensible package for work in PAD for the vein mode. Section 18.5 concludes this chapter.

18.2 Attack Recipes and Datasets

There are, currently, two public datasets for presentation attack vulnerability analysis and detection available in literature, both produced and distributed by the Idiap Research Institute in Switzerland. The VERA Spoofing Finger vein Database[1] contains attacks and *bona fide* samples of finger vein images captured with a transmission-based system [5]. By itself, this dataset can be used to train and evaluate PAD methods for finger vein recognition systems. When combined with the VERA Finger vein Database,[2] it is possible to study the vulnerability of recognition systems to presentation attacks. The VERA Spoofing Palm vein Database[3] contains attacks and *bona fide* samples of palm vein images captured with a reflection-based system [6]. This dataset can be used to develop and evaluate PAD methods for palm vein recognition systems. When combined with the VERA Palm vein Database,[4] one can study the vulnerability of palm vein recognition systems. Before introducing and detailing these datasets, we start by defining recipes for producing attacks to both

---

[1] https://www.idiap.ch/dataset/fvspoofingattack.

[2] https://www.idiap.ch/dataset/vera-fingervein.

[3] https://www.idiap.ch/dataset/vera-spoofingpalmvein.

[4] https://www.idiap.ch/dataset/vera-palmvein.

finger vein and palm vein recognition systems. Section 18.4 demonstrates that current state-of-the-art recognition systems are vulnerable to such attack instruments.

18.2.1 Attack Recipes

Building attacks for vein recognition systems are possible if one has access to images of vein networks of the identities to be attacked. Tome and others at [5, 6] suggest that, if it is the case, then presentation attack instruments may be easily created using laser/toner prints of objects (human fingers and palms). Starting from the images of existing datasets, they created a set of effective attacks to vein recognition systems. The motivation behind using printed images is based on its simplicity and the fact it does not require prior knowledge about the recognition system. Print attacks have been proved effective in the context of other biometric modalities such as face [8] or Iris [9].

The recipe for the generation of printed attacks is typically divided into four steps:

1. Original images from NIR sensors are preprocessed to improve their quality;
2. Preprocessed images are then printed on a piece of paper using a commercially available laser printer;
3. Contours are enhanced;
4. The printed image is presented to the sensor

In the case of finger veins, Tome and others [5] tested various configurations of paper and preprocessing to find an ideal combination of that works best:

1. Original images are preprocessed using histogram equalization and a Gaussian filtering of 10-pixel window. Before printing, proper rescaling (180×68 pixels) is performed so the printed fingers have the same size as the real ones. Furthermore, a background of black pixels is added around the finger to mask the outside regions of the finger during the acquisition, and finally, the image is flipped-up to handle the reflection of the internal mirror of the sensor.
2. Images are printed in a high-quality white paper (grammage: $200 \, g/m^2$). The thickness of the paper plays an important role in transmission systems since the amount of light that will pass through the paper should mimic that of a real finger;
3. The contours of the finger are enhanced using a black ink whiteboard marker;
4. Printed images are presented to the acquisition sensor at 2 cm of distance to the sensor as shown in Fig. 18.3f.

Palm vein presentation attack instruments (PAI) may be created in a similar manner as realized in [6]: palm vein images are reshaped to match *bona fide* comparable sizes and a padding array of black pixels is applied to the *bona fide* palm vein images to emulate the black background. Images are printed on a regular grammage white paper ($80 \, g/m^2$) with a commercial laser printer. Because palm vein attack instruments are usually meant for reflective-light systems, the grammage of the paper is less important than on transmission-based systems. Figure 18.4 (left) shows an example of the final printed image while the right part shows an example of the PAI acquisition process and how an attacker may position the printed palm. Because palm veins were acquired using a reflection-based sensor, a piece of white paper with grammage $100 \, g/m^2$ is applied to cover the NIR illumination from the sensor, in order to reduce the impact from illumination. This solution was adopted to improve the chances of successfully attacking even though authors note that without this resource, attacks are still successful to various recognition baselines.

Complete datasets for vulnerability analysis and presentation attack detection training and evaluation are discussed next. Images in these datasets were created using the recipes above.

18.2.2 Finger Vein Datasets

Motivated by the scarce literature in the area, the Idiap Research Institute published the first finger vein presentation attack database [5] in 2014. The dataset is composed of 880 image samples stored in PNG format, which are split into three distinct subsets for training (240), fine-tuning (240), and testing (400) PAD methods to finger vein recognition. The splits do not mix subjects. Samples that are on the training set come from a specific set of fingers which are not available in any other set on the database. The same holds true for the fine-tuning (development) and test (evaluation) sets. The training set should be used for training PAD statistical models, the development set to validate and fine-tune such models, and, finally, the test set should only be used to report the final PAD performance. The goal of this protocol is to evaluate the (binary classification) performance of countermeasures to presentation attacks.

The dataset contains two different sets of images following the beforementioned protocol: *full* printed images and *cropped* images (see Fig. 18.5). The so-called 'full' protocol contains untreated raw samples as perceived by the finger vein sensor. Such images include finger borders and have a size of 665 by 250 pixels. The 'cropped' protocol provides pre-cropped images of the subject fingers and discards the border, making them harder to discriminate when compared to images in the 'full' protocol. The intent behind this strategy is to be able to evaluate PAD methods with respect to their generalization capabilities when considering only parts of the images that would be effectively used to recognize individuals.

This dataset, when combined with the VERA Finger vein Database, can be used to study the vulnerability of recognition systems to presentation attacks. Such an approach is of high importance, as it enables one to assess how effective the attacks are in deceiving a finger vein recognition system and whether PA detection is indeed necessary. Section 18.4 contains reproducible results from the vulnerability analysis assessment of finger vein systems. In summary, more than 80% of the attacks in this database successfully bypass baseline algorithms.

As presented in Table 18.1, other datasets are mentioned in the literature [10, 11] but, up to date, none of them was made publicly available by authors.

18.2.3 Palm Vein Dataset

The only palm vein dataset for the analysis of vulnerability and development of presentation attack detection methods was published by the Idiap Research Institute in 2015 [6]. This dataset contains recaptured images from the first 50 subjects of the public VERA Palm vein Database. There are a total of 1000 recaptured palm images from both hands of all subjects, matching the original images available on the *bona fide* dataset.

The VERA Palm vein Database originally consists of 2,200 images depicting human palm vein patterns. Palm vein images were recorded from 110 volunteers for both left and right hands. For each subject, images were obtained in two sessions of five pictures each per hand. Palm vein images were acquired by the contactless palm vein prototype sensor comprised of an ImagingSource camera, a Sony ICX618 sensor and an infrared illumination of LEDs using a wavelength of 940 nm. The distance between the user hand and the camera lens is measured by an HC-SR04 ultrasound sensor and a led signal that indicates the user the correct position of the hand for the acquisition.

The *bona fide* palm vein samples have a resolution of 480×680 pixels and are stored in PNG format. The database is divided into two subdatasets: RAW and ROI-1 data. The *raw* folder corresponds to the full palm vein image and *roi* folder contains the region of interest (palm vein region) obtained automatically by the sensor during the acquisition process (see Fig. 18.6).

In order to test for the vulnerability of palm vein recognition systems to attacks in this dataset, we selected a baseline algorithm from the state-of-the-art in this field of research. The overall processing is divided into three stages: (i) segmentation and normalization, (ii) feature extraction and (iii) matching. In the segmentation process, the hand contour is localized by a binarization from greyscale palm vein images. Then, the hand landmarks (peaks and valleys) are extracted using the radial distance function (RDF) between the reference point (generally the starting of the wrist) and the contour points extracted [12, 13]. The palm region is extracted as a square region based on the located hand landmarks, and a geometric normalization (scaling and rotation) on the extracted palm vein region is performed. Finally, the palm veins are enhanced by using circular Gabor filters (CGF) [14]. Once palm vein of the region of interest (ROI-2) is extracted and normalized, local binary patterns (LBPs) are extracted as features [15] and the histogram intersection metric [16] is adopted as a similarity measure to compute the scores. The final score of the system per user is computed by the average of the scores of all enrollment samples for that user.

Since there are two different preprocessing configurations (none and CGF) and two regions of interest analysed (ROI-1 given by the database, and ROI-2), this finally leads to four different systems that can be evaluated for vulnerability. Figure 18.7 shows the results of vulnerability analysis for these four baseline variants using the VERA Palm vein Spoofing Database when combined with its *bona fide* sibling. In all cases, a large number of attacks bypass a naively tuned recognition system. In some configurations, the success rate is over 70%.

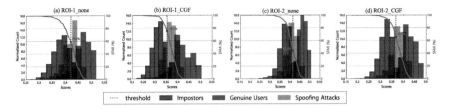

Fig. 18.7 Score distributions of palm vein recognition systems on the database for zero-effort impostors (red), genuine subjects (blue) and presentation attacks (grey). The solid green curve shows the incorrect attack presentation match rate (IAPMR) as a function of the recognition threshold. The dashed green line shows the equal error threshold (EER) for a system tuned naively, for a setup without presentation attacks

Up to date, palm vein presentation attack detection remains a yet-to-be-explored research domain, with no literature published. We focus our attention in finger vein PAD next, which is far more developed presently.

18.3 State of the Art in Finger Vein PAD

Great progress has been made in the field of biometric presentation attack detection (PAD) in recent years. But due to scarce number of databases, the potential attacks on finger vein systems focus on printed attacks. To the best of our knowledge, the first attempt to attack finger vein was conducted by Tsutomu Matsumoto and others, by using a overhead projector film on which finger vein images were printed as synthetically produced artefacts that mimic real finger veins. But it was not until years later, as a consequence of the development of non-commercial acquisition sensors [17–20], when a presentation attack that could successfully bypass a finger vein recognition system for a different number of identities was demonstrated [5], with an imposter attack presentation match rate (IAPMR) as high as 86%.

The current technology of finger vein sensors is based on 2D image acquisition, 3D finger vein sensors are still developing and they will bring the opportunity to ensure the robustness of this finger vein technology against presentation attacks. Hence, based on the 2D sensors, the PAD methods can be divided into two categories, i.e. texture-based [7, 10, 11, 21], where the texture analysis can be exploited to detect fake images and liveness-based [22, 23], where the motion in the image is studied for the same purpose.

Texture-based methods analyse differences in image quality between real and forged veins, which are primarily reflected in the texture resolution and noise level of the images. Table 18.1 summarizes the main PAD texture-based methods described in the literature. On the other hand, liveness-based methods determine whether a finger vein is real by detecting evidence of the liveness or vital signs of the finger, such as the blood movement across veins. These methods require more advanced sensor of image/video acquisition but its benefit can be more accurate and reliable than texture-based methods.

Following the literature [7] and available databases, it was demonstrated that texture-based methods perform well in preventing printed attacks because the forged fake finger vein images contain print artefacts, blurring and other noise originating from the printing, processing and reacquisition process, thus resulting in slightly different texture features between real and fake images.

As can be extracted from literature [5], one of the main problems when finger vein images are forged from a printed sample is the NIR illumination in the reacquisition process. The NIR light reflects in a different way on a paper than the human finger, generating acquisition noise artefacts, blurring effects, defocus and other similar factors. Most of these texture-based methods only exploit noise features or extract features directly from original images, neglecting other distinctive information as the blurriness. Therefore, by utilizing both these features, the discriminative power of texture-based methods can be further improved.

On the other side, liveness-based methods have higher implementation costs and require more sophisticated acquisition sensors. Some studies have been conducted in the literature, as an example, Qin et al. [22] explored a vital sign that involves the oxygen saturation of human blood and the heart rate, while Raghavendra et al. [23] measured the liveness of particular samples by magnifying the blood flow through the finger vein. Both of these approaches require capturing, storing and processing a sequence of infrared images, which requires significant time and less friendly acquisition processes.

Table 18.1 briefly describes an overview of main PAD texture-based methods available on literature. As it is shown, texture-based methods can be categorized into two groups: (i) methods implemented *without* decomposition, these methods extract features directly from original images, and thus, those implicit discriminative features in blurriness and noise components are degraded due to mutual interference between the two components. And (ii) methods implemented *with* decomposition, these methods first decompose original images into two components, but only one component is used for finger vein PAD, while some discriminative features determined from the other component are neglected.

Some of these methods implemented without decomposition are introduced by Nguyen et al. in 2013 [10]. They adopted three schemes to detect individual presentation attacks, (i) the Fourier spectral energy ratio (FSER), which explores the frequency information extracted from a vein image to construct features. It first transforms a vein image into the frequency domain using the 2-D FFT. Next, it calculates the ratio of high and low frequencies of image energy to determine whether the finger vein image is real. (ii) The discrete wavelet transform (DWT), in contrast to FSER, takes full advantage of both the spatial and frequency information obtained from an image. First, a one-level wavelet transform is used to divide the finger vein image into four sub-bands. Then the standard deviation of the pixel values in the HH sub-region (i.e. the high-frequency components in both the horizontal and vertical directions) is calculated as the PAD feature, since the printing noise that occurs in forged images is almost exclusively expressed more at high frequencies than at low frequencies.

Table 18.1 Summary of existing texture-based methods for finger vein presentation attack counter-measures and databases. HTER refers to half total error rate, EER to equal error rate, d to decidability or decision-making power and F-ratio, which relates the variances of independent samples

Method	Description	Database	Performance	
			Full	Cropped
Methods implemented without decomposition				
FSER [10]	Fourier transform		–	EER = 12.652%
HDWT [10]	Haar wavelet transform	330 fingers,	–	EER = 12.706%
DDWT [10]	Daubechies wavelet transform	5820 samples	–	EER = 14.193%
FSER-DWT [10]	Combining FSER, HDWT and DDWT by using SVM		–	EER – 1.176%
FSBE [7]	The average vertical energy of the fourier spectrum	IDIAP FVD	HTER = 0.00%, d = 11.17	HTER = 20.50%, d = 1.82
BSIF [7]	Binarized statistical image features		HTER = 4.00%, d = 4.47	HTER = 2.75%, d = 3.81
MSS [7]	Utilized monogenic scale space based global descriptors		HTER = 0.00%, d = 8.06	HTER = 1.25%, d = 5.54
LPQ-WLD [7]	Local phase quantization and Weber local descriptor		HTER = 0.00%, d = 8.03	–
RLBP [7]	LBP on the residual image		–	HTER = 0.00%, d = 5.20
Methods implemented with decomposition				
W-DMD [21]	Windowed dynamic mode decomposition	IDIAP FVD	EER = 0.08%, F-ratio = 3.15	EER = 1.59%, F-ratio = 2.14
SP [11]	Steerable pyramids - SVM	300 fingers, 12000 samples	–	HTER of Artefact 123 = 3.6%3.0%2.4%

Haar DWT (HDWT) and Daubechies DWT (DDWT) are discussed in [1] as possible methods of achieving better performance. And (iii) a combination of FSER and DWT (i.e. FSER-DWT), this scheme combines three of the abovementioned features, i.e. FSER, HDWT and DDWT, to strengthen the maximum classification performance by combining them with SVMs.

In 2014, the Idiap Research Institute made publicly available the first finger vein presentation attack database [5]. Given this resource, many new PAD methods for finger vein recognition have been proposed [7, 11, 21]. More recently, the first Competition on Counter Measures to Finger Vein Spoofing Attacks was organized in conjunction with the International Conference on Biometrics (ICB) in 2015 [7]. In this competition, five different PAD approaches were proposed, (*i*) Fourier spectral bandwidth energy (FSBE), which was used as a baseline PAD system in the competition. This method first extracts the average vertical energy of the Fourier spectrum, and then calculates the bandwidth of this average vertical signal using a -3 dB cut-off frequency as the criterion. (*ii*) Binarized statistical image features (BSIF), which are used to represent each pixel as a binary code constructed by computing its response to a filter learned using the statistical properties of natural images. (*iii*) Monogenic scale space (MSS), which extract global descriptors that represent local energy and orientation at a coarse level to distinguish between real and forged finger veins. (*iv*) Local phase quantization and Weber local descriptor (LPQ-WLD), this method first fuses local phase information generated by LPQ with the differential excitation and local orientation extracted by WLDs to form input features with simultaneous temporal and spectral cues for subsequent evaluation. And (*v*) the residual local binary pattern (RLBP), a residual version of a vein image is first generated using this method and a 3×3 integer kernel and then texture features can be extracted using LBP for PAD of cropped images. It is important to note that none of them could achieve 100% PAD accuracy on both cropped and full versions of the given test sets in the competition.

Following literature, Tirunagari et al. [21] proposed the Windowed Dynamic mode decomposition, focusing on light reflections, illumination and planar effect differences between real and forged images. By using W-DMD to effectively amplify such local variations, an increasing of the accuracy of PAD was achieved.

Finally, although its effectiveness has been demonstrated only on a private database, Raghavendra et al. [11] presented steerable pyramids (SP - SVM), a method to make full use of the texture information at different levels of a pyramid, based on combining a SP with support vector machines to detect presentation attacks.

18.4   Reproducible Experimental Results

This section contains reproducible experimental results demonstrating the vulnerability of finger vein recognition systems when exposed to the VERA Spoofing Finger vein Database alongside baselines for presentation attack detection based on the same dataset. All experiments in this section can be reproduced by downloading and installing the package bob.hobpad2.veins[5]

---

[5]Source code and results: https://pypi.org/project/bob.hobpad2.veins.

Fig. 18.8 Processing pipeline implemented by bob.bio.base and bob.bio.vein. The pipeline is composed of three stages for preprocessing, feature extraction and matching which can be implemented independently

18.4.1 Vulnerability Analysis

The experiments in this section were carried out using the open-source vein recognition framework called bob.bio.vein.[6] This framework is extensible and allows running complete finger vein recognition experiments, from the preprocessing of raw images (including segmentation) to the evaluation of biometric scores and performance analysis. The bob.bio.vein framework is itself based on another framework called bob.bio.base[7] which implements biometric recognition experiments based on a fixed pipeline composed of three components as shown in Fig. 18.8. Raw data are first preprocessed to extract the (finger) region of interest (RoI). In our experiments, we always used pre-annotated regions of interest. The finger image is then normalized by fitting a straight line between the detected finger edges, whose parameters (a rotation and a translation) are used to create an affine transformation [24]. Figure 18.9 illustrates this process.

Features from the RoI are subsequently extracted for each finger in the dataset according to one of three methods tested: Repeated Line Tracking [25] (rlt), Maximum Curvature [26] (mc) or Wide Line Detector [24] (wld). In the final stage of the pipeline, features from each finger are used to enrol and probe users respecting precise database protocols. Matching is performed using the correlation-based 'Miura Matching' algorithm introduced in [25].

18.4.1.1 Experimental Protocols

For the experiments, we consider each of the four verification protocols available with the dataset:

- 'Nom' (Normal Operation Mode): It corresponds to the standard verification scenario. For the VERA database, each of the 2 fingers for all 110 subjects is used for enrolling and probing. Data from the first session are used for enrollment, while data from the second session are used for probing. Matching happens exhaustively. In summary, this protocol encompasses 220 unique fingers and 440 unique images providing a total of 48180 zero-effort impostor protocols and 220 genuine scores.

---

[6]https://pypi.org/project/bob.bio.veins.

[7]https://pypi.org/project/bob.bio.base.

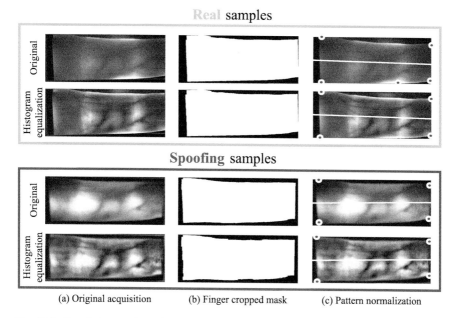

Fig. 18.9 Sample images from the database (665×250 pixels, **a**), as well as the finger cropped masks (**b**) and the normalized patterns (**c**) obtained after preprocessing these images

In order to analyse vulnerability of our baselines using this protocol, we also probe against each attack derived from session 2 data, making up an additional set of 220 presentation attack scores.

- 'Fifty': The 'Fifty' protocol is meant as a reduced version of the 'Nom' protocol, for quick check purposes. All definitions are the same, except we only use the first 50 subjects in the dataset (numbered 1 until 59). In this protocol, there are 100 unique fingers available and 200 unique images. There are 9900 zero-effort impostor scores and 100 genuine scores. Vulnerability analysis is done like for protocol 'Nom' which makes up an additional set of 100 presentation attack scores.

- 'B': This protocol was created to simulate an evaluation scenario similar to that from the UTFVP database [17]. In this protocol, only 108 unique fingers are used which makes up a total of 216 unique images. Matching happens exhaustively between all images in the dataset, including self-probing. With this configuration, there are 46224 zero-effort impostor scores and 432 genuine scores. For vulnerability analysis, an extra set of 432 presentation attack scores are derived matching presentation attack instructions generated from the equivalent probe set.

- 'Full': Each of the 2 fingers for all 110 subjects is used for enrolling and probing. Data from both sessions are used for probing and enrolling which makes this protocol slightly biased as in protocol 'B' above. Matching happens exhaustively. In summary, this protocol encompasses 220 unique fingers and 440 unique images providing a total of 192720 zero-effort impostor protocols and 880 genuine scores.

Table 18.2 Impostor attack presentation match rate (IAPMR) and equal error rate (EER, in parenthesis) for each of the baselines and verification protocols explored in this section. Values are approximate and expressed in percentage (%)

Protocol/Baseline	Maximum curvature	Repeated line tracking	Wide line detector
Nom	83 (2)	38 (19)	76 (7)
Fifty	77 (1)	35 (17)	70 (3)
B	86 (1)	32 (17)	75 (3)
Full	89 (1)	34 (11)	80 (4)

In order to analyse vulnerability of our baselines using this protocol, we also probe against each attack derived from session 2 data, making up an additional set of 880 presentation attack scores.

18.4.1.2 Experimental Results

Table 18.2 shows the results in terms of the equal error rate (EER) and the impostor attack presentation match rate (IAPMR) for the bona fide verification task and the vulnerability analysis, respectively. Vulnerability is reported on the threshold defined at the EER. As can be observed, explored baselines are vulnerable to attacks. In the case of some combinations of baseline techniques with protocols, the vulnerability at the EER threshold is nearly 90%, indicating 9 of 10 attacks to such a system would not be detected as such. As observed for other biometric modalities, better performing recognition (smaller error rates) loosely correlates with a more vulnerable system. For example, the system based on Miura's maximum curvature algorithm (mc) presents the lowest EER and the highest vulnerabilities.

Figure 18.10 shows the IAPMR against our mc pipeline with scores produced using the 'Nom' VERA finger vein database protocol. The decision threshold is fixed to reach a FNMR=FMR (i.e. EER) using *bona fide* samples (genuines and zero-effort impostors), and then the IAPMR of presentation attacks is computed. While the almost perfect separation of scores for *bona fide* users and impostors justifies the good verification performance, the presentation attacks appear optimal. This is demonstrated by the value of IAPMR as well as the percentage of presentation attacks that manage to bypass the system at the chosen threshold (i.e. an IAPMR of about 83% or higher is observed).

18.4.2 Presentation Attack Detection

The experiments in this section were carried out using the open-source vein presentation attack detection framework called bob.pad.vein.[8] This framework is extensible

---

[8]https://pypi.python.org/pypi/bob.pad.vein.

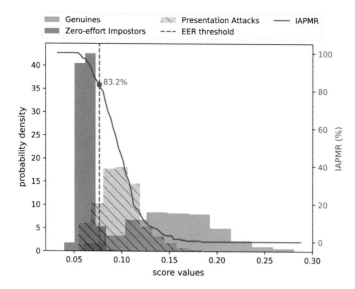

Fig. 18.10 Score distributions and IAPMR curve for a finger vein recognition system based on Miura's maximum curvature algorithm

and allows running complete finger vein presentation attack detection experiments, from the preprocessing of raw images (including segmentation) to the evaluation of scores and performance analysis. The algorithm described was used as baseline during the first competition on countermeasures to finger vein presentation attacks organized during the International Conference in Biometrics, 2015.

The baseline PAD system is a texture-based algorithm that exploits subtle changes in the finger vein images due to printed effects using the frequency domain. To recognize the static texture, the Fourier transform (FT) is extracted from the raw image after applying a histogram equalization. Once the FT is calculated and normalized on logarithmic scale, a window of 20 pixels centred vertically on the centre of the image is applied to extract the average vertical energy of the FT. Then, the bandwidth of this average vertical signal (Bw_v) at a cut-off frequency of -3 dB is calculated. The final score to discriminate between real-accesses and presentation attacks is this bandwidth normalized by the height of the image (h), i.e. $s = Bw_v/h$, resulting in a score normalized in the range $[0 - 1]$.

This method exploits the idea of the bandwidth of vertical energy signal on real finger vein images, which is weakly manifested on presentation attacks. The main reason of it is that the recaptured finger vein samples display a smooth texture with changes mainly vertical, changes translated as a horizontal energy in the Fourier domain. On the other hand, real finger vein images have better focus and sharpness on both horizontal and vertical directions, displaying both directions of energy in their Fourier transform. It is interesting to note that this approach does not need any kind of training or classifier to work.

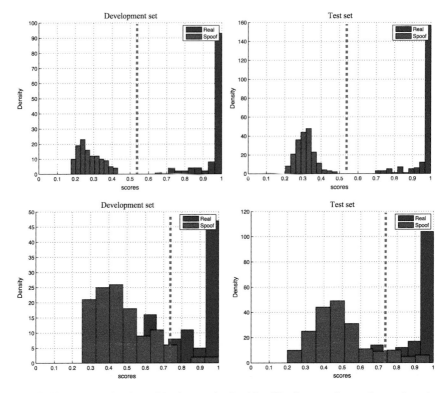

Fig. 18.11 Score distributions of the finger vein algorithm. The first two plots on the top show the performance of the PAD algorithm on the 'full' protocol (see also Fig. 18.5) while the histograms on the bottom show the performance on the 'cropped' protocol

Figure 18.11 shows the performance of the proposed PAD algorithm on the Spoofing-Attack Finger vein Database. The results are displayed in the form of histograms showing the class separation in this binary classification task. For the 'full' protocol, separation is clear between the two classes (presentation attacks and *bona fide*) with a half total error rate of 0% both on the development set and test set. For the cropped protocol, the performance degrades to about ∼23% EER with a matching ∼21% HTER on the test set. These results indicate that the baseline algorithm is likely using information from outside the finger vein region to detect the presence of attacks.

18.5 Conclusions

Human vein networks can be used to verify the identity of a person with high performance. Vein recognition is surrounded by an apparent sense of security as vein networks, unlike fingerprints, do not leave latent marks or are readily available on social networks. Yet, vein recognition systems are very vulnerable to presentation attacks. Instruments made out of simple toner prints of vein networks can successfully bypass naively tuned recognition systems.

Unfortunately, the number of available resources for the analysis of vulnerabilities or studying presentation attack detection remains limited to date. There are only two datasets available publicly and a handful of presentation attack detection algorithms were published. With this book chapter, associated software and scores for analysis, we hope to consolidate and seed further research in this area.

Acknowledgements The authors would like to thank the Swiss Centre for Biometrics Research and Testing and the Swiss Commission for Technology and Innovation (CTI) for supporting the research leading to some of results published in this book chapter.

References

1. Jain AK, Flynn P, Ross AA (eds) (2008) Handbook of biometrics. Springer, Berlin. https://doi.org/10.1007/978-0-387-71041-9
2. Finger vein authentication: white paper. Technical report, Hitachi, Ltd (2006)
3. Kono M, Ueki H, Umemura SI (2002) Near-infrared finger vein patterns for personal identification. Appl Opt 41(35):7429–7436. https://doi.org/10.1364/AO.41.007429
4. Kono M, Umemura S, Miyatake T, Harada K, Ito Y, Ueki H (2004) Personal identification system. US Patent 6,813,010. https://www.google.com/patents/US6813010
5. Tome P, Vanoni M, Marcel S (2014) On the vulnerability of finger vein recognition to spoofing. In: IEEE international conference of the biometrics special interest group (BIOSIG), vol 230
6. Tome P, Marcel S (2015) On the vulnerability of palm vein recognition to spoofing attacks. In: The 8th IAPR international conference on biometrics (ICB), pp 319–325. https://doi.org/10.1109/ICB.2015.7139056. http://ieeexplore.ieee.org/xpl/articleDetails.jsp?reload=true&arnumber=7139056
7. Tome P, Raghavendra R, Busch C, Tirunagari S, Poh N, Shekar BH, Gragnaniello D, Sansone C, Verdoliva L, Marcel S (2015) The 1st competition on counter measures to finger vein spoofing attacks. In: 2015 international conference on biometrics (ICB), pp 513–518. https://doi.org/10.1109/ICB.2015.7139067
8. Chingovska I, Anjos A, Marcel S (2012) On the effectiveness of local binary patterns in face anti-spoofing. In: Proceedings of the 11th international conference of the biometrics special interest group
9. Ruiz-Albacete V, Tome-Gonzalez P, Alonso-Fernandez F, Galbally J, Fierrez J, Ortega-Garcia J (2008) Direct attacks using fake images in iris verification. In: Proceedings of the COST 2101 workshop on biometrics and identity management, BIOID. LNCS, vol 5372. Springer, Berlin, pp 181–190
10. Nguyen DT, Park YH, Shin KY, Kwon SY, Lee HC, Park KR (2013) Fake finger-vein image detection based on Fourier and wavelet transforms. Digit Signal Process 23(5):1401–1413. https://doi.org/10.1016/j.dsp.2013.04.001
11. Raghavendra R, Busch C (2015) Presentation attack detection algorithms for finger vein biometrics: a comprehensive study. In: 2015 11th international conference on signal-image technology internet-based systems (SITIS), pp 628–632. https://doi.org/10.1109/SITIS.2015.74
12. Zhou Y, Kumar A (2011) Human identification using palm-vein images. IEEE Trans Inf Forensics Secur 6(4):1259–1274
13. Kang W, Wu Q (2014) Contactless palm vein recognition using a mutual foreground-based local binary pattern. IEEE Trans Inf Forensics Secur 9(11):1974–1985

14. Zhang J, Yang J (2009) Finger-vein image enhancement based on combination of gray-level grouping and circular Gabor filter. In: International conference on information engineering and computer science (ICIECS), pp 1–4
15. Mirmohamadsadeghi L, Drygajlo A (2014) Palm vein recognition with local texture patterns. IET Biom 1–9
16. Swain M, Ballard D (1991) Color indexing. Int J Comput Vis 7(1):11–32
17. Ton B (2012) Vascular pattern of the finger: biometric of the future? Sensor design, data collection and performance verification. Master's thesis, University of Twente
18. Ton B, Veldhuis R (2013) A high quality finger vascular pattern dataset collected using a custom designed capturing device. In: IEEE international conference on biometrics (ICB), pp 1–5
19. Xi X, Yang G, Yin Y, Meng X (2013) Finger vein recognition with personalized feature selection. Sensors 13(9):11243–11259
20. Raghavendra R, Raja KB, Surbiryala J, Busch C (2014) A low-cost multimodal biometric sensor to capture finger vein and fingerprint. In: IEEE international joint conference on biometrics, pp 1–7. https://doi.org/10.1109/BTAS.2014.6996225
21. Tirunagari S, Poh N, Bober M, Windridge D (2015) Windowed DMD as a microtexture descriptor for finger vein counter-spoofing in biometrics. In: 2015 IEEE international workshop on information forensics and security (WIFS), pp 1–6. https://doi.org/10.1109/WIFS.2015.7368599
22. Qin B, Pan J-F, Cao G-Z, Du G-G (2009) The anti-spoofing study of vein identification system. In: 2009 international conference on computational intelligence and security, vol 2, pp 357–360. https://doi.org/10.1109/CIS.2009.144
23. Raghavendra R, Avinash M, Marcel S, Busch C (2015) Finger vein liveness detection using motion magnification. In: 2015 IEEE 7th international conference on biometrics theory, applications and systems (BTAS), pp 1–7. https://doi.org/10.1109/BTAS.2015.7358762
24. Huang B, Dai Y, Li R, Tang D, Li W (2010) Finger-vein authentication based on wide line detector and pattern normalization. In: International conference on pattern recognition (ICPR), pp 1269–1272
25. Miura N, Nagasaka A, Miyatake T (2004) Feature extraction of finger-vein patterns based on repeated line tracking and its application to personal identification. Mach Vis Appl 15(4):194–203
26. Miura N, Nagasaka A, Miyatake T (2007) Extraction of finger-vein patterns using maximum curvature points in image profiles. IEICE Trans Inf Syst E90-D(8):1185–1194

Chapter 19
Presentation Attacks in Signature Biometrics: Types and Introduction to Attack Detection

Ruben Tolosana, Ruben Vera-Rodriguez, Julian Fierrez and Javier Ortega-Garcia

Abstract Authentication applications based on the use of biometric methods have received a lot of interest during the last years due to the breathtaking results obtained using personal traits such as face or fingerprint. However, it is important not to forget that these biometric systems have to withstand different types of possible attacks. This work carries out an analysis of different Presentation Attack (PA) scenarios for on-line handwritten signature verification. The main contributions of the present work are: (1) short overview of representative methods for Presentation Attack Detection (PAD) in signature biometrics; (2) to describe the different levels of PAs existing in on-line signature verification regarding the amount of information available to the attacker, as well as the training, effort and ability to perform the forgeries; and (3) to report an evaluation of the system performance in signature biometrics under different PAs and writing tools considering freely available signature databases. Results obtained for both BiosecurID and e-BioSign databases show the high impact on the system performance regarding not only the level of information that the attacker has but also the training and effort performing the signature. This work is in line with recent efforts in the Common Criteria standardization community towards security evaluation of biometric systems, where attacks are rated depending on, among other factors, time spent, effort and expertise of the attacker, as well as the information available and used from the target being attacked.

R. Tolosana (✉) · R. Vera-Rodriguez · J. Ortega-Garcia
Biometrics and Data Pattern Analytics - BiDA Lab, Escuela Politecnica Superior,
Universidad Autonoma de Madrid, 28049 Madrid, Spain
e-mail: ruben.tolosana@uam.es

R. Vera-Rodriguez
e-mail: ruben.vera@uam.es

J. Fierrez
Universidad Autonoma de Madrid, 28049 Madrid, Spain
e-mail: julian.fierrez@uam.es

J. Ortega-Garcia
e-mail: javier.ortega@uam.es

© Springer Nature Switzerland AG 2019   439
S. Marcel et al. (eds.), *Handbook of Biometric Anti-Spoofing*,
Advances in Computer Vision and Pattern Recognition,
https://doi.org/10.1007/978-3-319-92627-8_19

19.1 Introduction

Applications based on biometric user authentication have experienced a high deployment in many relevant sectors such as security, e-government, healthcare, education, banking or insurance in the last years [1]. This growth has been possible thanks to two main factors: (1) the technological evolution and the improvement of sensors quality [2], which have cut the prices of general purpose devices (smartphones and tablets) and therefore, the high acceptance of the society towards the use of them; and (2) the evolution of biometric recognition technologies in general [3–5]. However, it is important to keep in mind that these biometric-based authentication systems have to withstand different types of possible attacks [6].

In this work, we focus on different Presentation Attack (PA) scenarios for online handwritten signature biometric authentication systems. These systems have received a significant amount of attention in the last years thanks to improved signature acquisition scenarios (including device interoperability [7]) and writing inputs (e.g. finger [8]).

In general, two different types of impostors can be found in the context of signature verification: (1) *random (zero-effort or accidental)* impostors, the case in which no information about the user being attacked is known and impostors present their own signature claiming to be another user of the system, and (2) *skilled* impostors, the case in which impostors have some level of information about the user being attacked (e.g. image of the signature) and try to forge their signature claiming to be that user in the system.

Galbally et al. have recently discussed in [9] different approaches to report accuracy results in handwritten signature verification applying the lessons learned in the evaluation of vulnerabilities to Presentation Attacks (PAs). They considered skilled impostors as a particular case of biometric PAs that is performed against a behavioral biometric characteristic (referred to in some cases as *mimicry*). It is important to highlight the key differences between physical PAs and mimicry, while traditional PAs involve the use of some physical artefacts such as fake masks and gummy fingers (and therefore, can be detected in some cases at the sensor level), in the case of mimicry the interaction with the sensor is exactly the same followed in a normal access attempt. Galbally et al. in [9] modified the traditional nomenclature of impostor scenarios in signature verification (i.e. skilled and random) following the standard in the field of biometric Presentation Attack Detection (PAD). This way, the classical random impostor scenario was referred to as Bona Fide (BF) scenario, while the skilled impostor scenario was referred to as PA scenario. This new nomenclature is used in this chapter as well.

If those PAs are expected, one can include specific modules for PAD, which in the signature recognition literature are usually referred to as forgery detection modules. A survey of such PAD methods is out of the scope of the chapter. Here in Sect. 19.2, we only provide a short overview of some selected representative works in that area.

A different approach aimed at improving the security against attacks in signature biometrics different from including a PAD module is template protection [10].

Traditional on-line signature verification systems use very sensitive biometric data such as the X and Y spatial coordinates for the matching, storing this information as the user templates without any kind of protection. A compromised template, in this case, would easily provide an attacker with the X and Y coordinates along the time axis, making possible to generate very high-quality forgeries of the original signature. In [11], an approach for signature template generation was proposed not considering information related to X, Y coordinates and their derivatives on the biometric system, providing, therefore, a much more robust system against attacks, as this critical information would not be stored anywhere. Moreover, the results achieved had error rates in the same range as more traditional systems that store very sensitive information.

The main contributions of the present work are: (1) short overview of representative methods for PAD in signature biometrics; (2) to describe the different levels of PAs existing in on-line signature verification regarding the amount of information available to the attacker, as well as the training, effort and ability to perform the forgeries; and (3) to report an evaluation of the system performance in signature biometrics under different PAs and writing tools considering freely available signature databases.

The remainder of the chapter is organized as follows. The introduction is completed with a short overview of PAD in signature biometrics (Sect. 19.2). After that, the main technical content of the chapter begins in Sect. 19.3, with a review of the most relevant possible attacks, pointing out which type of impostors are included in various well-known public signature databases. Section 19.4 describes the on-line signature databases considered in the experimental work. Section 19.5 describes the experimental protocol and the results achieved. Finally, Sect. 19.6 draws the final conclusions and points out some lines for future work.

19.2   PAD in Signature Biometrics

Presentation Attack Detection (PAD) in signature biometrics can be traced back to early works by Rosenfeld et al. in the late 70s [12]. In that work, authors dealt with the detection of freehand forgeries (i.e. forgeries written in the forger's own handwriting without knowledge of the appearance of the genuine signature) on bank checks for off-line signature verification. The detection process made use of features derived from Eden's model [13], which characterizes handwriting strokes in terms of a set of kinematic parameters that can be used to discriminate forged from genuine signatures. Those features were based on dimension ratios and slant angles, measured for the signature as a whole and for specific letters on it. Finally, unknown signatures were classified as genuine or forgery on the basis of their distance from the set of genuine signatures. A more exhaustive analysis was later carried out in [14], performing skilled forgery detection by examining the writer-dependent information embedded at the substroke level and trying to capture unballistic motion and tremor information in each stroke segment, rather than as global statistics.

In [15], authors proposed an off-line signature verification and forgery detection system based on fuzzy modelling. The verification of genuine signatures and detection of forgeries was achieved via angle features extracted using a grid method. The derived features were fuzzified by an exponential membership function, which was modified to include two structural parameters regarding variations of the handwriting styles and other factors affecting the scripting of a signature. Experiments showed the capability of the system in detecting even the slightest changes in signatures.

Brault et al. presented in [16] an original attempt to estimate, quantitatively and a priori from the coordinates sampled during its execution, the difficulty that could be experienced by a typical imitator in reproducing both visually and dynamically that signature. To achieve this goal, they first derived a functional model of what a typical imitator must do to copy dynamically any signature. A specific difficulty coefficient was then numerically estimated for a given signature. Experimentation geared specifically to signature imitation demonstrated the effectiveness of the model. The ranking of the tested signatures given by the difficulty coefficient was compared to three different sources: the opinions of the imitators themselves, the ones of an expert document examiner, and the ranking given by a specific pattern recognition algorithm. They provided an example application as well. This work supposed one of the first attempts of PAD for on-line handwritten signature verification using a special pen attached to a digitizer (Summagraphic Inc. model MM1201). The sampling frequency was 110 Hz, and the spatial resolution was 0.025 inch.

More studies of PAD methods at feature level for on-line signature verification were carried out in [17, 18]. In [17], authors proposed a new scheme in which a module focused on the detection of skilled forgeries (i.e. PA impostors) was added to the original verification system. That new module (i.e. Skilled Forgeries Detector) was based on four parameters of the Sigma LogNormal writing generation model [19] and a linear classifier. That new binary classification module was supposed to work sequentially before a standard signature recognition system [20]. Good results were achieved using that approach for both skilled (i.e. PA) and random (i.e. BF) scenarios. In [18], Reillo et al. proposed PAD methods based on the use of some global features such as the total number of strokes and the signing time of the signatures. They acquired a new database based on 11 levels of PAs regarding the level of knowledge and the tools available to the forger. The results achieved in that work using the proposed PAD reduced the EER from a percentage close to 20.0% to below 3.0%.

19.3   Presentation Attacks in Signature Biometrics

This section aims to describe the different levels of skilled forgeries (i.e. PA impostors) that exist in the literature regarding the amount of information provided to the attacker, as well as the training, effort and ability to perform the forgeries. In addition, we consider the case of random forgeries (i.e. zero-effort impostors) although it belongs to the BF scenario and not to the PA scenario in order to review the whole range of possible impostors in handwritten signature verification.

Previous studies have applied the concept of Biometric Menagerie in order to categorize each type of user of the biometric system as an animal. This concept was initially formalized by Doddington et al. in [21], classifying speakers regarding how easy or difficult the speaker can be recognized (i.e. sheep and goats, respectively), how easily they can be forged (i.e. lambs) and finally, how good they are forging others (i.e. wolves). Yager and Dunstone have more recently extended the Biometric Menagerie in [22] by adding four more categories of users (i.e. worms, chameleons, phantoms and doves). Their proposed approach was investigated using a broad range of biometric modalities, including 2D and 3D faces, fingerprints, iris, speech and keystroke dynamics. In [23], Houmani and Garcia-Salicetti applied the concept of Biometric Menagerie for the different types of users found in the on-line signature verification task proposing the combination of their personal and relative entropy measures as a way to quantify how difficult it is a signature to be forged. Their proposed approach achieved promising classification results on the MCYT database [24], where the attacker had access to a visual static image of the signature to forge.

In [25], authors showed through a series of experiments that: (1) some users are significantly better forgers than others (these users would be classified as wolves in the previous user categorization); (2) forgers can be trained in a relatively straight-forward way to become a greater threat; (3) certain users are easy targets for forgers (sheep following the previous user categorization); and (4) most humans are relatively poor judges of handwriting authenticity, and hence, their unaided instincts cannot be trusted. Additionally, in that work, authors proposed a new metric for impostor classification: *naive*, *trained* and *generative*. They considered naive impostors as random impostors (i.e. zero-effort impostors) in which no information about the user to forge is available whereas they defined trained and generative impostors as skilled forgeries (i.e. PA impostors) when only the image or the dynamics of the signature to forge is available, respectively.

In [26], authors proposed a software tool implemented on two different computer platforms in order to generate forgeries with different quality levels (i.e. PA impostors). Three different levels of PAs were considered: (1) *blind forgeries*, the case in which the attacker writes on a blank surface having access just to textual knowledge (i.e. precise spelling of the user's name to forge); (2) *low-force forgeries*, where the attacker gets a blueprint of the signature projected on the writing surface (dynamic information is not provided), which they may trace; and (3) *brute-force forgeries*, in which an animated pointer is projected onto the writing pad showing the whole realization of the signature to forge. The attacker may observe the sequence and follow the pointer. Authors carried out an experiment based on the use of 82 forgery samples performed by four different users in order to detect how the False Acceptance Rate (FAR) is affected regarding the level of PA. They considered a signature verification system based on average quadratic deviation. Results obtained for four different threshold values confirmed a strong protection against attacks.

A more exhaustive analysis of the different types of forgeries possible in signature recognition was carried out in [27]. In that work, authors considered random or zero-effort impostors plus four different levels of PA impostors regarding the amount of information provided to the attacker and the tools used for the impostors in order to forge the signature:

- **Random or zero-effort forgeries**, in which no information of the user to forge is available and the impostor uses its own signature (accidentally or not) claiming to be another user of the system.
- **Blind forgeries**, in which the attacker has access to a descriptive or textual knowledge of the original signatures (e.g. the name of the person).
- **Static forgeries** (low-force in [26]), where the attacker has access to a visual static image of the signature to forge. There are two ways to generate the forgeries. The first one, the attacker can train to imitate the signature with or without time restrictions and blueprint, and then forge it without the use of the blueprint, leading to **static trained forgeries**. In the second one, the attacker uses a blueprint to first copy the signature of the user to forge and then put it on the screen of the device while forging, leading to **static blueprint forgeries**, more difficult to detect as they have the same appearance as the original ones.
- **Dynamic forgeries** (brute-force in [26]), where the attacker has access to both the image and also the whole realization process (i.e. dynamics) of the signature to forge. The dynamics can be obtained in the presence of the original writer or through the use of a video recording. In a similar way as the previous category, we can distinguish first **dynamic trained forgeries** in which the attacker can use dedicated tools to analyze and train to forge the genuine signature, and second, **dynamic blueprint forgeries** which are generated by projecting on the acquisition area a real-time pointer that the forger needs to follow.
- **Regained forgeries**, the case where the attacker has access only to the static image of the signature to forge and makes use of a dedicated software to regain its dynamics [28], which are later analyzed and used to create dynamic forgeries.

Figure 19.1 depicts examples of one genuine signature and three different types of forgeries (i.e. random, static blueprint and dynamic trained) performed for the same user. The image shows both the static and dynamic information with the X and Y coordinates and pressure.

Besides the forgery classification carried out in [27], Alonso-Fernandez et al. studied the impact of an incremental level of quality in the PAs against signature verification systems. Both off-line and on-line systems were considered using the BiosecurID database which contains both off-line and on-line signatures. For the off-line system, they considered a system based on global image analysis and a minimum distance classifier [29] whereas a system based on Hidden Markov Models (HMM) [30] was considered for the on-line approach. Their experiments concluded that the

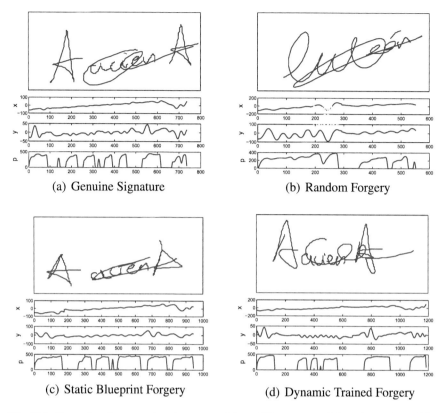

(a) Genuine Signature

(b) Random Forgery

(c) Static Blueprint Forgery

(d) Dynamic Trained Forgery

Fig. 19.1 Examples of one genuine signature and three different types of forgeries performed for the same user

performance of the off-line system is only degraded with the highest level of forgeries quality. On the contrary, the on-line system exhibits a progressive degradation with the forgeries quality, suggesting that the dynamic information of signatures is the one more affected by the considered increased forgeries quality.

Finally, Fig. 19.2 summarizes all different types of forgeries for both BF and PA scenarios regarding the amount of information available to the attacker, as well as the training, effort and ability to perform the attack. In addition, the most commonly used on-line signature databases are included in each PA group. It is important to highlight the lack of public on-line signature databases for the case of blind forgeries, as far as we know.

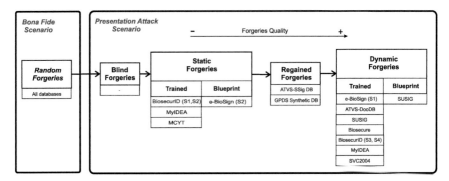

Fig. 19.2 Diagram of different types of forgeries for both BF and PA scenarios regarding the amount of information available to the attacker, as well as the training, effort and ability to perform the attack. The commonly used on-line signature databases are included in each PA group

19.4 On-Line Signature Databases

The following two databases are considered in the experiments reported here:

19.4.1 e-BioSign

For the e-BioSign database [8], we consider a subset of the freely available database[1] comprised of signatures acquired using a Samsung ATIV 7 general purpose device (a.k.a. W4 device). The W4 device has a 11.6-inch LED display with a resolution of 1920×1080 pixels and 1024 pressure levels. Data was collected using a pen stylus and also the finger in order to study the performance of signature verification in a mobile scenario. The available information when using the pen stylus is X and Y pen coordinates and pressure. In addition, pen-up trajectories are also available. However, for the case of using the finger as the writing tool, pressure information and pen-ups trajectories are not recorded. Regarding the acquisition protocol, the device was placed on a desktop and subjects were able to rotate the device in order to feel comfortable with the writing position.

Data were collected in two sessions for 65 subjects with a time gap between sessions of at least three weeks. For each user and writing tool, there are a total of eight genuine signatures and six skilled forgeries (i.e. PA impostors). Regarding skilled forgeries for the case of using the stylus as the writing tool, users were allowed during the first session to visualize a recording of the dynamic realization of the signature to forge as many times as they wanted whereas only the image of the signature to forge was available during the second session. Regarding skilled forgeries for the case of using the finger as the writing tool, in both sessions users

---

[1]https://atvs.ii.uam.es/atvs/eBioSign-DS1.html.

had access to the dynamic realization of the signatures to forge as many times as they wanted.

19.4.2   BiosecurID

For the BiosecurID database [31], we consider a subset [32] comprised of a total of 132 users.[2] Signatures were acquired using a Wacom Intuos 3 pen tablet with a resolution of 5080 dpi and 1024 pressure levels. The database comprises 16 genuine signatures and 12 skilled forgeries (i.e. PA impostors) per user, captured in four separate acquisition sessions. Each session was captured leaving a two-month interval between them, in a controlled and supervised office-like scenario. Signatures were acquired using a pen stylus. The available information within each signature is X and Y pen coordinates and pressure. In addition, pen-up trajectories are available.

The following PAs are considered in the database in order to analyze how the system performance differs regarding the amount of information provided to the attacker: (i) the attacker only sees the image of the signature once and tries to imitate it right away (session 1); (ii) the attacker sees the image of the signature and trains for a minute before making the forgery (session 2); (iii) the attacker is able to see the dynamics of the signing process three times, trains for a minute and then makes the forgery (session 3); and (iv) the dynamics of the signature are shown as many times as the attacker requests, being able to train for a minute and then sign (session 4).

19.5   Experimental Work

19.5.1   Signature Verification System

An on-line signature verification system based on time functions (a.k.a. local systems) is considered in the experimental work [33]. For each signature acquired using the stylus or the finger, only signals related to X and Y pen coordinates and their first- and second-order derivatives are used in order to provide reproducible results. Information related to pen angular orientation (azimuth and altitude angles) and pressure have been always discarded in order to consider the same set of time functions that we would be able to use in general purpose devices such as tablets and smartphones using the finger as the writing tool.

Our local system is based on DTW, which computes the similarity between the time functions from the input and training signatures. The configuration of the DTW algorithm described in [34].

---

[2]https://atvs.ii.uam.es/atvs/biosecurid_sonof_db.html.

19.5.2 Experimental Protocol

The experimental protocol has been designed to allow the study of both BF and PA scenarios on the system performance. Three different levels of impostors are analyzed: (1) random forgeries, (2) static forgeries (both trained and blueprint) and (3) dynamic forgeries. Additionally, for the e-BioSign subset, the case of using the finger as the writing tool is considered. All available users (i.e. 65 and 132 for e-BioSign and BiosecurID subsets, respectively) are used for the evaluation as no development of the on-line signature verification system is carried out.

For both databases, the four genuine signatures of the first session are used as reference signatures, whereas the remaining genuine signatures (i.e. 4 and 12 for the e-BioSign and BiosecurID databases, respectively) are used for testing. Skilled forgeries scores (i.e. PA mated scores) are obtained by comparing the reference signatures against the skilled forgeries (i.e. PA impostors) related to each level of attacker, whereas random forgeries scores (i.e. BF non-mated scores) are obtained by comparing the reference signatures with one genuine signature of each of the remaining users (i.e. 64 and 131 for the e-BioSign and BiosecurID databases, respectively). The final score is obtained after performing the average score of the four one-to-one comparisons.

19.5.3 Experimental Results

Tables 19.1 and 19.2 show the system performance obtained for each different type of impostor and database. Additionally, Fig. 19.3 shows the system performance in terms of DET curves for each impostor scenario and database.

First, we analyze the results achieved for the case of using the stylus as the writing tool. In this case, a system performance improvement can be observed for both BiosecurID (Table 19.1) and e-BioSign (Table 19.2) databases when the amount of information that the attacker has is reduced. For example, a 7.5% EER is obtained in Table 19.1 when the attacker has access to the dynamics and also the static information

Table 19.1 BiosecurID: system performance results (EER in %)

	Random forgeries	Static forgeries	Dynamic forgeries
Stylus	1.1	5.4	7.5

Table 19.2 e-BioSign: system performance results (EER in %)

	Random forgeries	Static forgeries	Dynamic forgeries
Stylus	1.0	11.4	12.3
Finger	0.4	8.9*	18.3

*Generated on new data captured after e-BioSign

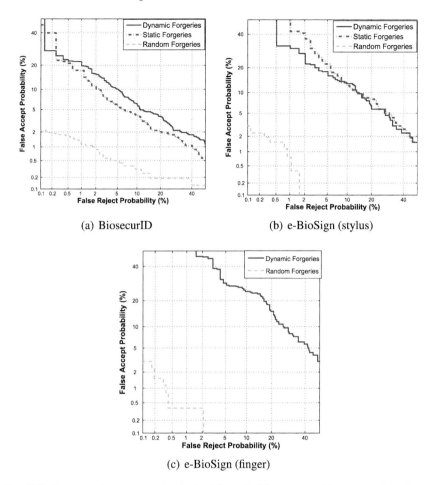

Fig. 19.3 System performance results obtained for each different type of impostor and database

of the signature to forge whereas this value is reduced to 5.4% EER when only the static information is provided to the forger.

We can also observe the impact of varying training and effort to perform the forgeries by comparing Tables 19.1 and 19.2. In general, higher errors are observed for the e-BioSign database for both types of skilled forgeries (i.e. dynamic and static) compared to the BiosecurID database. This is due to the fact that for dynamic forgeries, the attackers of the e-BioSign database had access to the dynamic realization of the signatures to forge as many times as they wanted and were also allowed to train without restrictions of time, whereas for the BiosecurID database the attackers had time restrictions, resulting in lower quality forgeries. For the case of static forgeries, the attackers of the e-BioSign database used a blueprint with the image of the signature to forge, placing it on the screen of the device while forging whereas for the BiosecurID database, the attackers just saw the image of the signatures to forge and trained before making the forgery without the help of any blueprint.

Finally, very similar good results are achieved in Tables 19.1 and 19.2 for random forgeries (i.e. zero-effort impostors) as the attackers have no information of the user to forge and present to the system their own signature.

Analyzing the case of using the finger as the writing tool, a high degradation of the system performance can be observed in Table 19.2 for the dynamic forgeries compared to the case of using the stylus as the writing tool. A recommendation for the usage of signature recognition on mobile devices would be for the users to protect themselves from other people that could be watching while signing, as this is more feasible to do in a mobile scenario compared to an office scenario. This way skilled forgers (i.e. PA impostors) might have access to the global shape of the signature but not to the dynamic information and results would be much better. For analyzing this scenario, we captured additional data after e-BioSign achieving a 8.9% EER (marked with * in Table 19.2, as the dataset in this case is not the same of the rest of the table), much better results compared to the 18.3% EER obtained for dynamic forgeries. For the case of random forgeries (i.e. zero-effort impostors), better results are obtained when the finger is considered as the writing tool compared to the stylus proving the feasibility of this scenario for random forgeries. Finally, it is important to remind that we are using a simple and reproducible verification system based only on X, Y coordinates and their derivatives. For a complete analysis of using the finger as the writing tool please refer to [8].

Finally, we would like to remark that the results obtained in this work should be interpreted in general terms as comparing different scenarios of attack. Specific results on operational setups can vary depending on the specific matching algorithm considered. An example of this can be seen in [35], where two different verification systems (i.e. Recurrent Neural Networks (RNNs) and DTW) were evaluated on the BiosecurID database for different types of attacks. The signature verification system based on RNNs obtained much better results than DTW for skilled forgeries, but DTW outperformed RNNs for random forgeries concluding that fusion of both systems could be a good strategy. Similar conclusions can be observed in previous studies [36, 37].

19.6 Conclusions

This work carries out an analysis of Presentation Attack (PA) scenarios [6] for on-line handwritten signature verification [33]. Unlike traditional PAs, which use physical artefacts (e.g. fake masks and gummy fingers), the most typical PAs in signature verification represent an attacker interacting with the sensor exactly in the same way followed in a normal access attempt, i.e. the presentation attack is a handwritten signature, in this case imitating to some extent the attacked identity. In such typical PA scenario, the level of knowledge that the attacker has and uses about the signature being attacked results crucial for the success rate of the attack.

The main contributions of the present work are: (1) short overview of representative methods for PAD in signature biometrics; (2) to describe the different levels of PAs existing in on-line signature verification regarding the amount of information available to the attacker, as well as the training, effort and ability to perform the forgeries and (3) to report an evaluation of the system performance in signature biometrics under different PAs and writing tools considering available signature databases.

Results obtained for both BiosecurID [31] and e-BioSign [8] databases show the high impact on the system performance regarding not only the level of information that the attacker has but also the training and effort performing the signature [27]. For the case of using the finger as the writing tool, a recommendation for the usage of signature recognition on mobile devices would be for the users to protect themselves from other people that could be watching while signing, as this is more feasible to do in a mobile scenario [38] compared to an office scenario. This way skilled forgers (i.e. PA impostors) might have access to the global shape of the signature but not to the dynamic information and results would be much better. This work is in line with recent efforts in the Common Criteria standardization community towards security evaluation of biometric systems, where attacks are rated depending on, among other factors: time spent, effort, and expertise of the attacker; as well as the information available and used from the target being attacked [39].

Acknowledgements This work has been supported by projects: Bio-Guard (Ayudas Fundación BBVA a Equipos de Investigación Científica 2017), UAM-CecaBank, and by TEC2015-70627-R (MINECO/FEDER). Ruben Tolosana is supported by a FPU Fellowship from the Spanish MECD.

References

1. Meng W, Wong DS, Furnell S, Zhou J (2015) Surveying the development of biometric user authentication on mobile phones. IEEE Commun Surv Tutor 17:1268–1293
2. Zhang DD (2013) Automated biometrics: technologies and systems. Springer Science & Business Media, Berlin
3. Jain AK, Nandakumar K, Ross A (2016) 50 Years of biometric research: accomplishments, challenges, and opportunities. Pattern Recognit Lett 79:80–105
4. Taigman Y, Yang M, Ranzato MA, Wolf L (2014) Deepface: closing the gap to human-level performance in face verification. In: The IEEE conference on computer vision and pattern recognition (CVPR)
5. Impedovo D, Pirlo G (2008) Automatic signature verification: the state of the art. IEEE Trans Syst Man Cybern Part C (Appl Rev) 38(5):609–635
6. Hadid A, Evans N, Marcel S, Fierrez J (2015) Biometrics systems under spoofing attack: an evaluation methodology and lessons learned. IEEE Signal Process Mag 32(5):20–30
7. Tolosana R, Vera-Rodriguez R, Ortega-Garcia J, Fierrez J (2015) Preprocessing and feature selection for improved sensor interoperability in online biometric signature verification. IEEE Access 3:478–489
8. Tolosana R, Vera-Rodriguez R, Fierrez J, Morales A, Ortega-Garcia J (2017) Benchmarking desktop and mobile handwriting across COTS devices: the e-Biosign biometric database. PLoS ONE 12(5):e0176792

9. Galbally J, Gomez-Barrero M, Ross A (2017) Accuracy evaluation of handwritten signature verification: rethinking the random-skilled forgeries dichotomy. In Proceedings of IEEE international joint conference on biometrics, pp 302–310

10. Gomez-Barrero M, Galbally J, Morales A, Fierrez J (2017) Privacy-preserving comparison of variable-length data with application to biometric template protection. IEEE Access 5:8606–8619

11. Tolosana R, Vera-Rodriguez R, Ortega-Garcia J, Fierrez J (2015) Increasing the robustness of biometric templates for dynamic signature biometric systems. In: Proceedings of 49th annual international Carnahan conference on security technology

12. Nagel RN, Rosenfeld A (1977) Computer detection of freehand forgeries. IEEE Trans Comput C-26;895–905

13. Eden M (1961) On the formalization of handwriting. Structure of language and its mathematical aspects (Proceedings of symposia in applied mathematics), vol 12. American Mathematical Society, Providence, pp 83–88

14. Guo JK, Doermann D, Rosenfeld A (2001) Forgery detection by local correspondence. Int J Pattern Recognit Artif Intell 15

15. Madasu VK, Lovell BC (2008) An automatic off-line signature verification and forgery detection system. In: Verma B, Blumenstein M (eds) Pattern recognition technologies and applications: recent advances. IGI Global, USA, pp 63–88

16. Brault JJ, Plamondon R (1993) A complexity measure of handwritten curves: modeling of dynamic signature forgery. IEEE Trans Syst Man Cybern 23:400–413

17. Gomez-Barrero M, Galbally J, Fierrez J, Ortega-Garcia J, Plamondon R (2015) Enhanced on-line signature verification based on skilled forgery detection using sigma-lognormal features. In: Proceedings of IEEE/IAPR international conference on biometrics. ICB, pp 501–506

18. Sanchez-Reillo R, Quiros-Sandoval HC, Goicochea-Telleria I, Ponce-Hernandez W (2017) Improving presentation attack detection in dynamic handwritten signature biometrics. IEEE Access 5:20463–20469

19. O'Reilly C, Plamondon R (2009) Development of a sigma-lognormal representation for on-line signatures. Pattern Recognit 42(12):3324–3337

20. Fierrez J, Morales A, Vera-Rodriguez R, Camacho D (2018) Multiple classifiers in biometrics. part 1: fundamentals and review. Inf Fusion 44:57–64

21. Doddington G, Liggett W, Martin A, Przybocki M, Reynolds D (1998) Sheeps, goats, lambs and wolves: a statistical analysis of speaker performance in the NIST 1998 speaker recognition evaluation. In: Proceedings of international conference on spoken language processing

22. Yager N, Dunstone T (2010) The biometric menagerie. IEEE Trans Pattern Anal Mach Intell 32(2):220–230

23. Houmani N, Garcia-Salicetti S (2016) On hunting animals of the biometric menagerie for online signature. PLoS ONE 11(4):e0151691

24. Ortega-Garcia J, et al. (2003) MCYT baseline corpus: a bimodal biometric database. IEE Proc Vis Image Signal Process Spec Issue Biom Internet 150(6):395–401

25. Ballard L, Lopresti D, Monroe F (2007) Forgery quality and its implication for behavioural biometric security. IEEE Trans Syst Man Cybern Part B 37(5):1107–1118

26. Vielhauer C, Zbisch F (2003) A test tool to support brute-force online and offline signature forgery tests on mobile devices. In: Proceedings of international conference on multimedia and expo, vol 3, pp 225–228

27. Alonso-Fernandez F, Fierrez J, Gilperez A, Galbally J, Ortega-Garcia J (2009) Robustness of signature verification systems to imitators with increasing skills. In: Proceedings of 10th international conference on document analysis and recognition, pp 728–732

28. Ferrer MA, Diaz M, Carmona-Duarte C, Morales A (2017) A behavioral handwriting model for static and dynamic signature synthesis. IEEE Trans Pattern Anal Mach Intell 39(6):1041–1053

29. Fierrez-Aguilar J, Alonso-Hermira N, Moreno-Marquez G, Ortega-Garcia J (2004) An off-line signature verification system based on fusion of local and global information. In: Proceedings of European conference on computer vision, workshop on biometric authentication, BIOAW, LNCS, vol. 3087. Springer, Berlin, pp 295–306

30. Tolosana R, Vera-Rodriguez R, Ortega-Garcia J, Fierrez J (2015) Update strategies for HMM-based dynamic signature biometric systems. In: Proceedings of 7th IEEE international workshop on information forensics and security, WIFS
31. Fierrez J, Galbally J, Ortega-Garcia J, et al. (2010) BiosecurID: a multimodal biometric database. Pattern Anal Appl 13(2):235–246
32. Galbally J, Diaz-Cabrera M, Ferrer MA, Gomez-Barrero M, Morales A, Fierrez J (2015) Online signature recognition through the combination of real dynamic data and synthetically generated static data. Pattern Recognit 48:2921–2934
33. Martinez-Diaz M, Fierrez J, Hangai S (2015) Signature features. In: Li SZ, Jain A (eds) Encyclopedia of biometrics. Springer, Berlin, pp 1375–1382
34. Martinez-Diaz M, Fierrez J, Hangai S (2015) Signature matching. In: Li SZ, Jain A (eds) Encyclopedia of biometrics. Springer, Berlin, pp 1382–1387
35. Tolosana R, Vera-Rodriguez R, Fierrez J, Ortega-Garcia J (2018) Exploring recurrent neural networks for on-line handwritten signature biometrics. IEEE Access 6:5128–5138
36. Liu Y, Yang Z, Yang L (2014) Online signature verification based on DCT and sparse representation. IEEE Trans Cybern 45:2498–2511
37. Diaz M, Fischer A, Ferrer MA, Plamondon R (2016) Dynamic signature verification system based on one real signature. IEEE Trans Cybern 48:228–239
38. Martinez-Diaz M, Fierrez J, Krish RP, Galbally J (2014) Mobile signature verification: feature robustness and performance comparison. IET Biom 3(4):267–277
39. Tekampe N, Merle A, Bringer J, Gomez-Barrero M, Fierrez J, Galbally J (2016) Toward Common Criteria evaluations of biometric systems. BEAT public deliverable D6.5. https://www.beat-eu.org/project/deliverables-public/d6-5-toward-common-criteria-evaluations-of-biometric-systems

Part VI
PAD Evaluation, Legal Aspects and Standards

Chapter 20
Evaluation Methodologies for Biometric Presentation Attack Detection

Ivana Chingovska, Amir Mohammadi, André Anjos and Sébastien Marcel

Abstract Presentation attack detection (PAD, also known as anti-spoofing) systems, regardless of the technique, biometric mode or degree of independence of external equipment, are most commonly treated as binary classification systems. The two classes that they differentiate are bona-fide and presentation attack samples. From this perspective, their evaluation is equivalent to the established evaluation standards for the binary classification systems. However, PAD systems are designed to operate in conjunction with recognition systems and as such can affect their performance. From the point of view of a recognition system, the presentation attacks are a separate class that need to be detected and rejected. As the problem of presentation attack detection grows to this pseudo-ternary status, the evaluation methodologies for the recognition systems need to be revised and updated. Consequentially, the database requirements for presentation attack databases become more specific. The focus of this chapter is the task of biometric verification and its scope is three-fold: first, it gives the definition of the presentation attack detection problem from the two perspectives. Second, it states the database requirements for a fair and unbiased evaluation. Finally, it gives an overview of the existing evaluation techniques for presentation attacks detection systems and verification systems under presentation attacks.

I. Chingovska · A. Mohammadi (✉) · A. Anjos · S. Marcel
Idiap Research Institute, rue Marconi 19, 1920 Martigny, Switzerland
e-mail: amir.mohammadi@idiap.ch

I. Chingovska
e-mail: ivana.cingovska@gmail.com

A. Anjos
e-mail: andre.anjos@idiap.ch

S. Marcel
e-mail: marcel@idiap.ch

© Springer Nature Switzerland AG 2019
S. Marcel et al. (eds.), *Handbook of Biometric Anti-Spoofing*,
Advances in Computer Vision and Pattern Recognition,
https://doi.org/10.1007/978-3-319-92627-8_20

457

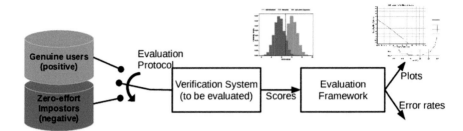

Fig. 20.1 Evaluation of a (unknown) verification system with regards to its capacity to discriminate genuine samples (positives) from zero-effort impostor samples (negatives). Data from each of the classes are fed into the verification system (treated as a black box) and the scores are collected. Collected scores are fed into an evaluation framework which can compute error rates and draw performance figures

20.1 Introduction

Biometric person recognition systems are widely adopted nowadays. These systems compare presentations of biometric traits to verify or identify a person. In the typical verification scenario, a biometric system matches a biometric presentation of a claimed identity against a prestored reference model of the same identity. The verification problem can be seen as a binary classification problem where presentations that are being matched against the same reference identity are considered positive samples (genuine samples) and the presentations that are being matched against another identity are considered negative samples (zero-effort impostor samples). Evaluation of verification systems as a binary classification problem is done using common metrics (error rates) and plots that are designed for binary classification problems. Figure 20.1 outlines such an evaluation framework.

Moreover, biometric systems are vulnerable to presentation attacks (PA, also known as spoofing). A printed photo of a person presented to a face recognition system with the goal of interfering with the operation of the system is an example of presentation attacks [1]. Presentation attack detection (PAD, also known as anti-spoofing) systems discriminate between bona-fide[1] (positives) and presentation attacks (negatives). The problem of PAs and PAD can be seen from different perspectives. As implied directly by the definition of the task of PAD systems, the problem is most often designed as a binary classification problem as outlined in Fig. 20.2.

On the other hand, presentation attacks are directed toward deceiving recognition systems,[2] regardless of whether there is a PAD algorithm to prevent them to do so, or not. From that perspective, the problem of PAs and PAD is not limited only to binary

---

[1] Bona-fide are also called real or live samples. Both genuine and zero-effort impostor samples are bona-fide samples. While zero-effort impostors are negative samples in a verification system, they are considered positive samples in a standalone PAD system (since they are not PAs).

[2] In this chapter, since we focus on the biometric recognition task, we will only consider PAs aiming to impersonate an identity and not to conceal (hide) an identity.

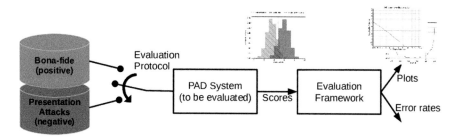

Fig. 20.2 Evaluation of a (unknown) PAD system with regards to its capacity to discriminate bona-fide samples (positives) from presentation attacks (negatives)

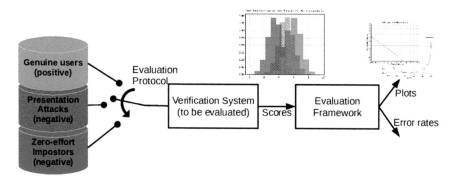

Fig. 20.3 Evaluation of a (unknown) verification system with regards to its capacity to discriminate genuine accesses from zero-effort impostors *and* presentation attacks

classification systems, as the isolated PAD systems are. It is of equal importance to transfer the problem understanding to the domain of biometric recognition systems (in particular, in this chapter, biometric verification systems).

Biometric verification under presentation attacks can be cast into a pseudo-ternary classification problem. While as binary classifiers, verification systems comply to typical evaluation methods, in this new perspective their concept and evaluation need to be changed accordingly. Figure 20.3 depicts these new settings. Instead of inputting a single set of negative examples, this new evaluation method requires two sub-classes of negative samples: samples coming from zero-effort impostors and the ones coming from presentation attacks.

This concept shift may influence the biometric verification systems at several levels. First of all, presentation attacks represent another class of input samples for the verification systems, which may cause changes in their internal algorithms to gain greater spoofing resilience. Two most prominent attempts for such changes are multimodal fusion [2–6] and fusion of a verification system with a PAD system [7–10]. Second, the problem restatement needs to modify the evaluation methods for verification systems. Finally, it may play a key role in the process of their parameterization.

While the first aspect of Presentation Attack (PA) and PAD problem under the umbrella of a verification system is out of the scope of this chapter, we will thoroughly inspect all the modifications that the evaluation methods need to undergo to accustom to the new setting. The main reason is that once the danger of presentation attacks is acknowledged, the verification performance of the biometric systems is not the only measurement of their quality. Important property to assess is their robustness to presentation attacks. Only in that case, one can say that the overall performance of the system is being estimated. In this context, by verification system, we could consider any type of system that can produce verification scores given a biometric sample as an input. No assumption on the mechanism the system employs for protection against presentation attacks, if any, is needed. The system may be solely any baseline biometric verification algorithm which disregards the hazard of presentation attacks, or a multimodal system or a fusion with a PAD algorithm. In any case, the system can be regarded as a black box, and the full evaluation can be done based on the verification scores it outputs for the input samples.

Mutual comparison of verification systems is the second matter of their evaluation with regards to presentation attacks. For example, it is of great importance to observe the performance change of a verification system before and after an integration with a PAD system. Blending in PAD into an existing verification system can increase its robustness to presentation attacks, but at the same time, it can affect its verification performance. The evaluation methodology which is going to be deployed should be able to assess the trade-off between these two effects.

Issues regarding the aspect of parameterization and tuning of the verification systems when presentation attacks have a non-negligible prior will be also touched upon in this chapter.

With the previous observations in mind, stating the problem of PAs from the perspective of a PAD system, as well as from the perspective of a verification system is the primary objective of this chapter (Sect. 20.2). Thorough review of the evaluation strategies for isolated presentation attack detection systems, as well as for verification systems commonly used in the literature will follow in Sect. 20.4. As a prerequisite, the concepts we are going to evaluate entail certain database structure that will be covered in Sect. 20.3.

20.2   Problem Statement

When treating PAD as a binary classification problem, designers are interested in determining the capacity of a given system to discriminate between bona-fide (positives) and presentation attacks (negatives).[3] These systems, which do not have any capacity to perform biometric verification, are only exposed to elements of these

---

[3]In this chapter, we shall treat examples in a (discriminative) binary classification system one wishes to keep as *positive class* or simply as *positives*, and, examples that should be discarded as *negative class* or *negatives*.

two classes. Figure 20.2 represents these settings in a block diagram. In order to evaluate a given system, one feeds data from each of the two classes involved on the assessment. Scores collected from the evaluated system are fed into an evaluation framework which can compute error rates or draw performance figures. This workflow, typical for evaluation of binary classification systems, is widely deployed by PAD developers as well [7, 11–16]. The database design and the evaluation of PAD systems comprise to the standards of general binary classification systems and will be revisited in Sects. 20.3.1 and 20.4.2, respectively.

A less considered perspective is how biometric verification systems treat presentation attacks. The classical approach puts biometric verification systems into the set of binary classifiers. Normally, such systems are designed to decide between two categories of verification attempts: bona-fide genuine users (positives) and the so-called bona-fide zero-effort impostors (negatives) [17]. Presentation attacks represent a new type of samples that can be presented at the input of this system. Considering that both presentation attacks and zero-effort impostors need to be rejected, it is still possible to regard the problem as a binary classification task where the genuine users are the positives, while the union of presentation attacks and zero-effort impostors are the negatives. Nevertheless, tuning of different properties of the verification system to make it more robust to presentation attacks may require a clearly separated class of presentation attacks. Furthermore, the correct ratio of presentation attacks and impostors in the negative class union is, at most times, unknown at design time. Applications in highly surveilled environments may consider that the probability of a presentation attack is small, while applications in unsurveilled spaces may consider it very high. Presentation attacks, therefore, should be considered as a third separate category of samples that the verification systems need to handle.

This viewpoint casts biometric verification into a pseudo-ternary classification problem as depicted in Fig. 20.3. Researchers generally simplify the pseudo-ternary classification problem so that it suits the binary nature of the verification systems. A common approach is to reduce it to two binary classification problems, each of which is responsible for one of the two classes of negatives. According to this, the verification system can be operating in two scenarios or operation modes: (1) when it receives genuine accesses as positives and only zero-effort impostors as negatives, and (2) when it receives only genuine accesses as positives and presentation attacks as negatives. Sometimes the first scenario is called a *normal operation mode* [18–20]. As it is going to be discussed in Sect. 20.4.3, it is beneficial to simplification that the positives (genuine accesses) that are evaluated completely match in both scenarios.

The workflow of the verification system confronted with presentation attacks, from the input to the evaluation stage, is represented in Fig. 20.3. The score histogram displays three distinctive groups of data: the positive class and the two negative ones. If the mixing factor between the negative classes is known at design time, system evaluation can be carried using known binary classification analysis tools. Since that is usually not the case, the evaluation tools for the verification systems need to be adapted to the new settings.

The new concept for verification systems explained above requires a database design and evaluation methodologies adapted to the enhanced negative class, regardless of the system's robustness to presentation attacks and how it is achieved. An overview of the research efforts in this domain will be given in Sects. 20.3.2 and 20.4.3, respectively.

20.3 Database Requirements

The use of databases and associated evaluation protocols allow for objective and comparative performance evaluation of different systems. As discussed on Sect. 20.2, the vulnerability (aka *spoofability*) of a system can be evaluated on isolated presentation attack detection systems, but also on fully functional verification systems. The simple evaluation of PAD requires only that database and evaluation protocols consider two data types: bona-fide and presentation attack samples. The evaluation of verification systems, merged with PAD or not, requires the traceability of identities contained in each presented sample, so that tabs are kept for probe-to-model matching and non-matching scenarios. The particular requirements for each of the two cases are given in Sects. 20.3.1 and 20.3.2. Databases for each of these two settings exist in the literature. An exhaustive listing of databases that allow for the evaluation of resilience against presentation attacks in isolated PAD or biometric verification systems is given by the end of this section, in Sect. 20.3.3.

20.3.1 Databases for Evaluation of Presentation Attack Detection Systems

The primary task of a database for evaluation of presentation attack detection systems is to provide samples of presentation attacks along with samples of bona-fide. The identity information of clients in each sample needs not to be present and can be discarded in case it is. The two sets of samples, which will represent the negative and the positive class for the binary classification problem, are just by themselves sufficient to train and evaluate a PAD system. It is a common practice that a database for binary classification provides a usage protocol which breaks the available data into three datasets [21]:

- *Training set* $\mathcal{D}_{train}$, used to train a PAD model;
- *Development set* $\mathcal{D}_{dev}$, also known as the validation set, used to optimize the decisions in terms of model parameters estimation or model selection;
- *Test set* $\mathcal{D}_{test}$, also known as the evaluation set, on which the performance is finally measured.

In the case of presentation attack databases, it is recommended that the three datasets do not contain overlapping client data to avoid bias related to client specific traits and to improve generalization [22]. A database with this setup completely satisfies the requirements of a two-class classification problem, as the isolated presentation attack detection is.

The process of generating presentation attacks requires bona-fide samples that will serve as a basis to create the fake copies of the biometric trait. These may or may not be the same samples as the bona-fide samples of the database. In any case, if they are provided alongside the database, it can be enhanced with new types of presentation attacks in future.

20.3.2 Databases for Vulnerability Analysis of Verification Systems

If a database is to serve for evaluation of a verification system, it needs to possess similar properties of a biometric database. Training and testing through biometric databases require (preferably) disjoint sets of data used for enrollment and verification of different identities. In practice, many databases also present a separation of the data in three sets as described above. Data from the training set can be used to create background models. The development data should contain enrollment (gallery) samples to create the user-specific models, as well as probe samples to match against the models. Similar specifications apply for the test set. The matching of the development probe samples against the user models should be employed to tune algorithms' parameters. Evaluation is carried out by matching probe samples of the test set against models created using the enrollment samples. The identity of the model being tested and the gallery samples are annotated to each of the scores produced so that the problem can be analyzed as a binary classification one: if model identity and probe identity match, the score belongs to the positive class (genuine client), otherwise, the score belongs to the negative class (zero-effort impostors). Usually, all identities in the three datasets are kept disjoint for the same reasons indicated in Sect. 20.3.1. Following this reasoning, a first requirement for a presentation attack database aspiring to be equally adapted to the needs of PAD and verification systems, is provision of separate enrollment samples, besides the bona-fide and presentation attack samples.

The pseudo-ternary problem of presentation attacks as explained in Sect. 20.2 imposes scenario for matching bona-fide genuine accesses, bona-fide zero-effort impostors, and presentation attacks against the models. In order to conform to this second requirement, the simplification of the pseudo-ternary problem introduced in Sect. 20.2 is of great help. In the case of the first scenario, or the normal operation mode, matching entries equivalent to the entries for genuine users and zero-effort impostors for a classical biometric verification database are needed. In the case of the second scenario, the provided entries should match the presentation attack samples to a corresponding model or enrollment sample.

To unify the terminology, we formalize the two scenarios of operation of the verification system as below:

Table 20.1 Creating licit and spoof scenarios out of the samples in a PA database. + stands for positives, − for negatives. L is for licit and S for spoof scenario. Note that the positives are the same for both L and S scenarios. Bona-fide enrollment samples will also be needed for each identity

Probe	Model for	A	B
A	Bona-fide	L+, S+	L−
	Presentation attack	S−	no match done
B	Bona-fide	L−	L+, S+
	Presentation attack	no match done	S−

- *Licit* scenario: A scenario consisting of genuine users (positives) and zero-effort impostors (negatives). The positives of this scenario are created by matching the genuine access samples of each client to the model or enrollment samples of the same client. The negatives can be created by matching the genuine access samples of each client to the model or enrollment samples of other clients. This scenario is suitable to evaluate a verification system in a normal operation mode. Evidently, no presentation attacks are present in this scenario;
- *Spoof* scenario: A scenario consisting of genuine users (positives) and presentation attacks (negatives). The positives of this scenario are created by matching genuine access samples of each client to the models or enrollment samples of the same client. The negatives are created by matching the presentation attacks of each client to the model or enrollment samples of the same client. No zero-effort impostors are involved in this scenario.

The licit scenario is necessary for evaluation of the verification performance of the system. The spoof scenario is necessary for evaluation of the system's robustness to PAs. If we follow a convention to match *all* the genuine access samples to the model or enrollment samples of the same client in both scenarios, we will end up having the same set of positives for the two scenarios. This agreement, as will be shown in Sect. 20.4.3, plays an important role in some approaches for evaluation of the verification systems.

To better illustrate how to create the scenarios out of the samples present in any presentation attack database, let us assume a simple hypothetical presentation attack database containing one bona-fide and one presentation attack of two clients with identities A and B. Let us assume that the database also contains bona-fide enrollment samples for A and B allowing computation of models for them. The matching of the samples with the models in order to create the positives and the negatives of the two scenarios is given in Table 20.1. To exemplify an entry in the table, L+ in the first row means that entries that match genuine accesses of client A to the model of client A belong to the subset of positives of the licit scenario. The same applies for L+ in the third row, this time for client B. Similarly, S− in the second row means that entries that match presentation attacks of client A to the model of client A belong to the subset of negatives in the spoof scenario.

Instead of creating a presentation attack database and then creating the licit and spoof scenario from its samples, an alternative way to start with is to use an existing biometric database which already has enrollment samples as well as data for the licit scenario. All that is needed is creating the desirable presentation attacks out of the existing samples. One should note, however, that the *complete* system used for the acquisition of samples, including the sensor, should be kept constant through all the recordings as differentiation may introduce biases. For example, consider a situation in which a speaker verification system is evaluated with data collected with a low-noise microphone, but in which attack samples are collected using noisier equipment. Even if attacks do pass the verification threshold, it is possible that potential PAD may rely on the additional noise produced by the new microphone to identify attacks. If that is the case, then such a study may be producing a less effective PAD system.

20.3.3   Overview of Available Databases for Presentation Attack Detection

Table 20.2 contains an overview of the existing PAD databases that are publicly available. The columns, that refer to properties discussed throughout this section, refer to the following:

- **Database**: the database name;
- **Trait**: the biometric trait on the database;
- **# Subsets**: the number of subsets in the database referring to existing separate set for training, developing and testing systems;
- **Overlap**: if there is client overlap between the different database subsets (training, development and testing);
- **Vulnerability**: if the database can be used to evaluate the vulnerability of a verification system to presentation attacks (i.e. contains enrollment samples);
- **Existing DB**: if the database is a spin-off of an existing biometric database not originally created for PAD evaluation;
- **Sensor**: If the sensors used to acquire the presentation attack samples are the same as those used for the bona-fide samples.

20.4   Evaluation Techniques

Several important concepts about evaluation of binary classification systems have been established and followed by the biometric community. Primarily, they are used to evaluate verification systems, which have a binary nature. They are also applicable to the problem of PAD as a binary classification problem.

Table 20.2 Catalog of evaluation features available on a few presentation attack databases available. For detailed column description, please see Sect. 20.3.3. This table is not an exhaustive list of presentation attack databases

Database	Trait	# Subsets	Overlap	Vulnerability	Existing DB	Sensor
ATVS-FFp[a] [20]	Fingerprint	2	No	No	No	Yes
LivDet 2009 [23]	Fingerprint	2	?	No	No	Yes
LivDet 2011 [16]	Fingerprint	2	?	No	No	Yes
LivDet 2013[b] [24]	Fingerprint	2	?	No	No	Yes
CASIA FAS[c] [25]	Face	2	No	No	No	Yes
MSU MFSD[d] [26]	Face	2	No	No	No	Yes
NUAA PI[e] [14]	Face	2	No	No	No	Yes
OULU-NPU[f] [27]	Face	2	No	No	No	Yes
Replay Attack[g] [28]	Face	3	No	Yes	No	Yes
Replay Mobile[h] [29]	Face	3	No	Yes	No	Yes
UVAD[i] [30]	Face	2	No	No	No	Yes
Yale Recaptured[j] [31]	Face	1	Yes	No	Yes	No
VERA Finger vein[k] [32–34]	Finger vein	2	No	Yes	No	Yes
VERA Palm vein[l] [35]	Palmvein	3	No	Yes	Yes	Yes
ASVSpoof 2017[m] [36]	Voice	2	No	Yes	Yes	No
AVSpoof[m] [37]	Voice	3	No	Yes	No	Yes
VoicePA[o] [38]	Voice	3	No	Yes	Yes	Yes

[a] http://atvs.ii.uam.es/atvs/ffp_db.html
[b] http://livdet.org/
[c] http://www.cbsr.ia.ac.cn/english/Databases.asp
[d] http://www.cse.msu.edu/rgroups/biometrics/Publications/Databases/MSUMobileFaceSpoofing/
[e] http://parnec.nuaa.edu.cn/xtan/data/nuaaimposterdb.html
[f] https://sites.google.com/site/oulunpudatabase/
[g] http://www.idiap.ch/dataset/replayattack
[h] http://www.idiap.ch/dataset/replaymobile
[i] http://ieeexplore.ieee.org/abstract/document/7017526/
[j] http://ieeexplore.ieee.org/abstract/document/6116484/
[k] https://www.idiap.ch/dataset/vera-fingervein
[l] https://www.idiap.ch/dataset/vera-palmvein
[m] http://dx.doi.org/10.7488/ds/2313
[n] https://www.idiap.ch/dataset/avspoof
[o] https://www.idiap.ch/dataset/voicepa

In Sect. 20.4.1, we revisit the basic notation and statistics for evaluation of any binary classification system. After that recapitulation, we give an overview of how the error rates and methodologies are adapted particularly for PAD systems in Sect. 20.4.2 and verification systems under presentation attacks in Sect. 20.4.3.

20.4.1 Evaluation of Binary Classification Systems

The metrics for evaluation of binary classification systems are associated to the types of errors and how to measure them, as well as to the threshold and evaluation criterion [39]. A binary classification system is subject to two types of errors: False Positive (FP) and False Negative (FN). Typically, the error rates that are reported are False Positive Rate (FPR), which corresponds to the ratio between FP and the total number of negative samples and False Negative Rate (FNR), which corresponds to the ratio between FN and the total number of positive samples.

Alternatively, many algorithms for binary classification report different error rates, but still equivalent to FPR and FNR. For example, True Positive Rate (TPR) refers to the ratio of correctly classified positives and can be computed as $1 -$ FNR. True Negative Rate (TNR) gives the ratio of correctly detected negatives, and can be computed as $1 -$ FPR.

To compute the error rates, the system needs to compute a decision threshold τ which will serve as a boundary between the output scores of the genuine accesses and presentation attacks. By changing this threshold one can balance between FPR and FNR: increasing FPR reduces FNR and vice-versa. However, it is often desired that an optimal threshold τ^* is chosen according to some criterion. Two well-established criteria are Minimum Weighted Error Rate (WER) and Equal Error Rate (EER) [39]. In the first case, the threshold τ^*_{WER} is chosen so that it minimizes the weighted total error rate as in Eq. 20.1 where $\beta \in [0, 1]$ is a predefined parameter which balances between the importance (cost) of FPR and FNR. Very often, they have the same cost of $\beta = 0.5$, leading to Minimum Half Total Error Rate (HTER) criteria. In the second case, the threshold τ^*_{EER} ensures that the difference between FPR and FNR is as small as possible (Eq. 20.2). The optimal threshold, also referred to as *operating point* should be determined using the data in the development set, denoted in the equations below as $\mathcal{D}_{dev}$.

$$\tau^*_{WER} = \arg \operatorname*{argmin}_{\tau} \beta \cdot \text{FPR}(\tau, \mathcal{D}_{dev}) + (1 - \beta) \cdot \text{FNR}(\tau, \mathcal{D}_{dev}) \qquad (20.1)$$

$$\tau^*_{EER} = \arg \operatorname*{argmin}_{\tau} |\text{FPR}(\tau, \mathcal{D}_{dev}) - \text{FNR}(\tau, \mathcal{D}_{dev})| \qquad (20.2)$$

Regarding the evaluation criteria, once the threshold τ^* is determined, the systems usually report the WER (Eq. 20.3) or its special case for $\beta = 0.5$, HTER (Eq. 20.4). Since in a real-world scenario the final system will be used for data which have not been seen before, the performance measure should be reported on the test set $\mathcal{D}_{test}$.

$$\text{WER}(\tau, \mathcal{D}_{test}) = \beta \cdot \text{FPR}(\tau, \mathcal{D}_{test}) + (1 - \beta) \cdot \text{FNR}(\tau, \mathcal{D}_{test}) \qquad (20.3)$$

$$\text{HTER}(\tau, \mathcal{D}_{test}) = \frac{\text{FPR}(\tau, \mathcal{D}_{test}) + \text{FNR}(\tau, \mathcal{D}_{test})}{2} \quad [\%] \qquad (20.4)$$

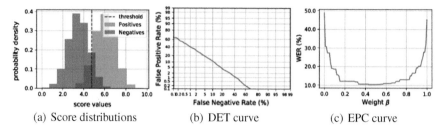

(a) Score distributions   (b) DET curve   (c) EPC curve

Fig. 20.4 Evaluation plots for a hypothetical binary classification system

Graphical Analysis

Important tools in evaluation of classification systems are the different graphical representations of the classification results. For example, one can get an intuition about how good the discriminating power of a binary classification system is by plotting its output score distributions for the positive and the negative class, as in Fig. 20.4a. Better separability between the two classes means better results in terms of error rates.

To summarize the performance of a system and to present the trade-off between FPR and FNR depending on the threshold, the performance of the binary classification systems are often visualized using Receiver Operating Characteristic (ROC) and Detection-Error Tradeoff (DET) [40] curves. They plot the FPR versus the FNR (or some of the equivalent error rates) for different values of the threshold. Sometimes, when one number is needed to represent the performance of the system in order to compare several systems, Area Under ROC curve (AUC) values are reported. Usually it is computed for ROC curves plotting FPR and TPR and in this case, the higher the AUC the better the system. Figure 20.4b illustrates the DET curve for a hypothetical binary classification system.

Unfortunately, curves like ROC and DET can only display a-posteriori performance. When reading values directly from the plotted curves, one implicitly chooses a threshold on a dataset and the error rates are reported on the same dataset. Although ROC and DET give a clear idea about the performance of a single system, as explained in [41], comparing two systems with these curves can lead to biased conclusions. To solve this issue, [41] proposes the so-called Expected Performance Curve (EPC). It fills in for two main disadvantages of the DET and ROC curves: 1. it plots the error rate on an independent test set based on a threshold selected a-priori on a development set; and 2. it accounts for the varying relative cost $\beta \in [0; 1]$ of FPR and FNR when calculating the threshold.

Hence, in the EPC framework, an optimal threshold τ^* is computed using Eq. 20.1 for different values of β, which is the variable parameter plotted on the abscissa. Performance for the calculated values of τ^* is then computed on the test set. WER, HTER or any other measure of importance can be plotted on the ordinate axis. The EPC curve is illustrated in Fig. 20.4c for a hypothetical classification system.

20.4.2 Evaluation of Presentation Attack Detection Systems

In the domain of PAD, bona-fide samples are the positive samples and presentation attacks are negative. Moreover, False Positive Rate (FPR) was renamed by the ISO standards [1] to Attack Presentation Classification Error Rate (APCER), and False Negative Rate (FNR) was renamed to Bona-Fide Presentation Classification Error Rate (BPCER). Before the ISO standardization, since the positives and the negatives are associated with the action of *acceptance* and *rejection* by the PAD system, False Accept Rate (FAR) and False Reject Rate (FRR) were used commonly in place of APCER and BPCER, respectively. Some publications utilize other synonyms which are listed in Table 20.3.

For a more general framework, where the system is specialized to detect any kind of suspicious or subversive presentation of samples, be it a presentation attack, altered sample or artifact, [44] has assembled a different set of notations for error measurements. Such a system reports False Suspicious Presentation Detection (FSPD) in the place of FNR and False Non-Suspicious Presentation Detection (FNSPD) in the place of FPR. To summarize the error rates into one value, some authors use accuracy [13, 31, 45], which is the ratio of the overall errors that the system made and the total number of samples. Finally, to graphically represent the performance of the PAD systems, score distribution plots [46], ROC, DET, and EPC curves are often used.

20.4.3 Evaluation of Verification Systems Under Presentation Attacks

The classical approach regards a biometric verification system as a binary classification system. In the scope of biometric verification systems, False Match Rate

Table 20.3 Typically used error rates in PAD and their synonyms

Error rate	Acronym	Synonyms
False positive rate	FPR	Attack presentation classification error rate (APCER), False accept rate (FAR), False spoof accept rate [9], False living rate (FLR) [15]
False negative rate	FNR	Bona-fide presentation classification error rate (BPCER), False reject rate (FRR), False alarm rate [11], False live rejection rate [9], False fake rate (FFR) [15]
True positive rate	TPR	True accept rate
True negative rate	TNR	True reject rate, detection rate [11, 12, 42] , detection accuracy [43]
Half total error rate	HTER	Average classification error (ACE) [15]

Table 20.4 Typically used error rates in biometric verification and their synonyms

Scenario	Error rate	Synonyms
Licit	Positive rate	False match rate (FMR), False accept rate (FAR) [9, 49], Pfa [7]
Spoof	False positive rate	Impostor attack presentation match rate (IAPMR), False accept rate (FAR) [18], Spoof false acceptance rate [3, 6], Liveness false acceptance rate [48], Success rate [19, 20], Attack success rate [49]
Both	False negative rate	False nonmatch rate (FNMR), False reject rate (FRR) [9, 49], Pmiss [7]
Both	True positive rate	True positive rate, True accept rate, Genuine acceptance rate [44, 50]
Union	False positive rate	Global false acceptance rate (GFAR) [9], System false acceptance rate (SFAR) [48]
	False negative rate	Global false rejection rate (GFRR)

(FMR) and False Nonmatch Rate (FNMR) are the most commonly used terms for the error rates FPR and FNR. FMR stands for the ratio of incorrectly accepted zero-effort impostors and FNMR for the ratio of incorrectly rejected genuine users. These and the equivalent error rates are often substituted with other synonyms which are different by different authors. The most common of them are listed in Table 20.4. Although not always equivalent [17], sometimes FMR and FNMR are substituted with FAR and FRR, respectively [47].

The simplification of ternary classification into two binary classification problems, as explained in Sect. 20.2, is the key step that set the standards for the evaluation of verification systems. Systems are usually evaluated separately in the two modes of operation associated with the two scenarios stated in Sect. 20.3.2. This section focuses on the error rates and plots typical for this evaluation.

While verification performance metrics are well established and widely used, metrics for PA evaluation is not as well defined and adopted. Some authors do not make a clear distinction between a presentation attack and a zero-effort impostor and refer to both types of samples as impostors. The nature of the sample can be concluded by the scenario in which it is being used: licit or spoof.

The importance of a clear distinction between the terminology for error rate reporting on misclassified zero-effort impostors and presentation attacks was outlined in [48]. Besides Liveness False Acceptance Rate (LFAR) as a ratio of presentation attacks that are incorrectly accepted by the system, [48] defines error rates connected to the total number of accepted negatives, regardless of whether they come from zero-effort impostors or presentation attacks. For example, the union of FAR in licit scenario and LFAR in spoof scenario is called System False Acceptance Rate (SFAR). However, since the introduction of [1], LFAR (also sometimes called Spoof False Accept Rate (SFAR)) was renamed to Impostor Attack Presentation Match Rate (IAPMR). A detailed overview of all the metrics utilized by various authors is given in Table 20.4. The table contains two metrics of error rates for negatives: for the licit and spoof scenario. It also reports the overall error rates that occur when both scenarios are considered as a union.

The adopted terminology in the remainder of this text is as follows:

- FNMR—ratio of incorrectly rejected genuine users (both licit and spoof scenario)
- FMR—ratio of incorrectly accepted zero-effort impostors (in the licit scenario)
- IAPMR—ratio of incorrectly accepted presentation attacks [3] (in the spoof scenario)
- GFAR—ratio of incorrectly accepted zero-effort impostors and presentation attacks.

Researchers generally follow three main methodologies for determining the effect of presentation attacks over the verification systems and obtaining the error rates. The differences between the three evaluation methodologies are in the way of computation of the decision threshold.

Evaluation Methodology 1

Two decision threshold calculations are performed separately for the two scenarios, resulting in two separate values of the error rate (HTER or EER) [3, 18, 51–53]. FNMR, FMR, and IAPMR are reported depending on the decision threshold obtained for the scenario they are derived from. One weak point of this type of evaluation is that it neglects that there is only one verification system at disposal and it should have only one operating point corresponding to one decision threshold. Furthermore, the decision threshold and the reported error rates of the spoof scenario are irrelevant in a real-world scenario. The problem arises because the spoof scenario assumes that all the possible misuses of the system come from spoofing attacks. It is not likely that any system needs to be tuned to operate in such a scenario. Therefore, the error rates depending on the threshold obtained under the spoof scenario are not a relevant estimate of the system's performance under presentation attacks. Furthermore, the error rates for the licit and spoof scenarios cannot be compared because they rely on different thresholds.

Evaluation Methodology 2

This methodology is adopted for more realistic performance evaluation. It takes advantage of the assumption that the licit and spoof scenarios share the same positive samples: a requirement mentioned to be beneficial in Sect. 20.3.2. In this case, the system will obtain the same FNMR for the both scenarios regardless of the threshold. Once the threshold of the system is chosen, FMR and IAPMR can be reported and compared. The threshold can be chosen using various criteria, but almost always using the licit scenario. Most of the publications report error rates for the two scenarios using a threshold chosen to achieve a particular desired value of FRR [2, 4–8, 19, 20, 49, 50, 54].

The issue that the evaluation methodology 2 oversees is that a system whose decision threshold is optimized for one type of negatives (for example, the zero-effort impostors), can not be evaluated in a fair manner for another type of negatives (the presentation attacks). If the system is expected to be exposed to two types of negatives in the test or deployment stage, it is fair that the two types of negatives play a role in the decision of the threshold in the development stage.

Evaluation Methodology 3

This methodology, introduced as Expected Performance and Spoofability (EPS) framework in [55], aims at filling in the gaps of the evaluation methodology 2 and establishes a criteria for determining a decision threshold which considers the two types of negatives and also the cost of rejecting positives. Two parameters are defined: $\omega \in [0, 1]$, which denotes the relative cost of presentation attacks with respect to zero-effort impostors; and $\beta \in [0, 1]$, which denotes the relative cost of the negative classes (zero-effort impostors and presentation attacks) with respect to the positive class. FAR_ω is introduced which is a weighted error rate for the two negative classes (zero-effort impostors and presentation attacks). It is calculated as in Eq. 20.5.

$$\text{FAR}_\omega = \omega \cdot \text{IAPMR} + (1 - \omega) \cdot \text{FMR} \tag{20.5}$$

The optimal classification threshold $\tau^*_{\omega,\beta}$ depends on both parameters. It is chosen to minimize the weighted difference between FAR_ω and FNMR on the development set, as in Eq. 20.6.

$$\tau^*_{\omega,\beta} = \arg\,\underset{\tau}{\text{argmin}}\ |\beta \cdot \text{FAR}_\omega(\tau, \mathscr{D}_{dev}) - (1 - \beta) \cdot \text{FNMR}(\tau, \mathscr{D}_{dev})| \tag{20.6}$$

Once an optimal threshold $\tau^*_{\omega,\beta}$ is calculated for certain values of ω and β, different error rates can be computed on the test set. Probably the most important is $\text{WER}_{\omega,\beta}$, which can be accounted as a measurement summarizing both the verification performance and the vulnerability of the system to presentation attacks and which is calculated as in Eq. 20.7.

$$\begin{aligned}\text{WER}_{\omega,\beta}(\tau^*_{\omega,\beta}, \mathscr{D}_{test}) &= \beta \cdot \text{FAR}_\omega(\tau^*_{\omega,\beta}, \mathscr{D}_{test}) \\ &\quad + (1 - \beta) \cdot \text{FNMR}(\tau^*_{\omega,\beta}, \mathscr{D}_{test})\end{aligned} \tag{20.7}$$

A special case of $\text{WER}_{\omega,\beta}$, obtained by assigning equal cost $\beta = 0.5$ to FAR_w and FNMR can be defined as HTER_ω and computed as in Eq. 20.8. In such a case, the criteria for optimal decision threshold is analogous to the EER criteria given in Sect. 20.4.2.

$$\text{HTER}_\omega(\tau^*_\omega, \mathscr{D}_{test}) = \frac{\text{FAR}_\omega(\tau^*_\omega, \mathscr{D}_{test}) + \text{FNMR}(\tau^*_\omega, \mathscr{D}_{test})}{2} \tag{20.8}$$

The parameter ω could be interpreted as relative cost of the error rate related to presentation attacks. Alternatively, it could be connected to the expected relative number of presentation attacks among all the negative samples presented to the system. In other words, it could be understood as the prior probability of the system being exposed to presentation attacks when it is deployed. If it is expected that there is no danger of presentation attacks for some particular setup, it can be set to 0. In this case, $WER_{\omega,\beta}$ corresponds to WER in the traditional evaluation scheme for biometric verification systems. When it is expected that some portion of the illegitimate accesses to the system will be presentation attacks, ω will reflect their prior and ensure they are not neglected in the process of determining the decision threshold.

As in the computation of WER in Sect. 20.4.2, the parameter β could be interpreted as the relative cost of the error rate related to the negative class consisting of both zero-effort impostors and presentation attacks. This parameter can be controlled according to the needs or to the deployment scenario of the system. For example, if we want to reduce the wrong acceptance of samples to the minimum, while allowing increased number of rejected genuine users, we need to penalize FAR_ω by setting β as close as possible to 1.

Graphical Analysis

Following the typical convention for binary classification system, biometric verification systems use score distributions, ROC or DET curves to graphically present their performance. The plots for a traditional biometric verification system regard the genuine users as a positive and the zero-effort impostors as a negative class. The details about these types of plots are given in Sect. 20.4.2.

When using graphical representation of the results, the researchers usually follow the evaluation methodology 2. This means that all the tuning of the algorithms, in particular in computation of the decision thresholds, is performed using the licit scenario, while the plots may represent the results for one of the scenarios or for the both of them.

When only the licit scenario is of interest, the score distribution plot contains the distributions only for the genuine users and the zero-effort impostors. If evaluation with regards to the vulnerability to presentation attacks is desired, the score distribution plot gets an additional distribution corresponding to the scores that the system outputs for the presentation attack samples in the spoof scenario. As a result, the score distribution plot presents three score distributions, which illustratively for a hypothetical verification system, are given in Fig. 20.5a.

An information about the dependence of IAPMR on the chosen threshold can be obtained directly from the score distribution plot. An example is shown in Fig. 20.5b, where the full red line represents how IAPMR varies with shifting the threshold, while the vertical dashed red line represents the threshold at a chosen operating point.

Typically, ROC and DET curves visualize the trade-off between FMR and FNMR for a biometric system with no danger of presentation attacks anticipated. The closest analogy to the ROC and DET curves when evaluating a system exposed to presentation attacks, can be found using the evaluation methodology 2. First, the curve using the licit scenario is plotted. Then, it can be overlaid with a curve for the spoof scenario. For the licit scenario, the horizontal axis represents FMR, while for the spoof scenario it represents IAPMR. However, meaningful comparison of the two curves is possible only if the number of genuine access samples in both licit and spoof scenario is the same. In such a case, a certain selected threshold will result in the same value of FNMR for the both scenarios. By drawing a horizontal line at the point of the obtained FNMR, one can examine the points where it cuts the curves for the licit and spoof scenario, and can compare FMR and IAPMR for the given system. Illustration of this analysis is given in Fig. 20.5c.

The drawback of the DET curve coming from its a-posteriori evaluation feature explained in [41] and obstructing fair comparison of two systems, is not a concern here. The plot does not compare different systems, but the same system with a single operating point under different set of negative samples.

As an alternative figure delivering similar information as DET, [2, 50] suggest to plot FMR Versus IAPMR. Thresholds are fixed in order to obtain all the possible values of FMR for the licit scenario and IAPMR is evaluated on the spoof scenario and plotted on the ordinate axis. By plotting the curves for different verification systems, the plot enables to compare which of them is less prone to spoofing given a particular verification performance. However, this comparison suffers from the same drawback as the DET: a-posteriori evaluation. As such, its fairness is limited. This plot is illustrated in Fig. 20.5d.

The logic for plotting the EPC curve is similar if one wants to follow the evaluation methodology 2. One has to vary the cost parameter β which balances between FMR and FNMR of the licit scenario and choose the threshold accordingly. Using the selected threshold, one can plot WER on the licit scenario. Afterwards, to see the method's vulnerability to presentation attacks depending on β, the WER curve can be overlaid with the IAPMR curve using the spoof scenario, as shown in Fig. 20.5e for a hypothetical system.

A graphical evaluation for the evaluation methodology 3 cannot be easily derived from the existing ROC or DET curves. The EPS framework computes error rates for a range of decision thresholds obtained by varying the parameters ω and β. The visualization of the error rates parameterized over two parameters will result in a 3D surface, which may not be convenient for evaluation and analysis, especially when one needs to compare two or more systems. Instead, plotting the Expected Performance and Spoofability Curve (EPSC) is suggested, showing $WER_{\omega,\beta}$ with respect to one of the parameters, while the other parameter is fixed to a predefined value. For example, we can fix the parameter $\beta = \beta_0$ and draw a 2D curve which plots $WER_{\omega,\beta}$ on the ordinate with respect to the varying parameter ω on the abscissa. Having in mind that the relative cost given to FAR_ω and FNMR depends mostly on the security preferences for the system, it is not difficult to imagine that particular values for β can be selected by an expert. Similarly, if the cost of IAPMR and FMR or the prior of presentation attacks with regards to the zero-effort impostors can be precisely estimated for a particular application, one can set $\omega = \omega_0$ and draw a 2D curve plotting $WER_{\omega,\beta}$ on the ordinate, with respect to the varying parameter β on the abscissa. Unlike for EPC, in the decision threshold calculation for EPSC both the licit and spoof scenario take place, because both FMR and IAPMR contribute with a certain weight.

The convenience of EPSC for evaluation of verification systems under presentation attacks is covered by several properties. Firstly, since it follows the evaluation methodology 3, it provides that both types of negatives participate in threshold

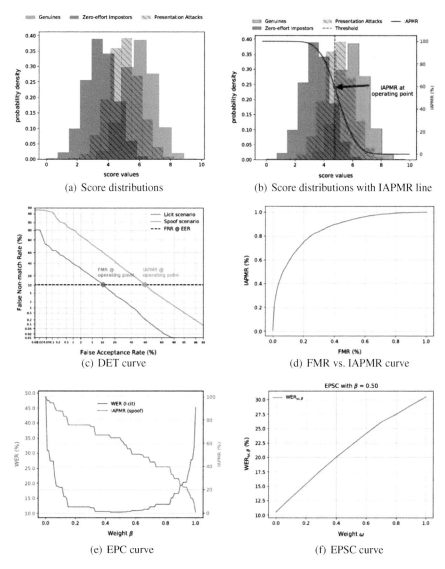

Fig. 20.5 Performance and spoofing vulnerability evaluation plots for hypothetical verification system

decision process. Second, it presents a-priori results: the thresholds are calculated on the development set, while the error rates are reported on the test set. This ensures unbiased comparison between algorithms. Furthermore, this comparison is enabled for a range of values for the cost parameters ω and β.

Besides $WEE_{\omega,\beta}$, other error rates of interest may be plotted on the EPSC plot, like IAPMR or FAR_ω.

20.5 Conclusions

Presentation attack detection systems in biometrics can rarely be imagined working as standalone. Their task is to perform an additional check on the decision of a biometric verification systems in order to detect a fraudulent user who possesses a copy of a biometric trait of a genuine user. Unless they have perfect detection rate, they inevitably affect the performance of the verification system they protect.

Traditionally, the presentation attack detection systems have been evaluated as binary classification systems, and in reason: by nature they need to distinguish between two classes - bona-fide and presentation attack samples. However, the above observation throws a light on the critical issue of establishing a methodology for evaluation of verification systems with regards to presentation attacks. This equally applies for verification systems with or without any mechanism for handling presentation attacks.

This task requires reformulation of the problem of biometric verification. They, as well, are, by definition, binary classification systems distinguishing between genuine accesses and zero-effort impostors. With the presentation attacks in play, the problem scales to pseudo-ternary classification problem, with two types of negatives: zero-effort impostors and presentation attacks.

As a result of the above observations, this chapter covers the problem of presentation attacks evaluation from two perspectives: evaluation of presentation attack detection systems alone and evaluation of verification systems with respect to presentation attacks. The evaluation in the first case means straightforward application of well-established evaluation methodologies for binary classification systems, in error rates (FAR, FRR, HTER, etc.), decisions on operating point (Minimum WER, EER, etc.) and graphical representation of results (ROC, DET, and EPC curves). The second perspective requires a simplification of the pseudo-ternary problem, in, for example, two binary classification problems. This, on the other hand, imposes certain database requirements, and presentation attacks databases which do not satisfy them can not be used for evaluation of biometric verification systems under presentation attacks. Depending on the steps undertaken to simplify the pseudo-ternary problem, the evaluation paradigm for the system differs. In particular, in this chapter, we discussed three evaluation methodologies, together with the error rates and the plots associated with them.[4]

As the interest for presentation attack detection in almost all biometric modes is growing both in research, but even more in industrial environment, a common fair criteria for evaluation of presentation attack detection systems and of verification systems under presentation attacks is becoming of essential importance. For the time being, there is a lot of inconsistency in the error rates conventions, as well as the evaluation strategies used in different publications.

---

[4]The software to reproduce the plots of this chapter is available in https://gitlab.idiap.ch/bob/bob.hobpad2.chapter20.

Acknowledgements The authors would like to thank the projects BEAT (http://www.beat-eu.org) and TABULA RASA (http://www.tabularasa-euproject.org) both funded under the 7th Framework Programme of the European Union (EU) (grant agreement number 284989 and 257289) respectively. The revision of this chapter was supported under the project on Secure Access Control over Wide Area Networks (SWAN) funded by the Research Council of Norway (grant no. IKTPLUSS 248030/O70). and by the Swiss Center for Biometrics Research and Testing.

References

1. Information technology – Biometric presentation attack detection – Part 3: Testing and reporting. Standard, International Organization for Standardization, Geneva, CH (2017). https://www.iso.org/standard/67381.html
2. Rodrigues RN, Ling LL, Govindaraju V (2009) Robustness of multimodal biometric fusion methods against spoofing attacks. J Vis Lang Comput 20(3):169–179
3. Johnson PA, Tan B, Schuckers S (2010) Multimodal fusion vulnerability to non-zero (spoof) imposters. In: IEEE international workshop on information forensics and security
4. Akhtar Z, Fumera G, Marcialis GL, Roli F. Robustness evaluation of biometric systems under spoof attacks. In: 16th international conference on image analysis and processing, pp 159–168
5. Akhtar Z, Fumera G, Marcialis GL, Roli F. Robustness analysis of likelihood ration score fusion rule for multi-modal biometric systems under spoof attacks. In: 45th IEEE international carnahan conference on security technology, pp 237–244
6. Akhtar Z, Fumera G, Marcialis GL, Roli F (2012) Evaluation of serial and parallel multibiometric systems under spoofing attacks. In: 5th IEEE international conference on biometrics: theory, applications and systems
7. Villalba J, Lleida E (2011) Preventing replay attacks on speaker verification systems. In: 2011 IEEE international carnahan conference on security technology (ICCST), pp 1–8
8. Marasco E, Johnson P, Sansone C, Schuckers S (2011) Increase the security of multibiometric systems by incorporating a spoofing detection algorithm in the fusion mechanism. In: Proceedings of the 10th international conference on Multiple classifier systems, pp 309–318
9. Marasco E, Ding Y, Ross A (2012) Combining match scores with liveness values in a fingerprint verification system. In: 5th IEEE international conference on biometrics: theory, applications and systems
10. Chingovska I, Anjos A, Marcel S (2013) Anti-spoofing in action: joint operation with a verification system. In: Proceedings of IEEE conference on computer vision and pattern recognition, workshop on biometrics
11. Pan G, Sun L, Wu Z, Lao S (2007) Eyeblink-based anti-spoofing in face recognition from a generic webcamera. In: IEEE 11th international conference on computer vision. ICCV 2007, pp 1–8
12. Bao W, Li H, Li N, Jiang W (2009) A liveness detection method for face recognition based on optical flow field. In: International conference on image analysis and signal processing. IASP 2009, pp. 233–236
13. yan J, Zhang Z, Lei Z, Yi D, Li SZ (2012) Face liveness detection by exploring multiple scenic clues. In: 12th international conference on control, automation, robotics and vision (ICARCV 2012), China
14. Tan X, Li Y, Liu J, Jiang L (2010) Face liveness detection from a single image with sparse low rank bilinear discriminative model. In: Proceedings of the european conference on computer vision (ECCV), LNCS 6316. Springer, Berlin, pp 504–517
15. Galbally J, Alonso-Fernandez F, Fierrez J, Ortega-Garcia J (2012) A high performance fingerprint liveness detection method based on quality related features. Future Gener Comput Syst 28(1):311–321

16. Yambay D, Ghiani L, Denti P, Marcialis G, Roli F, Schuckers S (2012) LivDet 2011 - fingerprint liveness detection competition 2011. In: 2012 5th IAPR international conference on biometrics (ICB), pp 208–215

17. Mansfield AJ, Wayman JL, Dr A, Rayner D, Wayman JL (2002) Best practices in testing and reporting performance

18. Galbally-Herrero J, Fierrez-Aguilar J, Rodriguez-Gonzalez JD, Alonso-Fernandez F, Ortega-Garcia J, Tapiador M (006) On the vulnerability of fingerprint verification systems to fake fingerprints attacks. In: IEEE international carnahan conference on security technology, pp 169–179

19. Ruiz-Albacete V, Tome-Gonzalez P, Alonso-Fernandez F, Galbally J, Fierrez J, Ortega-Garcia J (2008) Direct attacks using fake images in iris verification. In: Proceedings of the COST 2101 workshop on biometrics and identity management, BIOID, Springer, Berlin, pp 181–190

20. Galbally J, Fierrez J, Alonso-Fernandez F, Martinez-Diaz M (2011) Evaluation of direct attacks to fingerprint verification systems. Telecommun Syst, Spec Issue Biom 47(3):243–254

21. Hastie T, Tibshirani R, Friedman JH (2001) The elements of statistical learning: data mining, inference, and prediction: with 200 full-color illustrations. Springer, New York

22. Lui YM, Bolme D, Phillips P, Beveridge J, Draper B (2012) Preliminary studies on the good, the bad, and the ugly face recognition challenge problem. In: 2012 IEEE computer society conference on computer vision and pattern recognition workshops (CVPRW), pp 9–16

23. Marcialis GL, Lewicke A, Tan B, Coli P, Grimberg D, Congiu A, Tidu A, Roli F, Schuckers S (2009) First international fingerprint liveness detection competition – livdet 2009. In: Proceedings of the IAPR International Conference on Image Analysis and Processing (ICIAP), LNCS-5716, pp 12–23 (2009)

24. Ghiani L, Yambay D, Mura V, Tocco S, Marcialis G, Roli F, Schuckers S (2013) Livdet 2013 - fingerprint liveness detection competition. In: IEEE international conference on biometrics (ICB)

25. Zhiwei Z, Yan J, Liu S, Lei Z, Yi D, Li SZ (2012) A face antispoofing database with diverse attacks. In: Proceedings of the IAPR international conference on biometrics (ICB), pp 26–31

26. Wen D, Han H, Jain AK (2015) Face spoof detection with image distortion analysis. IEEE Trans Inf Forensics Sec 10(4):746–761. https://doi.org/10.1109/TIFS.2015.2400395

27. Boulkenafet Z, Komulainen J, Li L, Feng X, Hadid A (2017) OULU-NPU: A mobile face presentation attack database with real-world variations. In: 2017 12th IEEE international conference on automatic face and gesture recognition (FG 2017), IEEE, pp 612–618

28. Chingovska I, Anjos A, Marcel S (2012) On the effectiveness of local binary patterns in face anti-spoofing. In: Proceedings of the IEEE international conference of the biometrics special interest group (BIOSIG), pp 1–7

29. Costa-Pazo A, Bhattacharjee S, Vazquez-Fernandez E, Marcel S (2016) The REPLAY-MOBILE face presentation-attack database. In: 2016 International conference of the biometrics special interest group (BIOSIG), IEEE, pp 1–7. http://ieeexplore.ieee.org/abstract/document/7736936/

30. Pinto A, Schwartz WR, Pedrini H, de Rezende Rocha A (2015) Using visual rhythms for detecting video-based facial spoof attacks. IEEE Trans Inf Forensics Sec 10(5):1025–1038

31. Peixoto B, Michelassi C, Rocha A (2011) Face liveness detection under bad illumination conditions. In: 2011 18th IEEE international conference on image processing (ICIP), pp 3557–3560

32. Vanoni M, Tome P, El Shafey L, Marcel S (2014) Cross-database evaluation with an open finger vein sensor. In: IEEE workshop on biometric measurements and systems for security and medical applications (BioMS). http://publications.idiap.ch/index.php/publications/show/2928

33. Tome P, Vanoni M, Marcel S (2014) On the vulnerability of finger vein recognition to spoofing. In: IEEE international conference of the biometrics special interest group (BIOSIG). http://publications.idiap.ch/index.php/publications/show/2910

34. Tome P, Raghavendra R, Busch C, Tirunagari S, Poh N, Shekar BH, Gragnaniello D, Sansone C, Verdoliva L, Marcel S (2015) The 1st competition on counter measures to finger vein spoofing

attacks. In: The 8th IAPR international conference on biometrics (ICB). http://publications. idiap.ch/index.php/publications/show/3095

35. Tome P, Marcel S (2015) On the vulnerability of palm vein recognition to spoofing attacks. In: The 8th IAPR international conference on biometrics (ICB). http://publications.idiap.ch/index.php/publications/show/3096

36. Kinnunen T, Sahidullah M, Delgado H, Todisco M, Evans N, Yamagishi J, Lee KA (2017) The asvspoof 2017 challenge: assessing the limits of replay spoofing attack detection

37. Ergünay SK, Khoury E, Lazaridis A, Marcel S (2015) On the vulnerability of speaker verification to realistic voice spoofing. In: IEEE international conference on biometrics: theory, applications and systems (BTAS). https://publidiap.idiap.ch/downloads//papers/2015/KucurErgunay_IEEEBTAS_2015.pdf

38. Korshunov P, Goncalves AR, Violato RP, Simes FO, Marcel S (2018) On the use of convolutional neural networks for speech presentation attack detection. In: International conference on identity, security and behavior analysis

39. Poh N, Bengio S (2006) Database, protocols and tools for evaluating score-level fusion algorithms in biometric authentication. Pattern Recognition 2006

40. Martin A, Doddington G, Kamm T, Ordowski M (1997) The det curve in assessment of detection task performance. In: Eurospeech, pp 1895–1898

41. Bengio S, Keller M, Mariéthoz J (2003) The expected performance curve. Technical Report Idiap-RR-85-2003, Idiap Research Institute

42. Wang L, Ding X, Fang C (2009) Face live detection method based on physiological motion analysis. Tsinghua Sci Technol 14(6):685–690

43. Zhang Z, Yi D, Lei Z, Li S (2011) Face liveness detection by learning multispectral reflectance distributions. In: 2011 IEEE international conference on automatic face gesture recognition and workshops (FG 2011), pp 436–441

44. Johnson P, Lazarick R, Marasco E, Newton E, Ross A, Schuckers S (2012) Biometric liveness detection: Framework and metrics. In: International biometric performance conference

45. Gao X, Tsong Ng T, Qiu B, Chang SF (2010) Single-view recaptured image detection based on physics-based features. In: IEEE international conference on multimedia and expo (ICME). Singapore

46. Tronci R, Muntoni D, Fadda G, Pili M, Sirena N, Murgia G, Ristori M, Ricerche S, Roli F (2011) Fusion of multiple clues for photo-attack detection in face recognition systems. In: Proceedings of the 2011 international joint conference on biometrics, IJCB '11, IEEE Computer Society, pp 1–6

47. Jain AK, Flynn P, Ross AA (eds) (2008) Handbook of biometrics. Springer, Berlin

48. Adler A, Schuckers S (2009) Encyclopedia of biometrics, chap. security and liveness, overview, Springer, Berlin, pp 1146–1152

49. Galbally J, Cappelli R, Lumini A, de Rivera GG, Maltoni D, Fiérrez J, Ortega-Garcia J, Maio D (2010) An evaluation of direct attacks using fake fingers generated from iso templates. Pattern Recogn Lett 31(8):725–732

50. Rodrigues R, Kamat N, Govindaraju V (2010) Evaluation of biometric spoofing in a multimodal system. In: 2010 Fourth IEEE international conference on biometrics: theory applications and systems (BTAS)

51. Matsumoto T, Matsumoto H, Yamada K, Hoshino S (2002) Impact of artifical gummy fingers on fingerprint systems. In: SPIE proceedings: optical security and counterfeit deterrence techniques, vol 4677

52. Patrick P, Aversano G, Blouet R, Charbit M, Chollet G (2005) Voice forgery using alisp: indexation in a client memory. In: Proceedings of the IEEE international conference on acoustics, speech, and signal processing, 2005 (ICASSP '05), vol 1, pp 17–20

53. Alegre F, Vipperla R, Evans N, Fauve B (2012) On the vulnerability of automatic speaker recognition to spoofing attacks with artificial signals. In: 2012 Proceedings of the 20th european signal processing conference (EUSIPCO), pp 36–40

54. Bonastre JF, Matrouf D, Fredouille C (2007) Artificial impostor voice transformation effects on false acceptance rates. In: INTERSPEECH, pp 2053–2056
55. Chingovska I, Rabello dos Anjos A, Marcel S (2014) Biometrics evaluation under spoofing attacks. IEEE Trans Inf Forensics Sec 9(12):2264–2276 http://ieeexplore.ieee.org/xpls/abs_all.jsp?arnumber=6879440

Chapter 21
A Legal Perspective on the Relevance of Biometric Presentation Attack Detection (PAD) for Payment Services Under PSDII and the GDPR

Els J. Kindt

Abstract Payment applications turn in mass to biometric solutions to authenticate the rightful users of payment services offered electronically. This is due to the new regulatory landscape which puts considerable emphasis on the need of enhanced security for all payment services offered via internet or via other at-distance channels to guarantee the safe authentication and to reduce fraud to the maximum extent possible. The Payment Services Directive (EU) 2015/2366 (PSDII) which applies as of 13 January 2018 in the Member States introduced the concept of strong customer authentication and refers to 'something the user is' as authentication element. This chapter analyses this requirement of strong customer authentication for payment services offered electronically and the role of automated biometric presentation attack detection (PAD) as a security measure. PAD measures aid biometric (authentication) technology to recognize persons presenting biometric characteristics as friends or foes. We find that while PSDII remains vague about any obligation to use PAD as a specific security feature for biometric characteristics's use for authentication, PAD re-enters the scene through the backdoor of the General Data Protection Regulation (EU) 2016/679.

21.1 Introduction

1. Over the last years, (mobile) electronic online payments have considerably increased. Because such payments on a distance pose specific security risk, there is a growing urgency to implement appropriate security measures to secure such payment systems and to combat fraud. One of the risks is more specifically that

E. J. Kindt (✉)
KU Leuven – Law Faculty – Citip – iMec, Sint-Michielsstraat 6, 3000 Leuven, Belgium
e-mail: els.kindt@law.kuleuven.be

E. J. Kindt
Universiteit Leiden - Law Faculty - eLaw, Kamerlingh Onnes,
Steenschuur 25, 2311ES Leiden, The Netherlands

© Springer Nature Switzerland AG 2019
S. Marcel et al. (eds.), *Handbook of Biometric Anti-Spoofing*,
Advances in Computer Vision and Pattern Recognition,
https://doi.org/10.1007/978-3-319-92627-8_21

unauthorized persons request the execution of a payment transaction, typically using funds to their benefice but which are not their own.

2. Innovative systems thus come our way, attaching much importance on who we are and using our biometric characteristics, in addition to what we know and what we have. Some of these solutions are more original and newer than others. They range from smiling at the camera of your mobile phone, over using wearables deploying EKG readings tied to credit cards to swallowing eatable passwords which come in pill form dissolving in your stomach emitting a unique chemical signal that unlocks your account. All these solutions envisage to authenticate and to verify the identity of the user or the validity of the use of a specific payment instrument, and aim to enable authorized secure online payments. Biometric data use hereby will rapidly become mainstream. Mastercard, for example, announced the use of biometric data as 'the new normal' for safer online shopping.[1]

However, applications using biometric technology are prone to so-called 'spoofing'. Spoofing is a general fraudulent practice whereby one person presents falsified information, usually belonging to someone else, for gaining an illegitimate advantage.[2] In the context of payment services, the 'attacker' would attempt to authenticate as someone else. If the payment service uses a biometric solution, the attacker may attempt to spoof the system, for example, by presenting biometric characteristics of the rightful user.

3. PAD tools are security systems which aid biometric (authentication) technology to detect whether, when persons present biometric characteristics, such as a face or finger, there are indications that such characteristics are for example an artefact (e.g. a facial image on paper), have been altered (e.g. transplanted fingerprints) or were forced (e.g. by detecting voice emotion). This can be detected at the moment the characteristics are submitted to a sensor[3] but also at the level of a system. Such 'attacks', also known as biometric presentation attacks or 'spoofing', is now widely recognized and remains a weak point in biometric authentication applications.[4] It has restricted the take up of biometric authentication solutions

---

[1] See S. Clark, 'Mastercard to add biometric security to online transactions in Europe', 23. 1. 2018, available at https://www.nfcworld.com/technology/eu-payment-services-directive-ii-psd2/.

[2] Several kinds of information may hereby be falsified, such as a phone number someone is calling from, a URL for setting up a fraudulent website, but also an email address to mislead about the sender, etc.

[3] A sensor is also known as 'data capture subsystems'. The reason of such submission to a sensor is to be authenticated, e.g. at the border or for an electronic payment.

[4] About Presentation Attack, see also Ch. Busch e.a., *What is a Presentation Attack ? And how do we detect it ?*, 16. 1. 2018, Tel Aviv, also available at http://www.christoph-busch.de/about-talks-slides.html About standards for analyzing the effectiveness of direct attacks, countermeasures and more robust biometrics, see also the results of Tabula Rasa, a 7th framework research project funded by the EU Commission, at http://www.tabularasa-euproject.org and the results of the 7th Framework funded Biometrics Evaluation and Testing (BEAT) project in particular J. Galbally, J. Fierrez, A. Merle, L. Merrien and B. Leidner, *D4.6, Description of Metrics for the Evaluation of Vulnerabilities to Indirect Attacks*, BEAT, 27. 2. 2013, available at https://www.beat-eu.org/project/deliverables-public/d4.6-description-of-metrics-for-the-evaluation-of-vulnerabilities-to-indirect-attacks/view.

in applications which are not supervised or where biometric characteristics are collected over untrusted networks. Presentation Attack Detection (PAD) tools may therefore be a solution. At the same time, such PAD techniques vary in being effective, whereby trade-offs with respect to efficiency and security are made. Methods for assessing the performance of such tools have been discussed in international fora and standardization organization, including ISO, in order to lead to a general improvement of PAD tools.

4. In this chapter, we provide a first legal analysis as to the relevance and the need of security measures which detect presentation attacks to biometric data. Overall, technology will play an increasingly important role in finance, including in particular for personal data processing but also for biometric authentication. We focus on the specific domain of electronic (mobile) payments and the role of PAD for providing strong authentication. We hereby discuss PSDII and the GDPR (see below).[5]

21.2 PSDII and Presentation Attack Detection (PAD) for Electronic Payment Services

5. Over the last 15 years, the European legislator has issued common rules for payment services and payment service providers. These rules were needed to foster the common market for ecommerce and to increase competition. Those rules covered the rights and obligations of payment services providers and users while also aiming at harmonization of consumer protection legislation. It provided for a coherent legal framework for payment services in the European Union and in Iceland, Norway and Liechtenstein (European Economic Area). We take a closer look at this framework and whether it also imposes obligations when using biometric data for payment services.

21.2.1 Introduction into PSDII

6. The first Payment Services Directive 2007/64/EC was adopted in 2007, and is now modernized and replaced by a revised Directive on Payment Services (EU) 2015/2366 ('PSDII') which Member States shall implement in national law and apply by 13 January 2018.[6] This legislation is important because access to the payment market and services is *opened up to third parties*, other than the

---

[5]Our analysis does not include the eIDAS Regulation to the extent relevant.

[6]Directive (EU) 2015/2366 of the European Parliament and of the Council of 25 November 2015 on payment services in the internal market, amending Directives 2002/65/EC, 2009/110/EC and 2013/36/EU and Regulation (EU) No 1093/2010, and repealing Directive 2007/64/EC, *O.J.*, 23. 12. 2015, L 337/35 ('PSDII').

traditional banking institutions. PSDII hence applies to various kinds of payment institutions including not only credit institutions and electronic money institutions, but also third party payment service providers, which are hereby regulated.[7] Such third party payment service provider can be any company or business engaged in payment services, as listed and described in Annex 1 to PSDII, such as companies being engaged with the execution of payment transactions but also companies merely providing payment initiation services and account information services.[8] It means that a very broad range of businesses are covered by PSDII, and not only bank and credit institutions.[9] The security measures required under PSDII will be outlined in a Delegated Regulation. Such security measures are likely going to be applied as of mid 2019 (see below).

21.2.2 Strong Customer Authentication

7. PSDII deploys the notion of 'strong customer authentication' for requiring the security of payment transactions. Such authentication is needed to ensure the security of payments which are offered electronically and on a distance, for ensuring the protection of the users. The idea is to build and develop a sound environment for e-commerce where technologies are adopted to guarantee the safe authentication of the user and to reduce, to the maximum extent possible, the risk of fraud.[10]

8. 'Strong customer authentication' refers to the process of authentication and the longstanding theory that authentication can be ensured by using on, two or three factors: something someone *knows*, something one *has*, and something an individual *is*. 'Strong customer authentication' is defined in PSDII as 'authentication based on the use of *two or more* elements categorized as *knowledge* (something only the user knows), *possession* (something only the user possesses) and *inherence* (something the user is) that are independent, in that the breach of one does not compromise the reliability of the others, and is designed in such a way as to protect the confidentiality of the authentication data'.[11]

---

[7] About PSDII, see e.g. Ch. Riefa, 'Directive 2009/110/EC on the taking up, pursuit and prudential supervision of the business of electronic money institutions and Directive 2015/2366/EU on the control of electronic payments in the EU', in A. Lodder and A. Murray (eds.), *EU Regulation of e-Commerce*, Cheltenham, E. Elgar, 2017, 146–176.

[8] These companies are not necessarily engaged in payment operations. They could, e.g. merely combine and present information of different banking accounts to customers.

[9] For example, account information services (AIS), allowing customers and businesses to have a global view on their financial situation, for instance, by enabling consumers to consolidate the different payment accounts they may have with one or more banks in one (mobile) apps. These were before PSDII not specifically regulated.

[10] See recital 95 PSDII.

[11] Art. 4 (30) PSDII. See also the definition of 'authentication' in PSDII: 'authentication' means a procedure which allows the payment service provider to verify the identity of a payment service

'Inherence' or 'something the user is' hence refers to the use of biometric characteristics for authenticating the user of the payment service. While PSDII mentions explicitly nor imposes the use of biometric data, it refers indirectly to the
use of such data in the definition of strong customer authentication as a means
to produce strong customer authentication.[12] At the same time, and from the
definition it is clear that it is sufficient to use (at least) two elements (out of
the three) for strong customer authentication. Biometric data use is hence only
an option to obtain strong customer authentication. In these standards, security
features for 'devices and software that read elements categorized as inherence
(…)' are discussed, in particular, to mitigate the risks 'that those elements are
uncovered, disclosed to and used by unauthorized parties' (see below).

9. The methods and how strong customer authentication can be obtained, and
   whether for example a provider should include tools for biometric data use
   presentation attack detection, however, is not specified in PSDII as such. More
   detailed technical standards however are the subject of a delegated regulation,
   discussed below.

   The use of strong customer authentication will also play a critical role in the
   liability of parties involved. This will be further analyzed below.

10. There are exemptions to the need for strong customer authentication. Payments
    made through the use of an anonymous payment instrument for example are
    not subject to the obligation of strong customer authentication. However, if the
    anonymity of such instruments is lifted on contractual or legislative grounds, such
    payments become subject to the security requirements that follow from PSDII.
    Other exemptions exist for low-risk payments, such as low value contactless
    payments.

21.2.3 Delegation of Technical Standards on Security Aspects to the European Banking Authority

11. Further guidelines and technical standards on the security aspects, and in partic-
    ular with regard to strong customer authentication, needed to be adopted. PSDII
    empowered the Commission to adopt delegated and implementing acts to spec-
    ify how authorities and market participants shall comply with the obligations
    laid down in PSDII. The European legislator has subsequently delegated this to
    the European Banking Authority (EBA).

12. Of importance is that the technical standards have to be technology and business-
    model neutral. They should also not hamper innovation. Those regulatory tech-
    nical standards should be compatible with the different technological solu-

---

user or the validity of the use of a specific payment instrument, including the use of the user's
personalized security credentials;' (Art. 4(29) PSDII).

[12]The possibility to use biometric data is more prominent in the Regulatory Technical Standards
(see below).

tions available. When developing the Regulatory Technical Standards (RTS) on authentication and communication, PSDII also explains and requires that the privacy dimension should be systematically assessed and taken into account, in order to identify the risks associated with each of the technical options available and the remedies that could be put in place to minimize threats to data protection.[13] This is in our view also of interest if biometric solutions would be envisaged.

13. The EBA has then conducted in 2016 an open public consultation on the draft regulatory technical standards, analyzed the potential related costs and benefits and requested the opinion of the Banking Stakeholder Group established in accordance with Article 37 of Regulation (EU) No 1093/2010. The Banking Stakeholder Group published its final report in February 2017 with draft technical standards on *inter alia* strong customer authentication.[14] In this report, it made various recommendations for the technical standards, including with regard to the use of elements 'categorized as inherence'. We mention these also below.

21.2.4 Delegated Regulation on Reference Technical Standards

14. The EU Commission has published the Delegated Regulation on Regulatory Technical Standards for *inter alia* strong customer authentication and common and secure open standards on 27 November 2017 ('Regulation on RTS' or 'RTS').[15] The RTS came into force on 14 March 2018 and the requirements and standards will apply as from 14 March 2019 and as from 14 September 2019. In this regulation on RTS, it is stated that in order to ensure the application of strong customer authentication, it is also necessary to require *adequate security features* for the elements of strong customer authentication.

15. The aim is to mitigate the risk that those elements are 'uncovered, disclosed to and used by unauthorized parties'.[16] For 'elements categorized as inherence' and which, as stated, refer also to biometric information, Article 8 contains

---

[13]Recital 94 PSDII. About the need to respect data protection, see also e.g. Recital 89 and Recital 93 PSDII. See also *below*.

[14]European Banking Authority, *Final Report. Draft Regulatory Technical Standards on Strong Customer Authentication and common and secure communication under Article 98 of Directive 2015/2366 (PSD2)*, EBA/RTS/2017/02, 23. 2. 2017, 153 p., available at https://www.eba.europa.eu/documents/10180/1761863/Final+draft+RTS+on+SCA+and+CSC+under+PSD2+%20%28EBA-RTS-2017-02%29.pdf ('EBA, Final Report 2017').

[15]EU Commission, Commission Delegated Regulation (EU) 2018/389 of 27 November 2017 supplementing Directive 2015/2366 of the European Parliament and of the Council with regard to regulatory technical standards for strong customer authentication and common and secure open standards of communication, C(2017)7782final, *OJ L* 69, 13. 3. 2018, 23–43, available at ('Delegated Regulation RTS').

[16]Delegated Regulation RTS, Recital 6.

requirements of devices and software which are linked to such elements. It states the following:

'1. Payment service providers shall adopt *measures to mitigate the risk* that the authentication *elements* categorized as inherence and read by access devices and software provided to the payer *are uncovered by unauthorized parties*. At a minimum, the payment service providers shall ensure that those access devices and software *have a very low probability of an unauthorized party being authenticated as the payer*.

2. The use by the payer of those elements shall be subject to measures ensuring that those devices and the software *guarantee resistance against unauthorized use* of the elements through access to the devices and the software'. (emphasis added)

16. The text took several amendments which were suggested to the proposals of the standards into account (see below). Because of the amendments, the text is however difficult to read and to understand. What is meant by 'uncovered' ? This term seems to have not the same meaning as 'disclosed' ? In addition, while adopting these amendments, the original text and purposes have quite drastically changed. The original text stated that '…shall be characterized by security features including, but not limited to, (..) ensuring resistance against the risk of sensitive information related to the elements being disclosed to unauthorized parties'.[17] The text adopted presently requires security features in our view for any disclosure related to the elements and any use by an unauthorized party, and not only for sensitive information. As a minimum, and as a broader goal, it is more specifically required that providers ensure that those access devices and software have *a very low probability* of an unauthorized party being authenticated as the payer. This could refer to low false acceptance rates (FAR),[18] but it could also include the need for measures against spoofing.

17. During the public consultation, respondents asked the EBA to clarify the quality of biometric data, such as by reference to maximum/minimum false positive rates, mechanisms and procedures to capture biometric features, accepted biometric features (fingerprint, retina patter, facial recognition, etc.) and security measures used to store biometric data. The EBA, however, disagreed to add detail on the quality as a legal requirement. An important reason was the fear to undermine the objectives of technology neutrality and future proofing.[19]

18. Some respondents also asked the stakeholders group of the EBA to consider to define other features for biometric data, such as testing the strength of biometric data's use in authentication,[20] the fact that biometric characteristics could deteriorate over time, false positive and false negative parameters which could

---

[17]EBA, Final Report 2017, p. 61.

[18]About FAR and other technical aspects of biometric systems, see E. Kindt, *Privacy and Data Protection Issues of Biometric Applications. A Comparative Legal Analysis*, Dordrecht, Springer, 2013, 19–63 ("Kindt, Biometric Applications 2013").

[19]EBA, Final Report 2017, p. 61.

[20]It was suggested that this could be done by reference to the National Institute of Standards and Technology (NIST) draft focused on measuring Strength of Function for Authenticators – Biometrics (SOFA-B). See NIST, *Strength of Function for Authenticators – Biometrics (SOFA-B): Discussion Draft Open For Comments,* available at https://pages.nist.gov/SOFA/.

be adjusted (or attacked) to allow for impersonation, and mechanisms to re-calibrate biometric data (recapture fingerprint / face print, etc.). Others argued that behavioural biometrics should be accepted while one respondent, however, was of the view that using behavioural data as a standalone inherence element should be clearly excluded as an early technology which has to be monitored and tested in detail in combination with a specific threat model.[21] Important for this chapter is that it was also suggested to consider 'protection against presentation attacks'. The EBA, however, considered these specifications were too detailed and *feared* that these *could undermine future innovation*, and *the need to be technology- and business-model neutral*. The EBA suggested, however, that payment service providers may want to take such elements into consideration at the time of implementation.[22]

19. In general, it is fair to say that it is important to clarify and stress the general objective and more specifically, in this case, that limiting the misuse for the payment service purposes, in particular authentication by a perpetrator is aimed at. Too many technical requirements could hamper new business approaches. Overall, however, the technical guidance is quite below what can be useful and which can be enforced against the payment service provider.

A relevant question here is whether this leaves the payment service user with the burden to prove that enough nor appropriate security measures were in place to prevent fraudulent authentication and the misuse of funds, but also of the biometric data. Member States have to stipulate in national law that if a payment service user denies having authorized an executed payment transaction, the payment service *provider shall prove* that the payment transaction was authenticated, accurately recorded, entered in the accounts and *not affected by a technical breakdown or some other deficiency of the service* provided by the payment service provider.[23] If there is no specific obligation to protect biometric authentication by PAD, and no PAD installed at all, one may have difficulty to argue that non-installation of PAD by the provider is 'a technical breakdown'. Whether courts could consider it is some 'other deficiency' is to be awaited. While PSDII does also regulate liability for unauthorized or incorrectly executed payment transactions (see below), PSDII does not seem to regulate liability for losses in case of misuse of biometric data. The security, data protection by design and the liability rules under the GDPR, including the right of data subjects to lodge a complaint and to compensation, could therefore be helpful for the data subject here (see below). It also remains unsure what shall be understood by a 'very low probability'. The original text seemed to put the threshold less high and was

---

[21] Some respondents were also inviting to distinguishing between behavioural data in general and behavioural biometrics (such as typing recognition), arguing that the latter can very well be used. EBA, Final Report 2017, p. 61.

[22] *Ibid.* p. 63.

[23] Art. 72.1 PSDII. See further also Art. 72.2 PSDII.

less strict.[24] The preceding chapters in this book are therefore of particular relevance, setting the ways and methods to detect presentation attacks for various characteristics and how to test and compare PAD tools and rates.

20. The second paragraph of Article 8 of the Regulation on RTS with the requirements for hard- and software used for authentication is further quite broad. This could imply the need for access codes for accessing the biometric data, for example, but includes in our opinion also measures to resist perpetrators, such as during a spoofing attack, for example, by liveness detection but also by other PAD tools and ways. This will have to be further specified. The second paragraph of Article 8 could also be interpreted as requiring resistance against unauthorized access and use of biometric data as such. In this case, it would not be limited to security for authentication for purposes of the payment service, but could be far broader. But it is less clear whether it would include for example also the need for protection against access and the use for identity theft purposes.[25]

21. Finally, it is also required that each of the authentication factors are independent, so that the breach of one does not compromise the reliability of the others, in particular, when any of these elements are used through a multi-purpose device, namely, a device such as a tablet or a mobile phone which can be used both for giving the instruction to make the payment and in the authentication process.[26]

22. By way of summary, Article 8 Delegated Regulation RTS is quite general and does not contain more specific requirements for biometric data. Recital 6 Delegated Regulation RTS on the other hand provides some examples on how to avoid and to mitigate the risk that biometric data are 'uncovered, disclosed to and used by unauthorized third parties'. Recital 6 of the Delegated Regulation RTS reads as follows:

'In order to ensure the application of strong customer authentication, it is also necessary to require adequate security features for the elements of strong customer authentication, (…) for the devices and software that read elements categorized as "inherence" (something the user is) such as *algorithm specifications, biometric sensor and template protection features*, in particular to mitigate the risk that those elements are uncovered, disclosed to and used by unauthorized parties. (…)'.

In order to ensure the application of strong customer authentication, Recital 6 hence refers to adequate security features for biometric data such as algorithm specifications, biometric sensor and template protection features.[27] The wording is however somewhat confusing: does it refer to 'protection features' for 'biometric sensors' and 'templates' or to 'features' of 'biometric sensors' and

---

[24] The original text only required 'a sufficiently low likelihood of an unauthorized party being authenticated as the payer'. EBA, Final Report 2017, p. 61.

[25] Biometric data, already embedded in the EU ePassports of citizens who travel, will become increasingly object of desire and theft by those wanting to obtain identity documents or to engage in secure transactions but being deprived or already suspected.

[26] Delegated Regulation RTS, Recital 6.

[27] These examples were removed from the initial text of the draft articles of the Delegated Regulation RTS and mentioned in the Recitals.

'template protection'? And it is also not clear whether algorithm specifications would refer to PAD algorithms or mechanisms, while this is not excluded.[28]

23. Interpreting Recital 6 Delegated Regulation RTS in the way 'template protection' would be required, would in our view be favourable. However, it is not sure if this was intended. *Template protection* has a specific meaning in biometric data processing. It is an advanced form of protection of biometric data, which—rather than using 'raw' data (samples) or templates—the data is transformed in such way in that the data are *revocable, unlinkeable* and to some extent *irreversible*.[29] The EDPS and the Article 29 Working Party support the use of the biometric encryption technology, the latter stating in its Opinion 3/2012 in 2012 on new developments in biometric technologies that it '(…) has become sufficiently mature for broader public policy consideration, prototype development, and consideration of applications'.[30] Technology has in the meantime further developed.

21.2.5 Exceptions to the Need for Strong Customer Authentication and Other Provisions

24. There are a number of exceptions for RTS, including in case of low value payments below 30 euro and spending limits of 150 euro. If, for example, fingerprint is used for authorizing such payments below the threshold, the RTS, including Article 8 RTS, will not apply.
25. The fraud levels need to be reported to the competent authorities and also directly to EBA enabling it to conduct a review of the reference fraud rates within 18 months after the RTS enter into force.

Furthermore, the RTS specify the requirements for standards for secure communication between the institutions.

21.3 PAD: Re-Entry Through the 'Backdoor' of the General Data Protection Regulation ?

26. While PAD may not be explicitly mentioned and required for strong customer authentication solutions, one could question whether PAD should not be implemented under the provisions of the General Data Protection Regulation (EU)

---

[28] See *below* with regard to existing standard ISO/IEC 30107-1 and other parts adopted in 2016.

[29] This is also referred to sometimes as biometric information protection. See also ISO standard ISO/IEC 24745:2011 on biometric information protection.

[30] About template protection, see e.g. Kindt, Biometric Applications 2013, pp. 855–859.

2016/679, adopted in 2016 and applicable as of 25 May 2018 directly in all EU Member States ('General Data Protection Regulation' or 'GDPR').[31] The GDPR applies to the 'processing' of 'personal data', as defined, and hence will also apply if customer data are collected and used by payment services providers in (mobile) biometric payment solutions. In this view, we briefly discuss applicable privacy and data protection requirements hereunder. First, we mention how PSDII refers to data protection. We subsequently look at the GDPR.[32] We analyze in a succinct way to what extent the provisions of the GDPR may require PAD.

21.3.1 PSDII and Privacy and Data Protection in General

27. The legislator of PSDII realized that while being concerned for the security of electronic payments and the development of a competitive environment for e-commerce, more data of the payer will be needed, including for risk-assessments of unusual behaviour. For that reason, and as mentioned, PSDII pays specific attention to the need to implement and to respect data protection legislation and requires that the privacy dimension should be systematically assessed and taken into account, in order to identify the risks associated with each of the technical options available and the remedies that could be put in place to minimize threats to data protection.[33] This points directly to the need of a data protection impact assessment of personal data in specific cases as we will explain below, and in particular if biometric data would be used, for example, for authentication purposes.

28. PSDII also emphasizes the need to respect fundamental rights, including the right to privacy such as when competent authorities supervise the compliance of payment institutions. This should also be without prejudice to the control of national data protection authorities and in accordance with the Charter of Fundamental Rights of the European Union.[34]

---

[31] Regulation (EU) 2016/679 of the European Parliament and of the Council of 27 April 2016 on the protection of natural persons with regard to the processing of personal data and on the free movement of such data, and repealing Directive 95/46/EC (General Data Protection Regulation), *OJ L* 119, 4. 05. 2016, pp. 1–88, available at http://eur-lex.europa.eu/legal-content/EN/TXT/PDF/?uri=OJ:L:2016:119:FULL&from=NL ('Regulation (EU) 2016/679' or 'GDPR').

[32] Note that the GDPR also contains a new regime and a new definition (see Article 4 (14) GDPR) of biometric data. See also E. Kindt, 'Having Yes, Using No? About the new legal regime for biometric data', in *Computer Law and Security Report,* 523–538, 2018, available at http://authors.elsevier.com/sd/article/%20S0267364917303667.

[33] Recital 94 PSDII. See also recital 14 Delegated Regulation RTS.

[34] Recital 46 PSDII.

21.3.2 PSDII Requires Explicit Consent

29. Article 94.2 states that payment service providers shall only access, process
    and retain personal data which is necessary for the provision of their payment
    services, if the payment service users provide explicit consent. This article is
    general and somewhat confusing. The article on data protection is in our opinion
    too brief to avoid a good comprehension of its meaning. It seems at first sight
    also contradictory to the GDPR (see below). Under the GDPR, the necessity to
    process personal data under a contract to which the data subject is a party (or
    in order to take steps at the request of the data subject prior to entering into
    a contract) is a sufficient legal ground,[35] without the need for explicit consent
    in addition. The PSDII mentions both legal grounds as a double requirement?
    In addition, one could read that even with explicit consent, only the necessary
    personal data can be processed.[36] Furthermore, 'explicit consent' is more strict
    than consent, and is in the GDPR reserved and required for, for example, the
    processing of 'sensitive data' which includes the processing of biometric data for
    uniquely identifying, profiling and transfer to third countries. Would the same
    definition and principles in relation to consent and the GDPR apply?[37] This
    may well be possible as the GDPR is a *lex generalis*. We nevertheless plead for
    clarification.
30. Based on Article 94.2 PSDII, however, we understand that for access to biometric
    data for authentication purposes for a payment service, the explicit consent is
    required.

21.3.3 PAD and the GDPR

21.3.3.1 The General Data Protection Regulation and the Risk-Based Approach

31. The new General Data Protection Regulation has received much attention since
    its principles and obligations apply to almost all processing of personal data (safe
    specific restrictions[38]) for all sectors and, in case of non-compliance, allows for
    high administrative fines. While building further on the previous principles and

---

[35] See Art. 6.1(b) GDPR.

[36] See also Art. 66.3 (f) and (g) PSDII stating that payment initiation services 'shall not request
from the payment service user any data other than those necessary to provide the payment initiation
service' and 'not use, access or store any data for purposes other than for the provision of the
payment initiation service as explicitly requested by the payer'.

[37] See also Article 29 Working Party, Guidelines on Consent under Regulation 2016/679
(WP259rev0.1) July, 2018.

[38] See Art. 23 GDPR. The GDPR has a wide (material and territorial) scope, as detailed in the arts.
2–3 GDPR. It hence also applies as a general legislation for data processing activities to the payment
services sector.

obligations of data protection, it is important to retain that the GDPR stresses a specific approach in order to guarantee compliance with data protection requirements. In the first place, and which is also very relevant for this chapter, the new Regulation includes a so-called 'risk-based approach'. This implies that compliance with data protection legislation is not just a 'box-ticking exercise', but should really be about ensuring that personal data is sufficiently protected. It is not an entirely new concept. It means that data protection obligations for processing which is considered risky for the individuals concerned are to be complied with more rigour and are strengthened.[39] The risk-based approach is reflected not only in the obligation of security (Article 32) and the legal regime applicable to the processing of special categories of data (Article 9), which were already present in the previous legislation, but also in the obligation to carry out an impact assessment (Article 35) and other implementation measures such as the data protection by design principle (Article 25). The risk-based approach aligns with different levels of 'accountability obligations' *depending on the risk posed by the processing* in question, in particular, regarding data processing which, taking into account *the nature, scope, context, purposes* of the processing poses risks for data subjects. It should in any case *not* be seen as an alternative to well-established data protection rights and principles, but rather as a 'scalable and proportionate approach' to compliance.[40]

32. There is a general agreement that the use of biometric characteristics in applications poses specific risks for individuals for various reasons. Identity theft is just one but important example, which is as a risk also mentioned in recital 75 of the GPDR.[41] This implies, also in view of the explained risk-based approach, that specific scrutiny for the use of such data, also for payment services, will be required under the GDPR.

33. Furthermore, the GDPR in addition now also stresses as a general principle that controllers shall not only be responsible for compliance with the data protection obligations, but shall also be able to demonstrate compliance. This is referred to as the 'accountability' principle of controllers.[42] Payment services providers are—if controller—hence also bound by this principle. It requires that payment

---

[39]Control by authorities is likely to be more strict for such processing as compared to processing with relatively 'low risk'.

[40]See also the Article 29 Data Protection Working Party, *Statement on the role of a risk-based approach in data protection legal frameworks*, WP218, 30. 5. 2014, 4 p. available at http://ec.europa.eu/justice/data-protection/article-29/documentation/opinion-recommendation/files/2014/wp218_en.pdf. This document contains more clarifications on the not to be misunderstood risk-based approach.

[41]Some data protection authorities have mentioned the risk of identity theft and misuse of biometric data before. The Belgian Privacy commission, for example, mentioned the increased risk of identity theft in case biometrics are more commonly used as an authentication tool. CBPL, Opinion N17/2008 biometric data, 45–51. See also other studies, e.g. Teletrust, *White Paper zum Datenschutz*, 2008, 18–19. For an overview of the many risks, see Kindt, Biometric Applications 2013, 275–395.

[42]Art. 5.2 GDPR and art. 24.1 GDPR. The general principle is repeated in Article 24 as a specific obligation.

services providers as controllers, as well as their processors, shall implement appropriate technical and organizational measures. They shall hereby take the very nature of biometric data and the wide scope (for example, if applied to all payment service users) of the processing into account, as well as the risks of varying likelihood and severity for the rights and freedoms of natural persons. Furthermore, those measures shall be reviewed and updated where necessary.[43]

34. The GDPR hence requires when reviewing processing activities, to take into account the nature and risks. Operations with high risks hence require more protection measures. Mobile payment services deploying unique biometric characteristics which cannot be revoked for authentication should hence in our view be assessed, as we explain below.

21.3.3.2 The Use of Biometric Data for Strong Customer Authentication will Need a Data Protection Impact Assessment

35. One of the new obligations of the GDPR is the requirement for controllers to in addition conduct a data protection impact assessment (DPIA) whenever 'taking into account the nature, scope, context and purposes of the processing', such processing is 'likely to result in a high risk to the rights and freedoms of natural persons'. This applies the more if 'new technologies' are used.[44] Further guidelines for such DPIA have been provided by the Article 29 Working Party.[45] Biometric technology for authentication purposes, for example for payment services, is very likely to be considered such new technology.[46]

---

[43] Art. 24.1 GDPR. This is further reflected in several more specific obligations, stressing the burden of proof on the controllers and to keep records and evidence that they adhered to their obligations (for example, that an (explicit) consent was obtained). It is also reflected in the new obligation for controllers to make an impact assessment for processing operations 'if likely to result in high risks'.

[44] Art. 35.1 GDPR.

[45] See Article 29 Data Protection Working Party, *Guidelines on Data Protection Impact Assessment (DPIA) and determining whether processing is "likely to result in a high risk" for the purposes of Regulation 2016/679*, adopted on 4. 4. 2017 and last revised and adopted on 4. 10. 2017, WP 248rev.01, 21 p. ('WP 29 Guidelines on DPIA (WP248rev.01)'). The Article 29 Working Party will as of May 2018 be reformed in the European Data Protection Board ('EDPB'). Some national DPAs, such as in the United Kingdom and in France, have also provided more information and guidance on a DPIA in general, and in some case also specific for biometric data. See, e.g. France, CNIL, *Délibération n° 2016-187* of 30 June 2016 relating to the 'unique authorization' for access control to places, devices and computer applications in the workplace based on templates stored in a database (AU-053), 15 p., available at https://www.cnil.fr/sites/default/files/atoms/files/au-053. pdf , and *Grille D'Analyse*, 11 p., available at https://www.cnil.fr/fr/%20biometrie-un-nouveau-cadre-pour-le-controle-dacces-biometrique-sur-les-lieux-de-travail.

[46] This interpretation is in our view also suggested in the fore-mentioned Guidelines on DPIA. See WP 29 Guidelines on DPIA (WP248rev.01), pp. 9–10. The example where a DPIA is required is given therein: '8. Innovative use or applying technological or organizational solutions, like combining the use of fingerprint and face recognition for improved physical access control, etc.' (p. 9).

36. Moreover, a DPIA is in three specific cases always required, *inter alia* when special categories of data are processed *on a large scale*.[47] If biometric data are used for uniquely identifying, such processing is considered the processing of a special category of personal data, and consequently, such processing, upon the condition that it is *on a large scale*, and has a legal basis which permits such processing, would have to be submitted to an assessment exercise, named DPIA (formerly also known as a PIA).

37. As mentioned, conducting a DPIA is in accordance with the risk-based approach and global rationale of the GDPR that the controller is responsible and shall demonstrate compliance with the legislation as required by the new principle of being accountable for the processing (see above). The DPIA is herein an important tool. Such assessment shall consist of (a) a systematic description of the envisaged processing operations, in particular, the capture and use of the biometric data, the purposes of the processing, including, where applicable, the legitimate interest pursued by the controller; (b) an assessment of the necessity and proportionality of the processing operations in relation to the purposes; (c) an assessment of the risks to the rights and freedoms of data subjects; and (d) the measures envisaged to address the risks, including safeguards, security measures and mechanisms to ensure the protection of personal data and to demonstrate compliance with the GDPR taking into account the rights and legitimate interests of data subjects and other persons concerned.[48] As imposter attacks and identity theft belong to the risks, these shall be mentioned and appropriate measures, including but not limited to PAD for example, but also template protection mechanisms, should be considered and described.

38. Making an impact assessment under data protection legislation is a new obligation under the GDPR and creates lots of concern and questions. For example, as of when would a biometric use for uniquely identifying be on a large scale? To assess the scale, large-scale processing operations would aim at processing 'a considerable amount of personal data at regional, national or supranational level' and which 'could affect a large number of data subjects'.[49] This would be the case when a payment service provider would offer and implement to a large clientele biometric authentication.

39. The controller is responsible for the carrying out of the DPIA to evaluate, in particular, 'the origin, nature, particularity and severity' of the risk. The outcome of the assessment shall then be taken into account for determining the appropriate measures to be taken to mitigate the risks in order to demonstrate that the processing of the personal data complies with the Regulation. The GDPR hereby

---

[47]This is the second (explicit) scenario requiring a DPIA which expressly refers to the processing (a) on a large scale (b) of *special categories* of data of Article 9(1) or of Article 10. See Art. 35.3(b) GDPR.

[48]Art. 35.7 GDPR and recitals 84 and 90 GDPR. For addressing various risks, see also, e.g. Kindt, E., 'Best Practices for privacy and data protection for the processing of biometric data' in P. Campisi (ed.), *Security and Privacy in Biometrics*, 2013, Springer, pp. 339–367. For more guidelines on how to conduct such DPIA, see WP 29 Guidelines on DPIA (WP248rev.01).

[49]See Recital 91 GDPR.

takes up defined components of more general risk management processes, e.g. as known in ISO 31000 reviews. The international standard ISO/IEC 29134 will also provide for more guidelines on the methodology for such DPIA.[50] The DPIA shall always be done *before* the start of the processing. Where appropriate, the views of the data subjects shall also be sought.[51] Such DPIA for biometric data processing is an iterative process and each of the stages to be revisited multiple times before the DPIA can be completed. A DPIA is hence an important exercise which will require the necessary time, skills and insights in the biometric application, as well as the organizational and technical measures, but also a comprehension of the legal requirements. The controller is further free to publish the DPIA or not.

40. In case the DPIA indicates that the processing *would result in a high risk in the absence of or which the controller cannot mitigate by appropriate measures'* in terms of available technology and costs of im*plemen*tation', the controller will have to conduct a prior consultation with the supervisory authority.[52] Such prior consultation and authorization would only be required when *residual risks remain high*[53] and the data controller cannot find sufficient measures to cope with them. An example of an unacceptable high residual risk given by the Art. 29 WP is where 'the data subjects may encounter significant, or even irreversible, consequences, which they may not overcome, and/or when it seems obvious that the risk will occur.' In this context, the theft of biometric identity could be such risk. For available technology to cope with particular risks, one could think of for example the use of so-called 'protected biometric information' or 'protected templates'.[54]

21.3.3.3 Organizational and Technical Measures Required to Ensure Security Appropriate as Compared with the Risks

41. At the core of any personal data processing operation is the need and the obligation to ensure the security of the processing. Article 32 GDPR is explicit and requires on one hand that one shall take into account the 'state of the art' and 'the costs of implementation' of such security measures, but also the 'nature,

---

[50]See ISO/IEC 29134, Information technology—Security techniques —Privacy impact assessment—Guidelines, International Organization for Standardization (ISO).

[51]These views could be sought through a variety of means. See WP 29 Guidelines on DPIA (WP248rev.01), p. 13.

[52]Recital 84 GDPR. See Art. 36 GDPR.

[53]WP 29 Guidelines on DPIA (WP248rev.01), p. 18.

[54]See also EDPS, Opinion 1. 02. 2011 on a research project funded by the European Union under the 7th Framework Programme (FP 7) for Research and Technology Development (Turbine: TrUsted Revocable Biometric IdeNtitiEs), p. 3, available at http://www.edps.europa.eu/EDPSWEB/%20webdav/site/mySite/shared/Documents/Consultation/Opinions/%202011/11-02-01_FP7_EN.pdf; Kindt, Biometric Applications 2013, pp. 792–805. Recital 6 TRS (see above) may even require such protection.

scope, context and purposes of processing' and 'the risk of varying likelihood and severity for the rights and freedoms'. The aim is to ensure a level of appropriate protection taking into account the risk. We briefly mentioned that using biometric characteristics for authentication purposes remains subject to many risks, including the risk of (biometric) identity theft. It is hence required to protect against such risks. PAD in our view would be part of the required security measures. Standards for security evaluations of biometric systems have been or are developed on international level and are part of the 'state of the art'. Biometric sensor features should hence include *biometric-based attack detection* of impostors attempting to subvert the intended operation of the system. The ISO/IEC 30107 standard *part 1* presents the framework and describes the methods for detecting presentation attacks as well as obstacles, while *part 2* defines the data formats and *part 3* principles and methods for performance assessment of presentation attack detection algorithms or mechanisms. Although the ISO standards are not binding standards, they play an important role in adherence and compliance to the state of the art.

While PSDII hence may not be specific to the relevance, importance and requirement of PAD, the GDPR—because of the risk-based approach, the specific nature of biometric data and various specific obligations—would require in our view that PAD is—in accordance with the state of the art and costs—introduced and applied.

42. Furthermore, Article 32 GDPR explicitly mentions the use of pseudonymisation of the data (see also below), as well as processes for regular testing.[55] The latter points the more to the requirement to detect, evaluate and to test presentation attacks and to implement appropriate countermeasures.

43. Using PAD for payment services further to the GDPR is only but in line with the legislative goals and aims of PSDII. Various commentators have pointed to PSDII as an important factor in a wider and more general user acceptance of biometric data processing in electronic identity authentication and payment schemes.[56] PSDII requires that payment services offered by providers electronically shall be carried out *in a secure manner*, adopting technologies able to guarantee the safe authentication of the user and to reduce, to the maximum extent possible, the risk of fraud. The goals of PSDII and the RTS include explicitly ensuring appropriate levels of security based on effective and risk-based requirements but also ensuring the safety of the funds and personal data.[57] While PAD may not

---

[55]Art. 32.1(a) and (d) GDPR.

[56]See e.g. P. Counter, 'Unisys Says Biometrics Will Go Mainstream in 2018' in MobileID-World, 22. 1. 2018, available at https://mobileidworld.com/unisys-says-biometrics-mainstream-2018-901224/.

[57]Art. 98.2 PSDII.

be explicitly mentioned, it may be considered again upon review of the RTS.[58] In any case, we argue that in the meantime, the GDPR should play its role.

44. Payment providers from their side need to ensure in any case that the '*personalized security credentials*' are not accessible to parties other than the payment service user that is entitled to use the payment instrument. They shall also make sure that there are appropriate means available at all times to enable the payment service user to make a notification without undue delay on becoming aware of the loss, theft, misappropriation or unauthorized use of the payment instrument.[59] Personalized security credentials is defined as personalized features provided by the payment service provider to a user for the purposes of authentication. One could argue that biometric data could fall hereunder and hence—since many biometric characteristics could easily be captured in public places, that such information should specifically be protected, and PAD detected.

21.3.3.4   Data Minimization and Data Protection by Design and by Default

45. The GDPR imposes furthermore several other often detailed obligations, such as data minimization, including the use of pseudonyms, and the obligation to embed data protection into the design of data processing and to guarantee data protection by default. PSDII explicitly refers to this obligation. Recital 89 states that data protection by design and data protection by default should be embedded in all data processing systems developed and used within the framework of this Directive.

46. Following these obligations, PAD seems necessary in strong authentication solutions relying on biometric characteristics. The latter can easily be captured in ways unknown to the owner, and allowing for spoofing. Moreover, this could also imply in our opinion that not only PAD is needed as a security obligation, but also for example, template protection, as mentioned above. Such template protection allows to create *multiple biometric identities* for one and the same data subject, which can be regarded as pseudonyms,[60] reducing the possibility to link the biometric data to the data subject outside a particular domain or application but offering at the same time the advantage of enhanced claim or identity verification by using biometric characteristics.[61]

---

[58] See Art. 98.5 PSDII which requires the EBA to review and update the RTS on a regular basis.

[59] Art. 70 PSDII.

[60] See also the general comments in the Opinion 4/2007 of the Article 29 Working Party on pseudonymous and anonymous data, which are also relevant for the *processing* of biometric data.

[61] See also Kindt, Biometric Applications 2013, 792–806.

21.4 Liability

21.4.1 Liability Under PSDII

47. The use of strong customer authentication, including the enabling of PAD, is also relevant from a liability perspective.

    PSDII contains several provisions with regard to the liability for unauthorized or incorrectly executed payment transactions. In general, the payer's payment service provider is liable for unauthorized transactions. The payer's payment service provider shall *refund* the payer the amount of the unauthorized payment transaction immediately, and in any event no later than by the end of the following business day, after notification in the case of an unauthorized payment transaction. This does not apply, however, if the payment service provider *has reasonable grounds for suspecting fraud*. This requires also that the grounds be communicated by the latter to the relevant national authority in writing.[62] Other financial damages may be obtain under the national law which is applicable to the payment contract.

48. Member States may however opt for payers to bear the losses relating to any unauthorized payment transactions, up to a maximum of EUR 50,[63] resulting from the use of a lost or stolen payment instrument or from the misappropriation of a payment instrument. This is however only the case if the provider *did use strong customer authentication*. If the provider did not require and use strong customer authentication, for example because the amounts are low, the payer cannot be held to pay this amount and will not bear any financial losses (unless in case of fraud).[64]

    This liability regime could be an incentive for providers as well to use two or more factor authentication, which could include biometric authentication, as well as appropriate PAD.

21.4.2 High Administrative Sanctions Possible Under the GDPR and Right to Compensation

49. Another incentive to prevent unauthorized persons accessing payment services exists under the GDPR. Unauthorized access could for example be due to an incomplete or ineffective DPIA or inappropriate security measures, notwithstanding the obligation to provide such (see above).

---

[62] Art. 73 PSDII.

[63] This has been reduced as compared to Directive 2007/64/EC.

[64] Art. 74.2 PSDII.

In this case, the competent supervising authority may impose administrative fines up to 10 000 000, or in the case of an undertaking, up to 2 % of the total worldwide annual turnover of the preceding financial year, whichever is higher[65] or *double*, up to 20 000 000 euro or 4 % in case of non-compliance *inter alia* with the general principles or non observance of the orders of supervisory authorities.[66] The idea is however that any fines will be imposed from case to case, after assessments, and shall be 'effective, proportional and dissuasive'. The authorities shall further only impose fines which adequately respond to the 'nature, gravity and consequences of the breach', the results and measures taken.[67]

50. At the same time, data subjects who have suffered material or non-material damages, e.g. because of the use of biometric data in identity theft, and this is due to a breach of the GDPR, are entitled to receive compensation for the damages suffered.[68] Providers responsible for the processing of the data, could then only be exempted from liability if they prove that they are 'not in any way responsible for the event giving rise to the damage.'[69] It then remains to be seen whether payment services providers could validly invoke that PAD was not required.

21.5 Conclusions

51. We explained the need for adequate security features for online payments, including strong customer authentication. The so-called 'authentication elements' categorized as 'inherence' in the PSDII refer indirectly to the use of biometric characteristics. The regulation relating to PSDII, however, remains rather vague about any obligation to use existing and specific security features for such inherence elements, including features such as relating to the need for PAD.

52. The use of biometric characteristics in such authentication procedure, however, also qualifies as the use of biometric data as defined in the General Data Protection Regulation (EU) 2016/679. This Regulation imposes *inter alia* appropriate security measures, data protection impact assessments and data protection by design when new technologies are used, such as biometric technologies.

---

[65] Art. 83.4 GDPR.

[66] Arts. 83.5–83.6 GDPR.

[67] See art. 83 GDPR and also Rec. 150. Supervisory authorities must assess all the facts of the case in a manner that is consistent and objectively justified. About how such sanctions shall be applied, see also Article 29 Data Protection Working Party, *Guidelines on the application and setting of administrative fines for the purposes of the Regulation 2016/679*, WP253, 3. 10. 2017, 17 p.

[68] See art. 82 GDPR.

[69] Art. 82.3 GDPR.

We argued that because of the obligation to implement security measures required for 'inherence' (something the user is) such security obligation will bring along with it the need to review and implement privacy and personal data protections for biometric data used as well. This is positive. This 'security paradigm' may hence prove to be helpful this time to promote privacy. Nevertheless, it remains an indirect way to address the need for privacy and personal data protection for biometric data in such online payment environment. At the same time, it remains as welcome as any other way to point to the need for privacy and personal data protection for such very particular data. PAD, while not being explicitly required in PSDII, therefore re-enters through the backdoor of the GDPR, requiring security measures and a DPIA. Spoofing vulnerabilities as well as PAD shall hence from a legal point be adequately described and assessed in a DPIA under the GDPR.

Acknowledgements This article has been made possible in part by funding received from the European Union's 7th Framework Programme for research and innovation in the context of the EKSISTENZ project under grant agreement no 607049. The viewpoints in this article are entirely those of the author and shall not be associated with any of the aforementioned projects, persons or entities. The author nor the Commission may be held responsible for any use that may be made of the information herein contained.

Chapter 22
Standards for Biometric Presentation Attack Detection

Christoph Busch

Abstract This chapter reports about the relevant international standardization activities in the field of biometrics and specifically describes standards on presentation attack detection that have established a framework including a harmonized taxonomy for terms in the field of liveness detection and spoofing attack detection, an interchange format for data records and moreover a testing methodology for presentation attack detection. The scope and of the presentation attack detection multipart standard ISO/IEC 30107 is presented. Moreover, standards regarding criteria and methodology for security evaluation of biometric systems are discussed.

22.1 Introduction

Biometric systems are characterized by two essential properties. On the one hand, functional components or subsystems are usually dislocated. While the enrolment may take place as part of an employment procedure with the personal department or as part of an ePassport application in the municipality administration, the biometric verification takes place likely at a different location, when the data subject (e.g., the staff member) is approaching a certain enterprise-gate (or any other physical border gate) or when the citizen is traveling with his new passport. On the other hand, while the biometric enrolment is likely to be a supervised capture process and often linked with training on sensor interaction and guided by an operator, on the contrary, such supervision does not exist for the verification process. Further, the verification is often conducted based on a probe sample that was generated in a unsupervised capture process. In consequence of the verification not only usability of the sensor and

C. Busch (✉)
Hochschule Darmstadt and CRISP (Center for Research in Security
and Privacy), Haardtring 100, 64295 Darmstadt, Germany
e-mail: christoph.busch@h-da.de

© Springer Nature Switzerland AG 2019
S. Marcel et al. (eds.), *Handbook of Biometric Anti-Spoofing*,
Advances in Computer Vision and Pattern Recognition,
https://doi.org/10.1007/978-3-319-92627-8_22

ease of human–machine interaction is essential but also a measure of confidence that the probe sample was indeed collected from the subject that was initially enrolled and not from an biometric artefact that was presented by an attacker with the intention to pretend the presence of the enrollee. Such attacks are now becoming more relevant also for mobile biometric use cases and application like biometric banking transactions. Thus, industry stakeholders such as the FIDO-alliance as well as financial institutions, which are about to implement the security requirements of the European payment service directive (PSD2), are continuously concerned about unsupervised biometric authentication and robustness of biometric sensors.

In most cases, the comparison of a probe biometric sample with the stored biometric reference will be dislocated from the place of enrolment. Some applications store the reference in a centralized or decentralized database. More prominent are token-based concepts like the ICAO ePassports [1] since they allow the subject to keep control of his/her personal biometric data as the traveling individuals decide themselves, whether and when they provide the token to the controlling instance. The same holds for many smartphone-based biometric authentication procedures, where the reference is stored in secure memory.

The recognition task is likely to fail, if the biometric reference is not readable according to a standardized format. Any open system concept does, therefore, require the use of an open standard in order to allow that for the recognition task a component from a different supplier can be used. The prime purpose of a biometric reference is to represent a biometric characteristic. This representation must, on the one hand, allow a good biometric performance but at the same time, the encoding format must fully support the interoperability requirements. Thus, encoding of a biometric sample (i.e., fingerprint image or face image) according to the ISO/IEC Biometric data interchange format [2] became a prominent format structure for many applications.

For all data interchange formats, it is essential to store along with the representation of the biometric characteristic essential information (metadata) on the capture process and the generation of the sample. Metadata that is stored along with the biometric data includes information such as size and resolution of the image but also relevant data that impacted the data capturing process: Examples are the Capture Device Type ID that identifies uniquely the device, which was used for the acquisition of the biometric sample and also the certification block data that reports the certification authority, which had tested the capture device and the corresponding certification scheme that was used for this purpose. These data fields are contained in standardized interchange records [2–4].

An essential information that was furthermore considered helpful for the verification capture process is a measure to describe the reliability of the capture device against presentation attacks (a.k.a spoofing attacks). This becomes more pressing once the capture device and the decision subsystems are dislocated. Thus, ISO/IEC has developed in recent years a multipart standard that is covering this issue and will be introduced in this chapter. The chapter will outline the strategy behind this standardization process, cover the framework architecture and taxonomy that was established and will discuss the constraints that had to be considered in the respective standardization projects.

22.2   International Standards Developed in ISO/IEC JTC

International standardization in the field of information technology is driven by a Joint Technical Committee (JTC 1) formed by the International Organization for Standardization (ISO), and the International Electrotechnical Commission (IEC). An important part of the JTC1 is the Sub-Committee 37 (SC37) that was established in 2002. First standards developed in SC37 became available in 2005 and have found wide deployment in the meantime. More than 900 million implementations, according to SC37, standards are estimated to be in the field at the time of this writing. Essential topics that are covered by SC37 include the definition of a $Harmonized Biometric Vocabulary$ (ISO/IEC 2382-37) that removes contradictions in the biometric terminology [5, 6], a harmonized definition of a General Biometric System (ISO/IEC SC37 SD11) that describes the distributed subsystems, which are contained in deployed systems [7], a common programming interface $BioAPI$ (ISO/IEC 19784-1) that supports ease of integration of sensors and SDKs [8] and also the definition of data interchange formats. SC37 has over its first 15 years of work concentrated on the development of the ISO/IEC 19794 family, which includes currently the following 14 parts:

- Part 1: Framework
- Part 2: Finger minutiae data
- Part 3: Finger pattern spectral data
- Part 4: Finger image data
- Part 5: Face image data
- Part 6: Iris image data
- Part 7: Signature/Sign time series data
- Part 8: Finger pattern skeletal data
- Part 9: Vascular image data
- Part 10: Hand geometry silhouette data
- Part 11: Signature/Sign processed dynamic data
- Part 12: - void -
- Part 13: Voice data
- Part 14: DNA data
- Part 15: Palm crease image data

The framework part includes relevant information that is common to all subsequent modality specific parts such as an introduction of the layered set of SC37 standards and an illustration of a general biometric system with a description of its functional subsystems namely the capture device, signal processing subsystem, data storage subsystem, comparison subsystem, and decision subsystem [2]. Furthermore, this framework part illustrates the functions of a biometric system such as enrolment, verification, and identification and explains the application context of biometric data interchange formats. Part 2–Part 15 then detail the specification and provide modality related data interchange formats for both image interchange and template interchange on feature level. The 19794-family gained relevance, as the

International Civic Aviation Organization (ICAO) adopted image-based representations for finger, face, and iris in order to store biometric references in electronic passports. Thus, the corresponding ICAO standard 9303 [1] includes a normative reference to ISO/IEC 19794. ICAO estimated in June 2015 that there were over 700 million Electronic Passports issued by 112 member states of ICAO.

22.3   The Development of Presentation Attack Detection Standard ISO/IEC 30107

For more than a decade, along with the enthusiasm for biometric technologies the insight into potential risks in biometrics systems was developed and is documented in the literature [9–11]. Within the context of this chapter, the risks of subversive attacks on the biometric capture device became a major concern in unsupervised applications. Over the years, academic and industry research developed countermeasures in order to detect biometric presentation attacks that constitute a subversive activity. For a survey on attacks and countermeasures regarding face-, iris- and fingerprint recognition the reader is directed to [12–14].

From a general perspective, a presentation attack can be conducted from an outsider that interacts with a biometric capture device but could as well be undertaken from an experienced insider. However, the need to develop a harmonized perspective for presentation attacks that are conducted by biometric capture subjects became obvious. Thus, the motivation to develop a standard that is related to liveness detection and spoofing was supported from stakeholders of all three communities that are active in SC37, namely, from industry (essentially representatives from vendors working fingerprint-, vein-, face-, and iris-modality), from academia and research projects (e.g., European projects on liveness detection) as well as from governmental agencies (e.g., responsible for testing laboratories). The latter took the lead and have started the development. Since then experts from the biometric community as well as from the security community have intensively contributed to the multipart standard that is entitled "ISO/IEC Information Technology - Biometric presentation attack detection" [15–17]. The intention of this standard is to provide a harmonized definition of terms and a taxonomy of attack techniques, data formats that can transport measured robustness against said attacks and a testing methodology that can evaluate PAD mechanisms.

The objectives of a standardization project are best understood by analyzing the scope clause. For the Presentation Attack Detection (PAD) standard the scope indicates that it aims at establishing beyond the taxonomy, terms and definitions a specification and characterization of presentation attack detection methods. A second objective is to develop a common data format devoted to presentation attack assessments and a third objective is to standardize principles and methods for performance assessment of PAD-algorithms. This field of standardization work becomes sharpened, when topics that are outside of the scope are defined: Outside of this standardization project are definitions of specific PAD detection methods as well as detailed information about countermeasures that both are commonly valuable IPR of the industrial stakeholders. In addition, a vulnerability assessment of PAD is out of scope at this point in time.

22.4 Taxonomy for Presentation Attack Detection

Literature and science, specifically in a multidisciplinary community (as in biometrics), tends to struggle with a clear and noncontradicting use and understanding of its terms. Thus, ISO/IEC has undertaken significant efforts to develop a Harmonized Biometric Vocabulary (HBV) [6] that contains terms and definitions useful also in the context of discussions about presentation attacks. Without going into detail of the terminology definition process it is important to note that biometric *concepts* are always discussed in context (e.g., of one or multiple biometric subsystems) before a *term* and its *definition* for said concept can be developed. Thus, terms are defined in groups and overlap of groups ("concept clusters") and the interdependencies of its group members necessarily lead to revision of previously found definitions. The result of this work is published as ISO/IEC 2382-37:2017 [5] and is also available online [6]. It is of interest to consider here definitions in the HBV, as they are relevant for the taxonomy and terminology defined in ISO/IEC 30107 [15–17]. The following list contains definitions of interest:

- **biometric characteristic**: biological and behavioral characteristic of an individual from which distinguishing, repeatable biometric features can be extracted for the purpose of biometric recognition (37.01.02)
- **biometric feature**: numbers or labels extracted from biometric samples and used for comparison (37.03.11)
- **biometric capture subject**: individual who is the subject of a biometric capture process (37.07.03)
- **biometric capture process**: collecting or attempting to collect a signal(s) from a biometric characteristic, or a representation(s) of a biometric characteristic(s,) and converting the signal(s) to a captured biometric sample set (37.05.02)
- **impostor**: subversive biometric capture subject who attempts to being matched to someone else's biometric reference (37.07.13)
- **identity concealer**: subversive biometric capture subject who attempts to avoid being matched to their own biometric reference (37.07.12)
- **subversive biometric capture subject**: biometric capture subject who attempts to subvert the correct and intended policy of the biometric capture subsystem (37.07.17)
- **subversive user**: user of a biometric system who attempts to subvert the correct and intended system policy (37.07.18)
- **uncooperative biometric capture subject**: biometric capture subject motivated to not achieve a successful completion of the biometric acquisition process (37.07.19)

- **uncooperative presentation**: presentation by a uncooperative biometric capture subject (37.06.19)

In order to formulate a common understanding of attacks on biometric systems, the list of above terms was expanded with the following concepts that provided in ISO/IEC 30107-1 Biometric presentation attack detection—Part1: Framework [15] and in ISO/IEC 30107-3 Biometric presentation attack detection—Part3: Testing and reporting [16]:

- **presentation attack/attack presentation**: presentation to the biometric data capture subsystem with the goal of interfering with the operation of the biometric system
- **bona fide presentation**: interaction of the biometric capture subject and the biometric data capture subsystem in the fashion intended by the policy of the biometric system
- **presentation attack instrument (PAI)**: biometric characteristic or object used in a presentation attack
- **PAI species**: class of presentation attack instruments created using a common production method and based on different biometric characteristics
- **artefact**: artificial object or representation presenting a copy of biometric characteristics or synthetic biometric patterns
- **presentation attack detection (PAD)**: automated determination of a presentation attack

Note that the use of the above terms are recommended and similar terms such as $fake$ should be deprecated despite their intense previous use in the literature. In the development of ISO/IEC 30107 a framework was defined to understand presentation attack characteristics and also detection methods. Figure 22.1 illustrates the potential

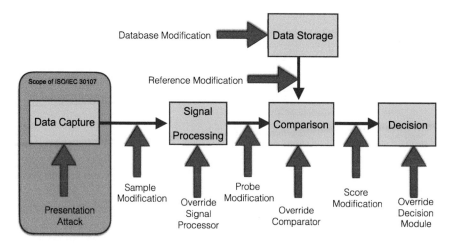

Fig. 22.1 Examples for points of attacks (following [15])

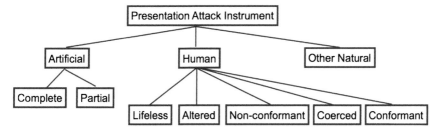

Fig. 22.2 Categories of attack instrument used in presentation attacks (following [15])

targets in a generic biometric system [7] that could be attacked. In this chapter, and moreover, in this book, we concentrate only on attack presentations that occur at the capture device.

The framework defined in [15] considers two types of attacks. On the one hand side, the *Active Impostor Presentation Attack* is considered, which attempts to subvert the correct and intended policy of the biometric capture subsystem and in which the attacker aims to be recognized as a specific data subject known to the system (e.g., an impersonation attack). On the other hand, the framework considers an *Identity Concealer Presentation Attack* as attempt of the attacker to avoid being matched to its own biometric reference in the system.

An attacker, be it an active impostor or an identity concealer, will use an object for his attack that is interacting with the capture device. Moreover, the potential of his attack will depend on his knowledge, the window of opportunity and other factors that we will discuss in Sect. 22.6. However, for the object that is employed the standard widens the scope from *gummy fingers* and considers various categories of objects that could be used in a presentation attack. Figure 22.2 illustrates that aside from artificial objects (i.e., *artefacts*) natural material could be used. When the expected biometric characteristic from an enrollee is absent and replaced by an *attack presentation characteristic* (i.e., the attack presentation object) this could be a human tissue from a deceased person (i.e., a cadaver part) or it could be an altered fingerprint [18], which is targeting on distortion or mutilation of a fingerprint— likely from an *Identity Concealer*. Moreover, an attacker might present his genuine characteristic but identification is avoided with nonconformant behavior with respect to the data capture regulations, e.g., by extreme facial expression or by placing the tip or the side of the finger on a sensor. But attack objects can also include other natural material such as onions or potatoes.

Detailed information about countermeasures (i.e., presentation attack detection techniques) to defend the biometric system against presentation attacks are out of scope of the standard in order to avoid conflicts of interests for industrial stakeholders. However, the standard does discuss a general classification in terms of detection on the level of a biometric subsystem (e.g., artefact detection, liveness detection, alteration detection, nonconformance detection) and through detection of noncompliant interaction in violation with system security policies (e.g., geographic or temporal exception)

Table 22.1 Selected data fields included in a PAD record according to ISO/IEC 30107-2

Field name	Valid values	Notes
PAD decision	0,1	Optional
PAD score	0–100	Mandatory
Level of supervision	1–5	

22.5 Data Formats

One of the objectives of the ISO/IEC 30107 multipart standard is to transport information about the presentation attack detection results from the capture device to subsequent signal processing or decision subsystems. The container to transmit such information is the open data interchange format (DIF) according to the ISO/IEC 19794 series [2]. This subsection outlines the conceptual data fields that are defined for a PAD record in ISO/IEC 30107-2 [16]. A selection of fields is illustrated in Table 22.1. It indicates that the result of the PAD functionality should be encoded as a scalar value in the range of 0 to 100 in analogy to the encoding of the sample quality assessment that is potentially also conducted by the capture device and stored as a quality score according to ISO/IEC 29794-1 [19].

The PAD decision is encoded with abstract values NO_ATTACK or ATTACK. The PAD score shall be rendered in the range between 0 and 100 provided by the attack detection techniques. Bona fide presentations shall tend to generate lower scores. Presentation attacks shall tend to generate higher scores. The abstract value FAILURE_TO_COMPUTE shall indicate that the computation of the PAD score has failed. The level of supervision expresses the surveillance during the capture process. Possible abstract values are UNKNOWN, CONTROLLED, ASSISTED, OBSERVED, UNATTENDED.

In the absence of standardized assessment methods, a PAD score would be encoded on a range of 0 (i.e., indicative of an attack) to 100 (i.e., indicative of a genuine capture attempt) the reliability that the transmitted biometric sample can be trusted. Any decision based on this information is at the discretion of the receiver. The described PAD data record is likely to become an integral part of the representation header in ISO/IEC 19794-x. Remaining challenges are to allow optional encoding since a capture device may or may not encode such additional information and further to achieve backwards compatibility with already deployed system that needs to parse DIFs according to 19794-1:2006 or 19794-1:2011.

22.6 Testing and Reporting

In order to evaluate the reliability of a PAD record that is transmitted two evaluations are foreseen. The first relevant information is to report for the capture device, which was encoding the interchange record, a meaningful performance testing result that was generated by an independent testing laboratory. Test procedures as such are well known since the biometric performance testing standards ISO/IEC 19795-1 was established in 2006 [20]. The framework for Biometric Performance Testing and Reporting was developed on the basis of established concepts such as the *Best Practices in Testing and Reporting Performance of Biometric Devices* [21] and it defines in which way algorithm errors such as false-match-rate (FMR) and false-non-match-rate (FNMR) as well as system errors such as false-accept-rate (FAR) and false-reject-rates (FRR) must be reported. For testing of presentation attack detection unfortunately such established concepts did not exist in the past. Thus, various national approaches have been proposed and were discussed as the standard ISO/IEC 30107 was developed. However, some metrics appear familiar to a testing expert and are indeed derived from biometric performance testing metrics.

An evaluation should determine whether artefacts with abnormal properties are accepted by the traditional biometric system and can result in higher-than-normal acceptance against bona fide enrolee references. This can be measured with the (1) Impostor Attack Presentation Match Rate (IAPMR): defined as: "in a full system evaluation of a verification system, the proportion of impostor attack presentations using the same PAI species in which the target reference is matched."

Moreover, when it comes to the testing of the detection component ISO/IEC 30107-3 introduces three levels of PAD evaluation namely: (1) PAD subsystem evaluation: This level evaluates only a PAD system, which is either hardware or software based (2) Data Capture subsystem evaluation: This will evaluate the data capture subsystem that may or may not include the PAD algorithms but is focused more on the biometric sensor itself (3) Full system evaluation: provided the end-to-end system evaluation.

Metrics for PAD Subsystem Evaluation

The PAD subsystem is evaluated using two different metrics, namely, [17]: (1) Attack Presentation Classification Error Rate (APCER): defined as the proportion of presentation attacks incorrectly classified as *Bona Fide* presentations (2) Bona Fide Presentation Classification Error Rate (BPCER): defined as the proportion of *Bona Fide* presentations incorrectly classified as presentation attacks.

The APCER can be calculated as follows:

$$APCER = \frac{1}{N_{PAIS}} \sum_{i=1}^{N_{PAIS}} (1 - RES_i) \tag{22.1}$$

where, N_{PAIS} is the number of attack presentations for the given Presentation Attack Instrument (PAI) [15]. RES_i takes the value 1 if the ith presentation is classified as an attack presentation and value 0 if classified as *Bona Fide* presentation.

While, the BPCER can be calculated as follows:

$$BPCER = \frac{\sum_{i=1}^{N_{BF}} RES_i}{N_{BF}} \tag{22.2}$$

where N_{BF} is the number of *Bona Fide* presentations. RES_i takes the value 1 if the ith presentation is classified as an attack presentation and value 0 if classified as *Bona Fide* presentation.

22.7 Conclusion and Future Work

This chapter introduces the standardization work that began a few years back in the area of presentation attack detection. Now that metrics are well established the published standards can contribute to mature taxonomy of presentation attack detection terms in future literature. The encoding details of the PAD interchange record are ready to be deployed in large-scale applications.

A challenge with the now established testing and reporting methodology is that unlike for biometric performance testing aka *technology testing* a large corpus of testing samples (i.e., PAI species) cannot be assumed to be available. Top national laboratories are in possession of no more than 60 PAI species for a fingerprint recognition system. In this case, it becomes essential that the proportion is computed not to a potentially large number of samples all of one single PAI species that are all of similar material properties and stemming from the same biometric source. At least the denominator should be defined by the number of *PAI species*. Note that one single artefact species would correspond to the set of fingerprint artefacts all made with the same *recipe* and the same materials but with different friction ridge patterns from different fingerprint instances. A complementary measure to the APCER is the BPCER, which should always be reported jointly.

An essential difference of PAD testing is that obviously there is beyond the mere statistical observations as expressed by APCER and BPCER metrics the need to categorize the attack potential itself. Such methodology is well established in the scope of Common Criteria testing that developed the Common Methodology for Information Technology Security Evaluation [22]. It might be desirable to replace the indication of a *defined level of difficulty* according to an attack potential attribute of a biometric presentation attack expressing the effort expended in the preparation and execution of the attack in terms of elapsed time, expertise, knowledge about the capture device being attacked, window of opportunity and equipment, graded as *no rating*, *minimal*, *basic*, *enhanced-basic*, *moderate* or *high*. Such gradings are established in Common Criteria testing and would allow a straightforward understanding of a PAD result for security purposes.

By separation of work tasks in ISO/IEC JTC1 discussion of security-related topics is not in scope of ISO/IEC 30107. However, the Common Criteria concept of attack potential should be seen as both a good categorization for the criticality of an attack and the precondition to conduct later a security evaluation based on the results of a ISO/IEC 30107 metric. However, this link needs to be established, and thus there is space for many activities as future work.

This work has started recently with the ISO/IEC 19989 multipart standard, which will be developed in the years ahead.

References

1. International Civil Aviation Organization NTWG: machine readable travel documents – Part 4 – specifications for machine readable passports (MRPs) and other TD3 size MRTDs (2015). http://www.icao.int/publications/Documents/9303_p4_cons_en.pdf
2. ISO/IEC JTC1 SC37 Biometrics: ISO/IEC 19794-1:2011 Information technology - biometric data interchange formats – Part 1: framework (2011). International organization for standardization
3. ISO/IEC JTC1 SC37 Biometrics: ISO/IEC 19794-4:2011 Information technology – biometric data interchange formats – Part 4: finger image data (2011). International organization for standardization
4. ISO/IEC JTC1 SC37 Biometrics: ISO/IEC 19794-5:2011. Information technology - biometric data interchange formats – Part 5: face image data (2011). International organization for standardization
5. ISO/IEC JTC1 SC37 Biometrics: ISO/IEC 2382-37 harmonized biometric vocabulary (2017). International organization for standardization
6. ISO/IEC JTC1 SC37 Biometrics: ISO/IEC 2382-37 harmonized biometric vocabulary (2017). http://www.christoph-busch.de/standards.html
7. ISO/IEC JTC1 SC37 Biometrics: ISO/IEC SC37 SD11 general biometric system (2008). International organization for standardization
8. ISO/IEC TC JTC1 SC37 Biometrics: ISO/IEC 19784-1:2006. Information technology – biometric application programming interface – Part 1: bioAPI specification (2006). International organization for standardization
9. Zwiesele A, Munde A, Busch C, Daum H (2000) Biois study - comparative study of biometric identification systems. In: IEEE computer society, 34th annual 2000 IEEE international carnahan conference on security technology (CCST), pp 60–63
10. Matsumoto T, Matsumoto H, Yamada K, Yoshino S (2002) Impact of artificial "Gummy" fingers on fingerprint systems. In: SPIE conference on optical security and counterfeit deterrence techniques IV, vol 4677, pp 275–289
11. Schuckers SA (2002) Spoofing and anti-spoofing measures. In: Information security technical report 4, Clarkson University and West Virginia University, USA
12. Raghavendra R, Busch C (2017) Presentation attack detection methods for face recognition systems: a comprehensive survey. ACM Comput Surv 50(1):8:1–8:37. https://doi.org/10.1145/3038924
13. Galbally J, Gomez-Barrero M (2016) A review of iris anti-spoofing. In: 4th international conference on biometrics and forensics (IWBF), pp 1–6
14. Sousedik C, Busch C (2014) Presentation attack detection methods for fingerprint recognition systems: a survey. IET Biom 3(1):1–15
15. ISO/IEC JTC1 SC37 Biometrics: ISO/IEC 30107-1. Information technology - biometric presentation attack detection – Part 1: framework (2016). International organization for standardization
16. ISO/IEC JTC1 SC37 Biometrics: ISO/IEC 30107-2. Information technology - biometric presentation attack detection – Part 2: data formats (2017). International organization for standardization

17. ISO/IEC JTC1 SC37 Biometrics: ISO/IEC 30107-3. Information technology - biometric presentation attack detection – Part 3: testing and reporting (2017). International organization for standardization

18. Ellingsgaard J, Busch C (2017) Altered fingerprint detection. Springer International Publishing, Cham, pp 85–123. https://doi.org/10.1007/978-3-319-50673-9_5

19. ISO/IEC JTC1 SC37 Biometrics: ISO/IEC 29794-1:2016 information technology - biometric sample quality – Part 1: framework (2016). International organization for standardization

20. ISO/IEC JTC1 SC37 Biometrics: ISO/IEC 19795-1:2017. Information technology – biometric performance testing and reporting – Part 1: principles and framework (2017). International organization for standardization and international electrotechnical committee

21. Mansfield T, Wayman J (2002) Best practices in testing and reporting performance of biometric devices. CMSC 14/02 Version 2.01, NPL

22. Criteria C (2012) Common methodology for information technology security evaluation - evaluation methodology. http://www.commoncriteriaportal.org/cc/

Glossary

ACE: Average Classification Error
Anti-Spoofing (term to be deprecated): countermeasure to spoofing
APCER: Attack Presentation Classification Error Rate
ASV: Automatic Speaker Verification
AUC: Area Under ROC

BEAT: Biometrics Evaluation and Testing
BPCER: Bona Fide Presentation Classification Error Rate

CFCCIF: Cochlear Filter Cepstral Coefficients with Instantaneous Frequency
CNN: Convolutional Neural Network
CQCC: Constant Q Cepstral Coefficients
CQT: Constant Q Transform

DET: Detection-Error Tradeoff

EER: Equal Error Rate
EPC: Expected Performance Curve
EPSC: Expected Performance and Spoofability Curve

FAR: False Accept Rate
FerrFake: Rate of misclassified fake fingerprints, see APCER
FerrLive: Rate of misclassified live fingerprints, see BPCER
FFR: False Fake Rate
FLR: False Living Rate
FMR: False Match Rate
FN: False Negative
FNMR: False Non-match Rate
FNR: False Negative Rate, FNSPD: False Non-suspicious Presentation Detection
FP: False Positive
FPR: False Positive Rate

© Springer Nature Switzerland AG 2019
S. Marcel et al. (eds.), *Handbook of Biometric Anti-Spoofing*,
Advances in Computer Vision and Pattern Recognition,
https://doi.org/10.1007/978-3-319-92627-8

FRR: False Reject Rate
FSPD: False Suspicious Presentation Detection

GFAR: Global False Accept Rate
GFRR: Global False Reject Rate
GMM: Gaussian Mixture Model
GRU: Gated Recurrent Unit

HTER: Half Total Error Rate

IAPMR: Impostor Attack Presentation Match Rate

LBP: Local Binary Pattern
LFAR: Liveness False Accept Rate
LFCC: Linear Frequency Cepstral Coefficients
LPQ: Local Phase Quantization
LSTM: Long Short-Term Memory
LWIR: Long-wave Infrared

MLP: Multi-layer Perceptron
MS-LBP: Multi-scale Local Binary Pattern

NIR: Near-Infrared

OCT: Optical Coherence Tomography

PA: Presentation Attack
PAD: Presentation Attack Detection
PAI: Presentation Attack Instrument

RNN: Recurrent Neural Network
ROC: Receiver Operating Characteristic
rPPG: Remote photo-plethysmography

Spoofing (term to be deprecated): attempt to impersonate a biometric system, see PA
SRC: Sparse Representation based Classifier
STFT: Short-Term Fourier Transform
SVM: Support Vector Machine
SWIR: Short-Wave Infrared

TABULA RASA: Trusted Biometrics under Spoofing Attacks
TPR: True Positive Rate

VC: Voice Conversion

WER: Weighted Error Rate

Index

Printed by Printforce, the Netherlands